T0142447

Neural Interface Engineering

Liang Guo

Editor

Neural Interface Engineering

Linking the Physical World and the Nervous System

 Springer

Editor
Liang Guo
Department of Electrical and Computer Engineering
The Ohio State University
Columbus, OH, USA

Department of Neuroscience
The Ohio State University
Columbus, OH, USA

ISBN 978-3-030-41856-4   ISBN 978-3-030-41854-0   (eBook)
https://doi.org/10.1007/978-3-030-41854-0

This Springer imprint is published by the registered company Springer Nature Switzerland AG
The registered company address is: Gewerbestrasse 11, 6330 Cham, Switzerland

Preface

Following recent worldwide advocations on brain-related research and bioelectric medicine, neurotechnologies have become one of the hottest scientific and technological frontiers attracting enormous academic and public interests. Not only are governments and private foundations generously investing in this field, but many industrial giants like Facebook, Google, and GlaxoSmithKline together with new startups like Neuralink and Kernel are also enthusiastically stepping into this venture. As the projected technological market is expanding unprecedentedly, interests in further learning the neurotechnological developments are growing fast in both the technical community and the general public.

In developing such body-machine symbiotic systems, the scientific community recognized the neural interfaces as the technological bottleneck hindering further advance of the field. As a result, tremendous efforts have been invested on neural interface engineering, leading to booming of this area over the past decade with a variety of exciting new developments. This book thus focuses on this important topic of neural interface engineering.

This book is targeted for graduate and advanced undergraduate students of bioengineering, biomedical engineering, applied physiology, biological engineering, applied physics, and related fields; for biomedical engineers, neuroscientists, neurophysiologists, and industry professionals wishing to take advantage of the latest and greatest in this emerging area; and for medical practitioners using products of this field. Readers in public services and government funding agencies may also find this book useful in learning the latest in the field.

This book provides an introduction to and summary of representative major neural interfacing technologies used to directly transmit signals between the physical world and the nervous system with the ultimate goals for repairing, restoring, and even augmenting body functions. It offers the readers a unique opportunity to obtain a panorama of this vibrant area in a handy format while elaborating the most important new developments. It covers classic noninvasive and invasive approaches for neural interfacing, as well as recent emerging techniques including advanced implantable neural electrodes and nanomaterial-assisted and genetically engineered neural interfaces.

Chapter authors on each topic are carefully selected among leading and practicing scientists. While it is not possible to cover all of the important approaches, for example, magnetothermogenetics and sonogenetics, due to unavailability of certain contributors, this book nonetheless strives to offer a comprehensive overview of the neural interfaces area to the readers, and it will be a valuable and convenient resource for grasping this specialized area comprehensively and in depth.

Columbus, OH, USA   Liang Guo

Contents

Chapter 1
Electroencephalography

Yalda Shahriari, Walter Besio, Sarah Ismail Hosni, Alyssa Hillary Zisk,
Seyyed Bahram Borgheai, Roohollah Jafari Deligani, and John McLinden

1.1 Introduction to Electroencephalography

Human electroencephalography (EEG) was first introduced by the German psychiatrist, Hans Berger, who first recorded EEG denoting the potential activity of the brain in 1924 (Haas 2003). His first description of EEG noted, "The electroencephalogram represents a continuous curve with continuous oscillations in which…one can distinguish larger first order waves with an average duration of 90 milliseconds and smaller second order waves of an average duration of 35 milliseconds." While EEG is one of the most common noninvasive approaches used to record the brain's electrical activities, invasive recordings can be obtained on the cortical surface that yields an electrocorticogram (ECoG) or in deeper structures yielding intracortical recordings, including local field potentials (LFPs) and single-unit recordings.

The recorded EEG is the superposition of thousands to millions of neuronal potentials within a volume–conductor medium. A single electrode's recording reflects a spatially smoothed version of the synchronized neural activities beneath a scalp surface on the order of 10 cm^2 (Nunez and Srinivasan 2006; Nunez 2000). When many dipoles (~60 million) in the same area discharge synchronously, the

Y. Shahriari (✉) · W. Besio
Department of Electrical, Computer & Biomedical Engineering,
University of Rhode Island (URI), Kingston, RI, USA

Interdisciplinary Neuroscience Program, URI, Kingston, RI, USA
e-mail: yalda_shahriari@uri.edu

S. I. Hosni · S. B. Borgheai · R. J. Deligani · J. McLinden
Department of Electrical, Computer & Biomedical Engineering, University of Rhode
Island (URI), Kingston, RI, USA

A. Hillary Zisk
Interdisciplinary Neuroscience Program, URI, Kingston, RI, USA

© Springer Nature Switzerland AG 2020
L. Guo (ed.), *Neural Interface Engineering*,
https://doi.org/10.1007/978-3-030-41854-0_1

1

superposition of their action potentials causes deflections in the cortical potential that can be detected through noninvasive recordings such as EEG as a macroscopic measure of a large population of synchronous neural spikes (Lopez-Gordo et al. 2014).

1.2   Introduction to Brain Anatomy

The human brain consists of two paired cerebral hemispheres covered with the cerebral cortex, which is a layered structure with a thickness that varies from 1.5 to 4 mm. It is a highly folded surface with gyri (ridges) and sulci (grooves) that enhance the processing capabilities of the brain while maintaining thickness. The cortex and a significant volume beneath it consist of the brain's gray matter structures (neural cell bodies), while deeper white matter structures connect different gray matter areas and carry nerve signals between neurons. The cerebral cortex includes four major lobes: frontal, parietal, occipital, and temporal lobes. The frontal lobe includes the prefrontal area, which is involved in higher-order executive functions, including cognitive workload, decision-making, planning, and personality. The central sulcus (CS) separates the frontal and parietal lobes. The primary motor area (M1) and somatosensory area (S1) are located anterior and posterior to the CS, respectively. The parietal lobe consists of the somatosensory area (S1), which is associated with somatosensory information processing, and the posterior parietal cortex (PPC), which is associated with different sensory inputs, including somatosensory, visual, and auditory information. The occipital lobe is associated with visual processing, and the temporal lobes are associated with auditory and memory processing. Each of these areas produces a different type of detectable EEG response applicable to the development of cutting-edge techniques in brain–computer interfaces (BCIs) and neuromodulation protocols (Wolpaw, J., & Wolpaw, E. W. Eds. 2012). In particular, EEG responses are typically used for various purposes, including controlling devices (e.g., a prosthetic arm), providing communication channels for patients lacking voluntary muscle control, and providing biomarkers for diagnostic applications and biofeedback for rehabilitation and treatment strategies. In human brain science, three principal anatomical planes are considered to describe the brain's anatomy, including the sagittal (longitudinal anteroposterior), coronal (vertical frontal or lateral), and transverse (axial or horizontal) planes. Typically, in discussions of animal neuroanatomy, such as that of rodents, the sections of the brain are named homologously to human brain sections. Figure 1.1 illustrates the major divisions of the human cortex in two views, transverse (top) and sagittal (bottom). This figure has also shown the four major lobes—frontal, parietal, occipital, and temporal—as well as several cortical functions

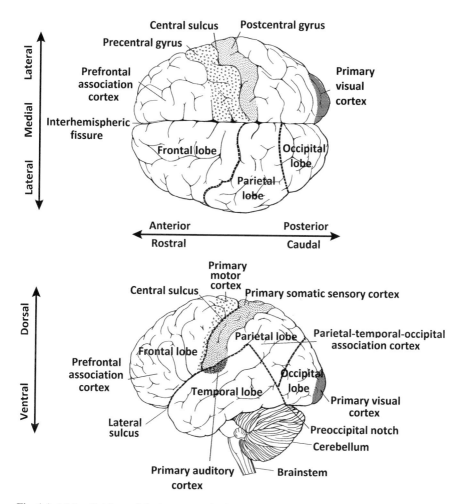

Fig. 1.1 Major divisions of the human cerebral cortex in two main views of transverse (top) and sagittal (bottom). The four major lobes of frontal, parietal, occipital, and temporal as well as several cortical functionalities (e.g., primary auditory cortex, primary visual cortex) also are shown. (*Adapted from* Kandel et al. (1991) (Kandel et al. 2000))

1.3 Brain Rhythms

EEG signals are typically described in terms of transient and oscillatory activities. Transient EEG features include sleep spindles, various components of event-related potentials (e.g., P200, P300, N100, and N200), and spikes corresponding to certain clinical conditions (e.g., seizures). Oscillatory activities are considered with respect to oscillatory frequency and are primarily divided into the delta, theta, alpha, beta, and gamma bands as explained below:

Fig. 1.2 Five major EEG oscillatory activities of delta, theta, alpha, beta, and gamma (from bottom to top) over 3-second segment

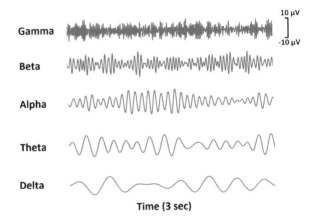

- *Delta* oscillations span frequencies up to 4 Hz and are normally associated with adult slow-wave sleep (Amzica and Steriade 1998) and attention-demanding tasks (Kirmizi-Alsan et al. 2006). *Delta* is also seen in infants' EEGs (Korotchikova et al. 2009). Typically, children's *delta* activity is greatest in the posterior cortical regions, while adult *delta* is strongest in the frontal regions.
- *Theta* oscillations span the 4–8 Hz range and can be seen in drowsiness or arousal (Daniel 1967), conflict error (Cohen and Donner 2013; van Driel et al. 2012), and mental workload (Käthner et al. 2014). *Theta* has also been associated with relaxation and creative states.
- Hans Berger named the *alpha* frequency band, which is associated with oscillations which span the 8–12 Hz range. This rhythmic activity largely is observed in the posterior regions during meditation, relaxation, and with closed eyes, while it is suppressed during mental tasks. A sensorimotor task-related Mu rhythm in this same frequency range also may be observed in sensory and motor cortical regions.
- *Beta* spans the 12–30 Hz range, and while it is strongly associated with motor tasks (Pfurtscheller et al. 1997), it also is seen during alert and anxious states (Kamiński et al. 2012).
- *Gamma* oscillations are 30 Hz or faster and are associated with a wide range of cognitive and motor functions (Fitzgibbon et al. 2004; Tallon-Baudry 2009; MacKay 1997).

These frequency bands have applications in various clinical conditions including neurodegenerative diseases (e.g., Parkinson's disease (PD)) (Weinberger et al. 2006; Heinrichs-Graham et al. 2014), neuro-psychiatric conditions (e.g., schizophrenia) (Kwon et al. 1999; Gotlib 1998), and trauma and brain injuries (Roche et al. 2004). Figure 1.2 illustrates the aforementioned five major EEG oscillatory activities over 3-second segments.

1.4 EEG Data Acquisition

1.4.1 EEG Sensors

The recorded scalp activity can be detected through EEG sensors (electrodes) that send relatively small recorded signals to an amplifier for amplification. While EEG electrodes can be made of various metals, the most common types are gold or Ag/Ag-Cl. Electrodes may be dry (gel-free) or wet (used with additional conductive material such as gel). Typically, dry electrodes use spiky contacts to minimize interface with the hair and outer skin. However, wet electrodes are considered the gold standard. A conducting electrolyte gel or paste is placed between the wet electrode and the skin to reduce skin–electrode impedance and thereby allow efficient current transduction. Although impedance is less than 5 KΩ ideally, impedance between 5 and 20 KΩ is considered acceptable (Nunez and Srinivasan 2006). One common problem with wet electrodes is that impedance deteriorates as the gel dries gradually, which makes these electrodes unsuitable for long-term use (Gargiulo et al. 2010).

Electrodes also may be active or passive. Passive electrodes connect the metal disk to the amplifier directly through a wire. As EEG signals are of relatively low amplitude, environmental factors, including movement and electromagnetic noise, can affect signal quality. Therefore, electrode locations may be rubbed with an abrasive paste to remove the outer layer of the skin, which reduces the signal quality. Active electrodes contain a built-in preamplifier that increases the signal's gain and signal-to-noise ratio (SNR). While these electrodes reduce possible environmental issues, they can amplify unwanted factors, such as input impedance or facial artifacts. Figure 1.3 shows an electrode cap on which active electrodes are mounted and an experimenter injecting conductive gel prior to a recording.

1.4.2 EEG Electrode Placement

Standard electrode montages use the 10–20, 10–10, and 10–5 international systems for EEG electrode positions. The most common landmarking methods are based on the bony parts of the skull beginning from the nasion (Nz) to inion (Iz) and left to right preauricular points (LPA and RPA) to determine the electrodes' placement on top of the head. The "20," "10," and "5" refer to interelectrode intervals 20%, 10%, or 5% of the total nasion–inion or left–right span of the head. The smaller the interelectrode interval, the higher the system's resolution. The standard 10–20 system consists of 21 electrodes, while the 10-10 system consists of 74 electrodes, and the 10-5 system consists of 142 electrodes (Oostenveld and Praamstra 2001). Each electrode name is a combination of letter and number that refers to a specific anatomical location ("Fp," frontal pole; "F," frontal; "T," temporal; "C," central; "P," parietal; and "O," occipital). The subscript "z" stands for zero for the midline

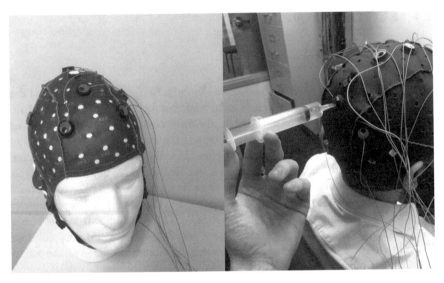

Fig. 1.3 (Right) An electrode cap with active electrodes mounted. (Left) The experimenter is injecting conductive gel prior to the recording

electrodes. Even numbers refer to the electrode positions in the right hemisphere and odd numbers to those in the left hemisphere. Smaller numbers are closer to the midline zero ("z"), and larger numbers represent more lateral electrodes. Figure 1.4 demonstrates the montages for the international 10–20 system as well as the extended 10–20 system.

1.4.3 Amplifiers

Hans Berger's initial human EEG recording used sensors and galvanometers that reside in museums now. However, the brain's potential today is detected using advanced amplifiers attached to fast computers for storage and analysis. EEG signals are relatively small (e.g., ~20 μv) and, therefore, must be amplified before any further processing. EEG amplifiers are differential amplifiers with two input terminals that output the amplified version of the voltage difference between the input terminals.

Thus, EEG amplifiers measure the potential difference and attenuate common signals that appear at both input terminals. As EEG has a very low amplitude, it is usually contaminated with electromagnetic interference from nearby instruments and power lines. The output of a real differential amplifier is defined as below:

$$V_{out} = A_d \left(V_+ - V_- \right) + \frac{1}{2} A_{cm} \left(V_+ + V_- \right) \tag{1.1}$$

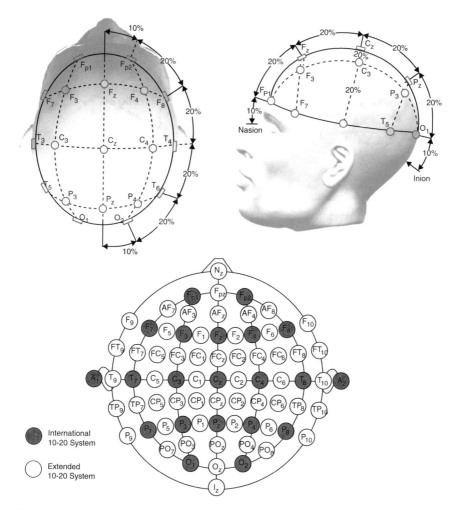

Fig. 1.4 International 10–20 system electrode placement on a 3-D head from two views of top (top left) and side (top right). The bottom figure shows 10–20 system (red electrodes) and the extended system (white electrodes) on a 2-D plot

where V_{out} is the output voltage, V_+ and V_- are the amplifier's two inputs, A_{cm} is the common-mode gain, and A_d is the differential gain, respectively.

EEG differential amplifiers have a high common-mode rejection ratio (CMRR) that amplifies the potential of interest and attenuates the interference from non-cerebral sources that appear simultaneously on both input terminals. Typically, the amplification factor is between 10^3 and 10^5, which results in a CMRR that ranges from 60 to 110 dB (Oostenveld and Praamstra 2001).

All EEG recordings measure the difference in potentials between two signals. Indeed, the output voltage (V_{out}) of an EEG amplifier with two inputs (V_+ and V_-) is the algebraic sum of the difference between two inputs minus the references

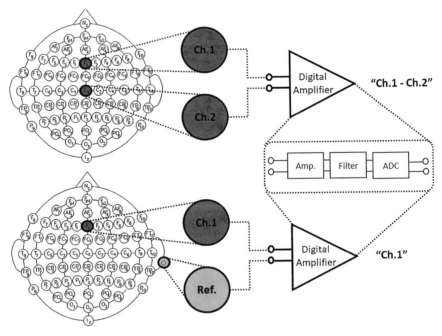

Fig. 1.5 Two main types of bipolar (top) and monopolar (bottom) recordings. For the bipolar recording, the differential potential of two channels (F_z and C_z) is the input of the amplifier to make one channel (F_z-C_z). For the monopolar recording, the differential potential of one channel (F_z) and the ear reference is the input of the amplifier to make one channel (F_z)

$(V_+ - V_{ref}) - (V_- - V_{ref})$. Conventional EEG recordings can be monitored either with *monopolar* or *bipolar* recording. In monopolar recordings, the electrode potential is measured with respect to a common reference electrode that is distant from the recording electrodes. Usually, this reference electrode is placed either on a mastoid or an earlobe for monopolar recording. Bipolar recordings use the difference between two electrode potentials to generate a recording channel. While bipolar recordings are less sensitive to common artifacts, they are more sensitive to localized brain activity (Oostenveld and Praamstra 2001)—this is the reverse for monopolar recording. However, as all channels in monopolar recordings have a common reference, further processing to make any montage desired can be achieved easily. Figure 1.5 shows a typical montage for two types of bipolar and monopolar recordings.

1.4.4 Digitization

Most amplifiers have an analog-to-digital converter (ADC) to digitize analog signals necessary for computer-based processing and storage. An ADC block in the amplifiers discretizes both amplitude and time and converts them into a series of numerical values. The sampling rate, expressed typically in Hz, or samples per second, indicates the frequency at which the data from the electrode is sampled in time. That is, a sampling rate of 256 Hz records 256 data points per second. According to the *Nyquist* criterion, to be able to fully reconstruct the information of interest from a sampled signal, the sampling rate should be at least twice the highest frequency of interest in the original signal. If the sampling rate does not meet the *Nyquist* criterion and the signal contains frequency components higher than half of the sampling rate, then aliasing will happen which will distort the digital signal. Under-sampling (aliasing) can lead to loss of information from the data that makes it impossible to fully reconstruct the signal when it is converted back to analog form. Thus, because of possible practical issues in anti-aliasing filters, the sampling rate should be several times higher than the highest frequency of interest. However, this is a trade-off, as higher sampling rates require greater data storage space.

The quantization block, which converts the analog amplitude into discrete form, is another important aspect of the digitization process. Binary bits are used to determine the quantization level, with 2^k possible values with k bits. Therefore, ADC amplitude resolution depends on the number of bits that represent the digital signal amplitudes. Most ADC blocks digitize signals using 16 bit (= 2^{16} = 65,536 levels), 24 bit (= 2^{24} = 16.8 million levels), or 32 bit (=2^{32} = 4.3 billion levels). Depending on the input voltage range and the number of bits, the ADC amplitude resolution is obtained as below:

$$V_{res} = \frac{V_{range}}{2^N} \tag{1.2}$$

where V_{res}, V_{range}, and N are the resolution, input range, and number of bits, respectively. For example, a 16-bit amplifier with an input voltage range of ±100 mV (range of 200 mV) has a 3 µV (=200 mV/2^{16}) resolution. This indicates that an amplifier with a 16-bit ADC can detect a signal as small as 3 µV for an input voltage of ±100 mV. However, as with increased sampling rates, higher resolution quantization requires more digital storage space.

1.4.5 Temporal Filtering

Because biological signals typically contain a large range of frequency components, generally they must be filtered to extract the desired activities. Filtering can remove certain unwanted activities including biological artifacts (e.g., electromyogram (EMG)), electrode-related noise (e.g., motion), and electromagnetic interference (e.g., mobile phones). Analog and digital filters establish the frequency components of the signal. Analog filters are implemented prior to the digitization block, and digital filters are implemented after digitization. While digital filters have no effect on source signals, analog filters do, and thus, the original unfiltered signal is no longer accessible. Anti-aliasing analog filters are required to avoid aliasing, which digital filters cannot accomplish, as aliasing occurs at the digital processing block.

Thus, to ensure that there is no frequency component above the *Nyquist* rate, anti-aliasing filters with a cutoff frequency equivalent to half of the sampling frequency (*Nyquist* rate) are applied to analog signals.

Most filters are described with respect to three main parameters: *filter order*, *phase*, and *cutoff frequency*. Filter order refers to the length of the filter, which determines its roll-off properties, i.e., the slope of the magnitude response in the transition bands. Sharp filters have narrow transition bands and steep roll-off with a longer response than do filters with a wide transition band. Filter phase refers to the frequency-dependent time displacement that causes delay at a particular frequency component. Group delay, which refers to general envelope delay, is among the filtering-related parameters important in EEG processing and results into two main classes, *linear phase* and *nonlinear phase*. The linear phase introduces a constant delay across all frequency components, while the nonlinear phase causes different delays at different frequency bands. The delay caused by linear phase filters can be corrected by filtering the filter output a second time in a backward direction. Typically, in broadband EEG components, nonlinear phase filters are undesirable, as they can distort the signal's temporal shape completely. The cutoff frequency, the frequency at which the signal is attenuated by 3-dB, refers to the transition frequency that separates the filter's passband and stopband. Depending on the filter type and frequency band of interest, the cutoff frequency should be accurately determined to pass the desired activities while blocking those unwanted. Four main types of filters include low pass (pass the low-frequency components), high pass (pass the high-frequency components), band pass (pass frequency components in a specific frequency range), and band stop (attenuate specific frequency components). Both band-pass and band-stop filters combine high- and low-pass filters to achieve the frequency range of interest.

1.5 Artifacts

EEG amplitude is small and, therefore, the signal is highly vulnerable to artifacts. The two primary artifact categories that affect EEG are biological (originating in the subject but from outside the brain) and nonbiological artifacts. Biological artifacts may include eye blinks, electrooculogram (EOG), EMG, and respiratory artifacts. Eye blink results from fast eyelid movement that generates changes in the dipole charge, which usually is observed as a strong, sharp deflection in the frontal EEG channels. EOG (ocular) also results from eye movement and can have symmetric or nonsymmetric polarity across channels, depending on whether the movement is vertical or horizontal. EOG artifacts are most profound in the frontal and frontal–temporal regions. Muscular artifacts (i.e., EMG) are caused typically by facial muscular (e.g., jaw or eyebrow) movements. EMG artifacts are most prominent at the temporal, frontal, and occipital peripheries. While EOG and eye blink artifacts have low-frequency spectral components (~1 Hz) largely and, therefore, are easier to remove, EMG artifacts have a broad frequency distribution and often are difficult to eliminate.

Fig. 1.6 Examples of EEG contaminated with various noises (eye blink, EMG, nearby instrument, and EMG) over a 3-second segment

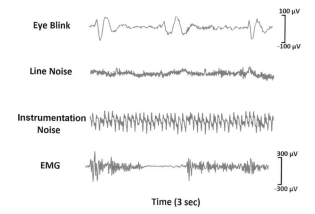

Nonbiological artifacts may include electromagnetic interference from nearby instruments and electrical power lines which use sinusoidal voltages with a frequency of 50 Hz (in Europe, Asia, Africa, and South America), or 60 Hz in North America, and electrode movement. Figure 1.6 shows examples of EEG contaminated with various biological (eye blink, EMG) and nonbiological (line noise, nearby instruments) artifacts over a 3-second segment.

Methods used commonly to remove artifacts include band-pass filtering, manual artifact rejection, and source decomposition techniques, such as independent component analysis (ICA) and principal component analysis (PCA). Band-stop (notch) filters that remove 58–62 Hz signals also may be used to remove power line interference if needed. Each type of artifact can affect the EEG signals differently in different frequency bands, and different processing methods are used depending on the frequency of interest. For example, if we are interested in EEG frequencies below 30 Hz, widespread EMG artifacts are more important to correct than 50 or 60 Hz line noise. Although using source decomposition methods, such as ICA and PCA, can be beneficial in many cases, selecting the optimal number of artifactual components can be challenging. Generally, rejecting contaminated EEG segments results in loss of data, and losing considerable information can damage the results and data interpretation. Thus, selecting proper artifactual components needs to be carefully investigated to minimize damage to the contents of the data.

1.6 Spatial Filtering

Considering EEG's low SNR, spatial filtering methods can improve source localization and increase SNR by making a particular channel more sensitive to certain sources and less sensitive to others (Oostenveld and Praamstra 2001). Typically, spatial filters use linear combinations of weighted channels with predefined geometrical patterns. Common spatial filters include common average reference (CAR) and surface Laplacian (small and large) filters (McFarland et al. 1997).

A CAR filter is implemented by subtracting the average activity across all digitized channels from each individual channel of interest as below:

$$V_i^{CAR} = V_i^{ER} - \frac{\sum\limits_{j=1}^{n} V_j^{ER}}{n} \tag{1.3}$$

where V_i^{ER} is the potential difference between the ith electrode and the reference and n is the number of electrodes in the montage. Typically, CAR filtering is used to reduce the effect of global artifacts that appear across all channels, such as line noise.

A surface Laplacian spatial filter is approximated by the second derivative of spatial voltage distribution, which is equivalent to subtracting the weighted sum of channels within a fixed distance from the channel of interest, as below:

$$V_i^{LAP} = V_i^{ER} - \sum\limits_{j \in S_i} g_{ij} V_j^{ER} \tag{1.4}$$

where $g_{ij} = 1/d_{ij} / \sum_{j \in S_i} 1/d_{ij}$. In this equation, S_i is the set of surrounding electrodes within the predetermined fixed distance from the electrode of interest (ith), and d_{ij} is the distance between the ith and jth electrodes, $j \in S_i$. Depending on the type of Laplacian filter (small or large), the electrode set S_i is defined as the set of nearest neighbor electrodes or next nearest neighbor electrodes for small and large Laplacian filters, respectively. The radius for small Laplacian is $d_{ij} < 3$ cm and for large Laplacian is 3 cm $< d_{ij} < 6$ cm. Depending on the spatial characteristics of the activity of interest, the filter's fixed distance should be determined before filtering. Laplacian spatial filters focus on the surrounding electrodes and thus localize the activity of interest by removing diffuse activity (unrelated to the task) that appears across neighboring electrodes. Figure 1.7 shows an example of electrode placement used in different types of spatial filtering, including small Laplacian, large Laplacian, and common average referencing (CAR) as well as conventional ear reference montage for EEG signal recorded from C_z (red electrode). For each type of spatial filter, the blue electrodes are the set of surrounding electrodes that are averaged and subtracted from electrode of interest (i.e., C_z in this example).

1.7 Tripolar Concentric Ring Electrodes

Tripolar concentric ring electrode (TCRE) sensors are one type of sensors that performs Laplacian filtering automatically at the hardware level (Besio et al. 2006). The potentials are recorded from closely spaced concentric electrodes and transferred to a t-interface preamplifier which then applies the tripolar Laplacian algorithm, $\{16 \times (M - D) - (O - D)\}$, in which M, O, and D are the TCRE sensor's potentials on the middle and outer rings, and central disc electrodes, respectively (Hjorth 1975). Figure 1.8 shows the tEEG setups, including the disk electrode, TCRE, and t-interface.

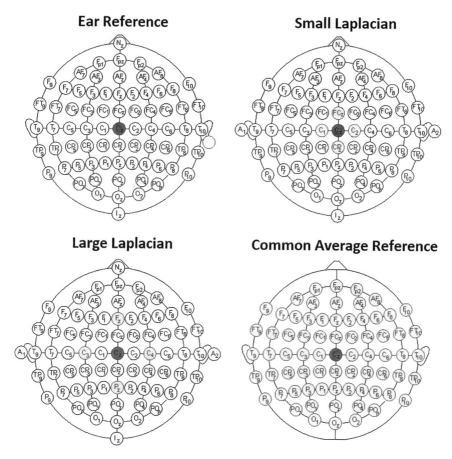

Fig. 1.7 Example of electrode placement used in different types of spatial filtering: small Laplacian, large Laplacian, and common average referencing (CAR) as well as conventional ear reference montage for EEG signal recorded from C_z (red electrode). The blue electrodes are the set of surrounding electrodes that are averaged and subtracted from the electrode of interest (i.e., C_z in this example) for each of the spatial filter types

As the electrodes within a sensor are millimeters apart, taking the differences from the t-interface cancels artifacts generated from high common mode rejection. This property leads to SNR approximately 374% higher than conventional EEG (Besio et al. 2013a). Because of these advantages, TCRE sensors have been used for clinical purposes including seizure detection. Figure 1.9 shows the onset of a tonic seizure recorded concurrently with conventional EEG (top) and tEEG (bottom) recorded with TCREs and the t-interface. tEEG localizes more independent sources (Besio et al. 2013a; Cao et al. 2009) and provides significantly better spatial resolution (approximately ten times better than conventional EEG using the same electrode size) (Besio et al. 2007). Further, it has been shown that high-frequency oscillations (HFOs), a promising biomarker for several neurological conditions

Fig. 1.8 tEEG setups. (**a**) Disc electrode (left) and TCRE (right). (**b**) T-interface as the preamplifier

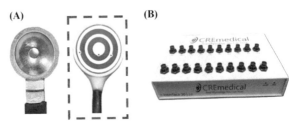

Fig. 1.9 Demonstration of seizure onset using EEG (top) and tEEG (bottom). While the EEG is severely contaminated with EMG that obscures brain signals, the signals are still evident in the tEEG (bottom) recorded concurrently from the same locations. The two panels are on different scales because the tEEG requires more amplification than does the EEG

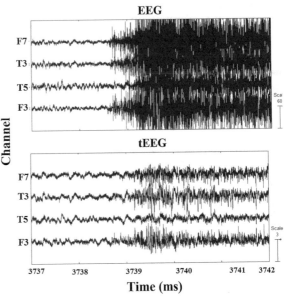

including epilepsy and schizophrenia, can be detected using TCRE recordings (Besio et al. 2013b). TCREs are also compatible with stimulation protocols, including transcranial focal stimulation (TFS), which can record from and stimulate the same region beneath the sensors (Besio et al. 2013a). This property is particularly desirable in cases such as seizures, in which seizure locations can be determined through recording and prediction algorithms and then TFS procedures can be used to stimulate the regions detected and prevent future seizures.

1.8 Conclusion

Since the discovery of EEG, scientists have developed EEG-based techniques to improve our understanding of brain functions and used this knowledge for multiple purposes, including diagnosis, treatment, communication, and rehabilitation. The rapid growth of research and development in this area reflects fertile ground for future studies of noninvasive brain recording techniques, including EEG. This chapter introduced the fundamental principles of EEG neural interfaces to provide a brief overview of challenges and possible solutions in the field. This chapter discussed signal acquisition, a major component of EEG neural interfaces that allows the signal to be detected and stored through hierarchical components, including sensors, amplifiers, filters, and digitization. Then, signal preprocessing, including artifact removal and temporal/spatial filtering, is also key to EEG neural interfaces that substantially extract responses of interest while removing unrelated activities. Finally, considering EEG's intrinsic challenges, including small amplitude, low SNR, and low spatial resolution, further advances in neural engineering are required to overcome the existing challenges and extend the boundaries of the field.

References

Amzica, F., & Steriade, M. (1998). Electrophysiological correlates of sleep delta waves. *Electroencephalography and Clinical Neurophysiology, 107*(2), 69–83.

Besio, G., Koka, K., Aakula, R., & Dai, W. (2006). Tri-polar concentric ring electrode development for Laplacian electroencephalography. *IEEE Transactions on Biomedical Engineering, 53*(5), 926–933.

Besio, W., Koka, K., & Cole, A. (2007). Feasibility of non-invasive transcutaneous electrical stimulation for modulating pilocarpine-induced status epilepticus seizures in rats. *Epilepsia, 48*(12), 2273–2279.

Besio, W. G., Makeyev, O., Medvedev, A., & Gale, K. (2013a). Effects of transcranial focal electrical stimulation via tripolar concentric ring electrodes on pentylenetetrazole-induced seizures in rats. *Epilepsy Research, 105*(1–2), 42–51.

Besio, W., Cuellar-Herrera, M., Luna-Munguia, H., Orozco-Suárez, S., & Rocha, L. (2013b). Effects of transcranial focal electrical stimulation alone and associated with a sub-effective dose of diazepam on pilocarpine-induced status epilepticus and subsequent neuronal damage in rats. *Epilepsy & Behavior, 28*(3), 432–436.

Cao, H., Besio, W., Jones, S., & Medvedev, A. (2009, September). Improved separability of dipole sources by tripolar versus conventional disk electrodes: A modeling study using independent component analysis. In *2009 Annual International Conference of the IEEE Engineering in Medicine and Biology Society* (pp. 4023–4026). IEEE.

Cohen, M. X., & Donner, T. H. (2013). Midfrontal conflict-related theta-band power reflects neural oscillations that predict behavior. *Journal of Neurophysiology, 110*(12), 2752–2763.

Daniel, R. S. (1967). Alpha and theta EEG in vigilance. *Perceptual and Motor Skills, 25*(3), 697–703.

Fitzgibbon, S. P., Pope, K. J., Mackenzie, L., Clark, C. R., & Willoughby, J. O. (2004). Cognitive tasks augment gamma EEG power. *Clinical Neurophysiology, 115*(8), 1802–1809.

Gargiulo, G., Calvo, R. A., Bifulco, P., Cesarelli, M., Jin, C., Mohamed, A., & van Schaik, A. (2010). A new EEG recording system for passive dry electrodes. *Clinical Neurophysiology, 121*(5), 686–693.

Gotlib, I. H. (1998). EEG alpha asymmetry, depression, and cognitive functioning. *Cognition & Emotion, 12*(3), 449–478.

Haas, L. F. (2003). Hans Berger (1873–1941), Richard Caton (1842–1926), and electroencephalography. *Journal of Neurology, Neurosurgery & Psychiatry, 74*(1), 9–9.

Heinrichs-Graham, E., Kurz, M. J., Becker, K. M., Santamaria, P. M., Gendelman, H. E., & Wilson, T. W. (2014). Hypersynchrony despite pathologically reduced beta oscillations in patients with Parkinson's disease: A pharmaco-magnetoencephalography study. *Journal of Neurophysiology, 112*(7), 1739–1747.

Hjorth, B. (1975). An on-line transformation of EEG scalp potentials into orthogonal source derivations. *Electroencephalography and Clinical Neurophysiology, 39*(5), 526–530.

Kamiński, J., Brzezicka, A., Gola, M., & Wróbel, A. (2012). Beta band oscillations engagement in human alertness process. *International Journal of Psychophysiology, 85*(1), 125–128.

Kandel, E. R., Schwartz, J. H., & Jessell, T. M. (Eds.). (2000). *Principles of neural science* (Vol. 4, pp. 1227–1246). New York: McGraw-hill.

Käthner, I., Wriessnegger, S. C., Müller-Putz, G. R., Kübler, A., & Halder, S. (2014). Effects of mental workload and fatigue on the P300, alpha and theta band power during operation of an ERP (P300) brain–computer interface. *Biological Psychology, 102*, 118–129.

Kirmizi-Alsan, E., Bayraktaroglu, Z., Gurvit, H., Keskin, Y. H., Emre, M., & Demiralp, T. (2006). Comparative analysis of event-related potentials during Go/NoGo and CPT: Decomposition of electrophysiological markers of response inhibition and sustained attention. *Brain Research, 1104*(1), 114–128.

Korotchikova, I., Connolly, S., Ryan, C. A., Murray, D. M., Temko, A., Greene, B. R., & Boylan, G. B. (2009). EEG in the healthy term newborn within 12 hours of birth. *Clinical Neurophysiology, 120*(6), 1046–1053.

Kwon, J. S., O'donnell, B. F., Wallenstein, G. V., Greene, R. W., Hirayasu, Y., Nestor, P. G., Hasselmo, M. E., Potts, G. F., Shenton, M. E., & McCarley, R. W. (1999). Gamma frequency–range abnormalities to auditory stimulation in schizophrenia. *Archives of General Psychiatry, 56*(11), 1001–1005.

Lopez-Gordo, M., Sanchez-Morillo, D., & Valle, F. (2014). Dry EEG electrodes. *Sensors, 14*(7), 12847–12870.

MacKay, W. A. (1997). Synchronized neuronal oscillations and their role in motor processes. *Trends in Cognitive Sciences, 1*(5), 176–183.

McFarland, D. J., McCane, L. M., David, S. V., & Wolpaw, J. R. (1997). Spatial filter selection for EEG-based communication. *Electroencephalography and Clinical Neurophysiology, 103*(3), 386–394.

Nunez, P. L. (2000). Toward a quantitative description of large-scale neocortical dynamic function and EEG. *Behavioral and Brain Sciences, 23*(3), 371–398.

Nunez, P. L., & Srinivasan, R. (2006). *Electric fields of the brain: The neurophysics of EEG*. USA: Oxford University Press.

Oostenveld, R., & Praamstra, P. (2001). The five percent electrode system for high-resolution EEG and ERP measurements. *Clinical Neurophysiology, 112*(4), 713–719.

Pfurtscheller, G., Neuper, C., Flotzinger, D., & Pregenzer, M. (1997). EEG-based discrimination between imagination of right and left hand movement. *Electroencephalography and Clinical Neurophysiology, 103*(6), 642–651.

Roche, R. A., Dockree, P. M., Garavan, H., Foxe, J. J., Robertson, I. H., & O'Mara, S. M. (2004). EEG alpha power changes reflect response inhibition deficits after traumatic brain injury (TBI) in humans. *Neuroscience Letters, 362*(1), 1–5.

Tallon-Baudry, C. (2009). The roles of gamma-band oscillatory synchrony in human visual cognition. *Frontiers in Bioscience, 14*, 321–332.

van Driel, J., Ridderinkhof, K. R., & Cohen, M. X. (2012). Not all errors are alike: Theta and alpha EEG dynamics relate to differences in error-processing dynamics. *Journal of Neuroscience, 32*(47), 16795–16806.

Weinberger, M., Mahant, N., Hutchison, W. D., Lozano, A. M., Moro, E., Hodaie, M., Lang, A. E., & Dostrovsky, J. O. (2006). Beta oscillatory activity in the subthalamic nucleus and its relation to dopaminergic response in Parkinson's disease. *Journal of Neurophysiology, 96*(6), 3248–3256.

Wolpaw, J., & Wolpaw, E. W. (Eds.). (2012). *Brain-computer interfaces: Principles and practice*. OUP USA.

Chapter 2
Functional Magnetic Resonance Imaging-Based Brain Computer Interfaces

Jeffrey Simon, Phillip Fishbein, Linrui Zhu, Mark Roberts, and Iwan Martin

2.1 Introduction to fMRI-BCI

Functional magnetic resonance imaging (fMRI) is a well-established, noninvasive neural imaging technique that has been used to describe cerebral hemodynamics. Comparing fMRI to electroencephalogram (EEG) and electrocorticography (ECoG) which monitor electrical activities associated with brain activities, fMRI monitors the hemodynamic responses and allows for neural activities to be localized to specific locations including parts mechanisms behind the learning in the procedure brain. The most prominently used fMRI technique is blood oxygen level-dependent (BOLD) contrast imaging, which indirectly monitors neural activities by determining the amount of oxyhemoglobin (oxygen-containing red blood cells) in each voxel (Buxton 2013). The fundamentals of fMRI imaging and this technique are described in Sect. 2.2.

Over the past decades, fMRI has been instrumental for neuroscience research exploring functionality and connectivity of the brain during different tasks, including speech and language comprehension. Additionally, it has been applied to abnormal psychology to study disorders such as bipolar disorder, schizophrenia, Alzheimer's disease, and eating disorders (Chen et al. 2011; Hafeman et al. 2012; Giraldo-Chica and Woodward 2016; Sheline and Raichle 2013; Val-Laillet et al. 2015; Price 2012; Mar 2011; Fedorenko and Thompson-Schill 2013; Pauls et al. 2001; Augath et al. 2006). Conventionally, fMRI data is taken during a study, and the majority of processing occurs offline. More recently, thanks to increased computational power and speed, fMRI data can be processed online (i.e., during the

J. Simon (✉) · P. Fishbein · L. Zhu · I. Martin
Department of Electrical and Computer Engineering, The Ohio State University, Columbus, OH, USA
e-mail: simon.546@osu.edu

M. Roberts
Department of Neuroscience, The Ohio State University, Columbus, OH, USA

© Springer Nature Switzerland AG 2020
L. Guo (ed.), *Neural Interface Engineering*,
https://doi.org/10.1007/978-3-030-41854-0_2

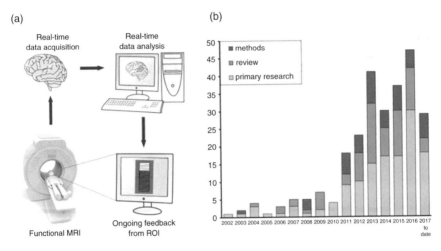

(a)

Real-time data acquisition → Real-time data analysis

Functional MRI → Ongoing feedback from ROI

(b)

■ methods
■ review
□ primary research

2002 2003 2004 2005 2006 2007 2008 2009 2010 2011 2012 2013 2014 2015 2016 2017 to date

Fig. 2.1 fMRI-BCI as an emerging research area. (**a**) Illustration of fMRI-BCI system components. (**b**) fMRI-BCI research is an emerging field that attracts a growing interest, as reflected by the increasing publications in the recent decade. (Images reproduced with permission from Thibault et al. (2018))

experiment). Furthermore, the processing time can be on the order of seconds allowing for the analysis to be used as feedback during a trial. The subject is then interacting with the computer via the fMRI (Fig. 2.1a). This is known as a brain–computer interface (BCI). This area of research has seen significant growth with increasing interests as indicated by the expanding literature. Figure 2.1b graphically describes the increase in primary and secondary research on fMRI-BCI.

Traditional BCI applications, such as controlling prostheses, motion devices, and communication interfaces, can be implemented with fMRI but are only practical for short-term applications under limited circumstances, because the hemodynamic response lags behind the neural activity. However, the unique capabilities of the fMRI-BCI enable novel BCI studies examining neuroplasticity, the treatment of clinical conditions, and the rehabilitation of lost motor functions. For a successful fMRI-BCI, efficient online characterization and classification algorithms must be implemented. The time frame to build a classification model by applying statistical methods is even more critical in fMRI-BCI than other BCIs because of the short duration of fMRI-BCI trials. The complexity of the characterization algorithms must be balanced with computation time and accuracy. Section 2.3 provides a qualitative introduction to these analysis methods.

There are a limited amount of fMRI-BCI studies that have focused on controlling an external device or communication interface. For this application, the limiting factors are the number of states the fMRI can classify and the inherent delays. We will describe the classification and the traditional device control applications in Sect. 2.4. The most common use of an fMRI-BCI is for neurofeedback applications that allow a patient to self-regulate brain activities. An overview of these applications will be provided in Sect. 2.5.

2.2 fMRI Physics, Technology, and Techniques

In comparison to conventional MRI, fMRI employs a fast imaging modality enabling real-time analysis of brain activities. Fortunately, technological developments in medical imaging during the past decade have reduced the acquisition time of imaging substantially. One of the most significant advances is the invention of the so-called echo-planar imaging (EPI) technique. The EPI technique obtains the spatial encoding information with only a single radio-frequency excitation, which is significantly faster than conventional MRI (Cohen 2001). This advantage enables EPI to image the brain within a second to acquire the brain signal in quasi-real time.

In this section, we briefly review the basic principles of fMRI including a qualitative description of the physics, EPI signal acquisition, fMRI imaging methods, and how fMRI can be used to monitor hemodynamic responses.

2.2.1 MRI and fMRI

fMRI is a type of MRI. fMRI coherently aligns, manipulates, and detects the spin of protons in the body, providing useful information about the structure. fMRI also provides information about the vascular architecture of the brain and is particularly useful when examining the blood flow in groups of three to five intracortical arteries. These arteries range in penetration depths from the middle cortex to the lower cortex (Kim and Ogawa 2012).

The primary requirement of an fMRI imaging system is to provide details about a specific physiological activity (e.g., blood flow in the brain) at a temporal and spatial resolution that allows for the response to be correlated with a stimulus or activity. The necessary resolutions depend specifically on the application. It must be understood that while cerebral fMRI imaging can be used to understand neural activities, the nature of observing primary and secondary effects leads to response delays at around 500 ms, and the measuring techniques result in low temporal resolutions of 1–3 seconds (Buxton 2013; Hillman 2014).

2.2.2 EPI

The echo-planar imaging (EPI) technique (also known as gradient spin technique) has been widely used in fMRI systems, including gradient echo and spin echo imaging. The raw data generated by EPI measurements is spin coherence information across different paths of a two-dimensional slice. The data is used to reconstruct a planar image. Gradient echo fMRI provides information about the spin dephasing caused by linear magnetic fields and random magnetic fields, while spin echo fMRI provides more information about the spin decoherence from random magnetic fields.

When a measurement is taken, the spins of water molecules in the slice are aligned in the longitudinal direction to a fixed magnetic field. A selective radio frequency (RF) pulse starts the spin precession of protons in a single transverse plane. Longitudinal magnetic fields with strengths that vary linearly across the transverse plane are applied to specifically adjust the precession frequency and thus selectively change the phase based on location. A time-correlated signal of the transverse magnetic field at each desired phase relation is acquired. Each sampled magnetic field provides information on the distribution of phase and strength of the spins at a specific point in k-space or reciprocal space. The data can be converted to a human-readable image through two-dimensional inverse Fourier analysis (Buxton 2013).

In the above introduction of the EPI technique, the decoherence of the spins in water molecules was assumed to be created only from the applied gradient magnetic field. There are other factors that cause the spin decoherence or free induction decay (FID). In fact, different magnetic components of intravenous blood cause FID. Before we discuss the specific understanding about hemodynamic responses that EPI provides, it is important to understand the parameters used to describe the spin decoherence and the techniques used to measure each.

2.2.3   T1, T2, and T2* Lifetimes

The coherence lifetimes in the longitudinal direction (T1) and transverse direction (T2, T2∗) describe the dephasing of each water molecule's relative spin. T1 describes the time constant of the spins aligning to the static longitudinal magnetic field. The magnetic field in the longitudinal direction can be described in Eq. 2.1, where B_l is the net magnetic field created by the static magnetic field $B_{l_{static}}$ and the magnetic field from the polarization of the spin $B_{l_{spin}}$. The typical lifetime of T1 is around a second at 3T (Buxton 2013).

$$B_l = B_{l_{static}} + B_{l_{spin}} \left(1 - e^{t/T1}\right) \qquad (2.1)$$

The coherence lifetime T2∗ in the transverse direction considers dephasing from the applied linear gradient and random magnetic fields in the brain, while T2 is a measure of the dephasing only from the random magnetic fields. The T2∗ lifetime is less than the T2 lifetime with the latter of gray matter being ~71 ms at 3T (Kim and Ogawa 2012). The magnetic field measured in the transverse direction decays exponentially over time according to a simple relation described in Eq. 2.2:

$$B_t = B_{t_{max}} e^{t/T2^*} \qquad (2.2)$$

2.2.4   Spin Echo and Gradient Echo Technique

The gradient echo technique is used to measure the linear dephasing of spins from the gradient magnetic field and the random dephasing from blood composition monitoring the FID. In contrast, the spin echo technique is used to measure the random dephasing of spins. This method is almost identical to the gradient spin technique except that the linear coherence caused by local magnetic fields is reversed via a spin flip. At a specific frequency dictated by the spin echo period, an RF pulse reverses the spin direction of the water molecules. The linear magnetic fields that caused the phase shift resulting in decoherence of spins will cause the phase shift to reverse and hence resulting in partial coherence. The signal measured can be thought of as an oscillating function that experiences an exponential decay. The oscillations arise from the spin flips, and the decay arises from the random dephasing (Buxton 2013). A graphical depiction of the spin reversal's effect on the fMRI signal is shown in Fig. 2.2.

2.2.5   fMRI Imaging of Hemodynamic Responses

As described above, fMRI imaging provides data about spin coherence of protons in water molecules. For the brain, different fMRI acquisition and data analysis techniques extract time-resolved information on vascular architecture and arterial and

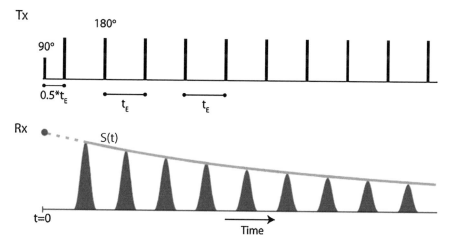

Fig. 2.2 Transmitted RF pulses (top) affect the collective magnetic field strength in the transverse direction (bottom) generated from the spin coherence. An RF pulse at 90° aligns the spins in the transverse direction. Between each RF pulse the spins decohere due to exponential free induction decay (dark gray), described by a lifetime of T2*. A spin flip induced by a 180° RF pule partially recoheres the spins. The recoverable magnetic field strength is described by an exponential decay, S(t), described by a lifetime of T2. (Image reproduced with permission from Walsh et al. (2013))

Fig. 2.3 (**a**) The cerebral
blood flow (CBF) and (**b**)
the BOLD responses are
shown to a stimulus of
finger tapping for 2 s. Both
responses exhibit a slight
dip in the signal during the
stimulus with a large
increase occurring ~5 s
after the stimulus begins.
Not all BOLD signals will
exhibit a noticeable dip
during stimulation, but the
peak is universal. (Image
reproduced from Buxton
(2010) under a Creative
Commons Attribution 4.0
International License
(http://creativecommons.
org/licenses/by/4.0/))

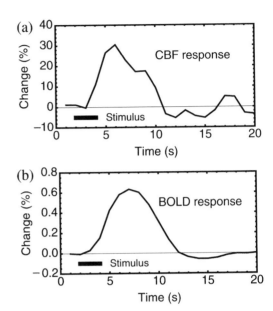

venous blood composition and flow. However, information about neuronal activities
is determined through a cascade of physiological responses. Therefore, the hemato-
logical response that fMRI monitors will have delays of around 500 ms with the
peak signal occurring at approximately 5 s later (Buxton 2013; Hillman 2014).
Despite this delay, the information can be used to temporally and spatially localize
brain activities.

Monitoring the BOLD signal, which provides information on oxyhemoglobin
and deoxyhemoglobin in the blood, is a popular technique for researchers. It enables
higher imaging resolution and contrast ratios. Less-practiced techniques seek to
isolate different properties of blood flow (e.g., cerebral blood flow and cerebral
blood volume) but suffer from a lower signal-to-noise ratio (SNR) and other
difficulties.

Each voxel, or 3D pixel, could include blood vessels, tissue, oxygenated blood,
and deoxygenated blood. The composite signal provides information about the neural activity. In an fMRI study, each image is compared with a baseline to determine
the change to the magnetic signal (Fig. 2.3).

2.2.6 BOLD

BOLD is the most common technique for analyzing neural activities in the brain
which relies on BOLD fMRI (Logothetis 2008). As described in the previous section, BOLD monitors the deoxyhemoglobin content in the blood (Logothetis 2008).
There are multiple factors that affect the deoxyhemoglobin concentration in the

blood during neural activities. Neuronal firing causes the depolarization across the cell membrane and a release of neurotransmitters. Adenosine triphosphate (ATP) is converted to adenosine diphosphate (ADP), releasing energy that is used to restore homeostasis. The ADP is converted back to ATP from the metabolism of glucose and oxygen, converting the oxygen-carrying molecule in the blood, hemoglobin, from oxyhemoglobin to deoxyhemoglobin (Buxton 2013).

Another major component of the BOLD response is the increase in cerebral blood flow (CBF) to the active area of the brain. In response to neuronal activities, the blood flow to the localized region increases. This increase, sometimes called an overshoot, concentrates oxyhemoglobin and carries away the deoxyhemoglobin in a way faster than it can be metabolized. The precise origin of this effect is still under debate, and several theoretical explanations exist. Experimental research and physiological models have attempted to understand the BOLD signal to neural activations to provide accurate quantitative models (Kim and Ogawa 2012; Dickson et al. 2011; Uludağ et al. 2009; Griffeth and Buxton 2011).

The changes in deoxyhemoglobin and oxyhemoglobinconcentration can be measured with spin echo fMRI approaches (Ogawa et al. 1990, 1992). The deoxyhemoglobin in the blood has a different structure than oxyhemoglobin and thus has different magnetic properties. Deoxyhemoglobin decreases the uniformity of the magnetic susceptibility, altering local magnetic fields that fMRI detects. The variation in the magnetic field in a local area causes the spin dephasing. Due to the non-uniform nature of the susceptibility, the reversal of spin in the spin echo technique cannot recover the local phase differences, causing a net decay in the signal. When deoxyhemoglobin is present, the $T2*$ lifetime is decreased. Additionally, the T2 lifetime measured in gradient echo lifetime is decreased from the spin decoherence (Buxton 2013; Kim and Ogawa 2012).

2.2.7 Other Methods

An alternate approach to understanding physiological activities is to utilize fMRI imaging methods to examine specific components that are affected by local neuronal activities (Huber et al. 2017). These components include CBF, oxygen extraction factor (OEF), cerebral metabolic rate of oxygen ($CMRO_2$), cerebral metabolic rate of glucose (CMRG1c), CBV, and arterial oxygen concentration ($[O_2]_a$) (Wong et al. 1998). fMRI imaging methods include arterial spin labeling (ASL) imaging for measuring CBF (Wong et al. 1997), vascular space occupancy (VASO) imaging for measuring CBV (Lu et al. 2013), and imaging methods that utilize different magnetic contrast agents (Huber et al. 2017). Despite the more direct information these individual components provide about neural activations in comparison to BOLD, the SNR, as well as the spatial and temporal resolution, is generally lower, preventing widespread adoption (Buxton 2013). Therefore, our primary discussion will involve the BOLD technique.

2.3   Real-Time fMRI Signal Analysis

Real-time imaging is a requirement for most BCI applications. If a system cannot update according to the data stream, it is then incapable of responding to the brain interfacing with it. In practice, several factors have impacts on brain signal acquisition, including magnetic field strength, echo time, spatial resolution and temporal resolution, etc. More specifically, adopting a high spatial resolution can decrease both the SNR and the temporal resolution. Therefore, practical fMRI-BCI system typically samples a small image size (e.g.,128 × 128) and 5-mm-slice thickness. On the other hand, a reduced spatial resolution helps to suppress the negative effect caused by body motion and intersubject variability (Weiskopf et al. 2004), and the echo time is typically chosen to be close to the relaxation time of the gray matter in the brain to maximize the functional sensitivity. Also, methods have been proposed to mitigate the signal attenuation in fMRI-BCI. Signal loss can be reduced by controlling the susceptibility-induced gradients in the EPI readout direction (Weiskopf et al. 2007).

To use fMRI as a basis for BCI, various preprocessing and signal analysis mechanisms exist to interpret the brain signals for feedback purposes. We will discuss these techniques as well as how they are used in fMRI-BCI in the following sections.

2.3.1   Signal Preprocessing in fMRI-BCI

The original fMRI signal is typically affected by, for example, head motion (Bandettini et al. 1992), respiratory and cardiac artifacts (Mathiak and Posse 2001), and spatial nonsmoothness. Therefore, several techniques are used to preprocess the signal to mitigate these undesirable effects. To elaborate, we will discuss two major techniques.

The head motion correction technique is applied to reduce the artifacts in the signal that are caused by unintended head motion, which can lead to convolution-like artifacts. Manual stabilization methods such as using padding or a bite bar can reduce head motion effects. Two advanced approaches for head motion correction are retrospective and prospective methods. The real-time retrospective algorithm applies a body motion correction of a complete multi-slice EPI dataset within a single repetition time (TR) cycle. More specifically, one of the first images is chosen as the reference image, to which subsequent images are realigned (Mathiak and Posse 2001). On the other hand, the prospective methods correct head motion before image acquisition by adjusting scanning parameters via tracking the moving anatomy. It measures the rotation and translation for each of the sagittal, axial, and coronal planes. The detected rotations and translations are further used to adjust the rotation matrix and the RF excitation frequency for the next acquisition (Tremblay et al. 2005). Both methods could be applied to fMRI-BCI provided that real-time adaptations of the methods are developed.

Another artifact is physiological noise, which is caused by the magnetic field fluctuation due to changes in the respiratory rhythm and blood volume (Thibault et al. 2018). Several offline approaches have been developed to reduce these physiological artifacts (Glover et al. 2000; Birn et al. 2006; Josephs et al. 1997). Recently, van Gelderen et al. designed a real-time shimming method to reduce respiration-induced fluctuations in the magnetic field. This approach can potentially be used in future implementations of fMRI- BCI for physiological artifacts and noise correction, especially when the magnetic field is higher (van Gelderen et al. 2007).

2.3.2   Brain Signal Analysis in fMRI-BCI

After proper preprocessing, the fMRI-BCI system performs further analysis to determine the regions of brain activities. Conventional methods, such as univariate analysis, are typically used, while machine learning-based multivariate methods of pattern recognition techniques have attracted much more interest recently. We will briefly review the univariate methods and provide some concrete illustration on the modern machine learning-based multivariate methods.

Univariate Signal Analysis

In univariate methods, one measures and records the brain activity at many brain locations and then compares the data at each location between consecutive time intervals. If a significant signal change is observed between two brain states at a given location, that location is used to help decode the brain state (Haynes and Rees 2006). Another commonly used method is correlation analysis, which evaluates the correlation between the time series of the reference vector that represents the hemodynamic response and the measurement vector of each voxel. The main advantage of this method is the flexibility of the shape of the reference vector that can match the hemodynamic response (Gembris et al. 2000).

Machine Learning–Based Multivariate Signal Analysis

In contrast to the univariate analysis which considers the brain signals at different locations separately, a recent study shows significant improvement in the quality of neuroimaging by exploiting the spatial pattern of brain activities in multivariate analysis (Mitchell et al. 2003; Polyn et al. 2005; Davatzikos et al. 2005; Norman et al. 2006; LaConte et al. 2007). It is shown that pattern-based methods use more information for more accurate detection of the current state from measurements of brain activities. In the last few years, there has been growing interest in the use of machine learning classifiers for analyzing fMRI data. A rapidly growing collection of literature has shown that machine learning-trained classifiers can efficiently extract information and automatically learn patterns from neuroimaging data. To be specific, recent advances in machine learning such as multilayer neural networks and support vector machines (SVMs) have proven their superior performance in classifying fMRI signals (Mitchell et al. 2003). We next briefly introduce the support vector machine approach for classifying brain activities based on fMRI data.

To train an SVM classifier, we are given a set of training data points, e.g., fMRI images and their corresponding activity labels. Based on the training dataset, we seek for multiple linear classifiers that can separate the data images in high dimensional space into corresponding clusters based on their labels. Moreover, the SVM approach will seek for the classifiers that provide the maximum classification margin among all possible choices, therefore enhancing the robustness and improving the generalization ability. These classifiers are usually obtained by optimization algorithms based on training data. Then, we apply the learned classifiers to the test fMRI images to predict the label of these images. Existing machine learning-based methods are implemented offline. More recently, Laconte et al. implemented a real-time pattern classification system that is applied to BCI. In their approach, they first train a classification model based on training fMRI data, which is used subsequently for classifying the brain states. One drawback of this approach is the limitation to binary-class classification of brain states (LaConte et al. 2007).

2.4 Real-Time fMRI-BCI: Enabling Communication Interface and Prosthetic Control

In the second part of this chapter, we focus on various applications of the BCIs to control external devices. Two primary applications exist: the first seeks to use existing neuronal activation patterns to control an external device with the aid of fMRI, and the second seeks to train the brain to react to certain stimulus in different ways for therapeutic purposes. This section will focus on the control of external devices.

Currently, BCIs that control external devices are targeted at nonhealthy people. Patients with varying forms of paralysis can use a BCI to control a prosthesis or a communication interface. A popular data acquisition technique for BCIs is to collect electrical signals that are related to brain activities. The most widely used techniques are the EEG, ECoG, and intracortical electrode arrays. For the use of BCIs in regular applications, these techniques work well; however, each technique has a varying level of invasiveness and effectiveness. Using fMRI for signal acquisition is noninvasive and can provide detailed information about brain activities across the entire brain (Bleichner et al. 2014).

The electrical activities of the brain have been shown to correlate with hemodynamic activities. To aid the placement of electrodes, fMRI could be used to localize brain activities which will be further analyzed by an electrode-based BCI to control an external device. The locations that are associated with different mental tasks vary on an individual basis, making the specific placement of electrodes critical, especially when surgery is required to place them on the surface or inside the brain (Andersson et al. 2012; 2013b; Bleichner et al. 2014; Yoo et al. 2018). Using a temporary fMRI-BCI setup, the feasibility of a permanent electrode-based BCI could be determined. Using multivariate and univariate statistical analysis of fMRI data when the patient is performing a certain task, a primary location or multiple locations of brain activities can be determined. Furthermore, classification accuracy can be assessed for different scenarios such as surface electrode placement, placement of one or multiple electrodes, and placement in one of two hemispheres of the brain. A determination can then be made about whether an implanted electrode BCI system would be suitable for a patient.

Communication BCIs using fMRI have been intended for patients with locked-in syndrome (LIS). LIS is characterized by the inability to move any muscle, sometimes with exception to limited eye motion. Despite the severe physical impairment, these individuals can have complete cognitive abilities and have all of their senses. The lack for a patient to be able to communicate with full brain functions often leads to conditions such as chronic depression. BCIs seek to alleviate such mental illnesses caused by the lack of socialization by restoring the patient's ability to communicate (Yoo et al. 2004; Sorger et al. 2012).

In order to differentiate between LIS and a vegetative state after severe trauma, fMRI can be used. A short-term BCI enabled by the fMRI is envisioned to allow patients to communicate critical information to doctors, consent to medical procedure, and messages to loved ones. Comparing fMRI-BCI to other BCI methods, classifiers can be trained faster because the fMRI provides a larger amount of information and better spatial localization of brain activities. Despite the slow hemodynamic responses causing a slower bit transfer rate, fMRI-BCI for communication could be effective (Sorger et al. 2009).

This section will be divided into two parts. First, we discuss signal classification using fMRI and the different regions of the brain which have provided the highest classification accuracy. Classification is the key challenge in creating a practical BCI. After mastering the classification, a BCI system is possible using current engineering techniques. Concluding the section, we discuss different applications where healthy individuals successfully interacted with an fMRI-BCI. These applications could theoretically be extended to nonhealthy individuals.

2.4.1 Signal Classification with fMRI

In an fMRI-BCI, classification strategies have been used to analyze different brain activities. These include distinguishing between different mental functions and similar mental functions. Different mental tasks activate distinctly different regions of the brain, and similar mental functions activate similar, more overlapping regions of the brain. Logically, higher classification accuracies are expected for classifying activities in distinct regions of the brain, but this may be a less practical approach. In studies classifying activities in different areas of the brain, different tasks such as mental calculations, mental imagery, inner speech, and motor imagery have been used (Sorger et al. 2009, 2012). Studies classifying similar activities have focused on spatial attention, imagined motion, and actuated motion.

When choosing a classification method and mental tasks to classify, many considerations are taken into account including classification time and the number of states that can be distinguished. Additionally, classification accuracy over time is a more complex factor. Some classification methods where participants switch between different mental tasks can lead to impressive initial classification accuracy but could lead to poor long-term performance caused by patient exhaustion. Furthermore, there are interests in using tasks intuitively related to the BCI application to make it easier to learn (Andersson et al. 2013b; Bleichner et al. 2014).

We will review different mental tasks that have been used for the successful classification of brain activities in an fMRI-BCI with regard to different factors, such as classification accuracy, ease of implementation, and the ability to be used with electrical recording techniques for BCI.

Varied Cognitive Function

The earliest studies examined the ability to classify mental states in real time using fMRI-examined subjects performing different mental tasks. Yoo et al. successfully distinguished the activation states of mental calculation, mental speech generation, right-hand motor imagery, and left-hand motor imagery. Exploiting the spatial separation of the brain activities associated with the mental activities shown in Fig. 2.4, they were able to distinguish states with an overall 90% accuracy. It took a total time of 135 s to classify each of the four states (Yoo et al. 2004).

While the experiment performed by Yoo et al. required the subject to alternate between performing mental tasks and resting for 15 s increments for over a minute, Sorger et al. demonstrated the ability to classify different mental states with the subject entering the mental state for a minimum time of 10 s. Furthermore, the researchers instructed each participant to perform the mental tasks for varying amounts of time and start the mental tasks at different offset times, each providing another classification variable. The classifier demonstrated an 82% success rate in distinguishing between 27 possible combinations of 3 activation offsets, 3 activation lengths, and 3 different mental tasks. Additionally, the classifier training time was in total 12 min faster than that of an EEG classifier (Sorger et al. 2009).

These studies examining varying cognitive functions demonstrated that it was possible to distinguish different mental tasks during online processing. However, more recent studies that explored the classification of brain states using fMRI focused on distinguishing information about more related mental tasks such as motor imagery and higher cognitive functions will be described in the following sections.

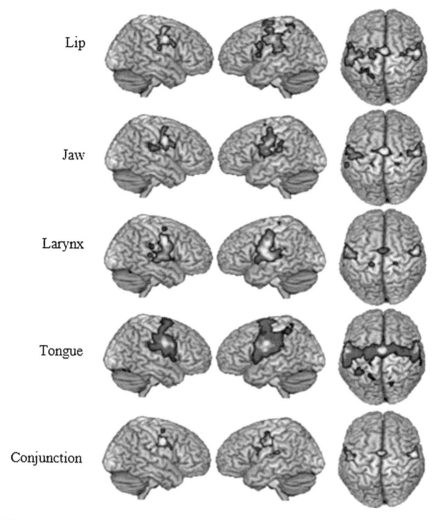

Fig. 2.4 Yoo et al. classified mental states based on spatially separated brain activities associated with each state. The ROI and brain activation areas are shown in the figure. Bilateral activation of the medial superior frontal gyrus correlated with mental calculations, activations of the left Broca's area and auditory association areas were associated with internal speech, activations of the left somatomotor areas were associated with right-hand motor imagery, and activations of the right somatomotor areas were associated with left-hand motor imagery. (Image reproduced with permission from Yoo et al. (2004))

Motion

A BCI based on the classification of imagined motion is practical for patients who are partially or completely paralyzed. Although the real-time fMRI (rt-fMRI) classification studies have taken place with healthy patients, the activation of the motor cortex during imagined motion in patients without the ability to move their limbs is stronger than in imagined motion of a healthy individual. This could be because a healthy patient must also inhibit the motion, while a patient in paralysis does not. Additionally, the patient can still activate their motor cortex even when they have lost functionality for an extended time (Bleichner et al. 2014).

Lee et al. conducted an rt-fMRI study based on the subject imaging clenching their right and left fist. After the classifier was trained on the motion-specific activation patterns in the hand motor areas of the primary motor cortex, the study instructed the subjects to control the amplitude of their BOLD response communicated using visual feedback. This task was more difficult than simply performing a cognitive function or imagining a motion which led to 25–50% success rate where chance success was 14% (Lee et al. 2009). In a study published in 2018, Yoo et al. explored the concept of systematically applying univariate and multivariate analyses for imagined motions: right ankle motion, left ankle motion, walk forward, and lean backward. Higher classification accuracies were obtained in multimodal versus unimodal analysis (Yoo et al. 2018).

Instead of examining imagined motion, Bleichner et al. took the approach of classifying actuated motion. In a study published in 2014, the team trained subjects to perform four letters in sign language. In monitoring a small area of the sensorimotor cortex that could be accessible with ECoG, a decoding accuracy of 63% was achieved with four commands possible per minute (Bleichner et al. 2014). This is lower than the theoretical 70% required for a spelling-based communication BCI. Although not as successful as hoped, this study demonstrated the ability to classify actual movements rather than imagined.

The motion of muscles related to speech has also been explored. In an fMRI study published in 2012, Grabski et al. determined activation patterns that correlate with lip, jaw, larynx, and tongue motion. The spatial separation of some of the regions of activation can be seen in Fig. 2.5. Additionally, statistical analysis indicated that the movements could be differentiated from each other (Grabski et al. 2012). Bleichner's team hypothesized that classification accuracy would be higher than in previous studies if mouth movements were classified and decoded. Bleichner et al. investigated the hypothesis in 2015. Applying multiple approaches to training the classifier including a pattern-based approach, the decoding accuracy could be determined using multiple different region-of-interest (ROI) selection criteria. This study, classifying four different mouth movements with a rate of four decoding events per second, had a high accuracy of around 90% for ROIs determined from the pattern-based approach and 82% for ROIs accessible by an ECoG electrode grid. The accuracy indicated that these mouth movements were suitable for use in a communication BCI. Furthermore, they believed that facial movements already associated with speech would allow for more intuitive control of a communication interface superior to other motions (Bleichner et al. 2015).

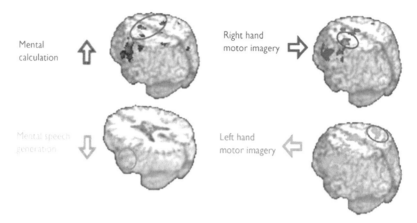

Fig. 2.5 The brain regions correlated with different lip, jaw, larynx, tongue, and conjunction movements are superimposed on a model of the brain. Each movement correlates to the sensorimotor and premotor cortices, the right inferior frontal gyrus, the supplementary motor area, the left parietal operculum and the adjacent inferior parietal lobule, the basal ganglia, and the cerebellum. The regions that can be used to differentiate these movements include the bilateral auditory cortices and the left ventrolateral sensorimotor cortex. (Image reproduced with permission from Grabski et al. (2012))

Higher-Order Cognitive Function

Spatial attention   Using fMRI for online classification of brain states based on spatial attention is promising, because of the mapping of different points in the visual field to cortical areas of the brain. However, fMRI classification has not received much attention in literature despite the accuracy of 95% that Andersson et al. reported with a pattern-based multivariate approach. Furthermore, Andersson et al. have demonstrated that regions close to the surface of the brain where ECoG electrode can be placed can be used for online classification of spatial attention. Restricting the surface ROIs to one hemisphere of the brain and removing small clusters, a classification accuracy of 82% was obtained (Andersson et al. 2012, 2013a, b). Other noninvasive and invasive BCIs based on spatial attention have been highly successful (Astrand et al. 2014).

Object attention   Researchers have also attempted to classify the higher-order cognitive functions such as object attention. Niazi et al. successfully used online fMRI analysis to classify whether participants were focused on a face or a place in a hybrid image (Niazi et al. 2014). Exanayake et al. distinguished which of the four classes of objects presented using online fMRI classification. Although these studies used a large number of ROIs, they demonstrated an increased interest in exploring classification with a top-down rather than a bottom-up approach (Ekanayake et al. 2018). Studying higher-order functions with fMRI-BCIs will be discussed in greater depth in the neurofeedback section.

2.4.2 fMRI-BCI for Controlling External Devices

The classification of mental states is a critical part and the limiting factor in an fMRI-BCI system used to control an external device; however, no discussion would be complete without a system-level description and exploration of specific applications. The fMRI-BCI for controlling external devices operates with feedback from the subject being monitored. The feedback occurs when the subject perceives the action, usually relayed visually through a screen, and responds by deciding the new mental state to enter. Refer to Fig. 2.1a for a visual representation of the system. The feedback loop takes place in a time frame on the order of multiple seconds due to the hemodynamic response and sampling time. Fundamentally, an fMRI-BCI controlling external devices is limited to short-term applications. However, it can still be used as a tool for proof-of-concept demonstrations.

The initial use of the fMRI-BCI to control external devices is rudimentary yet key milestones. Yoo et al., in a study published in 2004, allowed the subjects to control a cursor in a simple maze with their mind (Yoo et al. 2004). To demonstrate that a BCI was plausible to be used with simple prosthetics, Lee et al. used the magnitude of the BOLD response to control the relative position of a robotic arm (Lee et al. 2009), and Andersson et al. applied the fMRI-BCI for controlling a robot that moved forward, to the left and to the right (Andersson et al. 2013a).

Sorger et al. took a different approach using the fMRI-BCI as a communication interface. In a study published in 2009, they demonstrated the ability to chart the responses to multiple choice questions (Sorger et al. 2009). In a paper published in 2012, Sorger et al. reported an fMRI-BCI interface that allowed a participant to spell reporting-suitable accuracy to allow for contextual error correction when the classifier failed to correctly distinguish between letters (Sorger et al. 2012).

While studies have demonstrated the use of fMRI-BCIs to assist patients in communicating and performing motor functions, effectivity was limited by its slower response times than other BCI techniques. The use of the robotic arm can be used on its own to help rehabilitate chronic stroke patients, exploiting the brain's neuroplasticity to help return some of the lost functionalities (Lee et al. 2009). Studies also suggest that rt-fMRI could be more effective when applied in conjunction with faster imaging/recording modalities to improve BCI applications based on these faster modalities (Yoo et al. 2004).

2.5 rt-fMRI-BCI: Neurofeedback for Neuroscience Research and Therapeutic Applications

Currently, the most common use of fMRI-BCI systems is to study the upregulation and downregulation effects of various brain regions. This is a neurofeedback technique where a patient is taught to upregulate or downregulate a specific brain region or set of regions in response to a stimulus. Using the voluntary control of certain brain regions as the independent variable allows for the use of the neuroimaging of fMRI to study the neural response to this change. This enables the study of neuro-plasticity, pain, and emotional and language processing through the generation of functional maps during experiments. It has also been applied to decode brain states (Sitaram et al. 2008). This has been used to develop therapies to treat a wide variety of disorders. In the following section, we will discuss a selection of these experiments.

2.5.1 Neuroplasticity

Neuroplasticity is the process by which the brain undergoes long-lasting physical and chemical alterations to neural and glial cells. These changes are persistent and sometimes detectable by brain imaging (Chang 2014).

Neuroplasticity on the microscopic level occurs via synaptic transmission. Individual neurons exhibit Hebbian learning, as neural firing alters presynaptic and postsynaptic neural architecture (Mundkur 2005). Calcium ions play a critical role in this process. NMDA (N-methyl-D-aspartate) receptors allow calcium entry into postsynaptic neurons and cause a signaling cascade that causes changes in tran-scription and ultimately modifies the expression of postsynaptic receptor proteins and changes in synaptic architecture. Glial cells also play a crucial role in cellular and synaptic level neuroplasticity. Within synapses, glial cells can regulate both presynaptic and postsynaptic activities by the release of other neurotransmitters and signaling molecules (Duffau 2006).

Neuroplasticity also arises in large-scale brain structures. Following repeated training in juggling, requiring refinement of fine motor skills, subjects showed an increase in gray matter volume in the areas of the brain associated with visuomotor coordination (Schuierer et al. 2004). Furthermore, these large-scale changes appeared to be reversible, with subjects appearing to lose gray matter following discontinuation of the activity. Neuroplasticity, in this case, appears to be long-term, but the changes that occur appear to be reversible.

Using EEG and fMRI imaging, neuroplasticity and changes in brain structures can be tracked. Participants in a study were instructed on how to play the piano, who had no previous experience with the task. Following those participants over 5 weeks, EEG activities indicated an increase in motor cortex volume associated with finger movements. Furthermore, fMRI studies on other fine motor tasks indicated that learning of sequential finger movements led to a functional expansion of the motor cortex, followed by the striatum, cerebellum, and other areas of the brain responsi-ble for fine motor coordination (Chang 2014).

2.5.2 Application of fMRI-BCI to Neuroscience Research

fMRI-BCI is often used to study the interaction between the brain and behaviors. Specifically, fMRI-BCI can be used to examine the effects of lesions and changes to the brain tissue on different behaviors, as well as the effects that behaviors have on the brain. As this technique allows subjects to both learn new external behaviors, providing a route to noninvasively modify normal neural activities, as well as providing data in the way of fMRI, it is well suited to perform studies involving neuroplasticity, pain, language processing, as well as other neuroplastic processes (Sitaram et al. 2008).

fMRI-BCIs have been applied to study neuroplasticity of motor systems. Specifically, rt-fMRI feedback can be used to successively reactivate the affected regions of the brain. Sitaram et al. asked and trained four healthy volunteers to control their BOLD response of the supplementary motor area (SMA). Using offline analysis, they showed that there was significant activation of the SMA with enough training. Moreover, with such training they observed a reduction in activation in the surrounding areas, indicating that volitional control training focused activity in the ROI (Sheline and Raichle 2013).

Another application of fMRI-BCI is language processing. Rota et al. studied the self-regulation of the BOLD activity that was recorded in Broca's area (BA 45). They used linguistic tasks that consisted of reading and manipulating the syntactic structure of German sentences. They trained four healthy volunteers with thermometer feedback of activity from the ROI. Figure 2.5 shows brain activities during the training sessions. Before and after the feedback sessions, two linguistic tests were performed by the volunteers immediately. Their results showed that upregulation of the right BA 45 correlated with emotional prosody identification (Rota et al. 2006) (Fig. 2.6).

fMRI-BCI is also applied to study visual perception. Tong et al. applied fMRI to study binocular rivalry when a picture of a human face and a picture of a house were presented to different eyes. As the retinal stimulation remained constant, they found that subjects perceiving changes from house to face were accompanied by increased activity in the fusiform face area (FFA) and decreased activity in the parahippocampal place area (PPA). On the other hand, they found that subjects perceiving changes from face to house had the opposite change in activity (Tong et al. 1998)

2.5.3 Application of fMRI-BCI to Clinical Therapy

A promising application of fMRI-BCI is in clinical therapy. These therapies seek to treat a variety of different neurological and psychological disorders in clinical populations utilizing a combination of neurofeedback techniques and research on the neural bases for the targeted disorder. With a growing number of therapeutic design studies seeking clinically significant results, a standard methodology for these

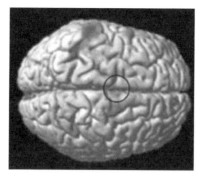

Overlay of localization and up-regulation

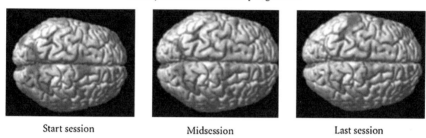

Start session   Midsession   Last session

Fig. 2.6 Applying neurofeedback shows the continuing focus of activity in the target region. (Image reproduced from Ranganatha Sitaram et al. (2007) under a Creative Commons Attribution 3.0 International License (http://creativecommons.org/licenses/by/3.0/))

studies needs to be determined (Thibault et al. 2018). A brief overview of important design considerations identified by Stoeckel et al. (2014) will preface examples of these studies for various disorders.

When designing a study for fMRI-BCI, one of the first considerations should be three properties associated with learning: dependent measures being analyzed, neural mechanisms behind the learning in the procedure, and duration of the learning and the therapy's effect. These properties allow for studies to be categorized for analysis and openings for further study found.

The next consideration concerns variables for designing training protocols. These include timing of the feedback (continuous or intermittent), whether a strategy is provided to the subjects, and the feedback modality itself. In terms of timing, both continuous and intermittent feedback have their strengths and weaknesses. Continuous feedback provides relatively quick response allowing modification of the regulation strategy in pseudo-real time but increases cognitive load and may hinder learning; intermittent feedback averages collected data to increase the SNR while minimizing cognitive load during regulation but won't allow for fine-tuning of individual strategies. Utilizing different feedback modalities may help minimize the weaknesses of the various feedback timing specifications by targeting different areas of the brain unrelated to the ROIs, and some subjects may be more receptive to different types of feedback. The provision of a cognitive strategy for control has yet to be decided since most studies either provide all participants with a strategy or none of the participants with a strategy. In the same vein, there needs to be an appropriate test for success of the training methodology, such as improved regulation with feedback or during a transfer run without feedback indicating true learning of regulation.

For approval in clinical practice, fMRI-BCI neurofeedback must undergo clinical trials to show meaningful clinical effects because of the therapy. To demonstrate the causality of the clinical effects, important control conditions must be implemented, such as giving no feedback, feedback from a brain area unrelated to the task, or feedback derived from another subject to name a few from past studies. These control groups and methods will help determine the overall effect of a neurofeedback therapy, which should then be compared with other existing therapies, particularly due to the high cost of fMRI.

Psychological Disorders

Eating disorders Eating disorders are often characterized by a disruption of the normal reward systems the brain uses to train the body to enjoy food. In people with obesity, dopamine reception is diminished (Wang et al. 2001). Dopamine plays a critical role in the brain's reward system, and the reduction of postsynaptic receptors may indicate that the pleasure people with eating disorders experience following eating may be modified in comparison to healthy subjects. fMRI-BCI neurofeedback can target various brain regions related to the reward system to return the patient's response to normal (Ihssen et al. 2017) (Fig. 2.7).

Treatments for eating disorders have also been studied. Take, for example, the studies by Ihssen et al. and Spetter et al. investigating reactions to food stimuli in obese patients (Spetter et al. 2017). Ihssen et al. found a statistically significant decrease in reported hunger after feedback training, targeting personalized areas having the strongest reaction to food stimuli (Ihssen et al. 2017). While they also saw a decrease in cravings, these were found to be mostly anecdotal with no statistical significance. There was also no change in reported satiety after training. While there was no change in snacking behavior after the training, it was attributed more to homeostatic hunger from the long time between tests. A unique design choice was employed (Fig. 2.8) to change the size of the stimulus cue itself, a picture of high-caloric food, to represent the BOLD feedback instead of another symbolic indicator like a thermometer, which minimizes potential dual-task interference from monitoring two stimuli concurrently.

Spetter et al. studied the effect of feedback training strengthening the functional connectivity between the dorsolateral and ventromedial prefrontal cortices, seen previously in obese subjects who had lost weight (Spetter et al. 2017). In strengthening these regions responsible for executive functions and reward processing, while unable to reduce hunger and cravings as in the previous study, patients were able to show an increased level of voluntary control of their craving for high-calorie

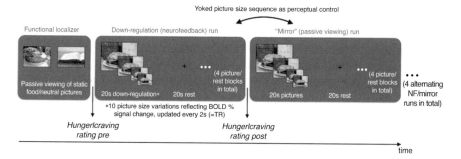

Fig. 2.7 Using the size of the stimulus as the feedback signal to avoid dual-task interference for improved subject performance. (Image reproduced from Ihssen et al. (2017) under a Creative Commons Attribution 4.0 International License (http://creativecommons.org/licenses/by/4.0/))

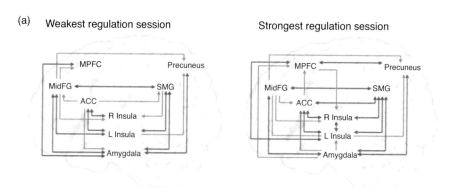

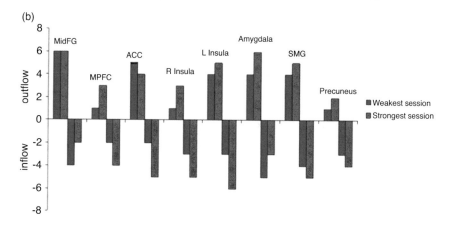

Fig. 2.8 Posttest analysis of data from a single-ROI study demonstrating increased functional connectivity between regions of the brain. (**a**) Shows changes in the connectivity between regions, with orange arrows indicating bidirectional and green arrows indicating unidirectional communication. (**b**) Shows changes in the magnitude of information flowing to and from each region. (Image reproduced with permission from Ruiz et al. (2014))

foods. This might help obese patients lose weight by allowing them to achieve satiety with fewer calories.

Schizophrenia Schizophrenia is a disease that affects both the cognitive function and perception of those with the disease. Often, individuals with the disease display reduced gray matter volume indicative of reduced cognitive performance, as well as increased volume of ventricles (Baaré et al. 2001). In addition, schizophrenics show activities below baseline in cingulate and frontal cortex, as well as a decrease in activation of the ventromedial prefrontal cortex (vmPFC), which plays a key role in the development of the self, and higher-order cognitive functioning, both of which could be targets for fMRI-BCI neurofeedback training (Kühn and Gallinat 2013).

While most studies have worked with the modulation of a single ROI, Ruiz et al. observed that regulation of a single ROI also strengthened the connections across the entire neural circuit in that region, represented in Fig. 2.8. In response, they developed a study to determine the effects of self-regulating multiple related areas, specifically the inferior frontal gyrus (IFG) and the superior temporal gyrus (STG), using healthy participants. This regulation was able to be learned and, using a priming task to determine behavioral modification, saw enhancement of the task after learned IFG-STG connectivity. This network modification was tested on schizophrenia patients as a potential treatment for the irregular circuit connectivity that led to symptoms. Using this therapy, schizophrenia patients were able to improve facial recognition symptoms. To help treatment of neurological disorders using network modulation, Ruiz et al. developed a subject-independent pattern classifier to allow the use of normal circuits as a basis for treating the abnormal circuits of patients (Ruiz et al. 2014).

Emotional disorders Major depressive disorder (MDD) can be treated with neurofeedback procedures. Distinct changes in cortical structure activation can be detected with fMRI. Often, the dorsolateral prefrontal cortex, involved in higher-level functioning, is often significantly downregulated in patients with MDD. In addition, changes in cingulate cortex associated with responses to positive stimuli also show a marked decrease in activity (Fitzgerald et al. 2008). fMRI BCI neurofeedback treatment can be used to increase or decrease the activation in these brain regions, developing changes in brain structure that are persistent and prevent recurrence of the disorder (Hammond 2005).

Inspired by the proposed functional dissociation between the right and left amygdala in emotional processing and the evidence suggesting increased response in the left amygdala to negative stimuli and attenuated response to positive stimuli being associated with major depressive disorder (MDD), Young et al. performed a double-blind experiment with volunteer patients diagnosed with MDD to test the viability of modulating the left amygdala as a treatment for MDD (Young et al. 2014). Experimental feedback was provided from the left amygdala, and control feedback was provided from the left horizontal segment of the intraparietal sulcus (HIPS). Using self-reporting emotional rating scales, a statistically significant increase in happiness was recorded between groups, as well as anger, anxiety, restlessness, and irritability; however, the observed decrease in depression ratings was not statistically significant. The ability of the experimental group to maintain increased BOLD regulation of the left amygdala without neurofeedback in the transfer run supports the clinical potential of this therapy and the potential for patients to maintain this self-control of the region in other, more general settings.

Another study had healthy participants modulate the anterior insula, involved in emotional processing. Upregulating this region resulted in a more negative emotional rating of fear-evoking pictures than downregulation. These results led the investigative team to attempt the same regulation with four psychopathic patients to determine the feasibility of using the training as a treatment. The success with these patients led the researchers to suggest this method for treatment to other emotional disorders (Rota et al. 2006).

Rehabilitation

Real-time feedback from fMRI-BCI has also been used to restore brain functions lost due to neural ischemia, typically caused by stroke. However, success varies significantly based on the area of the brain that was damaged in the patient. In studies that attempted to restore higher-order visual functions, very little to no difference was found in the brain following feedback training, while studies focusing on sensorimotor impairment following strokes were more successful in restoring function to patients (Wang et al. 2018).

One of the first studies on a clinical application of fMRI-BCI was conducted by de Charms et al. on the potential to treat chronic pain through modulation of the rostral region of the anterior cingulate cortex (rACC), a region implicated in pain processing and perception. This study was conducted with both healthy volunteers and chronic pain patients with four control groups, who were provided different instructions and/or feedback. Participants were instructed to both up- and down-regulate the ROI while receiving a painful thermal stimulus. The study reported a positive correlation between increased rACC activity and increased pain intensity, with 23% enhancement in controlling pain intensity and 38% enhancement in controlling pain unpleasantness in both healthy and chronic pain test groups (Fig. 2.9). Changes in the activation of the rACC also corresponded with increases in other brain areas, perhaps highlighting the other important areas in the network involved in pain processing (R. Christopher DeCharms et al. 2005).

In 2017, Sitaram et al. said that considering them as a clinical treatment, neurofeedback therapies are still in their early stages of development. There is a wide range of neural circuitry that has been modulated using these methods, but there is a need for more placebo-controlled trials to determine the real effect of these modulation therapies on the problems they are intended to address. Attempts to study neurofeedback treatment for ADHD and stroke rehabilitation have had mixed results. Cross-comparison suffers from differences in experimental design, difficulty identifying responses, and difficulty with nonhomogeneous populations being examined (Sitaram et al. 2017).

Fig. 2.9 Graph showing a positive correlation between increases in BOLD activation in the anterior cingulate cortex and increases in reported pain rating, indicating a possible change in pain responsiveness induced by neurofeedback training. (Figure reproduced from Christopher DeCharms et al. (2005). Copyright (2005) National Academy of Sciences, USA)

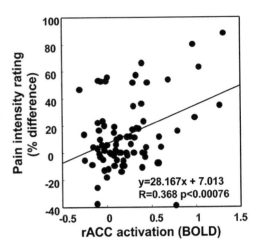

Enhancement in Healthy Subjects

The final applications discussed here can be applied to healthy subjects: improving the performance of working memory. Using a study of healthy subjects and a control group to control the left dorsolateral prefrontal cortex, both groups were able to improve working memory performance. However, they were able to demonstrate a larger differential improvement in the experimental group, who received real-time feedback, compared to the control group, who did not receive real-time feedback. The lasting effects of this training have not been studied, so the application of this cognitive enhancement is still unknown (Sherwood et al. 2016).

Up to this point, the studies discussed have exclusively filled their experimental and control groups with young adults. Rana et al. performed a study to evaluate the effectiveness of fMRI-BCI as a therapy to delay cognitive decline in the aging population. While they found improved regulation of the ROIs, it was much more moderate in comparison to younger populations. Many possible explanations for these moderate results were posited. A combination of fatigue and dual-task interference comprises some of them. Others looked at features of learning unique to the older adult population, such as knowing the background of a topic making it easier to learn a related skill. Further work in the optimization of these neurofeedback therapies for older populations must be done (Rana et al. 2016).

2.6   Limitations

An fMRI-BCI feedback loop is slow limiting the usefulness of applications. There are multiple-second delays in the physiological response after the neural activities occur, as well as delays in signal acquisition and processing of the fMRI image. Classification of the onset of the BOLD signal could reduce the feedback time.

Furthermore, accurately determining brain activities with as much specificity as possible is important for advancing the scope of rt-fMRI studies as well as their applications. Improvements in the SNR of the detector, computational neural models linking activity to physiological responses, and various statistical analysis methods are instrumental in improving the characterization of fMRI signals in real time.

Possible ways to improve fMRI sampling include utilizing different scanning patterns, adding multiple detector coils, and increasing the magnetic field strength (Sitaram et al. 2008; Ahn et al. 1986; Moeller et al. 2010; Dumoulin et al. 2018). Furthermore, advances in the understanding of how neural activities lead to the BOLD responses and the understanding of intrinsic physiological noise could provide more details about neurological activities. It is desirable to design real-time algorithms that can remove image artifacts in an efficient way. On the other hand, due to the massive development in machine learning techniques, it is promising to apply these models to learn the features of the fMRI images capturing and classifying the underlying brain activities in a way that traditional statistical approaches cannot. Moreover, it is possible to implement other more advanced signal-processing techniques such as independent component analysis to extract the BOLD response of interest.

The space requirement, the capital cost for acquiring or building an unit, and the operational costs prevent the fMRI-BCI from being implemented in a portable manner for long-term use. However, research in fMRI-BCIs can be used to support more realistic and portable BCI techniques. The neurofeedback from fMRI has been used to tune an EEG-BCI system (Hinterberger et al. 2004; Han and Im 2018). A new area of research is in functional near-infrared spectroscopy (fNIRS) which uses an infrared light source and detector to measure the BOLD signal in the brain (Chance et al. 1998). Similar principles as used in fMRI-BCI systems can be used for fNIRS systems. Despite being limited by the light penetration depth, a portable fNIRS-BCI system can be actuated. Furthermore, as with ECoG applications, fMRI could determine the ideal region of interest that the fMRI device should monitor and provide information on the signal characteristics to analyze.

2.7 Conclusion

fMRI used as a tool in BCIs has a unique set of applications that are different from an electrode-based BCI approach. fMRI provides details on the hemodynamic response to brain activities by monitoring changes in the magnetic susceptibility. The most widely used technique is BOLD. BOLD imaging provides details of the amount of oxyhemoglobin and deoxyhemoglobin in the blood. The two molecules have a differing magnetic signature that can be determined in fMRI. Monitoring the BOLD response allows for the localization of brain activities. The fMRI can monitor the entire brain at once with a relatively high spatial resolution and provide a unique set of information that can be used in a BCI. However, because the BOLD response is caused by a cascade of physiological events, there are intrinsic delays of multiple seconds before the peak response occurs.

It is necessary to analyze the fMRI data accurately and in a short amount of time. Univariate and multivariate methods can be used to analyze the data. Univariate analysis examines changes in signal strength in each voxel, while multivariate analysis finds distributed patterns in the signal across many voxels. A relation is determined in a training trial and classified in real time during the rest of the experiment.

Using an fMRI-BCI system to allow a patient to control an external device such as a robot or a communication interface is possible. Studies have classified different mental tasks and motion in real time with high accuracies and demonstrated mental control of external devices. Applications could help with the placement of electrodes in an electrode-based BCI and enable patients with LIS to communicate.

An emerging and promising area applying the fMRI-BCI focuses on therapeutic applications. While there are no set medical protocols to using neurofeedback techniques to treat neurological disorders, the fMRI-BCI along with other functional neurofeedback devices can alter larger-scale brain functions and provide persistent changes to the structure of the brain. Neurofeedback is noninvasive and provides long-lasting changes without the need for direct physical or chemical alterations to the brain tissue. fMRI-BCI neurofeedback techniques are largely used to alter cortical function in a specific area of the brain, allowing volitional control of a specific brain region responsible or contributing to a disorder. Future neurofeedback techniques seek to treat more global disruptions in neural functions.

Acknowledgments We thank Dr. Liang Guo from the Department of Electrical and Computer Engineering at The Ohio State University, Dr. Julie Golomb from the Department of Psychology at The Ohio State University, Dr. Yi Zhou from the Department of Electrical and Computer Engineering at Duke University, Richard Chan from the Department of Chemistry and Biochemistry at The Ohio State University, and Shwe Han from the Department of Computer Science and Engineering at The Ohio State University for valuable comments and suggestions that helped to improve the manuscript.

References

Ahn, C. B., Kim, J. H., & Cho, Z. H. (1986). High-speed spiral-scan echo planar NMR imaging-I. *TMI, 5*(1), 2–7. https://doi.org/10.1109/TMI.1986.4307732.

Andersson, J. P., Ramsey, N. F., Raemaekers, M., et al. (2012). Real-time decoding of the direction of covert visuospatial attention. *Journal of Neural Engineering, 9*(4), 045004. https://doi.org/10.1088/1741-2560/9/4/045004.

Andersson, J. P., Pluim, J. J., Viergever, M. M., et al. (2013a). Navigation of a telepresence robot via Covert Visuospatial Attention and real-time fMRI. *Brain Topogr, 26*(1), 177–185. https://doi.org/10.1007/s10548-012-0252-z.

Andersson, J. P., Ramsey, N. F., Viergever, M. A., et al. (2013b). 7 T fMRI reveals feasibility of covert visual attention-based brain–computer interfacing with signals obtained solely from cortical grey matter accessible by subdural surface electrodes. *Clinical Neurophysiology, 124*(11), 2191–2197. https://doi.org/10.1016/j.clinph.2013.05.009.

Astrand, E., Wardak, C., & Ben Hamed, S. (2014). Selective visual attention to drive cognitive brain-machine interfaces: From concepts to neurofeedback and rehabilitation applications. *Frontiers in Systems Neuroscience, 8*(144), 144. https://doi.org/10.3389/fnsys.2014.00144.

Augath, M., Logothetis, N. K., Shmuel, A., Oeltermann, A., (2006). Negative functional MRI response correlates with decreases in neuronal activity in monkey visual area V1. 9, 569–577. https://doi.org/10.1038/nn1675.

Baaré, W. F. C., van Oel, C. J., Hulshoff Pol, H. E., et al. (2001). Volumes of brain structures in twins discordant for schizophrenia. *Archives of General Psychiatry, 58*(1), 33–40. https://doi.org/10.1001/archpsyc.58.1.33.

Bandettini, P. A., Wong, E. C., Hinks, R. S., et al. (1992). Time course EPI of human brain function during task activation. *Magnetic Resonance in Medicine, 25*(2), 390–397. https://doi.org/10.1002/mrm.1910250220.

Birn, R. M., Diamond, J. B., Smith, M. A., et al. (2006). Separating respiratory-variation-related neuronal-activity-related fluctuations in fluctuations from fMRI. *Neuroimage, 31*(4), 1548. https://doi.org/10.1016/j.neuroimage.2006.02.048.

Bleichner, M., Jansma, J., Sellmeijer, J., et al. (2014). Give me a sign: Decoding complex coordinated hand movements using high-field fMRI. *Brain Topogr, 27*(2), 248–257. https://doi.org/10.1007/s10548-013-0322-x.

Bleichner, M. G., Jansma, J. M., Salari, E., et al. (2015). Classification of mouth movements using 7 T fMRI. *Journal of Neural Engineering, 12*(6), 066026. https://doi.org/10.1088/1741-2560/12/6/066026.

Buxton, R. B. (2013). The physics of functional magnetic resonance imaging (fMRI). Reports on progress in physics. *Physical Society (Great Britain), 76*(9), 096601. https://doi.org/10.1088/0034-4885/76/9/096601.

Buxton, R. B., (2010) Interpreting oxygenation-based neuroimaging signals: the importance and the challenge of understanding brain oxygen metabolism. 2, 8. https://doi.org/10.3389/fnene.2010.00008.

Chance, B., Anday, E., Nioka, S., et al. (1998). A novel method for fast imaging of brain function, non-invasively, with light. *Optics Express, 2*(10), 411. https://doi.org/10.1364/OE.2.000411.

Chang, Y. (2014). Reorganization and plastic changes of the human brain associated with skill learning and expertise. *Frontiers in Human Neuroscience, 8*, 35.

Chen, C., Suckling, J., Lennox, B. R., et al. (2011). A quantitative meta-analysis of fMRI studies in bipolar disorder. 13, 1–5. https://doi.org/10.1111/j.1399-5618.2011.00893.x.

Christopher DeCharms, R., Maeda, F., Glover, G. H., et al. (2005). Control over brain activation and pain learned by using real-time functional MRI. *Proceedings of the National Academy of Sciences of the United States of America, 102*(51), 18626–18631. https://doi.org/10.1073/pnas.0505210102.

Cohen, M. S. (2001). Real-time functional magnetic resonance imaging. *Methods, 25*(2), 201–220.

Davatzikos, C., Ruparel, K., Fan, Y., et al. (2005). Classifying spatial patterns of brain activity with machine learning methods: Application to lie detection. *Neuroimage, 28*(3), 663–668. https://doi.org/10.1016/j.neuroimage.2005.08.009.

Dickson, J. D., Ash, T. W. J., Williams, G. B., et al. (2011). Quantitative phenomenological model of the BOLD contrast mechanism. *Journal of Magnetic Resonance, 212*(1), 17–25. https://doi.org/10.1016/j.jmr.2011.06.003.

Duffau, H. (2006). Brain plasticity: From pathophysiological mechanisms to therapeutic applications. *Journal of Clinical Neuroscience, 13*(9), 885–897. https://doi.org/10.1016/j.jocn.2005.11.045.

Dumoulin, S. O., Fracasso, A., van der Zwaag, W., et al. (2018). Ultra-high field MRI: Advancing systems neuroscience towards mesoscopic human brain function. *Neuroimage, 168*, 345–357. https://doi.org/10.1016/j.neuroimage.2017.01.028.

Ekanayake, J., Hutton, C., Ridgway, G., et al. (2018). Real-time decoding of covert attention in higher-order visual areas. *Neuroimage, 169*, 462–472. https://doi.org/10.1016/j.neuroimage.2017.12.019.

Fedorenko, E., Thompson-Schill, S. L., (2013). Reworking the language network. 18, 120–126. https://doi.org/10.1016/j.tics.2013.12.006.

Fitzgerald, P. B., Laird, A. R., Maller, J., et al. (2008). A meta-analytic study of changes in brain activation in depression. *Human Brain Mapping, 29*(6), 683–695. https://doi.org/10.1002/hbm.20426.

Gembris, D., Taylor, J. G., Schor, S., et al. (2000). Functional magnetic resonance imaging in real time (FIRE): Sliding-window correlation analysis and reference-vector optimization. *Magnetic Resonance in Medicine, 43*(2), 259–268.

Giraldo-Chica, M., Woodward, N. D., (2016). Review of thalamocortical resting-state fMRI studies in schizophrenia. 180, 58–63. https://doi.org/10.1016/j.schres.2016.08.005.

Glover, G. H., Li, T. Q., & Ress, D. (2000). Image-based method for retrospective correction of physiological motion effects in fMRI: RETROICOR. *Magnetic Resonance in Medicine, 44*(1), 167. https://doi.org/10.1002/1522-2594(200007)44:1<162::aid-mrm23>3.0.co;2-e.

Grabski, K., Lamalle, L., Vilain, C., et al. (2012). Functional MRI assessment of orofacial articulators: Neural correlates of lip, jaw, larynx, and tongue movements. *Human Brain Mapping, 33*(10), 2306–2321. https://doi.org/10.1002/hbm.21363.

Griffeth, V. E. M., & Buxton, R. B. (2011). A theoretical framework for estimating cerebral oxygen metabolism changes using the calibrated-BOLD method: Modeling the effects of blood volume distribution, hematocrit, oxygen extraction fraction, and tissue signal properties on the BOLD signal. *Neuroimage, 58*(1), 198–212. https://doi.org/10.1016/j.neuroimage.2011.05.077.

Hafeman, D. M., Chang, K. D., Garrett, A. S., et al. (2012). Effects of medication on neuroimaging findings in bipolar disorder: an updated review. 14, 375–410. https://doi.org/10.1111/j.1399-5618.2012.01023.x.

Hammond, D. (2005). Neurofeedback treatment of depression and anxiety. *Journal of Adult Development, 12*(2), 131–137. https://doi.org/10.1007/s10804-005-7029-5.

Han, C., & Im, C. (2018). EEG-based brain-computer interface for real-time communication of patients in completely locked-in state. In: Anonymous IEEE (pp. 1–2).

Haynes, J., & Rees, G. (2006). Decoding mental states from brain activity in humans. *Nature Reviews Neuroscience, 7*(7), 523.

Hillman, E. M. C. (2014). Coupling mechanism and significance of the BOLD signal: A status report. *Annual Review of Neuroscience, 37*(1), 161–181. https://doi.org/10.1146/annurev-neuro-071013-014111.

Hinterberger, T., Weiskopf, N., Veit, R., et al. (2004). An EEG-driven brain-computer interface combined with functional magnetic resonance imaging (fMRI). *IEEE Transactions on Biomedical Engineering, 51*(6), 971–974. https://doi.org/10.1109/TBME.2004.827069.

Huber, L., Uludağ, K., & Möller, H. E. (2017). Non-BOLD contrast for laminar fMRI in humans: CBF, CBV, and CMRO2. *Neuroimage, 197*, 742–760.

Ihssen, N., Sokunbi, M. O., Lawrence, A. D., et al. (2017). Neurofeedback of visual food cue reactivity: A potential avenue to alter incentive sensitization and craving. *Brain Imaging and Behavior, 11*(3), 915–924. https://doi.org/10.1007/s11682-016-9558-x.

Josephs, O., Howseman, A. M., Friston, K., et al. (1997). Physiological noise modelling for multislice EPI fMRI using SPM. In *Anonymous Proceedings of the 5th Annual Meeting of ISMRM*, Vancouver (Vol. 1682).

Kim, S., & Ogawa, S. (2012). Biophysical and physiological origins of blood oxygenation level-dependent fMRI signals. *Journal of Cerebral Blood Flow & Metabolism, 32*(7), 1188–1206.

Kühn, S., & Gallinat, J. (2013). Resting-state brain activity in schizophrenia and major depression: A quantitative meta-analysis. *Schizophrenia Bulletin, 39*(2), 358–365. https://doi.org/10.1093/schbul/sbr151.

LaConte, S. M., Peltier, S. J., & Hu, X. P. (2007). Real-time fMRI using brain-state classification. *Human Brain Mapping, 28*(10), 1033–1044. https://doi.org/10.1002/hbm.20326.

Lee, J., Ryu, J., Jolesz, F. A., et al. (2009). Brain–machine interface via real-time fMRI: Preliminary study on thought-controlled robotic arm. *Neuroscience Letters, 450*(1), 1–6. https://doi.org/10.1016/j.neulet.2008.11.024.

Logothetis, N. K. (2008). What we can do and what we cannot do with fMRI. *Nature, 453*(7197), 869–878. https://doi.org/10.1038/nature06976.

Lu, H., Hua, J., & Zijl, P. C. M. (2013). Noninvasive functional imaging of cerebral blood volume with vascular-space-occupancy (VASO) MRI. *NMR in Biomedicine, 26*(8), 932–948. https://doi.org/10.1002/nbm.2905.

Mar, R. A., (2011). The Neural Bases of Social Cognition and Story Comprehension. 62, 103–134. https://doi.org/10.1146/annurev-psych-120709-145406.

Mathiak, K., & Posse, S. (2001). Evaluation of motion and realignment for functional magnetic resonance imaging in real time. *Magnetic Resonance in Medicine, 45*(1), 167–171. https://doi.org/10.1002/1522-2594(200101)45:1<167::AID-MRM1023>3.0.CO;2-M.

Mitchell, T. M., Hutchinson, R., Just, M. A., et al. (2003). Classifying instantaneous cognitive states from fMRI data. In *Anonymous AMIA Annual Symposium Proceedings* (Vol. 2003, p. 465). American Medical Informatics Association.

Moeller, S., Yacoub, E., Olman, C. A., et al. (2010). Multiband multislice GE-EPI at 7 tesla, with 16-fold acceleration using partial parallel imaging with application to high spatial and temporal whole-brain fMRI. *Magnetic Resonance in Medicine, 63*(5), 1144–1153. https://doi.org/10.1002/mrm.22361.

Mundkur, N. (2005). Neuroplasticity in children. *The Indian Journal of Pediatrics, 72*(10), 855–857. https://doi.org/10.1007/BF02731115.

Niazi, A. M., van den Broek, P. L. C., Klanke, S., et al. (2014). Online decoding of object-based attention using real-time fMRI. *European Journal of Neuroscience, 39*(2), 319–329. https://doi.org/10.1111/ejn.12405.

Norman, K. A., Polyn, S. M., Detre, G. J., et al. (2006). Beyond mind-reading: Multi-voxel pattern analysis of fMRI data. *Trends in Cognitive Sciences, 10*(9), 424–430. https://doi.org/10.1016/j.tics.2006.07.005.

Ogawa, S., Lee, T. M., Kay, A. R., et al. (1990). Brain magnetic resonance imaging with contrast dependent on blood oxygenation. *Proceedings of the National Academy of Sciences of the United States of America, 87*(24), 9868–9872. https://doi.org/10.1073/pnas.87.24.9868.

Ogawa, S., Tank, D. W., Menon, R., et al. (1992). Intrinsic signal changes accompanying sensory stimulation: Functional brain mapping with magnetic resonance imaging. *Proceedings of the National Academy of Sciences of the United States of America, 89*(13), 5951–5955. https://doi.org/10.1073/pnas.89.13.5951.

Pauls, J., Augath, M., Trinath, T., et al. (2001). Neurophysiological investigation of the basis of the fMRI signal. 412, 150–157. https://doi.org/10.1038/35084005.

Polyn, S. M., Natu, V. S., Cohen, J. D., et al. (2005). Category-Specific Cortical Activity Precedes Retrieval During Memory Search. *Science, 310*(5756), 1963–1966. https://doi.org/10.1126/science.1117645.

Price, C. J., (2012). A review and synthesis of the first 20 years of PET and fMRI studies of heard speech, spoken language and reading. 62, 816.

Rana, M., Varan, A. Q., Davoudi, A., et al. (2016). Real-time fMRI in neuroscience research and its use in studying the aging brain. *Frontiers in Aging Neuroscience, 8*, 239. https://doi.org/10.3389/fnagi.2016.00239.

Rota, G., Sitaram, R., Veit, R., et al. (2006). fMRI-neurofeedback for operant conditioning and neural plasticity investigation: A study on the physiological self-induced regulation of the BA 45. In *Anonymous Proceedings of the Cognitive Neuroscience Conference*.

Ruiz, S., Buyukturkoglu, K., Rana, M., et al. (2014). Real-time fMRI brain computer interfaces: Self-regulation of single brain regions to networks. *Biological Psychology, 95*, 4–20. https://doi.org/10.1016/j.biopsycho.2013.04.010.

Schuierer, G., May, A., Draganski, B., et al. (2004). Neuroplasticity changes in grey matter induced by training. *Nature, 427*(6972), 311–312. https://doi.org/10.1038/427311a.

Sheline, Y. I., & Raichle, M. E. (2013). Resting state functional connectivity in preclinical Alzheimer's disease. *Biological Psychiatry, 74*(5), 340. https://doi.org/10.1016/j.biopsych.2012.11.028.

Sherwood, M. S., Kane, J. H., Weisend, M. P., et al. (2016). Enhanced control of dorsolateral prefrontal cortex neurophysiology with real-time functional magnetic resonance imaging (rt-fMRI) neurofeedback training and working memory practice. *Neuroimage, 124*(Pt A), 214–223. https://doi.org/10.1016/j.neuroimage.2015.08.074.

Sitaram, R., Caria, A., Veit, R., et al. (2007). fMRI brain-computer interface: A tool for neuroscientific research and treatment. *Computational Intelligence and Neuroscience, 2007,* 25487–25410. https://doi.org/10.1155/2007/25487.

Sitaram, R., Weiskopf, N., Caria, A., et al. (2008). fMRI brain-computer interfaces. *IEEE Signal Processing Magazine, 25*(1), 95–106. https://doi.org/10.1109/MSP.2008.4408446.

Sitaram, R., Ros, T., Stoeckel, L., et al. (2017). Closed-loop brain training: The science of neurofeedback. *Nature Reviews. Neuroscience, 18*(2), 86–100. https://doi.org/10.1038/nrn.2016.164.

Sorger, B., Dahmen, B., Reithler, J., et al. (2009). Another kind of 'BOLD Response': Answering multiple-choice questions via online decoded single-trial brain signals. *Progress in Brain Research, 177,* 275–292.

Sorger, B., Reithler, J., Dahmen, B., et al. (2012). A real-time fMRI-based spelling device immediately enabling robust motor-independent communication. *Current Biology, 22*(14), 1333–1338. https://doi.org/10.1016/j.cub.2012.05.022.

Spetter, M. S., Malekshahi, R., Birbaumer, N., et al. (2017). Volitional regulation of brain responses to food stimuli in overweight and obese subjects: A real-time fMRI feedback study. *Appetite, 112,* 188–195. https://doi.org/10.1016/j.appet.2017.01.032.

Stoeckel, L. E., Garrison, K. A., Ghosh, S. S., et al. (2014). Optimizing real time fMRI neurofeedback for therapeutic discovery and development. *Neuroimage: Clinical, 5*(C), 245–255. https://doi.org/10.1016/j.nicl.2014.07.002.

Thibault, R. T., MacPherson, A., Lifshitz, M., et al. (2018). Neurofeedback with fMRI: A critical systematic review. *Neuroimage, 172,* 786–807. https://doi.org/10.1016/j.neuroimage.2017.12.071.

Tong, F., Nakayama, K., Vaughan, J. T., et al. (1998). Binocular rivalry and visual awareness in human extrastriate cortex. *Neuron, 21*(4), 753–759.

Tremblay, M., Tam, F., & Graham, S. J. (2005). Retrospective coregistration of functional magnetic resonance imaging data using external monitoring. *Magnetic Resonance in Medicine, 53*(1), 141–149. https://doi.org/10.1002/mrm.20319.

Uludağ, K., Müller-Bierl, B., & Uğurbil, K. (2009). An integrative model for neuronal activity-induced signal changes for gradient and spin echo functional imaging. *Neuroimage, 48*(1), 150–165. https://doi.org/10.1016/j.neuroimage.2009.05.051.

Val-Laillet, D., Aarts, E., Weber, B., et al. (2015). Neuroimaging and neuromodulation approaches to study eating behavior and prevent and treat eating disorders and obesity. 8, 1–31. https://doi.org/10.1016/j.nicl.2015.03.016.

van Gelderen, P., de Zwart, J. A., Starewicz, P., et al. (2007). Real-time shimming to compensate for respiration-induced B0 fluctuations. *Magnetic Resonance in Medicine, 57*(2), 362–368. https://doi.org/10.1002/mrm.21136.

Walsh, D., Turner, P., Grunewald, E., et al. (2013) A Small-Diameter NMR Logging Tool for Groundwater Investigations. 51, 914–926. https://doi.org/10.1111/gwat.12024.

Wang, G., Volkow, N. D., Logan, J., et al. (2001). Brain dopamine and obesity. *The Lancet, 357*(9253), 354–357. https://doi.org/10.1016/S0140-6736(00)03643-6.

Wang, T., Mantini, D., & Gillebert, C. R. (2018). The potential of real-time fMRI neurofeedback for stroke rehabilitation: A systematic review. *Cortex, 107,* 148–165. https://doi.org/10.1016/j.cortex.2017.09.006.

Weiskopf, N., Scharnowski, F., Veit, R., et al. (2004). Self-regulation of local brain activity using real-time functional magnetic resonance imaging (fMRI). *Journal of Physiology - Paris, 98*(4), 357–373. https://doi.org/10.1016/j.jphysparis.2005.09.019.

Weiskopf, N., Hutton, C., Josephs, O., et al. (2007). Optimized EPI for fMRI studies of the orbitofrontal cortex: Compensation of susceptibility-induced gradients in the readout direction. *Magnetic Resonance Materials in Physics, Biology and Medicine, 20*(1), 39. https://doi.org/10.1007/s10334-006-0067-6.

Wong, E. C., Buxton, R. B., & Frank, L. R. (1997). Implementation of quantitative perfusion imaging techniques for functional brain mapping using pulsed arterial spin labeling. *NMR in Biomedicine, 10*(4–5), 237–249. https://doi.org/10.1002/(sici)1099-1492(199706/08)10:4/5<237::aid-nbm475>3.0.co;2-x.

Wong, E. C., Buxton, R. B., & Frank, L. R. (1998). A theoretical and experimental comparison of continuous and pulsed arterial spin labeling techniques for quantitative perfusion imaging. *Magnetic Resonance in Medicine, 40*(3), 348–355. https://doi.org/10.1002/mrm.1910400303.

Yoo, S., Fairneny, T., Chen, N., et al. (2004). Brain–computer interface using fMRI: Spatial navigation by thoughts. *NeuroReport, 15*(10), 1591–1595. https://doi.org/10.1097/01.wnr.0000133296.39160.fe.

Yoo, P. E., Oxley, T. J., John, S. E., et al. (2018). Feasibility of identifying the ideal locations for motor intention decoding using unimodal and multimodal classification at 7 T-fMRI. *Scientific Reports, 8*(1), 1–15. https://doi.org/10.1038/s41598-018-33839-4.

Young, K. D., Zotev, V., Phillips, R., et al. (2014). Real-time fMRI neurofeedback training of amygdala activity in patients with major depressive disorder. *PLoS One, 9*(2), e88785. https://doi.org/10.1371/journal.pone.0088785.

Chapter 3
Transcranial Magnetic Stimulation

Gregory Halsey, Yu Wu, and Liang Guo

3.1 Introduction

Transcranial magnetic stimulation (TMS) is a noninvasive procedure which delivers a directed magnetic field into specified superficial regions of a patient's brain using an external current-carrying coil. TMS induces small electrical currents within the cortex of targeted brain areas to cause stimulation of the underlying neurons, which is believed to be the primary mode of action. TMS is a US Food and Drug Administration (FDA)-approved therapy for the treatment of major depressive disorder (MDD), obsessive compulsive disorder (OCD), and migraines due to its clinical successes in treating pharmacologically resistant patients. Despite this, the exact mechanism responsible for the effects of TMS is unclear, as magnetic fields have been shown to influence cellular responses and microenvironment in a variety of ways. In this chapter, we will review the history, potential mechanisms, current uses, and discuss future improvements for TMS-based therapy.

3.2 History of TMS

Here we discuss the history of TMS therapy, beginning with the discovery of fundamental electromagnetic properties which enabled the development of TMS.

G. Halsey
Department of Bioengineering, Clemson University, Clemson, SC, USA

Y. Wu · L. Guo (✉)
Department of Electrical and Computer Engineering, The Ohio State University, Columbus, OH, USA

Department of Neuroscience, The Ohio State University, Columbus, OH, USA
e-mail: guo.725@osu.edu

© Springer Nature Switzerland AG 2020
L. Guo (ed.), *Neural Interface Engineering*,
https://doi.org/10.1007/978-3-030-41854-0_3

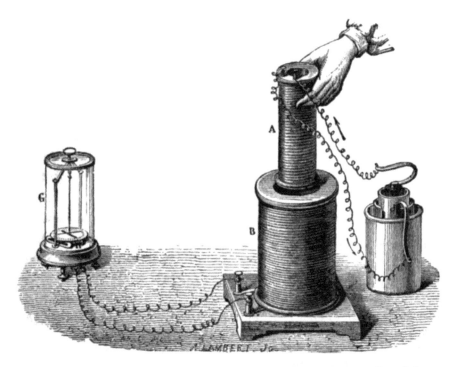

Fig. 3.1 Illustration of Michael Faraday's 1831 experiment demonstrating the effect of electromagnetic induction. The battery on the right provided a current that flew through the small wire coil A creating a magnetic field. Movement of the small coil within the larger coil B generated an electrical current which was detected by the movement of the needle within the galvanometer (Poyser 1892)

In 1831, Michael Faraday discovered the property of electromagnetic induction through experiments using an electromagnet and a tightly wound coil attached to a galvanometer. As depicted in Fig. 3.1, an electric potential was temporarily generated in the large coil and measured by the galvanometer. He then demonstrated that the same phenomenon occurred when a loop of wire was passed over a stationary magnet or when a magnet was passed through a stationary loop of wire (Barker and Shields 2017). The combination of these discoveries and the help of mathematician James Clerk Maxwell led to the creation of Faraday's law describing electromagnetic induction. This law became one of the four Maxwell equations used to explain electromagnetism.

Attempts to implement this effect in humans occurred near the turn of the nineteenth century when Jacques-Arsene d'Arsonval demonstrated the presence of magnetophosphenes (a visual sensation), when a subject was exposed to alternating magnetic fields and when Sylvanius P. Thompson (Fig. 3.2) attempted to use magnetic fields to stimulate his brain (Lovsund et al. 1980). These latter efforts by Thompson were largely unsuccessful as the technology to generate large and rapidly changing fields had yet to be invented. However, Thompson was able to corroborate d'Arsonval's earlier magnetophosphene findings. After little work in the

Fig. 3.2 Sylvanus
P. Thompson with
prototype magnetic brain
stimulation device (1910).
(Figure adapted with
permission from
Noakes 2007)

area was done during the Great Wars period, Kolin et al. showed in 1959 that it was possible to elicit a contractile response from frog muscle when it was exposed to intensity-altering magnetic fields (Kolin et al. 1959). This result showed promise in that action potentials could be produced in muscles using just magnetic fields.

At the University of Sheffield in 1974, then doctoral candidate Anthony Barker created an early magnetic stimulator with a "C"-shaped core attached to a 200 V capacitor bank. When he placed his wrist in the air gap of the electromagnet and fired the stimulation pulse, he reported feeling a "sensation and slight muscular contraction of the hand" (Polson et al. 1982). This finding led to the start of a program in 1976 at the Royal Hallamshire Hospital and the University of Sheffield "with the specific goal of stimulating nerves using currents induced by short-duration magnetic field pulses such that the resultant electrophysiological response could be recorded" (Barker and Shields 2017). In 1982 Barker and colleagues, pictured in Fig. 3.3, were able to produce reactions in the hand and wrist with a further developed stimulator coil, when it was placed over the ulnar nerve (Polson et al. 1982). Finally, in 1985, the preliminary prototype of a TMS machine was created.

The first stimulator coil was a closed-loop system that was able to deliver a peak value of 4000 amps after 110 microseconds once every 3 seconds, which at the time was a considerably higher frequency than any other group was able to produce (Barker et al. 1985). In his initial publication, Barker touted the ability of his device to produce musculature reactions with the absence of pain, unlike electrical stimulation that had preceded his findings. He was also able to demonstrate the use of his device in accessing deep nerves, such as the ulnar nerve mentioned previously,

Fig. 3.3 Anthony
T. Barker (right), Ian
L. Freeston (middle), and
Reza Jalinous (left)
presenting a TMS device in
1985. (Courtesy of
Anthony Barker)

whereas electrical methods at the time were unable to reach without considerable pain to the patient. Barker was able to record action potentials produced by the affected muscles using recording electrodes and thus theorized that it could be used as "a neurodiagnostic tool to produce an evoked potential in muscle tissue by activating neurons in the motor cortex" (Barker et al. 1985). The initial design of a round loop coil proved to have its challenges in localizing stimulation within the desired area of the brain. Ueno et al. was able to navigate around this difficulty in 1988 by proposing a "figure eight" coil that was able to produce opposing magnetic fields (Ueno et al. 1990a). This allowed the coil to have a much more focused point of application and produce a stronger reaction within the brain tissue. Soon after this proposed design was conceived, the achievable resolution of motor cortex stimulation had been drastically reduced to 5 mm, and in 1990 the functional map of the human motor cortex had been mapped using this method (Ueno et al. 1990b).

Based on the results of his 1985 publication, Dr. Barker's technology received considerable interest from the medical and scientific communities, especially since the device could be shown to create responses with little to no pain. The Sheffield group produced six of Dr. Barker's devices for distribution to research groups to "demonstrate proof of concept" and "explore possible clinical applications" (Barker and Shields 2017). Instead of patenting the new technology, Dr. Barker decided to promote his creation to manufacturers in hopes that they could help mass produce his device. By the late 1990s, there were at least three stimulators in widespread use, and by 2016 there were more than ten manufacturers whose devices were used in both clinical and research settings (Barker and Shields 2017). Since the initial use of neurodiagnostics, TMS has been used in a variety of settings including "stimulating peripheral nerves and brain tissue in studies encompassing motor conduction in human development, motor control, movement disorders, swallowing, vision, attention, memory, speech and language, epilepsy, depression, stroke, pain and

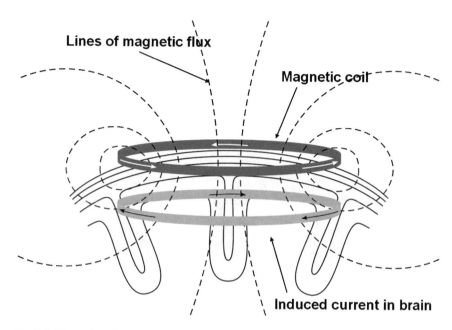

Fig. 3.4 Illustration of induction of electrical currents in the brain. The magnetic coil above the scalp produces a magnetic field that induces an electric field perpendicular to the magnetic field. The electric field will cause current to flow in loops parallel to the plane of the coil. TMS ordinarily does not activate corticospinal neurons directly; rather it activates them indirectly through synaptic inputs. (Figure adapted with permission from Hallett 2000)

plasticity" (O'Shea and Walsh 2007). Recent developments in the field have given insights into how TMS can alter brain activities by "virtue of inducing long-term changes in excitability and connectivity of the stimulated brain networks" (Ziemann 2017). Figure 3.4 below demonstrates how a magnetic field flux induces an electrical current within the brain.

3.3 Current Applications of TMS

In this section we provide an overview of the current uses of TMS including clinical applications as well as diagnostic purposes. TMS is an FDA-approved therapy for a variety of neurological disorders. Evidence has shown that TMS has potential to relieve MDD symptoms in antidepressant-resistant patients (Groppa et al. 2012).

According to the World Health Organization, major depressive disorder is the leading contributor to the overall burden of diseases worldwide, with approximately 16.2 million adults in the USA (6.7% of the population) suffering from depression in 2016 (Organization 2018, Health 2019). Of depression patients treated with commonly prescribed antidepressants, 20–30% never fully recover, and continued treatment becomes increasingly ineffective due to resistance (Brigitta 2002, NiCE 2014).

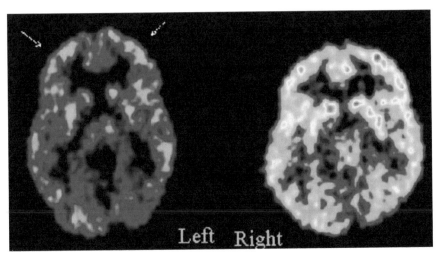

Fig. 3.5 PET scan study in clinically depressed patient (left) and matched control (right). Blue represents less glucose metabolism and hypoactivity, which are noted throughout the different areas of the brain in the depressed patient. (Figure adapted with permission from Faria 2013)

The primary neural circuit involved in MDD is most likely the limbic-cortical-striatal-pallidal-thalamic tract, which is connected to the prefrontal cortex, basal ganglia, and hippocampus among other regions (Price and Drevets 2012). Additionally, chronic stress has been associated with atrophy of certain regions of the hippocampus, and overall volume decrease was observed in MRI scans of MDD patients (Bremner et al. 2000). Positron emission tomography (PET) studies have also shown that in patients suffering from MDD, the frontal lobe and basal ganglia have reduced blood flow, as illustrated in Fig. 3.5 below (Bench et al. 1992). TMS therapy has been successful in stimulating areas lacking electrical activity and blood flow, helping to relieve and sometimes completely mitigate the symptoms of MDD. Patients typically report improvement in mood for over a year after treatment, and it is common for patients to return for subsequent rounds of treatment (Jeon and Kim 2016).

Though most commonly used as a treatment for MDD, TMS may also effectively treat migraines. Approximately 38 million Americans suffer from migraines annually, which are commonly characterized by substantial head pain, hypersensitivity to sensory stimuli, nausea, vomiting, and disability. Many pharmacological treatments are available, but only 2 hours of pain relief is experienced for approximately 60% of users, while continuing pain freedom after 2 hours decreases to approximately 30%. Prophylactic migraine medications, or medications intended to prevent migraines, have shown the ability to reduce migraine attack by approximately 50% in 20–40% of patients. Due to the inability for pharmacological and prophylactic treatments to completely treat and prevent migraines, single-pulse TMS has been tested as a treatment option and has become approved by the FDA for acute treatment of migraines. Similarly, repetitive TMS has shown promise as a therapy for migraine prophylaxis due to its effects on neurotransmitters and reduction of cortical excitability (Schwedt and Vargas 2015).

In August 2018, TMS was approved by the FDA for the treatment of OCD, an often-chronic psychological disorder in which a person experiences uncontrollable, recurring thoughts and behaviors. According to the National Institute of Mental Health, about 1% of adults in the USA suffered from some form of OCD in the past year. Like many other psychological diseases, treatment often consists of a combination of medication and psychotherapy, though these are frequently found to be ineffective. In a recent FDA approved study, approximately 38% of patients positively responded to OCD treatment via TMS (Administration 2018).

Although TMS therapy is currently limited to use in FDA-approved applications, research indicates there may be more potential uses. Low-frequency repetitive TMS has shown promising results as an antiepileptic treatment due to its inhibitory effect on spastic neuronal activity associated with the disease. It has also shown the capability of suppressing seizures in patients with neocortical epilepsy and interrupting ongoing seizures in status epilepticus (Gersner et al. 2016). Similarly, it has provided symptom relief for patients suffering from schizophrenia (Barker and Shields 2017).

Despite the potential benefits of TMS, it may introduce some adverse side-effects. Patients are at a slight risk for fainting or experiencing seizures. Additional common complaints include minor to moderate stimulation site pain and discomfort, jaw pain, facial pain, muscle pain, spasms and twitching, as well as neck pain. Transient memory and hearing losses have also been reported, as well as appearance of magnetophosphenes in the vision field. Most of these side-effects resolved shortly after the treatment had concluded and became more common with application of higher-frequency TMS (Rossi et al. 2009).

3.4 Potential Mechanisms of TMS Therapy

TMS uses the principle of magnetic induction to deliver localized electric currents to pyramidal neurons within the cortex. The physics behind this concept is based on Faraday's law, which states that the induced current is proportional to the changing rate of the stimulating magnetic field. A typical TMS device has a circular or figure eight-shaped coil as the magnetic field generator, which is attached by a cable to a box, containing a signal generator and a power supply. As a large time-varying current runs through the coil, a changing magnetic field can be formed. If the coil is placed close to the scalp, the magnetic field can penetrate the skull and deliver enough energy to stimulate the brain tissues. Although the principle of magnetic induction was extensively studied approximately 200 years ago, the intracranially induced electrical field is still not fully understood due to the varying conductivity of the cranial contents. According to Ohm's law, the current has the tendency to follow the path with the highest conductance. Areas with a high proportion of ventricles may have unpredictable current distributions, where the current may concentrate

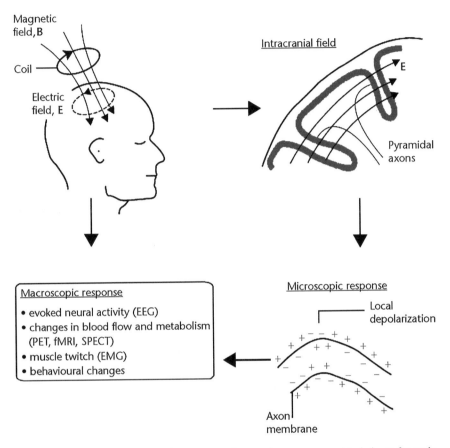

Fig. 3.6 Overview of TMS-induced responses and respective measurement techniques for analysis of effects, including electroencephalogram (EEG), PET, functional magnetic resonance imaging (fMRI), single-photon emission computed tomography (SPECT), and electromyography (EMG). (Figure adapted with permission from Butler and Wolf 2007)

in less resistive cerebrospinal fluid and travel to other parts of the brain (Wassermann and Zimmermann 2012). Figure 3.6 below shows potential mechanisms for the TMS therapy and methods for measuring their effects.

Magnetic field exposure has been shown to influence the nanostructure of the cellular microenvironment which may contribute to the beneficial effects observed from TMS therapy. Collagen is an important extracellular matrix (ECM) protein which makes up a large portion of connective tissues in the brain and has been demonstrated to be magnetically active (Torbet and Ronziere 1984). As demonstrated in Fig. 3.7, when exposed to a constant magnetic field, collagen fibrils began to align perpendicular to the direction of the field, while Schwann cells aligned parallel (Eguchi et al. 2015). The alignment of collagen has been shown to promote neuronal axonal regeneration and neural stem cell migration along the length of the fibril by acting as a guiding "track" for the cells to adhere to (Eguchi et al. 2015). There

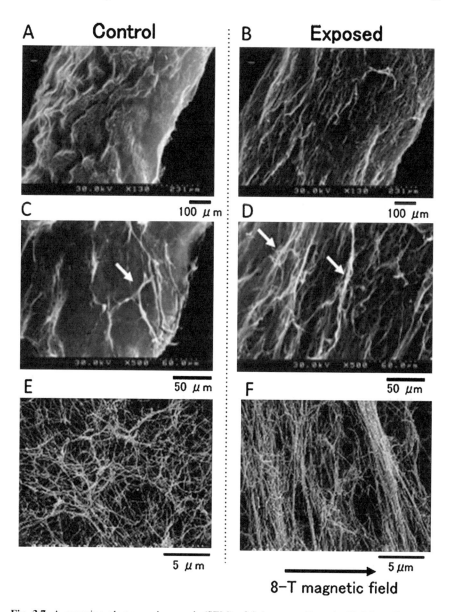

Fig. 3.7 A scanning electron micrograph (SEM) of Schwann cells embedded in collagen gel within a silicone tube. After 2 h of magnetic exposure, collagen fibers aligned perpendicular to the magnetic field, forming microbundles from 100 nm fibrils (**f**), while the control group fibers remained randomly oriented (**e**). In the exposed group, Schwann cells aligned parallel to the axis of the silicone tube, perpendicular to the magnetic field (**b, d**). In contrast, Schwann cells were randomly oriented in the control group (**a, c**). (Figure adapted with permission from Eguchi et al. 2015)

is also evidence that neural stem cells which are resident to the central nervous system will migrate directionally toward a cathode in response to an applied electric field. Interestingly, human-induced pluripotent stem cells migrate in the opposite direction toward the anode (Feng et al. 2012). Since MDD is linked to neuronal cell death and modifications to axonal connections between certain brain regions, guidance of axons and neural stem cells by induced magnetic fields is an attractive potential mechanism to explain the beneficial effects of TMS therapy for certain patients. This theory is supported by the fact that TMS often takes many sessions to achieve noticeable effects which would coincide with changes in the microenvironment being slowly altered by magnetic forces. Applications of magnetic collagen gel alignment are currently being explored for use in neural tissue engineering and CNS regenerative efforts (Plant et al. 2009; Eguchi et al. 2015).

Alternatively, the bulk movement of ECM proteins by an applied magnetic field may cause mechanically induced biochemical changes due to the activation of mechanosensitive signaling systems. Collagen fibrils bound to integrins on the surface of cells will stretch and realign perpendicularly to the field and subsequently exert a tensile or compressive force. This force can cause the activation of certain ion channels which alter polarization of a neuron or trigger signaling cascades which could result in alternate gene expression (Plant et al. 2009, Ranade et al. 2015, Johnson 2017).

TMS has been shown to alter cerebral blood flow (CBF), which is an important factor in the development of neuropathology, as decreased blood flow in critical areas of the brain can induce hypoxia and nerve death resulting in diseased states (Mesquita et al. 2013). Evidence suggests that there is an apparent modification of the fluid properties of blood when placed in a magnetic field. It has been shown in vitro and in vivo that a static magnetic field can decrease blood flow rate by increasing the apparent viscosity of blood potentially due to the exertion of magnetic torque forces on ferromagnetic iron in hemoglobin (Ichioka et al. 2000, Haik et al. 2001). This effect is apparently contradictory, as there are studies which found that repetitive TMS therapy increased CBF in the primary motor cortex (Conca et al. 2002).

3.5 TMS Stimulation Protocol and Device Design Considerations

In this section we discuss the effect of varying stimulation frequency, magnitude, and duration regarding TMS therapy protocol considerations, as well as the importance of applicator coil design.

A single-pulse TMS (sTMS) can induce depolarization in neurons, which can disrupt the normal encoding process temporarily. After sTMS is applied to a neuron, a large spike can be registered on the EMG, followed by a 150–200 ms of nearly "flat" line, which is known as the silent period (Wilson et al. 1993). During the silent period, the neural activity is suppressed comparing to the baseline readings. After the silent period, the neuron regains its normal activity, and no long-term effect is observed (Valero-Cabre and Pascual-Leone 2005). Also, multiple loosely spaced sTMS induce uniform silent periods, which confirms there is no effect outside of the silent period that can be summed up. In contrast, a burst of tightly spaced sTMS produces a longer silent period, which indicates that the interference caused by TMS may be a function of TMS frequency and duration (Valero-Cabre et al. 2019).

At the tissue level, a train of repetitive TMS (rTMS) can induce various neural behaviors with different frequencies in the stimulation area (Valero-Cabre et al. 2007). When high-frequency (f = 20 Hz) stimulation is delivered to the cerebral cortex in animal studies, the local cortical activity is suppressed as soon as the rTMS is turned on due to disrupting effects exerted by the induced current. When the rTMS is off, the activity recovers almost immediately, followed by a noticeable overshoot that lasts for a significant period. However, when stimulation is at a low-frequency pattern (f = 1 Hz), the local region experiences a less intensive suppression during stimulation and an extended suppression even after the TMS is off. This difference in local activities in an offline rTMS period is also observed in clinical experiments with human subjects. High-frequency rTMS is often used as an excitatory method for left dorsolateral prefrontal cortex (DLPFC) in MDD treatments, while low-frequency rTMS with inhibitory properties is sometimes used to suppress the right DLPFC and achieve similar antidepressant effects (Wassermann and Zimmermann 2012, Janicak and Dokucu 2015).

Coil design is an important factor in TMS therapy considerations as it alters the induced magnetic field's magnitude and direction. The size of the paddle is directly correlated with the size of the magnetic field induced, and larger coil designs can elicit stimulations deeper in the brain. However, the increasing magnetic field size also hinders the ability to direct the field toward specific targets, and thus, TMS is typically restricted to superficial cortical targets typically ranging from 2 to 3 cm below the surface of the scalp. Though the figure eight pattern is the most common paddle shape, there are many other configurations that can allow for deeper cortical stimulation to reach specific areas. As shown in Fig. 3.8, by applying different paddle sizes and shapes, it is easy to see how different cortical areas can be stimulated. Often, depth of stimulation and the focality of the stimulation area are inversely related (Groppa et al. 2012, Deng et al. 2014).

3.6 Future Perspectives

TMS is a relatively new technology with great promise in many areas but also room for improvements. The inability to target regions of the brain deeper than 2–3 cm without stimulating unsafe volumes of neural tissue is a major limiting factor in the feasible applicability of TMS to many neural diseases. Simulations of different stimulating coil geometries have revealed that the tradeoff between stimulated brain

G. Halsey et al.

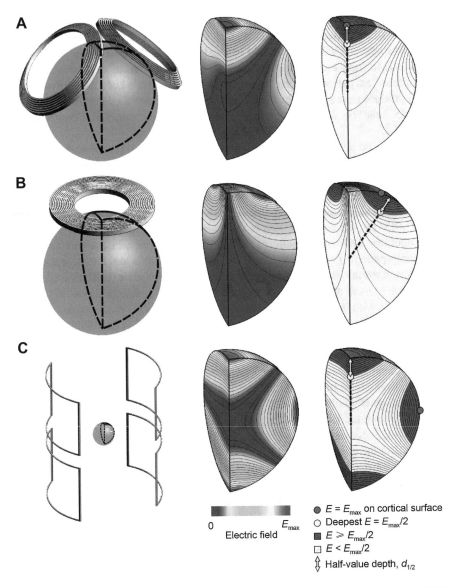

Fig. 3.8 Characterization of the electric field of (**a**) double-cone, (**b**) 90-mm circular, and (**c**) MRI gradient coils. From left to right is the respective coil geometry with a sphere representing the brain, electric field strength contour and color maps on the quarter-sphere segment of the brain, and the location of the maximum induced electric field on the brain surface, E_{max} (green circle), and the location of the maximum depth where electric field strength is $E_{max}/2$ (yellow circle) (Deng et al. 2013)

volume and depth cannot be avoided due to the intrinsic directional nature of magnetic fields induced by electrical currents in a coil. The improvement of TMS therapy hardware is effectively capped by our current understanding of electromagnetic physics, and thus, to enhance the use of TMS for clinical treatments, better patient-specific stimulation protocols and understanding of the underlying mechanisms of the effects of TMS are warranted.

The effects of magnitude and pulse duration are not well defined and vary widely among TMS studies making clinical protocol design for specific conditions challenging. Additionally, the effects of TMS to ease many neuropathological conditions may be due to a variety of mechanisms including macroscopic changes in blood flow as well as microscopic changes such as electrical depolarization of neurons and reorganization of tissue nanostructures. Cross-disciplinary coordinated studies involving experts in the fields of both neuroscience and electromagnetic physics need to be conducted to further investigate the effects of magnetic fields in vivo and determine how to take advantage of these to modulate the brain and treat neural diseases. Given the benefits that rTMS has shown for the treatment of MDD, migraines, and OCD, there is incredible potential for this technology to be applied to treat other neurological conditions which affect similar neural cross-circuits. Conditions such as Parkinson's and Huntington's diseases act on the same signaling systems in the brain as MDD; therefore, TMS may potentially be effectively repurposed to treat these conditions given an improved understanding of the feedback mechanisms and varying stimulus effects (Drevets et al. 2008).

The more recently developed transcranial direct current stimulation (tDCS) technique may overcome some design challenges faced with TMS; however, currently, tDCS is considered an investigational therapy and not approved for clinical treatments. tDCS works via the emission of a weak electrical current by electrodes attached to the subject's head with conducting jelly between the scalp and electrode surface. In the original tDCS design configuration, one electrode is known as the anodal target electrode and attached to the scalp, while the reference electrode may be placed anywhere on the body. Multiple electrode systems that can stimulate numerous parts of the brain simultaneously have greatly increased in popularity as well (Kropotov 2016).

As demonstrated in Fig. 3.9, a small direct electric current of 1–2 mA is applied to the scalp which flows by Ohm's law subsequently depolarizing or hyperpolarizing cortical pyramidal cells at their basal membrane depending on the current direction. tDCS currents are not powerful enough to evoke action potentials but, rather, change overall collective neural activities. tDCS allows for better spatial resolution of stimulation at a depth than TMS therapy, but the ability to define exact protocols for electrode placement and stimulation frequency is diminished due to limitations in determining the exact direction and magnitude of current densities spreading throughout the neural tissue (Kropotov 2016).

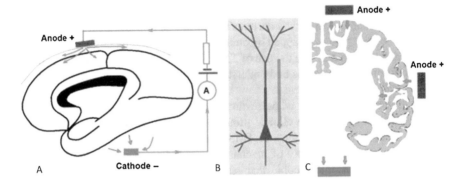

Fig. 3.9 (**a**) Two-electrode tDCS system, anode (+) and cathode (−). (**b**) In cortical layers oriented perpendicular to the current, pyramidal cells are depolarized at their basal membrane. (**c**) Anode location and cortical folding can influence depolarization (green arrows) and hyperpolarization (blue arrows) or have no effect on average potential (white arrows). (Adapted with permission from Kropotov 2016)

3.7 Summary

TMS technology has shown efficacy in treating MDD, migraines, and OCD, with great further potential to treat a wide variety of other neurological disorders. However, despite clinical approval and advancements in research, the exact mechanisms and direct protocols for treating different diseases remain not well defined. Further studies into the direct effects of TMS with varying stimulus protocols may increase the utility of this technology, which could be potentially combined with traditional surgical and pharmacological approaches to treat neurological disease.

The development of improved methods for tDCS has provided another means to achieve similar effects to those of the TMS therapy; however, this technology has similar issues with the requirement for developing patient- or condition-specific protocols. While tDCS has improved the spatial stimulation resolution compared to TMS, the tissue density, volume, and shape of the brain vary from patient to patient and affect the spread of the applied current. Significantly more efforts towards understanding the distribution of intracranial electrical currents are needed to advance this emerging technology to maturity.

Acknowledgments This work was partially supported by the Chronic Brain Injury Program of The Ohio State University through a Pilot Award. The authors thank Nick Heydinger, Daniel J. Robinson, and Hao Chen for valuable discussions on the topic.

References

Administration, U. S. F. a. D. (2018). *Press Announcements – FDA permits marketing of transcranial magnetic stimulation for treatment of obsessive compulsive disorder.* From https://www. fda.gov/NewsEvents/Newsroom/PressAnnouncements/ucm617244.htm

Barker, A. T., & Shields, K. (2017). Transcranial magnetic stimulation: Basic principles and clinical applications in migraine. *Headache, 57*(3), 517–524.

Barker, A. T., Jalinous, R., & Freeston, I. L. (1985). Non-invasive magnetic stimulation of human motor cortex. *Lancet, 1*(8437), 1106–1107.

Bench, C. J., Friston, K. J., Brown, R. G., Scott, L. C., Frackowiak, R. S., & Dolan, R. J. (1992). The anatomy of melancholia–focal abnormalities of cerebral blood flow in major depression. *Psychological Medicine, 22*(3), 607–615.

Bremner, J. D., Narayan, M., Anderson, E. R., Staib, L. H., Miller, H. L., & Charney, D. S. (2000). Hippocampal volume reduction in major depression. *The American Journal of Psychiatry, 157*(1), 115–118.

Brigitta, B. (2002). Pathophysiology of depression and mechanisms of treatment. *Dialogues in Clinical Neuroscience, 4*(1), 7–20.

Butler, A. J., & Wolf, S. L. (2007). Putting the brain on the map: Use of transcranial magnetic stimulation to assess and induce cortical plasticity of upper-extremity movement. *Physical Therapy, 87*(6), 719–736.

Conca, A., Peschina, W., Konig, P., Fritzsche, H., & Hausmann, A. (2002). Effect of chronic repetitive transcranial magnetic stimulation on regional cerebral blood flow and regional cerebral glucose uptake in drug treatment-resistant depressives. A brief report. *Neuropsychobiology, 45*(1), 27–31.

Deng, Z. D., Lisanby, S. H., & Peterchev, A. V. (2013). Electric field depth-focality tradeoff in transcranial magnetic stimulation: Simulation comparison of 50 coil designs. *Brain Stimulation, 6*(1), 1–13.

Deng, Z. D., Lisanby, S. H., & Peterchev, A. V. (2014). Coil design considerations for deep transcranial magnetic stimulation. *Clinical Neurophysiology, 125*(6), 1202–1212.

Drevets, W. C., Price, J. L., & Furey, M. L. (2008). Brain structural and functional abnormalities in mood disorders: Implications for neurocircuitry models of depression. *Brain Structure & Function, 213*(1–2), 93–118.

Eguchi, Y., Ohtori, S., Sekino, M., & Ueno, S. (2015). Effectiveness of magnetically aligned collagen for neural regeneration in vitro and in vivo. *Bioelectromagnetics, 36*(3), 233–243.

Faria, M. A. (2013). Violence, mental illness, and the brain – A brief history of psychosurgery: Part 3 – From deep brain stimulation to amygdalotomy for violent behavior, seizures, and pathological aggression in humans. *Surgical Neurology International, 4*, 91.

Feng, J. F., Liu, J., Zhang, X. Z., Zhang, L., Jiang, J. Y., Nolta, J., & Zhao, M. (2012). Guided migration of neural stem cells derived from human embryonic stem cells by an electric field. *Stem Cells, 30*(2), 349–355.

Gersner, R., Oberman, L., Sanchez, M. J., Chiriboga, N., Kaye, H. L., Pascual-Leone, A., Libenson, M., Roth, Y., Zangen, A., & Rotenberg, A. (2016). H-coil repetitive transcranial magnetic stimulation for treatment of temporal lobe epilepsy: A case report. *Epilepsy Behav Case Rep, 5*, 52–56.

Groppa, S., Oliviero, A., Eisen, A., Quartarone, A., Cohen, L. G., Mall, V., Kaelin-Lang, A., Mima, T., Rossi, S., Thickbroom, G. W., Rossini, P. M., Ziemann, U., Valls-Sole, J., & Siebner, H. R. (2012). A practical guide to diagnostic transcranial magnetic stimulation: Report of an IFCN committee. *Clinical Neurophysiology, 123*(5), 858–882.

Haik, Y., Pai, V. N., & Chen, C. J. (2001). Apparent viscosity of human blood in a high static magnetic field. *Journal of Magnetism and Magnetic Materials, 225*(1–2), 180–186.

Hallett, M. (2000). Transcranial magnetic stimulation and the human brain. *Nature, 406*(6792), 147–150.

Health, N. I. o. M. (2019). *Major depression*. From https://www.nimh.nih.gov/health/statistics/major-depression.shtml

Ichioka, S., Minegishi, M., Iwasaka, M., Shibata, M., Nakatsuka, T., Harii, K., Kamiya, A., & Ueno, S. (2000). High-intensity static magnetic fields modulate skin microcirculation and temperature in vivo. *Bioelectromagnetics, 21*(3), 183–188.

Janicak, P. G., & Dokucu, M. E. (2015). Transcranial magnetic stimulation for the treatment of major depression. *Neuropsychiatric Disease and Treatment, 11*, 1549–1560.

Jeon, S. W., & Kim, Y. K. (2016). Molecular neurobiology and promising new treatment in depression. *International Journal of Molecular Sciences, 17*(3), 381.

Johnson, W. A. (2017). Two views of the same stimulus. *eLife, 6*, e30191.

Kolin, A., Brill, N. Q., & Broberg, P. J. (1959). Stimulation of irritable tissues by means of an alternating magnetic field. *Proceedings of the Society for Experimental Biology and Medicine, 102*, 251–253.

Kropotov, J. D. (2016). Transcranial direct current stimulation. *Functional Neuromarkers for Psychiatry: Applications for Diagnosis and Treatment* 1st ed. Vol. 1, 273–280.

Lovsund, P., Oberg, P. A., Nilsson, S. E., & Reuter, T. (1980). Magnetophosphenes: A quantitative analysis of thresholds. *Medical & Biological Engineering & Computing, 18*(3), 326–334.

Mesquita, R. C., Faseyitan, O. K., Turkeltaub, P. E., Buckley, E. M., Thomas, A., Kim, M. N., Durduran, T., Greenberg, J. H., Detre, J. A., Yodh, A. G., & Hamilton, R. H. (2013). Blood flow and oxygenation changes due to low-frequency repetitive transcranial magnetic stimulation of the cerebral cortex. *Journal of Biomedical Optics, 18*(6), 067006.

NiCE. (2014). *Transcranial magnetic stimulation for treating and preventing migraine, Guidance and guide-lines*. From https://www.nice.org.uk/guidance/ipg477/chapter/3-The-procedure

Noakes, R. (2007). Cromwell Varley FRS, electrical discharge and Victorian spiritualism. *Notes and Records of the Royal Society of London, 61*, 5.

Organization, W. H. (2018). *Depression*. Fact sheets. From http://www.who.int/news-room/fact-sheets/detail/depression

O'Shea, J., & Walsh, V. (2007). Transcranial magnetic stimulation. *Current Biology, 17*(6), R196–R199.

Plant, A. L., Bhadriraju, K., Spurlin, T. A., & Elliott, J. T. (2009). Cell response to matrix mechanics: Focus on collagen. *Biochimica et Biophysica Acta, 1793*(5), 893–902.

Polson, M. J., Barker, A. T., & Freeston, I. L. (1982). Stimulation of nerve trunks with time-varying magnetic fields. *Medical & Biological Engineering & Computing, 20*(2), 243–244.

Poyser, A. W. (1892). *Magnetism and electricity: A manual for students in advanced classes*. New York: Longmans, Green, & Company.

Price, J. L., & Drevets, W. C. (2012). Neural circuits underlying the pathophysiology of mood disorders. *Trends in Cognitive Sciences, 16*(1), 61–71.

Ranade, S. S., Syeda, R., & Patapoutian, A. (2015). Mechanically activated ion channels. *Neuron, 87*(6), 1162–1179.

Rossi, S., Hallett, M., Rossini, P. M., Pascual-Leone, A., & T. M. S. C. G. Safety. (2009). Safety, ethical considerations, and application guidelines for the use of transcranial magnetic stimulation in clinical practice and research. *Clinical Neurophysiology, 120*(12), 2008–2039.

Schwedt, T. J., & Vargas, B. (2015). Neurostimulation for treatment of migraine and cluster headache. *Pain Medicine, 16*(9), 1827–1834.

Torbet, J., & Ronziere, M. C. (1984). Magnetic alignment of collagen during self-assembly. *The Biochemical Journal, 219*(3), 1057–1059.

Ueno, S., Matsuda, T., & Fujiki, M. (1990a). Functional mapping of the human motor cortex obtained by focal and vectorial magnetic stimulation of the brain. *IEEE Transactions on Magnetics, 26*(5), 1539–1544.

Ueno, S., Matsuda, T., & Hiwaki, O. (1990b). Localized stimulation of the human brain and spinal-cord by a pair of opposing pulsed magnetic-fields. *Journal of Applied Physics, 67*(9), 5838–5840.

Valero-Cabre, A., & Pascual-Leone, A. (2005). Impact of TMS on the primary motor cortex and associated spinal systems. *IEEE Engineering in Medicine and Biology Magazine, 24*(1), 29–35.

Valero-Cabre, A., Payne, B. R., & Pascual-Leone, A. (2007). Opposite impact on C-14-2-deoxyglucose brain metabolism following patterns of high and low frequency repetitive transcranial magnetic stimulation in the posterior parietal cortex. *Experimental Brain Research, 176*(4), 603–615.

Valero-Cabre, A., Amengual, J. L., Stengel, C., Pascual-Leone, A., & Coubard, O. A. (2019). Transcranial magnetic stimulation in basic and clinical neuroscience: A comprehensive review of fundamental principles and novel insights (vol 83, pg 381, 2017). *Neuroscience and Biobehavioral Reviews, 96*, 414–414.

Wassermann, E. M., & Zimmermann, T. (2012). Transcranial magnetic brain stimulation: Therapeutic promises and scientific gaps. *Pharmacology & Therapeutics, 133*(1), 98–107.

Wilson, S. A., Lockwood, R. J., Thickbroom, G. W., & Mastaglia, F. L. (1993). The muscle silent period following transcranial magnetic cortical stimulation. *Journal of the Neurological Sciences, 114*(2), 216–222.

Ziemann, U. (2017). Thirty years of transcranial magnetic stimulation: Where do we stand? *Experimental Brain Research, 235*(4), 973–984.

Chapter 4
Intracortical Electrodes

Meijian Wang and Liang Guo

4.1 Introduction

Neural signals, produced and transmitted through the networks of nervous systems, account for how we feel, move, reason, learn, and emote. Cerebral cortex is the outer surface of the brain including the gray matter, which consists of the neural somas, and is a part of intense electrophysiological research interests. Implantable neural interfaces have been designed for revealing the mechanisms of signal processing in the brain and treating related neurological diseases over the past 70 years (Hodgkin and Katz 1949; Hubel and Wiesel 1959, 1962). Intracortical electrodes are a class of devices inserted in the cerebral cortex to record electrical signals generated by cortical neurons in order to investigate the neural mechanisms or interface with external artificial systems (Fig. 4.1). A wealth of different neural electrodes have been developed for intracortical recording. They usually contain conductive materials, in shape of wires or strips for relaying bio-electrical signals, and insulated materials, covering the electrode shanks and exposing the electrodes for sensing nearby electric signals. These electrodes are capable of recording different signal types, including single-unit, multiunit, and local field potentials with better spatial and temporal resolutions comparing to other less invasive cortical recording approaches (Fig. 4.1). Intracortical recording can be performed with a single micro-wire electrode, a grid of micro-wire electrodes, or a micromachined electrode array, whose implantation can be either acute or chronic.

The recording quality and biocompatibility of intracortical electrodes are being continuously improved with introduction of new materials, designs, and fabrication techniques, since intracortical recording became a tool for the neuroscientists

M. Wang · L. Guo (✉)
Department of Electrical and Computer Engineering, The Ohio State University,
Columbus, OH, USA

Department of Neuroscience, The Ohio State University, Columbus, OH, USA
e-mail: guo.725@osu.edu

© Springer Nature Switzerland AG 2020
L. Guo (ed.), *Neural Interface Engineering*,
https://doi.org/10.1007/978-3-030-41854-0_4

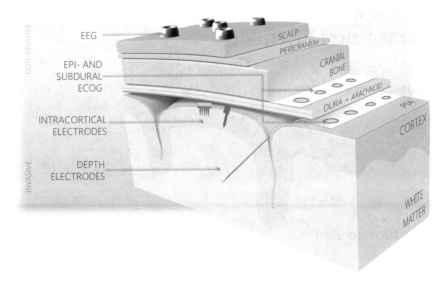

Fig. 4.1 Different types of electrodes for recording the brain's electrical activities. Invasive approaches provide better resolution compared to less invasive ones. (The figure is adapted from Szostak et al. (2017) with permission)

(Fig. 4.2). Contemporary intracortical electrodes can simultaneously record from hundreds of neurons over many weeks, months, or even years and have a good performance not only in rodents and monkeys but also in humans. Development and application of these intracortical electrodes are promoting new understanding in cerebral circuits and their disease models, as well as new therapies for neurological or psychiatric disorders.

Brain–machine interfaces (BMIs) or brain–computer interfaces (BCIs) are a creative concept that has captured much interests from the scientific community. BMIs aim at building a direct information exchange channel between the brain and an external artificial system and can provide innovative bypassing capabilities in restoration of lost brain or communication functions. For example, there have been successful BMIs for improving the independence of individuals with severe motor impairments (Velliste et al. 2008; Hochberg et al. 2006, 2012). In these science-fiction-like applications, intracortical electrodes are an essential component responsible for extracting relevant information from the brain (Brandman et al. 2017).

Even though intracortical electrodes have achieved many successes, there are still many challenges limiting their capability and stability (Sommakia et al. 2014; Groothuis et al. 2014). Among these challenges, long-term viability and biocompatibility have been reported as the most important problem to date. Therefore, it is essential to understand the factors that lead to foreign body responses and create targeted interventions to control them.

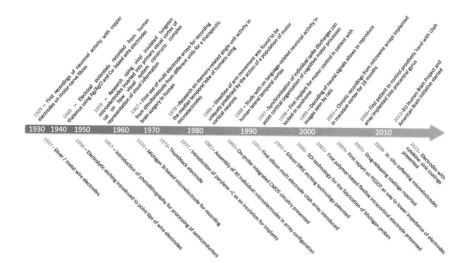

Fig. 4.2 Milestones in the history of brain recording. (The figure is adapted from Szostak et al. (2017) with permission)

4.2 Types of Intracortical Electrodes

4.2.1 Wire-Based Intracortical Electrodes

Wire-based electrodes are made of insulated 10–200 μm diameter metal wires with exposed tips to detect the electrical activities of neurons in their vicinity. Because metal wires were easily acquirable, wire-based electrodes emerged very early, dating back to the studies using silver probes in the early twentieth century (Rheinberger and Jasper 1937). A number of features of conventional metal wires are ideally suited for single neuron recording in the mammalian cortex (Garner et al. 1972; Kaltenbach and Gerstein 1986). Up to now, wire-based intracortical electrodes are still widely used in neural electrophysiological studies.

The selection of materials for wire-based intracortical electrodes depends on the location of target cortex to be recorded, requirement for signal quality, time course of implantation, and whether subject's movement is to be restrained. Metallic wires used for the electrodes include tungsten, stainless steel, nichrome, iridium, platinum, platinum-iridium alloys, etc. Their stiffness, conductivity, and corrosion resistance are often considered in the selection (Lehew and Nicolelis 2008). Insulation layers are coated onto the wires to ensure electrical isolation, protect them from biofluid corrosion, and increase their biocompatibility. Common materials for insulation include glass, Teflon, resins, polyimide (PI), Parylene, formvar, etc. Wires for

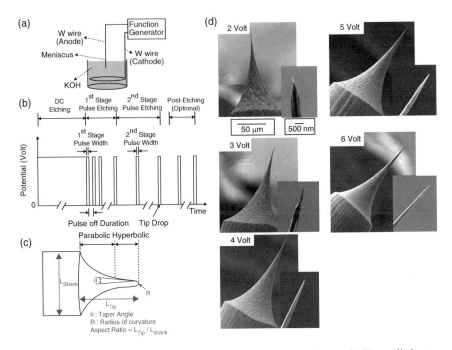

Fig. 4.3 Electrochemical etching of a tungsten tip. (**a**) Experimental setup. (**b**) The applied potential comprises DC and subsequent pulsatile potentials. (**c**) Tip profile includes tip length, radius of curvature at the apex, and taper angle. (**d**) SEM images of etched tungsten tips. Higher potentials decrease the surface roughness. (The figure is adapted from Chang et al. (2012) with permission)

the electrodes are generally obtained off the shelf, pre-coated with an insulation layer. Proper methods, such as cutting, breaking, and grinding, can produce conductive tips on these wires. If the wires are not pre-coated, insulation deposition would be needed after the tips are processed. In general, the fabrication of wire-based electrodes includes straightening and cutting the wire, shaping and smoothing the tip, insulation deposition, tip de-insulation, array assembling, etc.

Without post-processing, smooth tips of electrodes can be cut by sharp blade, surgical scissors, or lasers and can be directly used for neural recording. With post-processing, tapered micrometer–diameter tips can be obtained and used for high-quality recording. The quality and shape of the tip are a key factor in determining the performance of neural recording (Palmer 1978; Hubel 1957). According to early studies, tips of less than 5 μm in diameter are much more satisfactory; tips of about 20 μm in diameter may at times be adequate for extracellular single-unit recording; and tips of less than 1 μm in diameter are usually demanded for intracellular recording (Hubel 1957).

Electrochemical etching is the most predominant technique for post-processing electrode tips (Hubel 1957; Grundfest et al. 1950). In the etching, the metal wires are dipped into an electrolyte, such as 50% sodium hydroxide with 30% sodium hydroxide for platinum–iridium wires (Wolbarsht et al. 1960), saturated aqueous

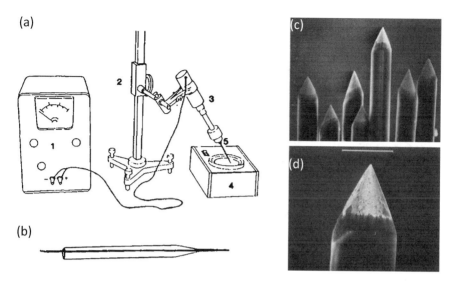

Fig. 4.4 Setup for grinding a tapered tip on a pre-insulated tungsten wire. (**a**) (1) Power supply, (2) pipette mount with vertical adjust, (3) motor-driven chuck, (4) rotating disk with a thin layer of diamond, and (5) a micropipette centered with a tungsten wire as shown in (**b**). (**b**) Enlargement of the tungsten wire's position in the pipette. (**c, d**) SEM images of ground tips. Scale bar: 25 μm. (The figures are adapted from Kaltenbach and Gerstein (1986) with permission)

potassium nitrite solution for tungsten wires (Hubel 1957), and an acid solution for stainless steel wires (Grundfest et al. 1950). A potentiostat applies a DC potential to the electrode wire as an anode (Fig. 4.3). A meniscus of solution is formed around the wire at the surface of the electrolyte. Its height depends on the diameter of the wire and decreases as the etching proceeds. As a result, a tapered tip is obtained. In addition, the wire can be repeatedly immersed and withdrawn from the electrolyte to produce a variety of tip shapes. Since the etching usually requires the electrode not to be insulated, an insulation deposition is needed afterward. The coating methods include dip coating, heat shrink, physical vapor deposition, electrodeposition, fluidized bed, etc.

Sharpening the tip on a pre-insulated wire can be achieved by grinding without a further process for insulation deposition and tip exposure (Kaltenbach and Gerstein 1986). One example of such a method is shown in Fig. 4.4. A pre-insulated wire was threaded into a micropipette and fixed by vinyl base red utility wax. The micropipette was pulled with as nearly a centered tip as possible, because this would contribute to symmetry of the ground tip. The wire was cut to leave 2–3 mm extending beyond the micropipette tip. The micropipette with wire was mounted in a motor-driven rotating chuck. A rotating plate coated with diamond powder was used to grind the tip. To prevent the grinding from heating that might damage the insulation, water was poured to cover the entire surface of the rotating plate. Then, the wire was lowered to touch the grinding surface with the aid of a dissection microscope. According to prior experience, an incident angle of 40° produced a 65–70° tapered

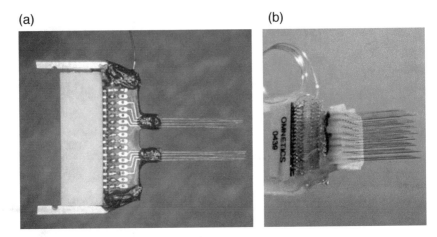

Fig. 4.5 Examples of MWAs assembled by layered and discretely wired approaches, respectively. (**a**) Layered MWA assembled onto a PBC. (**b**) Discrete MWA assembled onto connectors. (Figures are adapted with permissions from Lehew and Nicolelis (2008))

tip. By observing under the dissection microscope, the chuck was raised up when the desired sharpness was achieved. After the grinding, the pipette was disconnected from the chuck and then the wax was melted in boiling water to free the wire.

A single wire electrode can be used to record one neuron at a time. To record the activities of a population of neurons at the same time, a bundle or array of wire-based electrodes, organized in different geometrical configurations, have been designed. There are various methods to assemble the microwire arrays (MWAs). Those methods can be classified into layered or discretely wired approaches (Lehew and Nicolelis 2008). The layered approach involves producing a printed circuit board (PCB) to provide a platform for mounting the array components (Fig. 4.5a). The discretely wired approach mechanically bonds the electrodes in a desired pattern with customized jigs and spacers for routing to a connector (Fig. 4.5b).

4.2.2 *Micromachined Intracortical Electrodes*

Microfabrication is a technique with which micro-scale structures are fabricated on the surface of a substrate material or patterned into the substrate material itself (Ainslie and Desai 2008). Micromachined intracortical electrodes are a type of silicon-based electrode arrays created by the microfabrication techniques that can pattern substrates' shape and add/remove conducting or insulating layers. In the 1960s, the invention of photolithography promoted the progress of micromatching technologies, which in turn benefited the development of the new generation of intracortical electrodes (Jules 1964). The key advantage over the wire-based electrodes is the precise location of electrodes in the array across the same design, which provides more accurate spatial representations of population neural activities.

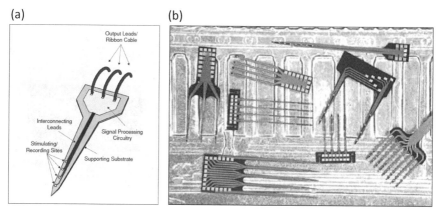

Fig. 4.6 Michigan electrode array. (**a**) General structure of a Michigan electrode array. (**b**) Different variants of Michigan-style microelectrode arrays on the back of a US Penny. (The figures are adapted with permission from Wise (2005))

Materials for the substrate of the electrode include glass, semiconductor, and metal. Materials for the conductor and insulator layers vary and mainly include those commonly used in CMOS and MEMS industries, such as oxides, nitrides, polymers, and metals. Most materials in MEMS are considered biocompatible and nonirritant according to ISO 10993 biocompatible test, such as silicon, polysilicon, silicon nitride, silicon carbide, silicon thermal oxide, titanium, and SU-8 (Kotzar et al. 2002).

Micromachined intracortical electrodes have a variety of designs and are still evolving in face of emerging new techniques. Two classic designs are the Michigan electrode arrays with two-dimensional recording sites (Fig. 4.6) and the Utah electrode array (UAE) with three-dimensional multineedles (Fig. 4.8).

Wise and colleagues at the University of Michigan developed a type of silicon-based electrodes by the then newly developed microfabrication technique, which was referred to as the Michigan electrode array (Wise 2005; Wise et al. 1970). This type of electrode arrays was first designed specifically for extracellular neural recording in the brain and consisted of a gold wire insulated by a thin layer of silicon dioxide and formed on a silicon carrier. The arrays had multiple recording sites placed along a single or multiple planar shanks and was capable of recording from well-controlled depths (Fig. 4.6).

Developments of microfabrication techniques, including etch stops, deep reactive ion etching (DRIE), and anisotropic silicon etching, have dramatically advanced the designs of Michigan electrode array from its first version. The basic microfabrication processes are shown in Fig. 4.7 (Wise 2005). A thermal oxide mask is used to create a diffused boron etch stop that defines the substrate's dimensions. These stops allow for defining the probe thickness to less than 15 μm. Then, a dielectric layer, usually consisting of a silicon oxide/silicon nitride stack, is deposited to insulate the backside of the electrode. Up on the dielectrics, a series of conductive traces

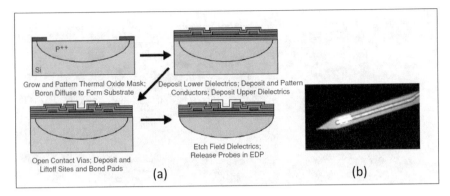

Fig. 4.7 Fabrication of Michigan electrode array. (**a**) The basic process flow. (**b**) An example fabricated electrode. (The figures are adapted with permission from Wise (2005))

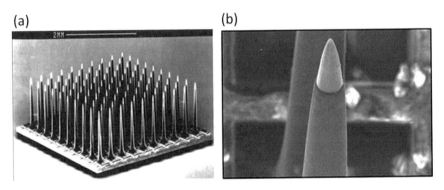

Fig. 4.8 Example UEA. (The figures are adapted with permission from Normann and Fernandez (2016))

are patterned to the length of the electrodes for connecting the recording sites to bonding pads, which are then created from metals. The common choices for the recording sites include gold, platinum, and iridium (iridium oxide). To insulate upper face of the conductive traces, another dielectric layer of stress-compensated silicon dioxide/silicon nitride is deposited. In addition, to increase the biocompatibility and protect the dielectrics from dissolution in in vivo conditions, an insulating polymer layer such as Parylene-C or Epoxylite has been adopted.

In parallel with the development of the Michigan electrode array, Normann and colleagues at the University of Utah developed another type of silicon-based electrode array, which was referred to as the UEA (Campbell et al. 1991). Instead of patterning recording sites on a planer surface, a grid of microneedles, which were first cut with a dicing saw, were processed by a series of steps such as etching and deposition into the electrodes with recording tips in the UEA (Fig. 4.8).

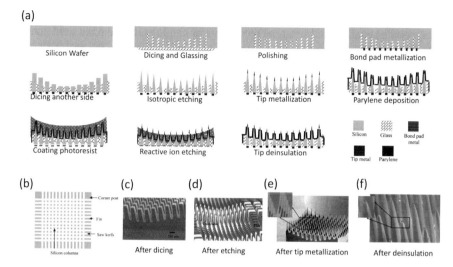

Fig. 4.9 Fabrication of the UEA. (**a**) Process flow for fabricating a diverse-height UEA. (**b**) Top view of the array dicing. (**c–f**) SEM images of the array during fabrication. (The figures are adapted with permission from Bhandari et al. (2008))

The typical UEA structure is shown in Fig. 4.8 (Normann and Fernandez 2016). The example UEA in Fig. 4.8a consists of a 4 mm × 4 mm substrate plate that projects out 100 (10 × 10 in a grid) penetrating microneedles, each 1.5 mm in length. The microneedles are electrically isolated from each other by a "moat" of glass surrounding at the needle roots. On the back surface of the substrate, 100 bonding pads are deposited to connect with each microneedle, and 100 insulated wires are used to connect these pads to an externally mounted connector. The tip of each microneedle is deposited with iridium oxide to facilitate ionic-to-electronic signal transduction. Except the tip of the microneedles, a biocompatible Parylene-C is used to insulate the entire array.

Generally, the basic fabrication processes of UEA include dicing, glassing, grinding, etching, and depositing conductive and insulative layers (Fig. 4.9) (Jones et al. 1992; Green et al. 2008; Bhandari et al. 2010). A grid of kerfs are cut by a diamond dicing saw on the back side of a chunk of doped silicon wafer, and the cut's gap and depth depend on the design of specific UEA. The kerfs are filled with an insulating material of glass. The excess glass layer on the wafer top is removed by a grinding process. Then, back-side metalization is conducted to form metal bonding pads on the back side of each electrode. To produce the columns that will eventually be etched to microneedles, a dicing saw is used again to cut the front side of the wafer. The square columns are subsequently etched into rounded and sharp needles. To facilitate charge transfer between the neural tissue and electrode, iridium oxide film (AIROF) is deposited on the microneedle tips. To form an insulative and biocompatible layer, low-pressure chemical vapor deposition is used to deposit Parylene-C on the array. Similar to the wire-based electrodes, the insulation layer on the electrode tip must be removed to form recording sites. A wafer-scale tip deinsulation can be achieved by a combination of photoresist masking and oxygen plasma etching.

The unique architecture of the UEA with a hundred, few-millimeters-long microneedles enables multichannel recordings of single units and local field potentials (LFPs) with high spatial and temporal resolutions. Therefore, the correlation or communication between a large number of neurons within a cortical volume that the array covers can be analyzed. The capability of the UEA has been demonstrated in multiple cortical recordings, such as in visual and auditory cortices (Warren et al. 2001; Kim et al. 2006). In addition, the UEA has achieved great success in BMIs after being approved by FDA for human testing (Hochberg et al. 2006; Cyberkinetics 2005).

4.3 Intracortical Recording

4.3.1 Placement of Intracortical Electrodes

Intracortical electrodes have been applied in animal studies for a long time and in human clinical treatment preliminarily. Here, we describe the basic procedures of placing intracortical electrodes into interested regions of the brain in the view of animal research.

Intracortical electrodes can be placed into the cortex either chronically or acutely. For the acute placement, the electrodes are inserted into the cortical regions of interest at each time of recording. If the animal is kept alive for consecutive recording sessions, a recording chamber is usually implanted on its skull by a surgery. In some cases, the surgery also performs craniotomy and implant guiding tubes or grids on the skull inside the recording chamber. In other cases, the craniotomy and electrode placement are performed just before each recording session. The electrodes selected for acute implantation are usually easy to insert, such as a single tungsten wire electrode. A microdriver is applied, manually or electrically, to advance the electrode to a desired depth with a micrometer precision. For chronic implantation, a surgery is performed to implant the electrodes in the brain, and the electrodes would be kept for long-time usage. Their connectors are placed in a connector chamber, with which the electrodes can be connected to an external system. The electrodes chosen for chronic implantation are usually arrays, such as the UEA, because they can be used for a long time with enough available recording channels. Sometimes, the depth of chronic electrodes are also adjustable by microdrivers.

In animal studies, the surviving surgeries are performed in a sterile environment (Carter and Shieh 2010). The surgery room is cleaned and sterilized with UV lights just before surgery. The tools and probes that will contact with the inside of the animal body need complete sterilization. There are several ways to sterilize. A bead sterilizer is a small and convenient device for sterilization, filled with tiny glass beads that allow heat to evenly surround the tools. An autoclave is a commonly used device to kill microorganisms by high-temperature steam in a high-pressure container. Most utilities, including stainless tools, gauzes, cotton swabs, suture lines,

(a) (b) (c)

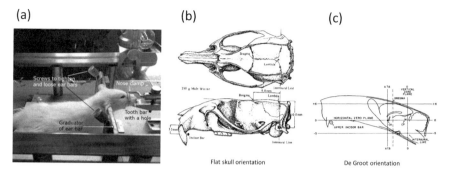

Fig. 4.10 A stereotaxic instrument (**a**) and head orientations (**b** and **c**) for stereotaxic coordinates in rats. ((**a**) is adapted with permission from (Zhang and Xiong 2014);(**b**) is adapted with permission from (Paxinos and Watson 2007); and (**c**) is adapted with permission from (De Groot 1959))

needles, saline, glass wares, etc., can be sterilized by the autoclave. But, some electrodes and tools cannot tolerate the high heat. A relatively lower-temperature method, ethylene oxide sterilization, commonly called EtO, is suitable for sterilizing these things. If the EtO is not available, these heat-sensitive tools can be soaked in 70% ethanol or other disinfectant for sterilization before use.

Once the surgery room and tools are ready, the animals can be anesthetized by either gas anesthetics, such as isoflurane, or pharmacological agents, such as ketamine. If it is a long-time surgery for a large animal, such as a nonhuman primate and cat, gas anesthetics with anesthetic machine is preferred, and some of the pharmacological agents through intravenous infusion are also suitable. In addition, pre-anesthesia by intramuscular or intraperitoneal injection with pharmacological agents is usually applied before the long-time anesthesia. Other agents, such as atropine for restraining saliva and dexamethasone for reducing brain swelling, might also be injected before the anesthesia. The fur on the head is shaved after the anesthetization. To keep body temperature during the surgery, heat pads are usually used. To monitor the animal's vital signs, including body temperature, blood oxygen saturation, heart rate, respiratory rate, and end-tidal CO_2, a vital sign alert system might be applied, especially for large animals. Intravenous infusion and oxygen might also be provided to keep the animal in a good condition during the surgery.

The properly anesthetized animal is placed on a stereotaxic instrument, which is designed to hold the animal's head in a precise orientation and uses a micromanipulator to measure and access targets with about 1 μm precision (Fig. 4.10a). First, ear bars are inserted into the ear canals. Second, a bite bar is place under the teeth, which can be adjusted to allow the head tilt up and down. Third, a nose brace is pressed on the nose, or a pair of eye braces is pressed on the eyes' lower orbits to restrict the upward movement of the head's anterior part. The head orientation can be adjusted by moving the bite bar and the braces.

It is usually necessary to have a sterile field around the operated part of the animal body. The sterile field can cover the contaminated part of the animal and serve as a resting place for tools. Some autoclaved strips of gauze or commercially available papers can be arranged to a sterile field. The scalp that was shaved can be sterilized by wiping with iodine or peroxide.

Localization of the recording site in the brain is one essential step for the placement of intracortical electrodes. Stereotaxic technique of brain is a minimally invasive method that utilizes a three-dimensional coordinate system and refers extra- and intracranial landmarks to insert a cannula or electrode into a target location of the brain. The brain atlases providing the three-dimensional coordinates of brain structures are available for a variety of animals, including rodents, cats, bats, birds, and primates. Here, we describe an example of rat to illustrate the stereotaxic coordinates (Paxinos and Watson 2007). To be consistent with the three-dimensional coordinates of the published brain atlases, flat skull orientation (Fig. 4.10b) or De Groot orientation (Fig. 4.10c) is commonly applied. The top surface between bregma and lambda is flat in the former one and is tilted about 68 degrees in the latter one. The rat stereotaxic coordinates can be determined by three values corresponding to dorsal–ventral (D–V), midline–lateral (M–L), and anterior–posterior (A–P) coordinates. The D–V coordinate is a vertical axis, the A–P coordinate is an axis from left to right, and the A–P coordinate is an axis from anterior to posterior parts of the rat skull. Their values would be different depending on which skull orientation is set and how to define the reference plane. Therefore, in the research articles, the coordinate values would be noted by which system was applied. For animals with limited resources, structural MR images can be taken before and/or after the implantation surgery to help the localization (Fiebelkorn et al. 2019).

With the stereotaxic technique, a craniotomy is performed by a dental drill to expose the brain. Its size and shape are determined by the requirements of the implanted electrodes. A section of dura is often removed for most delicate electrodes. After the brain is exposed, the intracranial landmarks, including blood vesicles, sulcus, and gyrus, are visible for localization of the recording sites. For some large areas of the surface of the brain that can be easily accessed to, several of extra- and intracranial landmarks might be enough for the localization without aiding by the stereotaxic coordinates. For deep or folded areas, the coordinate system might be needed.

Once the target location determined, the electrode can be delivered to the brain using a micromanipulator during the surgery, if it is for chronic implantation. In some cases, the surgery does not place the electrode but only implants some device that allows long-term access to the brain, such as a sealable chamber, cannula, and cranial window, and then the electrode is placed in assistance with this device before each recording session. Titanic screws, dental acrylic, and bone cement are commonly applied to hold the implants in place.

After implantation, sterile sutures or tissue glues are used to seal the skin incision. The scalp can be lightly treated with iodine solution to clean the affected area. It will take about a week for the animal to recover from the surgery. During the recovery time, the animal should be inspected for any signs of discomfort and pain, including a lack of eating, hunched posture, and slow movement. Injections of an analgesic and antibiotics are used to prevent infection. After recovery, the intracortical electrode can be tested for neural recordings.

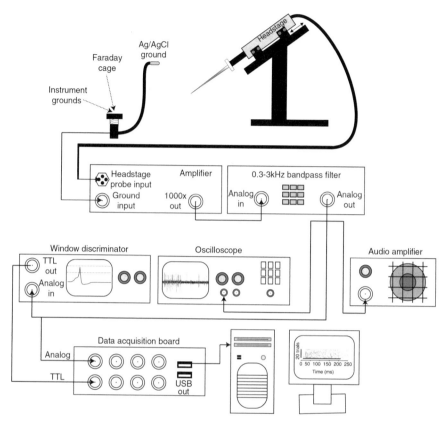

Fig. 4.11 A basic extracellular electrophysiology setup. (The figure is adapted with permission from Coleman and Burger (2015))

4.3.2   Data Acquisition and Analyses

The classic intracortical electrodes are usually used to record extracellular neural signals in research experiments and clinical treatments. The setup for intracortical recording includes the following key components: (1) recording electrodes, (2) a ground wire, (3) a headstage, (4) a differential amplifier, (5) a filter, (6) some means of signal monitoring, (7) an AD converter, and (8) a computer with acquisition software. An example of the setup is shown in Fig. 4.11 (Coleman and Burger 2015). There are many ways to incorporate these key components into a recording setup to suit to different experimental needs and personal preferences. Nowadays, a variety of multichannel acquisition systems are commercially available, such as the systems from Neuralynx, Tucker–Davis Technologies, Plexon, Blackrock Microsystems, Ripple, etc.

In general, the recorded signal from the electrode is preamplified by a small circuit board named headstage. The preamplified signal, which becomes less susceptible to surrounding noises than the original signal, is transmitted through a tether to an amplifier for further amplification. The digitalization might be set at the time of either preamplification or post-amplification according to different setups. The wideband signal is filtered to a low-frequency part for LFP analysis and/or a high-frequency part for spike analysis. Some commercially available software is typically designed to monitor and record multichannel signals from electrode array or several single electrodes.

The electrode itself and the recorded cortical region are two main factors affecting the quality of the signal collected from the electrode and should be carefully controlled during the recording session. For some electrodes, such as the UEA, their locations are determined from the implantation; for others, such as the single tungsten electrodes (Pardo-Vazquez et al. 2009), their locations can be adjusted by a microdriver. Because the target neurons usually have specific electrophysiological properties that differ from neurons in the nearby cortical areas, in most cases, especially when two cortical areas folded together, the experimenters can drive the electrodes to a proper depth with a desired signal quality by observing the signal in the acquisition software and the electrode depth in the microdriver software. Because the observed signal might not be stable due to flexibility of the brain tissue, advancing of electrodes needs to be slow and sometimes wait for several minutes to allow the brain tissue to rebound.

For further analysis, the low and high frequencies are often extracted from the raw signal by a high-pass and/or a low-pass filter (Fig. 4.12). The high-frequency part can be analyzed for fast electrical changes, specifically the spikes of neurons. On the other hand, the low-frequency part can be analyzed for slow electrical changes, specifically the LFPs, electrical fluctuations within a cortical area resulted from a population of neurons. The filters can be implemented in either software or hardware. There is no universal definition on the frequency limits for the low- and high-frequency parts. The parameters of the filters are set depending on the requirements of specific experiments and the type of neurons. The low-frequency part is usually obtained under 200 Hz and the high-frequency part above 300 Hz. The types of digital filter algorithms used in software implementation include Bessel, Butterworth, or elliptic filters. An example of such a flow of data processing, which is readable and customizable in that software interface, is shown in Fig. 4.12.

Spike sorting is a process to extract and categorize spikes from the high-frequency part of recording from the cortex (Fig. 4.13). The spike sorting can be applied during or after the recording, that is, online or offline, respectively. To save the computing resource and guarantee the speed of real-time monitoring and recording, online spike sorting is usually based on a simple sorting algorithm, such as setting a threshold for detecting spikes. On the other hand, in the absence of time constraint, offline sorting can scrutinize the details of the spikes based on complex algorithms, such as principle component analyses of the spike's waveform.

Because intracortical electrodes are extracellularly placed in the cortex and more than one neuron surrounds it, the high-frequency signal may contain activities from many nearby neurons. Online spike sorting is performed by setting a threshold to preliminarily detect spikes, usually containing both single-unit activities (SUAs)

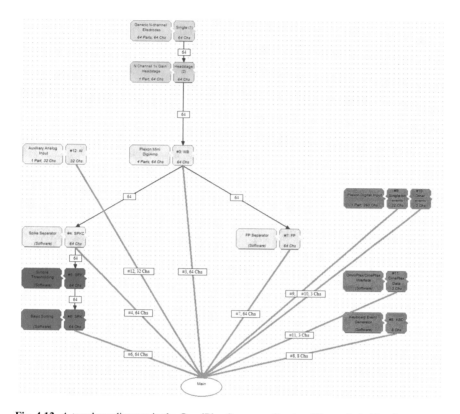

Fig. 4.12 A topology diagram in the OmniPlex Server application (Plexon Inc). The diagram can be viewed by the users in the OmniPlex software interface. It shows the flow of data from hardware devices into software-processing modules and eventually flowing into the Main Data pool at the bottom of the diagram. Users can change the setups for the data acquisition using this interface. The colors of the diagram rectangles are based on the types of data: green, analog signals; blue, continuously digitized sample data; and pink, digital event data. WB, wideband; SPKC, spike-filtered continuous signal; SPK, extracted spike waveform; FP, field potentials; EVT, events; AI, continuously digitized non-neural data; KBD, Keyboard Digital events; CPX, data generated by a CinePlex System. (The figure is adapted with permission from the User Guide for an OmniPlex system, Plexon Inc.)

and multiunit activities (MUAs). The SUA signal is supposed to be the activities from a single neuron, and the MUA is from a bunch of neurons at a moderate distance to the electrode. Given the small scale of the electrode tip, the activities of those nearby neurons differ in the signal amplitude and waveform due to their different distances to the electrode tip, and usually only one neuron that is close enough to the electrode tip can evoke a clear spike signal with a classic waveform of an extracellular action potential. Based on the differences, offline spike sorting can continue to separate the SUAs from the MUAs. As shown in Fig. 4.13d, yellow signals are the sorted SUAs and gray are the MUAs. If the online sorted signal is not good and the wideband signal is recorded, offline sorting can also redo the threshold

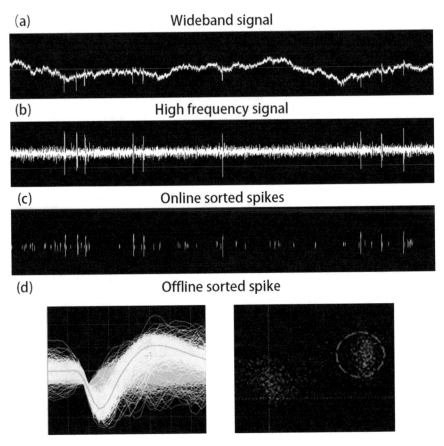

Fig. 4.13 High-frequency signal filtered from a wideband recording of a V1 neuron in monkey and the spike sorting. (The figures are from the authors' unpublished data)

sorting beginning from the wideband signal. Because the spike is a potential change within a few milliseconds, a high temporal resolution is required with a sampling rate typically of 40 kHz set for digitalizing the high-frequency signal.

The SUA signal is often preferred in research because it is a more reliable result with the highest spatial and temporal resolutions to reflect the spike activities of one single neuron. However, recording this signal needs more rigorous requirements, such as a small conductive area on the electrode tip, which is characterized by high impedance, and a proper distance of the neuron to the tip. Sometimes, the MUA and LFP can also be used in analyses when the SUA is not available.

Because LFP is relatively easy to obtain with intracortical electrodes, it is commonly found in research and clinical application (Lashgari et al. 2012). Even though LFP cannot specify one single neuron's activity as the SUA does, its spatial resolution is still higher than signals obtained from most of the noninvasive approaches. LFPs were initially thought to combine signals within a cortical area of several millimeters (Mitzdorf 1985), and later research indicated that the signals could be

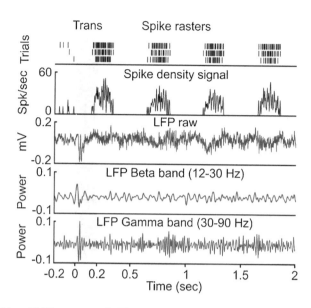

Fig. 4.14 SUA and LFPresponses of a V1 neuron to a grating stimulus. The first two panels are spike analyses, and the lower three panels are LFP analyses. (Figure adapted with permission from Lashgari et al. (2012))

restricted within a few hundred micrometers with a small electrode tip (Katzner et al. 2009). Moreover, further filtering the LFP into different frequency bands can increase the signal specifications (Fig. 4.14) (Lashgari et al. 2012) and address more aspects of the brain activities (Wang et al. 2012). The frequency bands usually include delta (1–4 Hz), theta (4–8 Hz), alpha (8–13 Hz), beta (13–30 Hz), and gamma (30–100 Hz). Despite many investigations in the past, researchers are still exploring new analysis methods to dig more valuable information from the LFPs.

4.3.3 BMI

BMI is a real-time bidirectional link between a living brain and artificial actuators (Lebedev and Nicolelis 2017). Though the conceptual propositions and proofs date back to the early 1960s, BMI studies only took off in the late 1990s (Lebedev and Nicolelis 2017). It integrates principles, techniques, and approaches derived from engineering, computer science, neuroscience, and neurology. The direct interface between the brain and devices could one day restore lost sensory and motor functions in patients, as well as revolutionize the way we interact with the computers and other smart environments (Nicolelis 2001).

The modern era of BMI emerged with the development of chronic intracortical electrode arrays (Lebedev and Nicolelis 2017) (Fig. 4.15). Using neural signals recorded from the brain to control external devices is one important type of

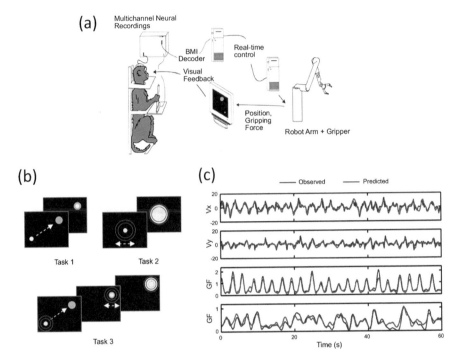

Fig. 4.15 The first BMI for reaching and grasping. (**a**) Diagram setup of the experiment in monkeys. The extracellular electrical activity was recorded from multiple cortical areas and then decoded to control the movements of a virtual or real robotic arm. (**b**) Schematics of three tasks for the monkeys. Task 1, monkeys moved a cursor over a target with a joystick. Task 2, the joystick was fixed, and the monkeys grasped a virtual object by gripping the joystick. Task 3, the monkeys moved the cursor over the target then produced a gripping force. (**c**) The observed movements (blue lines) and the movement predicted by a decoder of the recorded neural activities (red lines). Hand velocity in reference to x-axis or y-axis (Vx, Vy) and gripping force (GF). (Figure adapted with permission from Carmena et al. (2003))

BMI. Notwithstanding the countless innovations, this type of BMI includes three essential components: a sensor (e.g., intracortical electrode arrays) to sample large-scale brain activities, a decoder to convert the recorded signals into a command signal, and an effector to execute the command.

The possibility of the motor cortex serving as an information source of BMI has been evidenced in several aspects: First, the recordings from a small number of intracortical electrodes implanted in the primary motor cortex have the potential to sample a variety of parameters for voluntary limb movements. Second, in primates including humans, the hand motor cortex area is identifiable and accessible easily for the electrode implantation surgery. Third, neural information sufficient for the BMI can be recorded from the motor cortex despite the long-standing injury of its downstream neurons, such as in people with amyotrophic lateral sclerosis. As a series of original studies on intracortical electrodes were conducted in rat, monkey, and humans, it soon became apparent that BMIs could serve as a new generation of

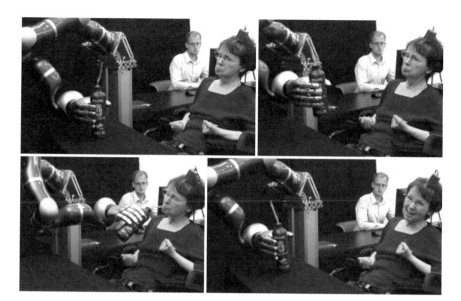

Fig. 4.16 A tetraplegic patient using a BMI to control a DLR robotic arm. (Figure adapted with permission from Hochberg et al. (2012))

prosthesis to restore mobility in patients severely paralyzed due to trauma, notably spinal cord injuries (SCIs) or neurodegenerative diseases (Hochberg et al. 2012; Bouton et al. 2016) (Fig. 4.16).

4.4   Challenges

Intracortical electrodes often fail in several months to years after implantation, likely due to foreign body responses (FBRs) (Stiller et al. 2018; Burns et al. 1974). The FBR progress is related to aspects of both neuronal dieback and electrode degradation. Its extent depends on the biocompatibility of the electrode. To maintain a stable recording from the chronically implanted electrode, it is essential to control the development of FBRs. Strategies for controlling FBRs can roughly be categorized into two types: interventions by materials science and bioactive treatments (Groothuis et al. 2014).

4.4.1   FBR

FBR is the body's self-defense mechanism and can be described as any tissue reaction in response to the presence of a foreign object implantation. It starts immediately after the electrode implantation and evokes a stream of events accountable for deteriorating the electrode and forming scar tissue around it (Grand et al. 2010; Kozai et al. 2015). The progress of FBR can be divided into three phases: acute, early, and chronic (Groothuis et al. 2014).

The acute phase takes place within several hours after the implantation. The electrode insertion causes damages to cells and vasculature along the implant trajectory. The blood–brain barrier is disrupted, and the blood constituents, including monocytes and other blood-borne cells, influx into the surroundings. The cell damage leads to acute cytokine response and microglia activation within about 50 μm around the injury site (Davalos et al. 2005).

The early phase typically takes place within the first week following the acute phase. During this phase, a loose glial sheath is formed around the electrode. Initially, microglia are activated by cytokines released by the damaged cells. Then, the activated microglia induce proliferation and migration of other microglia and astrocytes to form a lamellipodia-based encapsulation. The activated microglia and astrocytes are enriched in a large area ranging to several hundred micrometers around the electrodes (Szarowski et al. 2003).

The chronic phase may take over 6 weeks to several months after the early phase. During this phase, a dense sheath populated with fibrotic and glial cells is formed around the electrode (Fig. 4.17). Reactive microglia remain close to the electrode, indicating a presence of "frustrated phagocytosis," which may aggravate the neuron loss (Groothuis et al. 2014). Reactive astrocytes form a compact layer that is mostly separated from the microglia (Szarowski et al. 2003). The sheath also contains some extracellular matrix molecules inhibiting neuronal growth, such as chondroitin sulfate proteoglycans (CSPGs) secreted by the reactive microglia. Eventually, it results in changes in cellular architecture and increases the distance between the electrode and neurons.

4.4.2   Materials Science in Controlling FBR

Biocompatibility is an implant property that allows for the materials to be compatible with living cells and tissues. This implies that a biocompatible implant can remain functional for a long term in the organism without evoking considerable FBR and harming enzymes, cells, or tissues (Groothuis et al. 2014; Heiduschka and Thanos 1998). The degree of biocompatibility of an intracortical electrode depends on its material properties, geometric characteristics, and microstructures (Williams 2008).

The materials selected for the substrate, recording site, and encapsulation of the electrode should be biocompatible for chronic implantation. Metals are the most common material applied as the substrate and recording sites of traditional intracortical electrodes. Silver, copper, iron, cobalt, and palladium are found to provoke severe immune responses and not safe for chronic usage. Tungsten, platinum, iridium, and gold are overall safe and regularly used for chronic implantation (Szostak et al. 2017). Formation of galvanic cell structure should be avoided when metals,

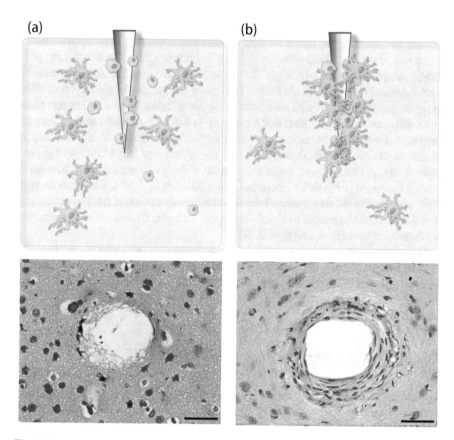

Fig. 4.17 Formation of glial sheath on an intracortical electrode. (**a**) Acute injury after inserting an electrode activates nearby microglia and astrocytes to venture to the area of injury. (**b**) Chronic reactions form a dense glial encapsulation containing fibroblasts, macrophages, and astrocytes. Top, illustrations; bottom, histological images with the cell nuclei stained blue. (Figures adapted with permission from Marin and Fernández (2010))

especially their alloy, are used. Otherwise, it would result in the faster corrosion of metals in a saline environment. Conductive polymers (CPs) have been used in the newer generation of electrode designs. The CPs have advantages of biocompatibility, fast charge transfer, conductive tuning, and nano- or drug-releasing structures (Chen et al. 2013). The most popular CPs that have been used for intracortical electrodes include polypyrrole (PPy), Poly(3, 4-ethylenedioxythiophene) (PEDOT), and polyaniline (PANI). They are usually blended with hydrogels, elastomers, and nanomaterials.

Mechanical moduli of most electrode materials mismatch to those of the soft brain tissues by two to eight orders of magnitude. The stiff electrodes exert continuous stress and cause irritation to the microtissue environment around the electrodes under the constant agitations of the brain (Lee et al. 2005). The stress may lead to gaps and shift the electrodes away from active zones. The FBRs were found more severe with a rigid electrode comparing to a soft one (Köhler et al. 2015). Therefore, flexible materials are preferred for intracortical electrodes.

The geometry of electrodes depends on the fabrication technology, biological model, and target brain structure. The stems of electrodes are usually needle-like with a conical tip or of a blade shape. Tip size of the electrode is believed to affect tissue responses, but it is still controversial about whether sharp or blunt tips cause more affection. Some suggests that a sharp tip is better because it can penetrate the heterogeneous tissue layers with a smaller displacement and a lower possibility of tearing blood vessel walls than a blunt one. Others suggest that a blunt tip can push tissue away and cause fewer traumas, whereas a sharp one would cut it. Moreover, a sharp tip might persistently injure the tissue after implantation. Research on electrode sizes indicated that a size of the tip comparable to that of the neuronal soma (5–10 μm) would be proper to reduce implantation trauma, tissue displacement, and mechanical mismatch (Ludwig et al. 2011).

Roughening the electrode surface by nanomaterials is found to increase the cellular attachment and neuronal growth. Such nanomaterials can provide a surface similar to that of the extracellular matrix environment (Silva 2006). These nanomaterials include carbon nanotubes, platinum black, platinum grass, and CPs.

4.4.3   Bioactive Intervention in Controlling FBR

In addition to optimizing the materials and mechanical characters of intracortical electrodes, bioactive interventions to the tissue responses are available for controlling the FBRs in recent years. Some compounds and other treatments that inhibit inflammation or promote neuronal regrowth can be used for this purpose.

It has been reported that dexamethasone (DEX), a corticosteroid anti-inflammatory compound, can restrict astrocytic responses after inserting intracortical electrodes. In the initial experiments, DEX was administered as subcutaneous injections during the first 6 days after inserting the electrode (Spataro et al. 2005). The results indicated that the DEX administration attenuated both the early phase and sustained phase of the FBR. However, another problem of side effects would be caused by long-term injection of DEX. Therefore, the following studies looked for solutions to locally deliver DEX, such as coating the electrode with DEX (Zhong and Bellamkonda 2007) or incorporating DEX with other coating materials (Mercanzini et al. 2010; Kim and Martin 2006). According to the research on DEX coating, the FBR was significantly lowered, and the neuronal loss was reduced at both 1 and 4 weeks (Zhong and Bellamkonda 2007). Moreover, the in vivo impedance of the electrode was also improved by this local delivery of DEX.

Studies found that laminin, a common extracellular matrix protein, can be used for controlling FBR in the electrode implantation. It is hypothesized that a layer of laminin on the electrode would stabilize the interface between the electrode and the tissue (Grill et al. 2009). Although microglia activity was increased by the laminin coating at the beginning (within about 1 week) of implantation, it later decreased to a lower level than without using the coating at 4 weeks (He et al. 2006). This study suggested that the initial boost in microglia activation triggered by the laminin coating might act to remove the necrotic debris, then the tissue responses would be attenuated after the cleanup (He et al. 2006).

The loss and growth inhibition of neurons at the interface during FBR are one of the factors affecting the recording quality. Therefore, protecting the neurons from loss and growth inhibition would increase the performance of chronic electrodes. Drugs or treatments used for recovering nerve injury have the potential effects in this purpose. A possible option is to disrupt the inhibition caused by chondroitin sulfate proteoglycans (CSPGs), an extracellular proteoglycans in the glial sheath at the electrode surface. Chondroitinase ABC (ChABC) had some success in treating spinal cord injury. It can remove the sugar chains that activate CSPGs' inhibitory functions (Filous et al. 2010).

4.5 Summary

Intracortical electrodes provide a valuable tool for scientists to extract neural information from the brain. This chapter reviewed the traditional intracortical electrodes. According to their fabrication and architecture, intracortical electrodes include two main classes, the wire-based and micromachined. After decades of developments, they can acquire both SUAs and LFPs with a high recording quality over a long-term implantation. Moreover, methods for signal analysis are improving to offer more capabilities, such as decoding voluntary movements from neural signals of motor cortex (Carmena et al. 2003) and investigating attention mechanisms (Chen et al. 2008). Therefore, the traditional intracortical electrodes, which can be used in vivo in animals, have achieved many successes in the research as well as clinical areas. In acute research using anesthetized animals, basic mechanisms of cortical information processing have been interpreted (Hubel and Wiesel 1959). In chronic research, higher cognitive functions have been studied as the animal can be trained with a behavior task (Reppas et al. 2002; O'Keefe and Burgess 1996). Based on the successes of chronic intracortical electrode arrays, BMIs emerged, and their pioneering clinical studies have shown the potential to restore lost motor functions in paralyzed patients (Hochberg et al. 2012). Chronic implantation of intracortical electrodes can provide more perceived values, whereas evidences show a great challenge to keep the quality of electrode recording for a long term. Many studies have been devoted to understanding the mechanisms underlying the chronic degradation of recording quality and to improving the long-term viability. Factors relating to biocompatibility and FBRs are supposed to account for this problem, which can be controlled with interventions by materials science and bioactive treatments.

Acknowledgments This work was supported by the Defense Advanced Research Projects Agency through Grant # D17AP00031 of the USA and the Chronic Brain Injury Program of The Ohio State University through a Pilot Award. The views, opinions, and/or findings contained in this article are those of the author and should not be interpreted as representing the official views or policies, either expressed or implied, of the Defense Advanced Research Projects Agency or the Department of Defense.

References

Ainslie, K. M., & Desai, T. A. (2008). Microfabricated implants for applications in therapeutic delivery, tissue engineering, and biosensing. *Lab on a Chip, 8*(11), 1864–1878. https://doi.org/10.1039/b806446f.

Bhandari, R., Negi, S., Rieth, L., Normann, R. A., & Solzbacher, F. (2008). A novel method of fabricating convoluted shaped electrode arrays for neural and retinal prostheses. *Sensors and Actuators A: Physical, 145–146*(1–2), 123–130. https://doi.org/10.1016/j.sna.2007.10.072.

Bhandari, R., Negi, S., & Solzbacher, F. (2010). Wafer-scale fabrication of penetrating neural microelectrode arrays. *Biomedical Microdevices, 12*(5), 797–807. https://doi.org/10.1007/s10544-010-9434-1.

Bouton, C. E., Shaikhouni, A., Annetta, N. V., Bockbrader, M. A., Friedenberg, D. A., Nielson, D. M., Sharma, G., Sederberg, P. B., Glenn, B. C., Mysiw, W. J., Morgan, A. G., Deogaonkar, M., & Rezai, A. R. (2016). Restoring cortical control of functional movement in a human with quadriplegia. *Nature, 533*(7602), 247–250. https://doi.org/10.1038/nature17435.

Brandman, D. M., Cash, S. S., & Hochberg, L. R. (2017). Review: Human intracortical recording and neural decoding for brain-computer interfaces. *IEEE Transactions on Neural Systems and Rehabilitation Engineering, 25*(10), 1687–1696. https://doi.org/10.1109/TNSRE.2017.2677443.

Burns, B. D., Stean, J. P., & Webb, A. C. (1974). Recording for several days from single cortical neurons in completely unrestrained cats. *Electroencephalography and Clinical Neurophysiology, 36*(3), 314–318. https://doi.org/10.1016/0013-4694(74)90175-8.

Campbell, P. K., Jones, K. E., Huber, R. J., Horch, K. W., & Normann, R. A. (1991). A silicon-based, three-dimensional neural interface: Manufacturing processes for an intracortical electrode array. *IEEE Transactions on Biomedical Engineering, 38*(8), 758–768. https://doi.org/10.1109/10.83588.

Carmena, J. M., Lebedev, M. A., Crist, R. E., O'Doherty, J. E., Santucci, D. M., Dimitrov, D. F., Patil, P. G., Henriquez, C. S., & Nicolelis, M. A. (2003). Learning to control a brain-machine interface for reaching and grasping by primates. *PLoS Biology, 1*(2), E42. https://doi.org/10.1371/journal.pbio.0000042.

Carter, M., & Shieh, J. C. (2010). Stereotaxic surgeries and in vivo techniques. In *Guide to research techniques in neuroscience* (pp. 73–90). https://doi.org/10.1016/B978-0-12-374849-2.00003-3.

Chang, W. T., Hwang, I. S., Chang, M. T., Lin, C. Y., Hsu, W. H., & Hou, J. L. (2012). Method of electrochemical etching of tungsten tips with controllable profiles. *The Review of Scientific Instruments, 83*(8), 083704. https://doi.org/10.1063/1.4745394.

Chen, Y., Martinez-Conde, S., Macknik, S. L., Bereshpolova, Y., Swadlow, H. A., & Alonso, J. M. (2008). Task difficulty modulates the activity of specific neuronal populations in primary visual cortex. *Nature Neuroscience, 11*(8), 974–982. https://doi.org/10.1038/nn.2147.

Chen, S., Pei, W., Gui, Q., Tang, R., Chen, Y., Zhao, S., Wang, H., & Chen, H. (2013). PEDOT/MWCNT composite film coated microelectrode arrays for neural interface improvement. *Sensors and Actuators, A: Physical, 193*, 141–148. https://doi.org/10.1016/j.sna.2013.01.033.

Coleman, W. L., & Burger, R. M. (2015). Extracellular single-unit recording and Neuropharmacological methods. In *Basic electrophysiological methods*. Oxford University Press. https://doi.org/10.1093/med/9780199939800.003.0003

Cyberkinetics. (2005). NeuroPort cortical microelectrode array system (Neuroport Electroe) 510K Summary. https://www.accessdata.fda.gov/cdrh_docs/pdf4/K042384.pdf.

Davalos, D., Grutzendler, J., Yang, G., Kim, J. V., Zuo, Y., Jung, S., Littman, D. R., Dustin, M. L., & Gan, W. B. (2005). ATP mediates rapid microglial response to local brain injury in vivo. *Nature Neuroscience, 8*(6), 752–758. https://doi.org/10.1038/nn1472.

De Groot, J. (1959). The rat hypothalamus in stereotaxic coordinates. *The Journal of Comparative Neurology, 113*(3), 389–400.

Fiebelkorn, I. C., Pinsk, M. A., & Kastner, S. (2019). The mediodorsal pulvinar coordinates the macaque fronto-parietal network during rhythmic spatial attention. *Nature Communications, 10*(1), 215. https://doi.org/10.1038/s41467-018-08151-4.

Filous, A. R., Miller, J. H., Coulson-Thomas, Y. M., Horn, K. P., Alilain, W. J., & Silver, J. (2010). Immature astrocytes promote CNS axonal regeneration when combined with chondroitinase ABC. *Developmental Neurobiology, 70*(12), 826–841. https://doi.org/10.1002/dneu.20820.

Garner, H. E., Amend, J. F., Rosborough, J. P., Geddes, L. A., & Ross, J. N. (1972). Electrodes for recording cortical electroencephalograms in ponies. *Laboratory Animal Science, 22*(2), 262–265.

Grand, L., Wittner, L., Herwik, S., Göthelid, E., Ruther, P., Oscarsson, S., Neves, H., Dombovári, B., Csercsa, R., Karmos, G., & Ulbert, I. (2010). Short and long term biocompatibility of NeuroProbes silicon probes. *Journal of Neuroscience Methods, 189*(2), 216–229. https://doi.org/10.1016/j.jneumeth.2010.04.009.

Green, R. A., Lovell, N. H., Wallace, G. G., & Poole-Warren, L. A. (2008). Conducting polymers for neural interfaces: Challenges in developing an effective long-term implant. *Biomaterials, 29*(24–25), 3393–3399. https://doi.org/10.1016/j.biomaterials.2008.04.047.

Grill, W. M., Norman, S. E., & Bellamkonda, R. V. (2009). Implanted neural interfaces: Biochallenges and engineered solutions. *Annual Review of Biomedical Engineering, 11*, 1–24. https://doi.org/10.1146/annurev-bioeng-061008-124927.

Groothuis, J., Ramsey, N. F., Ramakers, G. M. J., & van der Plasse, G. (2014). Physiological challenges for intracortical electrodes. *Brain Stimulation, 7*(1), 1–6. https://doi.org/10.1016/j.brs.2013.07.001.

Grundfest, H., Sengstaken, R. W., Oettinger, W. H., & Gurry, R. W. (1950). Stainless steel microneedle electrodes made by electrolytic pointing. *Review of Scientific Instruments, 21*(4), 360–361. https://doi.org/10.1063/1.1745583.

He, W., McConnell, G. C., & Bellamkonda, R. V. (2006). Nanoscale laminin coating modulates cortical scarring response around implanted silicon microelectrode arrays. *Journal of Neural Engineering, 3*(4), 316–326. https://doi.org/10.1088/1741-2560/3/4/009.

Heiduschka, P., & Thanos, S. (1998). Implantable bioelectric interfaces for lost nerve functions. *Progress in Neurobiology, 55*(5), 433–461. https://doi.org/10.1016/s0301-0082(98)00013-6.

Hochberg, L. R., Serruya, M. D., Friehs, G. M., Mukand, J. A., Saleh, M., Caplan, A. H., Branner, A., Chen, D., Penn, R. D., & Donoghue, J. P. (2006). Neuronal ensemble control of prosthetic devices by a human with tetraplegia. *Nature, 442*(7099), 164–171. https://doi.org/10.1038/nature04970.

Hochberg, L. R., Bacher, D., Jarosiewicz, B., Masse, N. Y., Simeral, J. D., Vogel, J., Haddadin, S., Liu, J., Cash, S. S., van der Smagt, P., & Donoghue, J. P. (2012). Reach and grasp by people with tetraplegia using a neurally controlled robotic arm. *Nature, 485*(7398), 372–U121. https://doi.org/10.1038/nature11076.

Hodgkin, A. L., & Katz, B. (1949). The effect of sodium ions on the electrical activity of giant axon of the squid. *The Journal of Physiology, 108*(1), 37–77.

Hubel, D. H. (1957). Tungsten microelectrode for recording from single units. *Science, 125*(3247), 549–550. 125/3247/549 [pii].. https://doi.org/10.1126/science.125.3247.549.

Hubel, D. H., & Wiesel, T. N. (1959). Receptive fields of single neurones in the cat's striate cortex. *The Journal of Physiology, 148*, 574–591. https://doi.org/10.1113/jphysiol.1959.sp006308.

Hubel, D. H., & Wiesel, T. N. (1962). Receptive fields, binocular interaction and functional architecture in the cat's visual cortex. *The Journal of Physiology, 160*, 106–154. https://doi.org/10.1113/jphysiol.1962.sp006837.

Jones, K. E., Campbell, P. K., & Normann, R. A. (1992). A glass/silicon composite intracortical electrode array. *Annals of Biomedical Engineering, 20*(4), 423–437. https://doi.org/10.1007/bf02368134.

Jules, A. (1964). Fabrication of Semiconductor Devices. U.S. Patent No. 3,122,817. Washington, DC: U.S. Patent and Trademark Office.

Kaltenbach, J. A., & Gerstein, G. L. (1986). A rapid method for production of sharp tips on preinsulated microwires. *Journal of Neuroscience Methods, 16*(4), 283–288. 0165-0270(86)90053-1 [pii].

Katzner, S., Nauhaus, I., Benucci, A., Bonin, V., Ringach, D. L., & Carandini, M. (2009). Local origin of field potentials in visual cortex. *Neuron, 61*(1), 35–41. https://doi.org/10.1016/j.neuron.2008.11.016.

Kim, D. H., & Martin, D. C. (2006). Sustained release of dexamethasone from hydrophilic matrices using PLGA nanoparticles for neural drug delivery. *Biomaterials, 27*(15), 3031–3037. https://doi.org/10.1016/j.biomaterials.2005.12.021.

Kim, S. J., Manyam, S. C., Warren, D. J., & Normann, R. A. (2006). Electrophysiological mapping of cat primary auditory cortex with multielectrode arrays. *Annals of Biomedical Engineering, 34*(2), 300–309. https://doi.org/10.1007/s10439-005-9037-9.

Köhler, P., Wolff, A., Ejserholm, F., Wallman, L., Schouenborg, J., & Linsmeier, C. E. (2015). Influence of probe flexibility and gelatin embedding on neuronal density and glial responses to brain implants. *PLoS One, 10*(3), e0119340. https://doi.org/10.1371/journal.pone.0119340.

Kotzar, G., Freas, M., Abel, P., Fleischman, A., Roy, S., Zorman, C., Moran, J. M., & Melzak, J. (2002). Evaluation of MEMS materials of construction for implantable medical devices. *Biomaterials, 23*(13), 2737–2750. https://doi.org/10.1016/s0142-9612(02)00007-8.

Kozai, T. D. Y., Jaquins-Gerstl, A. S., Vazquez, A. L., Michael, A. C., & Cui, X. T. (2015). Brain tissue responses to neural implants impact signal sensitivity and intervention strategies. *ACS Chemical Neuroscience, 6*(1), 48–67. https://doi.org/10.1021/cn500256e.

Lashgari, R., Li, X., Chen, Y., Kremkow, J., Bereshpolova, Y., Swadlow, H. A., & Alonso, J. M. (2012). Response properties of local field potentials and neighboring single neurons in awake primary visual cortex. *The Journal of Neuroscience, 32*(33), 11396–11413. https://doi.org/10.1523/JNEUROSCI.0429-12.2012.

Lebedev, M. A., & Nicolelis, M. A. (2017). Brain-machine interfaces: From basic science to neuroprostheses and neurorehabilitation. *Physiological Reviews, 97*(2), 767–837. https://doi.org/10.1152/physrev.00027.2016.

Lee, K., Massia, S., & He, J. (2005). Biocompatible benzocyclobutene-based intracortical neural implant with surface modification. *Journal of Micromechanics and Microengineering, 15*(11), 2149–2155. https://doi.org/10.1088/0960-1317/15/11/022.

Lehew, G., & Nicolelis, M. A. L. (2008). State-of-the-art microwire array design for chronic neural recordings in behaving animals. NBK3901 [bookaccession].

Ludwig, K. A., Langhals, N. B., Joseph, M. D., Richardson-Burns, S. M., Hendricks, J. L., & Kipke, D. R. (2011). Poly(3,4-ethylenedioxythiophene) (PEDOT) polymer coatings facilitate smaller neural recording electrodes. *Journal of Neural Engineering, 8*(1), 014001. https://doi.org/10.1088/1741-2560/8/1/014001.

Marin, C., & Fernández, E. (2010). Biocompatibility of intracortical microelectrodes: Current status and future prospects. *Frontiers in Neuroengineering, 3*. https://doi.org/10.3389/fneng.2010.00008.

Mercanzini, A., Reddy, S. T., Velluto, D., Colin, P., Maillard, A., Bensadoun, J. C., Hubbell, J. A., & Renaud, P. (2010). Controlled release nanoparticle-embedded coatings reduce the tissue reaction to neuroprostheses. *Journal of Controlled Release, 145*(3), 196–202. https://doi.org/10.1016/j.jconrel.2010.04.025.

Mitzdorf, U. (1985). Current source-density method and application in cat cerebral cortex: Investigation of evoked potentials and EEG phenomena. *Physiological Reviews, 65*(1), 37–100. https://doi.org/10.1152/physrev.1985.65.1.37.

Nicolelis, M. A. (2001). Actions from thoughts. *Nature, 409*(6818), 403–407. https://doi.org/10.1038/35053191.

Normann, R. A., & Fernandez, E. (2016). Clinical applications of penetrating neural interfaces and Utah Electrode Array technologies. *Journal of Neural Engineering, 13*(6), 061003. https://doi.org/10.1088/1741-2560/13/6/061003.

O'Keefe, J., & Burgess, N. (1996). Geometric determinants of the place fields of hippocampal neurons. *Nature, 381*(6581), 425–428. https://doi.org/10.1038/381425a0.

Palmer, C. (1978). A microwire technique for recording single neurons in unrestrained animals. *Brain Research Bulletin, 3*(3), 285–289. https://doi.org/10.1016/0361-9230(78)90129-6.

Pardo-Vazquez, J. L., Leboran, V., & Acuna, C. (2009). A role for the ventral premotor cortex beyond performance monitoring. *Proceedings of the National Academy of Sciences of the United States of America, 106*(44), 18815–18819. https://doi.org/10.1073/pnas.0910524106.

Paxinos, G., & Watson, C. (2007). *The rat brain in stereotaxic coordinates* (6th ed.). Amsterdam; Boston: Academic Press/Elsevier.

Reppas, J. B., Usrey, W. M., & Reid, R. C. (2002). Saccadic eye movements modulate visual responses in the lateral geniculate nucleus. *Neuron, 35*(5), 961–974. https://doi.org/10.1016/s0896-6273(02)00823-1.

Rheinberger, M. B., & Jasper, H. H. (1937). Electrical activity of the cerebral cortex in the unanesthetized cat. *American Journal of Physiology, 119*(1), 186–196.

Silva, G. A. (2006). Neuroscience nanotechnology: Progress, opportunities and challenges. *Nature Reviews Neuroscience, 7*(1), 65–74. https://doi.org/10.1038/nrn1827.

Sommakia, S., Lee, H. C., Gaire, J., & Otto, K. J. (2014). Materials approaches for modulating neural tissue responses to implanted microelectrodes through mechanical and biochemical means. *Current Opinion in Solid State & Materials Science, 18*(6), 319–328. https://doi.org/10.1016/j.cossms.2014.07.005.

Spataro, L., Dilgen, J., Retterer, S., Spence, A. J., Isaacson, M., Turner, J. N., & Shain, W. (2005). Dexamethasone treatment reduces astroglia responses to inserted neuroprosthetic devices in rat neocortex. *Experimental Neurology, 194*(2), 289–300. https://doi.org/10.1016/j.expneurol.2004.08.037.

Stiller, A. M., Usoro, J., Frewin, C. L., Danda, V. R., Ecker, M., Joshi-Imre, A., Musselman, K. C., Voit, W., Modi, R., Pancrazio, J. J., & Black, B. J. (2018). Chronic intracortical recording and electrochemical stability of thiol-ene/acrylate shape memory polymer electrode arrays. *Micromachines (Basel), 9*(10). https://doi.org/10.3390/mi9100500.

Szarowski, D. H., Andersen, M. D., Retterer, S., Spence, A. J., Isaacson, M., Craighead, H. G., Turner, J. N., & Shain, W. (2003). Brain responses to micro-machined silicon devices. *Brain Research, 983*(1–2), 23–35. https://doi.org/10.1016/s0006-8993(03)03023-3.

Szostak, K. M., Grand, L., & Constandinou, T. G. (2017). Neural interfaces for intracortical recording: Requirements, fabrication methods, and characteristics. *Frontiers in Neuroscience, 11*, 665. https://doi.org/10.3389/fnins.2017.00665.

Velliste, M., Perel, S., Spalding, M. C., Whitford, A. S., & Schwartz, A. B. (2008). Cortical control of a prosthetic arm for self-feeding. *Nature, 453*(7198), 1098–1101. https://doi.org/10.1038/nature06996.

Wang, L., Saalmann, Y. B., Pinsk, M. A., Arcaro, M. J., & Kastner, S. (2012). Electrophysiological low-frequency coherence and cross-frequency coupling contribute to BOLD connectivity. *Neuron, 76*(5), 1010–1020. https://doi.org/10.1016/j.neuron.2012.09.033.

Warren, D. J., Fernandez, E., & Normann, R. A. (2001). High-resolution two-dimensional spatial mapping of cat striate cortex using a 100-microelectrode array. *Neuroscience, 105*(1), 19–31. https://doi.org/10.1016/s0306-4522(01)00174-9.

Williams, D. F. (2008). On the mechanisms of biocompatibility. *Biomaterials, 29*(20), 2941–2953. https://doi.org/10.1016/j.biomaterials.2008.04.023.

Wise, K. D. (2005). Silicon microsystems for neuroscience and neural prostheses. *IEEE Engineering in Medicine and Biology Magazine, 24*(5), 22–29. https://doi.org/10.1109/memb.2005.1511497.

Wise, K. D., Angell, J. B., & Starr, A. (1970). An integrated-circuit approach to extracellular microelectrodes. *IEEE Transactions on Biomedical Engineering, BME-17*(3), 238–247. https://doi.org/10.1109/tbme.1970.4502738.

Wolbarsht, M. L., Macnichol, E. F., Jr., & Wagner, H. G. (1960). Glass insulated platinum micro-electrode. *Science, 132*(3436), 1309–1310. https://doi.org/10.1126/science.132.3436.1309.

Zhang J., Xiong H. (2014) Brain Stereotaxic Injection. In *Current Laboratory Methods in Neuroscience Research*. Springer. https://doi.org/10.1007/978-1-4614-8794-4_2.

Zhong, Y., & Bellamkonda, R. V. (2007). Dexamethasone-coated neural probes elicit attenuated inflammatory response and neuronal loss compared to uncoated neural probes. *Brain Research, 1148*, 15–27. https://doi.org/10.1016/j.brainres.2007.02.024.

Chapter 5
Peripheral Nerve Electrodes

Yu Wu and Liang Guo

5.1 Introduction

With the goal of restoring muscle functions or recording neural signals, peripheral nerve electrodes have been extensively studied for decades. For restoration of lost motor function, in most cases, multiple groups of muscle fibers need to be electrically activated. The straightforward approach is to use muscle-targeted electrodes (either epimysial or intramuscular electrodes (Navarro et al. 2005)), by implanting at least one electrode on/in each muscle (Veraart et al. 1993). Apparently, due to the sparse distribution of implanted electrodes, the wiring scheme is not only complicated for surgical implantation procedure and maintenance but also fragile to limb movements. Therefore, within the options of muscular or neural interfaces, targeting peripheral nerves has the advantage of much lower activation thresholds, a smaller number of required electrodes, and higher recruiting accuracy.

In contrast to the complexity in the central nervous system, the relatively simple anatomical structure of peripheral nerves makes them easier to be accessed by electrodes. Typically, neurons in the peripheral nervous system have their somas located in or around the spinal cord while extending their axons all the way to target organs. While eventually ending with branches and forming neuromuscular junctions with skeletal muscle fibers, the axons, for the most part of their lengths, are grouped together in fascicles that are wrapped by three protective layers: epineurium, perineurium, and endoneurium (Fig. 5.1). Such a compact and clearly mapped structure

Y. Wu
Department of Electrical and Computer Engineering, The Ohio State University, Columbus, OH, USA

L. Guo (✉)
Department of Electrical and Computer Engineering, The Ohio State University, Columbus, OH, USA

Department of Neuroscience, The Ohio State University, Columbus, OH, USA
e-mail: guo.725@osu.edu

© Springer Nature Switzerland AG 2020
L. Guo (ed.), *Neural Interface Engineering*,
https://doi.org/10.1007/978-3-030-41854-0_5

Fig. 5.1 Cross-sectional view of a typical peripheral nerve where nerve fibers are bundled into perineurium-ensheathed fascicles. (Adapted with permission from Yoshida et al. 2010)

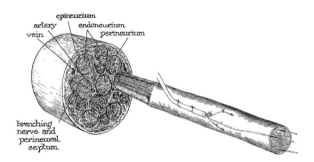

not only minimizes electrode size and quantity but also leaves adequate space for high-resolution interfaces such as intrafascicular electrodes.

The advances of peripheral nerve electrodes have been summarized in multiple excellent comprehensive reviews (Navarro et al. 2005; Yoshida et al. 2010; Ortiz-Catalan et al. 2012; Patil and Thakor 2016) which organize their discussions based on different types of devices. Instead, by focusing on challenges and corresponding design motivations, this chapter covers the efforts made in the field to improve issues of selectivity and noise reduction, electrode-neural interfaces, and surgical implantations. To provide an overview on the working principles of peripheral nerve electrodes, Sect. 5.2 will briefly introduce three major types of electrodes. The three major challenges and corresponding strategies will then be discussed in detail in Sect. 5.3.

5.2 Peripheral Nerve Electrodes

In this section, we discuss three types of conventional peripheral nerve electrodes used to stimulate peripheral nerves or record neural activity: cuff electrodes, intra-fascicular electrodes, and regenerative electrodes.

5.2.1 Cuff Electrodes

Among peripheral electrodes, cuff electrodes are perhaps the most studied and investigated toward clinical applications. They work by completely encircling the nerve with an insulated tubular sheath and using two or more electrode contacts on the inner surface to either stimulate the nerve or record neural activity. These electrodes can either completely encircle the nerve (circumferential) or come in contact with just a section of it (differential). Circumferential contacts have mainly been used for recording purposes, while differential contacts have been found to provide better stimulation (McNeal et al. 1989; Sweeney et al. 1990; Grill and Mortimer

1998). Furthermore, differential contacts offer a variety of designs as they can be placed in a bipolar, tripolar, or short-circuit tripolar configuration to reduce noise or current leaks. However, the use of differential contacts for recording has been rarely reported in the current literature.

The flexibility of cuff electrodes allows them to avoid the problems of mechanical stress and displacement that are common in muscle-based electrodes, thereby reducing the possibility of electrode or lead failure. They also have a higher accuracy when recording data because they increase the resistance of the extracellular return path, which increases the amplitude of the recorded signals (Sahin et al. 1997; Struijk et al. 1999).

Although several issues with cuff electrodes have been observed, they were often remedied in subsequent investigations. For example, it was found that cuff electrodes caused a loss of myelinated fibers, but these fibers regenerated to a smaller diameter over time without any signs of control or strength loss (Larsen et al. 1998). It was also found that the self-sizing spiral cuff electrode could cause demyelination and axon losses, perineurium thickening, increased intraneural tissue, or axonal swelling (Naples et al. 1988). However, a different study found that these effects could be circumvented by placing the cuff farther away from a joint (Romero et al. 2001).

5.2.2 Intrafascicular Electrodes

An intrafascicular electrode is directly inserted into the peripheral nerve for direct contact with the nerve tissue that they are intended to activate or record. This direct contact enhances the selectivity and increases the signal-to-noise ratio (SNR) of the recording. Stimulation on a certain nerve fascicle has less effect on adjacent ones. More than one intrafascicular electrode can be implanted to stimulate multiple areas along the nerve.

Longitudinally implanted intrafascicular electrodes (LIFEs) are implanted parallel to the nerve fibers and are inserted a few millimeters into the endoneurium before exiting. The active section inside the nerve can have multiple contacts placed in different orientations to improve selectivity comparing to the transverse intrafascicular multichannel electrodes (TIME).

Studies of metal and polymer LIFE show no damage caused by the electrode or biocompatibility issues and good selectivity for stimulation and multiunit extracellular recording (Nannini and Horch 1991; Goodall et al. 1993). They also have long-term reliability, being used for studies for over 6 months (Goodall et al. 1991).

Furthermore, the electrical stimulation produced by LIFE is able to elicit sensations of touch, joint movement, and position, potentially allowing amputees to have prosthetic limbs with a more natural feel and control. However, the selectivity is challenging to achieve, as it is difficult to implant multiple LIFEs in different fascicles to stimulate the appropriate muscle groups (Dhillon et al. 2004a, b).

5.2.3 Regenerative Electrodes

Unlike cuff and intrafascicular electrodes that passively access intact nerves, regenerative electrodes employ a completely different strategy in which nerve fibers of interest are severed first and then regenerate through micro via-holes or channels to reconstruct neural connections. Probes or stimulation electrodes are integrated in the channels to directly interface with regenerated fascicles, giving rise to highly intimate electrode–tissue interface as well as excellent selectivity to individual fibers.

Due to its highly aggressive nature, however, many factors should be considered during design to ensure successful nerve regeneration. For example, appropriate channel dimensions should be customized so that regenerated nerve fibers will not be damaged by compression force. Additionally, it may also require trophic factors to facilitate regeneration.

5.3 Challenges and Strategies on Electrode Design

5.3.1 Toward Better Selectivity and SNR

Selectivity and SNR are the two most important properties of peripheral neural interfaces regarding to functionality. For both stimulation and recording purposes, it is always desired to recruit or record a smaller group of nerve fibers with less interference from others. Higher selectivity in stimulation enables finer and more coordinated control over many muscle fibers, yielding a more effective motor prosthesis process.

Although recording peripheral nerve activities cannot improve neural modulation directly, it provides important information on the functionality and mechanism of peripheral nervous system and can be used for controlling an external prosthesis. In addition, neural recording is an indivisible part of feedback control that enhances the neural modulation effectiveness. Therefore, the goal of recovering action potentials while suppressing noise from muscle activities and stimulation artifacts has been pursued for any type of electrodes.

Selectivity

Steering electric fields with cuff electrodes Selectively activating nerve fascicles can be realized in two ways: (1) manipulating the electric field inside a nerve trunk and (2) directly accessing individual fascicles by electrodes. The first method, also referred as electric field steering, was developed for cuff electrodes which have no direct access to axons inside a nerve. And it can be considered as a remote control of electric field distribution by electrodes wrapping around the nerve trunk.

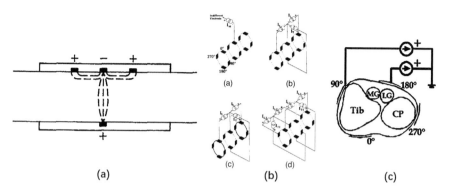

Fig. 5.2 Electric field-steering techniques for cuff electrodes. (**a**) Schematic view of longitudinal tripolar cuff with field steering where the transverse current serves as steering current. (Adapted with permission from Sweeney et al. 1990). (**b**) Four longitudinal configurations based on field steering. Adapted with permission from Veraart et al. (1993). (**c**) The configuration of four-contact transversal cuff electrode. (Adapted with permission from Tarler and Mortimer 2004)

This technique is based on two findings: (1) the excitability of a myelinated axon is determined by longitudinal electric field rather than transverse field (Rushton 1927; Coburn 1980; Rattay 1986); and (2) a transverse current could be used to steer the longitudinal current, restricting the region of excitation (Sweeney et al. 1990). As shown in the longitudinal tripolar configuration (Fig. 5.2a), by passing a transverse current flow at a subthreshold level from a "steering" electrode (anode) 180° opposite to the central cathode, the longitudinal current can be repelled away from the steering electrode and into the region near the cathode. In this way, only a portion of the nerve trunk is stimulated rather than the entire region. Based on this longitudinal tripolar configuration, further modifications have been proposed and investigated (Fig. 5.2b) (Veraart et al. 1993), in which double tripolar configuration shows the best selectivity to activate fascicles that could not be activated selectively by a single tripole, at the cost of more electrodes and lead wires that complicate the system. To achieve the best selectivity for a given muscle, it is necessary to modulate the amplitude of both longitudinal and transverse current, which means the stimulation conditions are not predictable and highly sensitive to electrodes' relative position to nerve fascicles. In clinical practices, however, such a trial-and-error process may result in unstable implantation because changes in electrodes' position may invalidate the preset stimulation parameters.

As an alternative approach, arranging electrodes transversely can achieve the same selectivity without compromising on robustness and simplicity. Such concept originates from studies of longitudinal configurations where spatial selectivity is maximized when the transverse current constitutes a high proportion of the total current (Goodall et al. 1996; Deurloo et al. 1998). Accordingly, this suggests that 100% transverse current would result in maximum spatial selectivity. In that case, the two anodes from the longitudinal tripole are effectively eliminated (Goodall et al. 1996). This has been tested and confirmed in modeling works (Parrini 1997; Deurloo et al. 1998) and cat sciatic nerve implantations (Tarler and Mortimer 2003;

Tarler and Mortimer 2004). A representative device is the four-contact cuff electrode (Fig. 5.2c) (Tarler and Mortimer 2004). By modulating anodic and cathodic steering current to hyperpolarize undesired fascicles and excite the target fascicle, respectively, each of the four motor fascicles in the cat sciatic nerve can be recruited selectively and independently. However, since nerve fibers are more sensitive to longitudinal electric fields, the activation threshold is significantly higher for transverse configurations than for longitudinal ones (Goodall et al. 1996).

Reshaping Nerve Trunk with Flat Interface Nerve Electrode (FINE) While selective recruitment of cat nerve fascicles can be achieved by a field-steering cuff interface, the circular distribution of electrode sites reduces their ability to selectively access central axon populations due to a significant distance to inner fascicles (Veltink et al. 1988, 1989a, b; Altman and Plonsey 1990). Different from the simple structure of cat sciatic nerve, human peripheral nerves typically contain more fascicles and may not have a cylindrical cross-section. For example, a human femoral nerve (12 mm wide, 4 mm thick) is composed of at least 20 fascicles (Schiefer et al. 2008), making it difficult for field-steering cuffs to work effectively. As an evolution, a noninvasive FINE was proposed to form a rectangular cross-section. By slowly squeezing the nerve trunk into an elongated and flattened oval, fascicles are lined up on a two-dimensional flat surface instead of a three-dimensional cylinder (Fig. 5.3a, b), resulting in more exposure of smaller and inner fascicles under stimulating electrodes. The nerve reshaping process may take from hours to days, depending on the force applied on nerve trunk.

In comparison with conventional cuff electrodes, this design exploits the oblong cross-section of the nerve and takes advantage of the nerve's ability to reshape (Tyler and Durand 2002). Since electrodes are more intimate to individual fascicles, they are theoretically more effective to recruit a certain fascicle without causing cross-talks to other fascicles (Choi et al. 2001; Schiefer et al. 2005). Using simple monopolar stimulation without field steering, acute experiments on cat and beagle have shown high degree of selectivity for major fascicles (Fig. 5.3c) (Tyler and

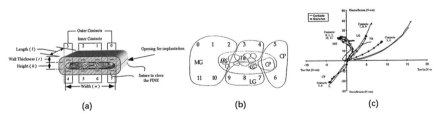

(a)   (b)   (c)

Fig. 5.3 FINE. (**a**) Schematic cross-section of a FINE on a nerve. (**b**) Relative position of the electrodes and the nerve fascicles. Electrical contacts giving functionally equivalent torque outputs are circled. (**c**) Example of selective recruitment of four fascicles within the sciatic nerve. The lines with symbols show recruitment curves of individual branches having electrodes distal to the FINE. The recruitments from the FINE contacts are shown with the thin lines. The contacts are listed next to their respective recruitment trajectories. (Adapted with permission from Tyler and Durand 2002)

Durand 2002; Leventhal and Durand 2004; Yoo et al. 2004), and even subfascicular recruitment can be achieved with optimized electrode design (Leventhal and Durand 2003). Moreover, studies on human peripheral nerves also proved the effectiveness of FINE design. Modeling and simulation predicted that eight-contact FINE is enough to selectively stimulate each of the six muscles innervated by the proximal femoral nerve (Schiefer et al. 2008), which was further confirmed for femoral, tibial, and peroneal nerves in human subject implantation carried by the same group (Schiefer et al. 2010; Schiefer et al. 2013).

However, with a flat interface to nerve trunk, the high selectivity and the simple monopolar stimulation without complicated field steering are actually achieved at the cost of greater possibility of nerve damage by pressure. Although chronic implantation on cat sciatic nerves up to 3 months suggested that moderate reshaping caused no axonal damage (Daniel et al. 2006), the chronical biocompatibility of FINE still needs to be confirmed with human subject test.

Subfascicular Selectivity with Intrafascicular Electrodes To further enhance the selectivity, either field-steering cuff or FINE has seemed to be reaching their limits due to the inherent disadvantage of remote stimulation. The need for finer neural modulation interfaces promotes intrafascicular electrodes that are placed within individual fascicles to directly access the nerve fibers, instead of manipulating electric fields outside the nerve fascicles. These electrodes are usually composed of an insulated conducting wire with small openings at stimulation sites. Essentially, it is the proximity to nerve fibers and restricted current flow in space that give rise to their excellent selectivity and low-activation threshold.

A variety of designs have been developed for intrafascicular recording and modulation. Based on the implanting orientation, they can be categorized into LIFEs and TIMEs. As indicated by the name, LIFEs are designed to be implanted in parallel to the nerve fibers (Fig. 5.4a) (Malagodi et al. 1989). Basically, it is a conductive wire wrapped by an insulation layer, with several stimulation sites where insulation is removed to expose the conductive core. During the surgical implantation, an insertion of micro-needle will penetrate the nerve tissue and lead the wire through the nerve trunk (Fig. 5.4b). Since stimulating sites are physically in contact with individual nerve fibers within a certain fascicle, this interface has shown high selectivity with low-activation threshold (Yoshida and Horch 1993; Micera et al. 2008; Jordi et al. 2011) and little cross-talk to adjacent fascicles (Navarro et al. 2005) in acute animal studies, as opposed to field-steering cuff electrodes.

The major disadvantage of LIFEs, however, is the challenging task to access multiple fascicles with just one device. Based on the guiding-needle method of implantation, it is difficult to implant a few wires in different fascicles without causing substantial tissue damage or device malfunction (Navarro et al. 2005). Therefore, TIMEs are designed to address this issue and further enhance the selectivity. By inserting the wire transversally to the nerve trunk, a single TIME is able to interface with several fascicles (Fig. 5.4a), thus reducing the number of implanted electrodes required by LIFEs (Boretius et al. 2010). Aside from positive confirmation from computational modeling (Raspopovic et al. 2011), acute studies on rat sciatic nerve

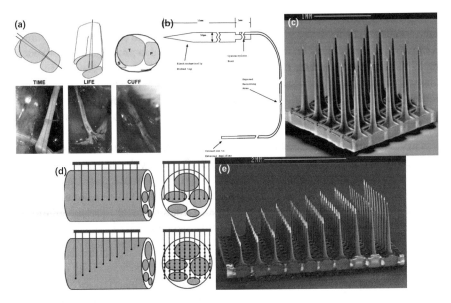

Fig. 5.4 Longitudinal and transversal intrafascicular electrodes. (**a**) Schematic diagrams and sciatic nerve implantations of TIME, LIFE, and cuff electrodes. (Adapted with permission from Jordi et al. 2011). (**b**) Structure of a longitudinal intrafascicular electrode. (Adapted with permission from Malagodi et al. 1989). (**c**) Electron microscopic picture of the Utah electrode array with 25 probes on a 2 mm by 2 mm base. (Adapted with permission from Branner and Normann 2000). (**d**) Schematic comparison of planar UEA and USEA. The varying length of electrodes allows more nerve fibers to be recruited. (Adapted with permission from Branner et al. 2001). (**e**) Scanning electron microscope image of the USEA. (Adapted with permission from Branner et al. 2001)

and human amputee have also shown remarkable selectivity of TIME, despite the small distance between active sites and the small size of the nerve (Boretius et al. 2010; Jensen et al. 2010; Badia et al. 2011; Boretius et al. 2012; Kundu et al. 2014; Harreby et al. 2015).

With similar principles of TIMEs but much more recording/stimulating channels, the Utah electrode array (UEA) can be customized for peripheral nerve interfacing, though originally developed for use in the cerebral cortex (Branner and Normann 2000). Benefiting from microfabrication technologies, 25 probes can be fabricated on a single device (Fig. 5.4c), significantly improving the accessibility to nerve fibers comparing to LIFEs and TIMEs. Furthermore, the planar array was improved by the Utah slanted electrode array (USEA) to avoid a nerve fiber being recruited by many electrodes along the axis (Fig. 5.4d, e) (Branner et al. 2001). The three dimensional structure of USEA with electrodes of varying length can provide access to more individual fibers in each fascicle and enhance graded recruitment of force in muscle groups in a highly selective fashion (Branner et al. 2001).

Axonal Selectivity with Regenerative Sieve Electrodes Regenerative electrodes are designed to precisely interface with each axon in a nerve fascicle, which reaches the highest resolution a peripheral nerve electrode can get. Ideally, as mentioned in

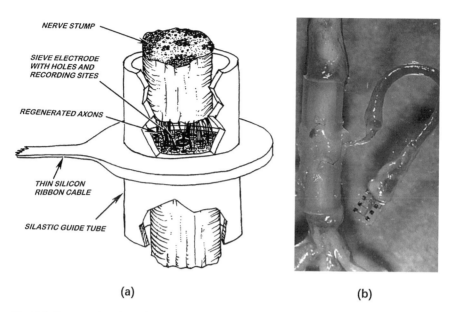

Fig. 5.5 Regenerative sieve electrodes. (**a**) Schematic diagram. (Adapted with permission from Bradley et al. 1997). (**b**) Photo of implantation of the sieve electrode. (Adapted with permission from Lago et al. 2005)

Sect. 5.2, by guiding transected axons to grow through an array of microscale via-holes, individual axons can be selectively stimulated or recorded by the conductive layer deposited around the holes (Fig. 5.5). Practically, since the nerve trunk has to be severed before implantation, this method actually reorganizes nerve structure rather than simply accessing target axons. Therefore, successful application of regenerative electrodes depends on robust axonal regeneration that is remarkably affected by the dimensions of via-holes. Excluding the feasibility of the ideal one-axon-one-hole design (Navarro et al. 2005), most chronic tests on animals found the optimal diameter to be 30–60 μm (Navarro et al. 1996; Wallman et al. 2000) for rat sciatic nerve. However, 5 μm designs with higher selectivity have also been realized and proved successful in recording afferent signals from a rat glossopharyngeal nerve (Bradley et al. 1992, 1997).

Although regenerative electrodes offer the highest selectivity among peripheral nerve devices, their traumatic implantation procedure brings more risks than others. Such design requires high regenerative ability of peripheral nerves, and it takes a long period of time for severed nerve fibers to regenerate and recover. Moreover, it is challenging to perfectly and coherently match the number, size, and layout of via-holes with the intact nerve structure, resulting in functional loss and instability from case to case. For example, thermal and taste responses were lost (Bradley et al. 1997) using the nearly same device that successfully recorded tactile, thermal, and taste signals (Bradley et al. 1992). Also, due to the variety in neuron types, each nerve fiber does not have the same or even similar regenerating capability. Implantations in rat sciatic nerve reveal that regeneration is much less successful for myelinated axons than unmyelinated ones, that motor axons regenerate more poorly than sensory axons, and that some subclasses of sympathetic fibers regenerate better than others (Negredo et al. 2004; Castro et al. 2008).

SNR

For cuff electrodes, noise rejection can be realized by optimizing electrode configuration and dimensions or an external shielding layer. Based on the simple principle of Ohm's law, the innovative design of tripole configuration has significantly improved the quality of recorded signals (Stein et al. 1975; Hoffer and Marks 1976). Instead of detecting the voltage between two electrodes, three equally spaced electrodes are placed along the nerve trunk (Fig. 5.6a) (Stein et al. 1975), and the two terminal ones are electrically shorted. In principle, such a design eliminates the potential gradient for external sources outside the cuff, allowing noises to bypass the device. In the meanwhile, action potentials that propagate inside the nerve trunk

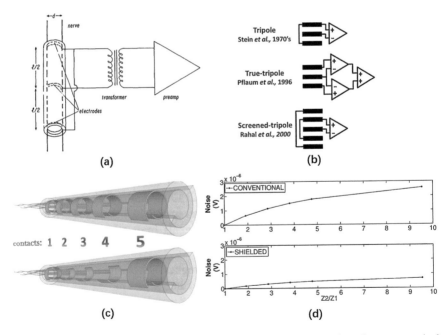

Fig. 5.6 Noise rejection for cuff electrodes. (**a**) Original tripolar configuration where two terminal electrodes are shorted. (Adapted with permission from Stein et al. 1975). (**b**) Variations based on the original tripolar configuration. (Adapted with permission from Ortiz-Catalan et al. 2013). (**c**) Schematic diagram of split-contact design. (Adapted with permission from Ortiz-Catalan et al. 2013). (**d**) Noise amplitude comparison between conventional cuff and a shielded cuff. (Adapted with permission from Sadeghlo and Yoo 2013)

are still able to be detected using the central contact. However, this tripolar configuration is based on the assumption that the potential gradient of noise is linearly distributed along the cuff, which is often compromised in clinical applications due to cuff imbalance and connective tissue (Rahal et al. 2000a; Triantis and Demosthenous 2008; Sadeghlo and Yoo 2013). Therefore, several variations on wiring scheme with amplifiers have been reported to achieve a robust recording performance (Fig. 5.6b) (Pflaum et al. 1996; Rahal et al. 2000b; Demosthenous et al. 2004; Jun-Uk et al. 2012). In addition to focusing on electrode configurations, studies on electrode dimensions found that splitting the ring-shaped central contact into discrete electrode pads (Fig. 5.6c) leads to considerable enhancement on SNR (Ortiz-Catalan et al. 2013). Moreover, inspired by the principle of electromagnetic shielding, a conductive shielding layer has shown its effectiveness on external noise rejection (Fig. 5.6d) (Perez-Orive and Durund 2000; Sadeghlo and Yoo 2013) and can simplify the circuit complexity by working with bipolar cuffs (Parisa et al. 2017). Conductive shielding has also been integrated with thin-film LIFE (tf-LIFE) to reject electromyogram (EMG) artifacts (Djilas et al. 2007).

Compared to cuff electrodes, intrafascicular electrodes typically have a higher SNR due to the insulating effect from the perineurium. The currents of EMG from nearby muscles tend to go around the fascicle rather than into it (Clark and Plonsey 1968). However, conventional monopolar scheme that records potential difference between intrafascicular space and extrafascicular space can still result in substantial EMG pickup. Based on the concept of differential recording, bipolar recording was proposed on LIFEs where axonal signals were detected between two intrafascicular electrodes (Yoshida and Stein 1999), eliminating the interference from extrafascicular sources. Moreover, the interelectrode spacing of 2 mm was found to provide an optimal SNR (Fig. 5.7) (Yoshida and Stein 1999).

The recording of sieve electrodes relies on close proximity of the ring electrode to the axons and the corresponding nodes of Ranvier; therefore, not all electrodes record equally well and are subject to cross-talk (Lago et al. 2007a). Microchannel regenerative interface combines the electrode–axon proximity of sieve electrodes and the snug enclosure of cuff electrodes. Instead of going through via-holes, the regenerated nerve fibers are enclosed by electrically insulated microchannels (Fig. 5.8) (Ivan et al. 2012) that restrict the amount of extracellular fluids. Such an encircle of axons works as a natural amplifier due to the restriction of extracellular current return path (Ivan et al. 2012). In addition, an axon is intimately interfaced by multiple recording sites. Consequently, it is easy to pick up weak axonal signals and obtain a high SNR up to four (Ivan et al. 2012; Musick et al. 2015).

5.3.2   Toward a Better Tissue–Device Interface

Clinical applications require not only high selectivity and noise suppression but also a good tissue–device interface that maintains the stability over long-term implantation. The implanted neural interface should be able to adapt to local motions and minimize neural damage without significant degradation on the stimulation/recording

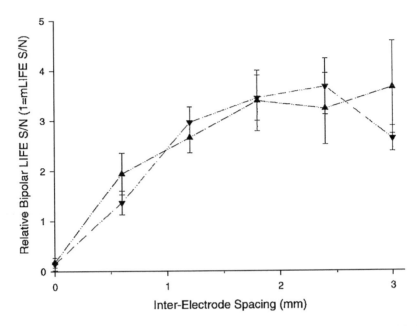

Fig. 5.7 Relative bipolar LIFE SNR as a function of interelectrode spacing (Yoshida and Stein 1999). (Adapted with permission from Yoshida and Stein 1999)

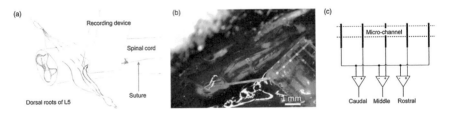

Fig. 5.8 Microchannel regenerative electrodes (Ivan et al. 2012). (**a**) Schematic diagram of the microchannel device interfaced with root L5. (**b**) Photo of the implanted device during recording. (**c**) Circuit diagram of recording sites and amplifiers. (Adapted with permission from Ivan et al. 2012)

capability. Taking advantages on microfabrication techniques and material advances, the quality of tissue–device interface can be improved through structural optimization, flexible base materials, and high-performance electrode materials.

Innovations on Structural Design

tf-LIFE and ACTIN tf-LIFE (thin-film longitudinal intrafascicular electrode) was proposed to improve not only selectivity but also mechanical matching of the neural interface (Yoshida et al. 2000). Using micromachining process, electrode sites can be fabricated on a flat ribbon of polyimide substrate (Fig. 5.9a, b). Comparing to

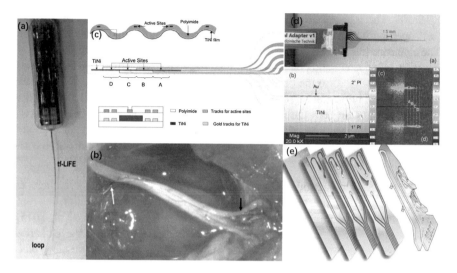

Fig. 5.9 tf-LIFE, ACTIN, and SELINE designs. (**a, b**) tf-LIFE and its implantation on a rat sciatic nerve. (Adapted with permission from Lago et al. 2007b). (**c, d**) Corrugated structure of ACTIN and its thermal responses under applied voltages. (Adapted with permission from Bossi et al. 2009). (**e**) The wing-shaped anchoring structure of SELINE. (Adapted with permission from Cutrone et al. 2015)

metal wire LIFEs, such a 2D structure allows for more recording sites and a much smaller interfacial area, which significantly enhances the selectivity without applying intensive spike-sorting algorithms (Mirfakhraei and Horch 1994). More importantly, the flexibility of the polymer substrate relieves the mechanical stress between the implanted electrodes and endoneural tissue (Yoshida et al. 2010). While a fibrous response is inevitable, tf-LIFEs produced less axonal damage than conventional LIFEs or polyLIFEs during a 3-month implantation (Lago et al. 2007b).

To address the issue of formation of encapsulation layer in chronic implantation, the concept of "movable interface" has been proposed and tested by embedding microactuators in the core of a polyLIFE device, referred to as actuated intraneural (ACTIN) electrodes (Bossi et al. 2009) (Fig. 5.9c, d). The ACTIN is actuated by the deforming of a TiNi thin film sandwiched by top and bottom polyimide layers. When applying voltage to the TiNi actuator, the bending force of the TiNi core (due to heating) causes the entire interface to turn from flat to corrugated. In addition, the memorized shape of TiNi microactuator is optimally designed so that its peaks coincide with the active sites of the interface, maximizing the displacement of active sites without undesired movements of the rest of the structure. With displacement of up to 60 μm, such micro-adjustment after implantation is very helpful to recover the lost connection between electrodes and nerve tissue. However, in vivo test of this design has not been reported yet, and due to the cytotoxicity of pure Ni, the biocompatibility of TiNi microactuator still remains unknown.

SELINE The issues with LIFEs and TIMEs inspire the development of SELINE (self-opening intrafascicular neural interface), an evolution of LIFEs and TIMEs. The electrode has a main shaft with two lateral wings on each side (Fig. 5.9e). After insertion into a nerve and gently being pulled back, the two wings open transversely and remain anchored to the nerve tissue, making the implanted device more stable over long-term implantation (Cutrone et al. 2015). In addition, since there are active stimulation sites on both the main shaft and opening wings, axons from different sub-fascicles can be accessed in a three-dimensional way. Comparing to LIFEs and TIMEs, this design significantly improves the mechanical stability of the tissue–device interface, while inheriting the outstanding selectivity from TIMEs.

For regenerative electrodes that have more intimate contact with nerve fibers, an implantable interface must be able to facilitate healthy axonal regeneration and maturation while maintaining the close contact with axons (Srinivasan et al. 2015). Moreover, the mixed motor and sensory fibers after transection make it even harder to separately record motor signals and stimulate sensory nerves. Rigid sieve electrodes fall short on these requirements due to the difficulty of designing a suitable layout for the complicated axon populations (Thompson et al. 2016). Even with the flexible polyimide polymer, the thin-plate structure can generate substantial compressive force on a small area to cause damage to nerve fibers. Likewise, the low-channel density also limits chronic applications of the flat microchannel electrode. Therefore, several three-dimensional microchannel interfaces have been proposed to promote the chronic performance by combining sieve and microchannel structures. Rolling a flat microchannel device for implantation not only increases channel density significantly (Fig. 5.10a) (Srinivasan et al. 2015), but also forms a more friendly interface with the nerve fibers, as its tubing structure provides more support to relieve localized compressive force and mimics the natural cross-sectional architecture of a nerve. With modifications, such tubing microchannel structure can also realize guided growth of motor and sensory fibers, separately. As shown in Fig. 5.10b, the Y-shape tubing together with nerve growth factor (NGF) and neurotrophin-3 (NT-3) has preferentially enticed specific axonal populations into separate compartments (Lotfi et al. 2011). Alternatively, multiple nerve fibers have been guided into wider channels first and then grow into bifurcated channels with diminished sizes (Fig. 5.10c) (Irina et al. 2013), which facilitates the separation of the regenerating axonal bundles in a more gradual and effective way than sieve electrodes.

Expanding the principle of in situ amplification of microchannel electrodes, a biohybrid regenerative peripheral nerve interface (RPNI) has been proposed by inclusion of living muscle grafts (Fig. 5.11) (Stieglitz et al. 2002; Urbanchek et al. 2011; Kung 2014; Langhals et al. 2014a; b; Urbanchek et al. 2016). This design integrates muscle grafts and recording electrodes together and then interfaces with transected nerves. Consequently, nerve fibers will grow toward these muscle grafts and regenerate neuromuscular junctions to form new innervations. In this way, action potentials from regenerated nerves give rise to contraction of muscle cells whose membrane potentials are picked up by recording probes. Comparing to sieve and microchannel devices, RPNI innovatively exploits the tendency of neurons to form neuromuscular junctions, which serve as biological amplifiers as muscular

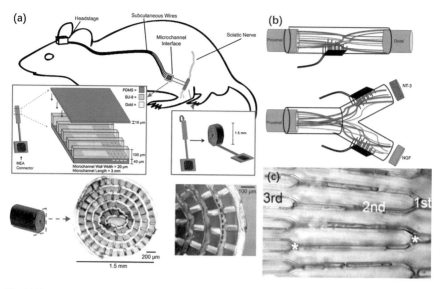

Fig. 5.10 Microchannel electrodes as an improved neural interface. (**a**) Schematic view of micro-channel structure consisting of PDMS substrate, SU8 walls, and gold electrodes (up); cross-sectional view of the device rolled for implantation (down). (Adapted with permission from Srinivasan et al. 2015). (**b**) Mixed nature of regenerative nerve in the absence of any molecular cues (up); specific growth factors attract a subtype of neurons to the modality-specific compart-ment (down). (Adapted with permission from Lotfi et al. 2011). (**c**) Three stages of bifurcated microchannels with diminished sizes for axons to grow through. (Adapted with permission from Irina et al. 2013)

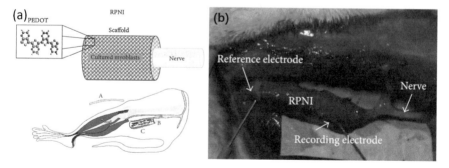

Fig. 5.11 Biohybrid muscle graft regenerative electrode. (Adapted with permission from Urbanchek et al. 2016). (**a**) Schematic view of RPNI in which the peripheral nerve is wrapped by acellular muscle with PEDOT and the device is populated with cultured myoblasts (**b**) RPNI after 4 months of implantation on rat peroneal nerve

ionic currents are much larger in amplitude than those of neurons. In addition, since axons are interfaced by the RPNI in a natural way rather than going through a pas-sive structure, foreign body reactions and traumas are significantly alleviated. However, nerve fibers innervating artificial muscle grafts cannot engage again with natural muscles, thus this strategy only applies to amputees whose original muscles are already beyond repair.

Innovations on Materials

The development of more stable and biocompatible materials is crucial to improving tissue–device interface and often serves as the basis of novel design on electrode structures. For peripheral nerve interfaces, material innovations are often used for either the skeleton of devices or recording/stimulation sites, in which the former focuses on mechanical and biological properties while the latter on electrical characteristics.

For chronic implantations, the high stiffness of conventional Pt-Ir LIFEs compared to surrounding neural tissue often leads to relative motions of electrodes, causing nerve damage and fibrous encapsulation (Lefurge et al. 1991). To minimize the mechanical mismatch, flexible polymers are used as core structures to replace pure metal wires (Yoshida et al. 2010). Based on the metalization techniques (McNaughton and Horch 1996), metalizing a 12-μm Kevlar fiber and insulating it with silicone (Fig. 5.12a) can make a polyLIFE device 60 times more flexible than Pt-Ir electrodes, without a significant loss in the recording capability (Lawrence et al. 2003). Similarly, the tf-LIFE mentioned before uses flexible polyimide as the substrate material to reduce the structural rigidity. These efforts have demonstrated reduced nerve damage and high degree of biocompatibility in chronic implantations (Malmstrom et al. 1998; Lawrence et al. 2002; Dhillon et al. 2004a, b; Dhillon and Horch 2005; Dhillon et al. 2005). In addition, flexible polyimide has shown better long-term stability in sieve electrodes and promotes faster axonal regeneration than traditional silicon-based devices (Navarro et al. 1998; Lago et al. 2005). To further increase the flexibility, liquid GaIn alloy as electrode material has been proposed to fabricate stretchable peripheral nerve interfaces (Rui and Jing 2017) similar to cuff electrodes (Fig. 5.12b). While liquid metal cannot be surpassed by solid materials in terms of mechanical flexibility, whether it is suitable for intrafascicular applications is still unclear.

The critical stimulation/recording sites that electrically interact with nerve fibers also benefit from the advances of biomaterials. Among four popular electrode materials, Pt, IrO_x, poly(3,4-ethylenedioxythiophene) (PEDOT), and platinum black (Pt black), acute and long-term stimulations with cuff electrodes have demonstrated Pt black as the best candidate for chronically implantable electrodes, due to its excellent performance on charge delivery capacity, charge injection capacity, interfacial impedance, and most importantly the stable electrochemical properties (Lee et al. 2016). Although tested on cuff electrodes, the application of Pt black can also be extended to smaller and more invasive interfaces that demand higher charge delivery capability for stimulating sites.

In contrast, the application of ion-selective membrane (ISM) stands out of the traditional electron–ion interfaces by directly modulating ion concentrations along the nerve (Fig. 5.12c) (Song et al. 2011). Since the essence of action potential is the motion of different ions into or out of plasma membrane, such innovative design actually works the same way as neurons and eliminates the electron–ion conversion process at the neural interface, resulting in 40% reduction of stimulation threshold on a frog sciatic nerve.

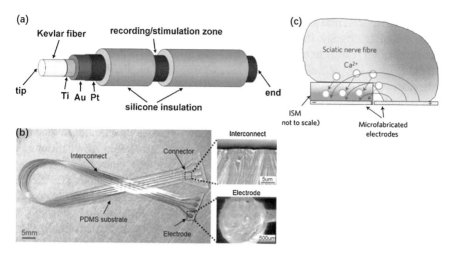

Fig. 5.12 Innovative neural interfaces based on advanced materials. (**a**) General design of polymer-based intrafascicular electrode. (Adapted with permission from Lawrence et al. 2003). (**b**) Flexible cuff electrode using liquid alloy as the electrode core. (Adapted with permission from Rui and Jing 2017). (**c**) Neural modulation using ion-selective membrane. (Adapted with permission from Song et al. 2011)

5.3.3 Toward Easier Surgical Operation and Implantation

Surgical operation is not only the very first step of peripheral nerve electrode implantation but also determines the performance and stability of implanted devices. Since this process is very likely to introduce neural trauma and displacement of electrodes, it is critical to make surgical protocols as simple and robust as possible when designing a neural interface. Therefore, this section will discuss the strategies for easier surgical implantation, from the improvement on electrodes and device wirings.

Modification of Electrodes

Since cuff, intrafascicular, and regenerative electrodes are significantly distinguished from each other in both structure and working principles, the corresponding surgical procedures should also be discussed individually. As discussed before, a conventional split-cylinder cuff interface is made of silicon rubber with metal contacts on its inner side, which requires sutures to secure the device around the target nerve during implantation. Not only does the suturing complicate the surgical installation but causes a non-intimate interface as well (Naples et al. 1988). Spiral cuff and shape memory alloy (SMA) cuff are therefore proposed to eliminate the suturing process using self-closed scheme. By bonding two pieces of silicone sheet with different resting lengths, stress is stored in the spiral cuff, and the device can spontaneously coil into a spiral tube once released (Fig. 5.13a) (Naples et al. 1988). With a similar strategy, the SMA cuff utilizes the shape memory effect and super elasticity of NiTi alloy to make the device self-closed at 37 °C while self-opening at 10 °C (Fig. 5.13b) (Crampon et al. 1998; Crampon et al. 1999). Benefitted from advances on microfabrication and flexible materials, various new designs have been reported under the concept of self-closed cuff (see the table in Kang et al. (2015)). A typical example is the microfabricated parylene-based cuff electrode (Fig. 5.13c), which further reduces mechanical mismatch and increases the number of channels to 16 (Kang et al. 2015). Compared to the original design, these modifications help the cuff wrap snugly around nerve trunk and adapt to size changes, significantly enhancing the quality of the neural interface and its recording/stimulating efficiency.

Compared to cuff electrodes that only need to wrap the entire nerve trunk, the surgical installation of intrafascicular electrodes is much more complicated and easier to cause tissue damage. According to the basic design of intrafascicular electrodes (Malagodi et al. 1989), the device can be introduced into a nerve fascicle by an electrosharpened tungsten needle that is cut off after implantation (Fig. 5.14a). However, this apparently leaves the electrode unsecured and subject to longitudinal movements. This was addressed by anchoring the device on a silicone tube (Lefurge et al. 1991). Specifically, medical-grade Silastic adhesive was injected into the tubing to anchor the electrode wires and suture (Fig. 5.14b). Furthermore, since electrodes are glued with the introducing needle, the entry terminal of a wire LIFE has to be cut off when removing the needle, and this may cause further longitudinal and rotatory motion of already positioned electrode, as well as the exposure of the metal core to body fluids. The creative dual electrode from a single insulated wire can solve this issue by eliminating the need of gluing (Fig. 5.14c, d). By breaking the

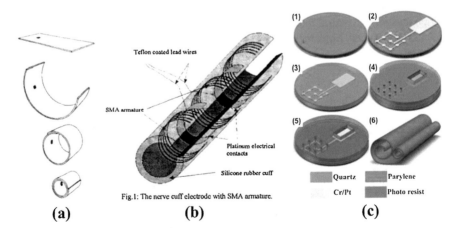

(a)   (b)   (c)

Fig. 5.13 Self-closed cuff electrodes. (**a**) Spiral cuff electrode is fully closed when stress is released, forming snug and adaptive wrapping on a nerve trunk. (Adapted with permission from Naples et al. 1988). (**b**) Structure of the shape memory alloy electrode. (Adapted with permission from Crampon et al. 1998). (**c**) Fabrication procedure of self-closed parylene cuff electrode. (Adapted with permission from Kang et al. 2015)

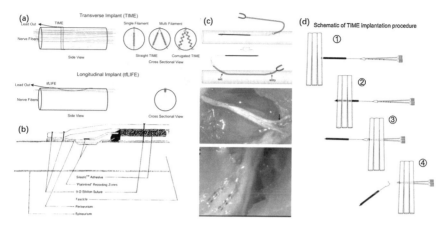

Fig. 5.14 Surgical implantation of intrafascicular electrodes. (**a**) Schematic representation of implantation process for LIFE and TIME. (Adapted with permission from Yoshida et al. 2010). (**b**) Anchoring of LIFE using a tube filled with Silastic adhesive. (Adapted with permission from Lefurge et al. 1991). (**c, d**) Implantation of LIFE (Lago et al. 2007b) and TIME (Boretius et al. 2010) without gluing to the introducer needle (Lago et al. 2007b). (Adapted with permission from Lago et al. (2007b) and Boretius et al. (2010))

metal contact between two interfacial sites and stretching the insulation layer, the device can be folded into a loop and linked with a sacrificial polyaramid loop glued to the needle. In this way, the needle can be released after implantation by simply cutting off the sacrificial loop rather than the electrode body.

Modification of Device Wirings

In addition to the implantation of the device itself, wiring and accessories of a peripheral nerve interface, especially those with multiple channels, also constitute substantial liability to surgical operation as well as chronic implantation after surgery. On the one hand, it is highly desired to minimize the overall dimensions of the implanted device using integrated circuit technology. For FINE, despite plenty of leads required for electrode contacts, the wiring scheme could be significantly simplified by integrating a multiplexer (Lertmanorat et al. 2009) with the electrodes. Similar integration design can also be implemented in sieve electrodes to enable chronic recordings (Bradley et al. 1997).

On the other hand, wireless transmission has been proposed to completely eliminate the need for tethered communication as well as powering of implanted devices. Based on electromagnetic energy coupling and communication, many systems have been developed using radiofrequency (RF) electromagnetic transmission. Wireless powering for in vivo stimulation can be realized by direct antenna harvesting (Park et al. 2015), resonant cavity (Montgomery et al. 2015), or midfield regime transfer (Tanabe et al. 2017). Through capacitive coupling between an adjacent serpentine antenna, RF power can be harvested to provide electricity for LEDs that are implanted to optogenetically stimulate rat sciatic nerves (Fig. 5.15a). Alternatively, electromagnetic energy can be uniformly localized on an aluminum resonant cavity to power LEDs implanted on mouse peripheral nerve endings (Fig. 5.15b). In addition to optogenetic probes, RF transmission has also been applied to cuff electrodes on the vagus nerve using midfield regime that significantly enhanced transmission efficiency (Fig. 5.15c).

However, further miniaturization of electromagnetic wireless electronics is bottlenecked by their poor efficiency at dimensions lower than 5 mm due to the inefficient RF coupling within tissue (Rabaey et al. 2011). To overcome this inherent limitation, ultrasonic transmission has been proposed to replace the electromagnetic strategy. With much smaller wavelength and less attenuation within tissue, ultrasonic wave is able to achieve higher spatial resolution and penetration depth, yielding excellent power efficiency at smaller dimensions compared to its electromagnetic counterparts (Seo et al. 2015). By integrating a piezocrystal to convert ultrasonic energy into electricity, the recording transistor can be electrically powered, and neural potentials can be detected by the transistor's gate, modulating the current flowing through the piezocrystal (Fig. 5.15d) (Seo et al. 2016). In turn, this current modulation affects the vibration of the crystal and the reflected ultrasonic wave that is then reconstructed externally. This technology, named Neural Dust, is well known for its sub-mm size and high efficiency, which makes the surgical operation easier and reduces trauma caused by micromotion of the electrodes during chronic implantation.

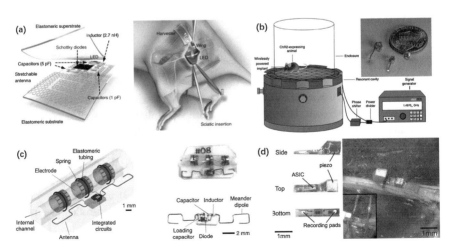

Fig. 5.15 Electromagnetic wireless powering for peripheral nerve electrodes. (**a**) Soft and wireless optogenetic stimulation device based on capacitive coupling through serpentine antenna. (Adapted with permission from Park et al. 2015). (**b**) Resonant cavity-based light delivery system and implanted devices (subfigure). (Adapted with permission from Montgomery et al. 2015). (**c**) Schematic diagram (left) and photos (right) of the wireless cuff, which consist of electrode sites, a meandered antenna, and integrated circuits. (Adapted with permission from Tanabe et al. 2017). (**d**) Neural Dust mote implanted on rat sciatic nerve. A piezoelectric crystal, a transistor, and two recording electrodes are assembled on a flexible PCB. (Adapted with permission from Seo et al. 2016)

5.4 Conclusion

Directly acting on nerve fibers, peripheral nerve interfaces can modulate muscular activity with less activation energy and more compact structure. Decades of extensive studies on electrode-based interfaces have witnessed significant progress on selectivity, noise rejection, tissue–device interface, and surgical implantation. Although the challenges are discussed individually here, they are inherently correlated to each other, particularly between the functionality and tissue–device interface. Both higher resolution of recruitment and lower noise level require more intimate interface between the implanted probes and nerve fibers, which inevitably brings issues of scar encapsulation, neural trauma, and instability of recording/stimulation during chronic implantation. While difficult to be perfectly eliminated, these side effects have been much alleviated through more stabilized device structures, the application of flexible materials, and robust implantation procedures. Moreover, the emergence of muscle-graft interface and ion-selective membrane has opened up the possibility of accessing peripheral nerves in a more natural way than conventional electrodes.

Acknowledgments This work was supported by the Defense Advanced Research Projects Agency through Grant # D17AP00031 of the USA and the Chronic Brain Injury Program of The Ohio State University through a Pilot Award. The views, opinions, and/or findings contained in this article are those of the author and should not be interpreted as representing the official views or policies, either expressed or implied, of the Defense Advanced Research Projects Agency or the Department of Defense.

References

Altman, K. W., & Plonsey, R. (1990). Point source nerve bundle stimulation: Effects of fiber diameter and depth on simulated excitation. *IEEE Transactions on Biomedical Engineering, 37*(7), 688–698.

Badia, J., Boretius, T., et al. (2011). Biocompatibility of chronically implanted transverse intrafascicular multichannel electrode (TIME) in the rat sciatic nerve. *IEEE Transactions on Biomedical Engineering, 58*(8), 2324–2332.

Boretius, T., Badia, J., et al. (2010). A transverse intrafascicular multichannel electrode (TIME) to interface with the peripheral nerve. *Biosensors and Bioelectronics, 26*(1), 62–69.

Boretius, T., Yoshida, K., et al. (2012). *A transverse intrafascicular multichannel electrode (TIME) to treat phantom limb pain — Towards human clinical trials.* 2012 4th IEEE RAS & EMBS International Conference on Biomedical Robotics and Biomechatronics (BioRob).

Bossi, S., Kammer, S., et al. (2009). An implantable microactuated intrafascicular electrode for peripheral nerves. *IEEE Transactions on Biomedical Engineering, 56*(11), 2701–2706.

Bradley, R. M., Smoke, R. H., et al. (1992). Functional regeneration of glossopharyngeal nerve through micromachined sieve electrode arrays. *Brain Research, 594*(1), 84–90.

Bradley, R. M., Cao, X., et al. (1997). Long term chronic recordings from peripheral sensory fibers using a sieve electrode array. *Journal of Neuroscience Methods, 73*(2), 177–186.

Branner, A., & Normann, R. A. (2000). A multielectrode array for intrafascicular recording and stimulation in sciatic nerve of cats. *Brain Research Bulletin, 51*(4), 293–306.

Branner, A., Stein, R. B., et al. (2001). Selective stimulation of cat sciatic nerve using an array of varying-length microelectrodes. *Journal of Neurophysiology, 85*(4), 1585–1594.

Castro, J., Negredo, P., et al. (2008). Fiber composition of the rat sciatic nerve and its modification during regeneration through a sieve electrode. *Brain Research, 1190*(Supplement C), 65–77.

Choi, A. Q., Cavanaugh, J. K., et al. (2001). Selectivity of multiple-contact nerve cuff electrodes: A simulation analysis. *IEEE Transactions on Biomedical Engineering, 48*(2), 165–172.

Clark, J., & Plonsey, R. (1968). The extracellular potential field of the single active nerve fiber in a volume conductor. *Biophysical Journal, 8*(7), 842–864.

Coburn, B. (1980). Electrical stimulation of the spinal cord: Two-dimensional finite element analysis with particular reference to epidural electrodes. *Medical and Biological Engineering and Computing, 18*(5), 573–584.

Crampon, M. A., Sawan, M., et al. (1998). *New nerve cuff electrode based on a shape memory alloy armature*. Proceedings of the 20th annual international conference of the IEEE Engineering in Medicine and Biology Society. Vol.20 Biomedical Engineering Towards the Year 2000 and Beyond (Cat. No.98CH36286).

Crampon, M.-A., Sawan, M., et al. (1999). New easy to install nerve cuff electrode using shape memory alloy armature. *Artificial Organs, 23*(5), 392–395.

Cutrone, A., Valle, J. D., et al. (2015). A three-dimensional self-opening intraneural peripheral interface (SELINE). *Journal of Neural Engineering, 12*(1), 016016.

Daniel, K. L., Mark, C., et al. (2006). Chronic histological effects of the flat interface nerve electrode. *Journal of Neural Engineering, 3*(2), 102.

Demosthenous, A., Taylor, J., et al. (2004). Design of an adaptive interference reduction system for nerve-cuff electrode recording. *IEEE Transactions on Circuits and Systems I: Regular Papers, 51*(4), 629–639.

Deurloo, K. E. I., Holsheimer, J., et al. (1998). Transverse tripolar stimulation of peripheral nerve: A modelling study of spatial selectivity. *Medical and Biological Engineering and Computing, 36*(1), 66–74.

Dhillon, G. S., & Horch, K. W. (2005). Direct neural sensory feedback and control of a prosthetic arm. *IEEE Transactions on Neural Systems and Rehabilitation Engineering, 13*(4), 468–472.

Dhillon, G. S., Lawrence, S. M., et al. (2004a). Residual function in peripheral nerve stumps of amputees: Implications for neural control of artificial limbs. *The Journal of Hand Surgery, 29*(4), 605–615. discussion 616-608.

Dhillon, G. S., Lawrence, S. M., et al. (2004b). Residual function in peripheral nerve stumps of amputees: Implications for neural control of artificial limbs1 1No benefits in any form have been received or will be received from a commercial party related directly or indirectly to the subject of this article. *The Journal of Hand Surgery, 29*(4), 605–615.

Dhillon, G. S., Krüger, T. B., et al. (2005). Effects of short-term training on sensory and motor function in severed nerves of long-term human amputees. *Journal of Neurophysiology, 93*(5), 2625–2633.

Djilas, M., Yoshida, K., et al. (2007). *Improving the signal-to-noise ratio in recordings with thin-film longitudinal intra-fascicular electrodes using shielding cuffs*. 2007 3rd international IEEE/EMBS conference on neural engineering.

Goodall, E. V., Lefurge, T. M., et al. (1991). Information contained in sensory nerve recordings made with intrafascicular electrodes. *IEEE Transactions on Biomedical Engineering, 38*(9), 846–850.

Goodall, E. V., Horch, K. W., et al. (1993). Analysis of single-unit firing patterns in multi-unit intra-fascicular recordings. *Medical and Biological Engineering and Computing, 31*(3), 257–267.

Goodall, E. V., de Breij, J. F., et al. (1996). Position-selective activation of peripheral nerve fibers with a cuff electrode. *IEEE Transactions on Biomedical Engineering, 43*(8), 851–856.

Grill, W. M., & Mortimer, J. T. (1998). Stability of the input-output properties of chronically implanted multiple contact nerve cuff stimulating electrodes. *IEEE Transactions on Rehabilitation Engineering, 6*(4), 364–373.

Harreby, K. R., Kundu, A., et al. (2015). Subchronic stimulation performance of transverse intrafascicular multichannel electrodes in the median nerve of the Göttingen Minipig. *Artificial Organs, 39*(2), E36–E48.

Hoffer, J. A., & Marks, W. B. (1976). Long-term peripheral nerve activity during behavior in the rabbit. In R. M. Herman, S. Grillner, P. S. G. Stein, & D. G. Stuart (Eds.), *Neural control of locomotion* (pp. 767–768). Boston: Springer.

Irina, I. S., van Wezel, R. J. A., et al. (2013). In vivo testing of a 3D bifurcating microchannel scaffold inducing separation of regenerating axon bundles in peripheral nerves. *Journal of Neural Engineering, 10*(6), 066018.

Ivan, R. M., Daniel, J. C., et al. (2012). High sensitivity recording of afferent nerve activity using ultra-compliant microchannel electrodes: An acute in vivo validation. *Journal of Neural Engineering, 9*(2), 026005.

Jensen, W., Micera, S., et al. (2010). *Development of an implantable transverse intrafascicular multichannel electrode (TIME) system for relieving phantom limb pain.* 2010 annual international conference of the IEEE engineering in medicine and biology.

Jordi, B., Tim, B., et al. (2011). Comparative analysis of transverse intrafascicular multichannel, longitudinal intrafascicular and multipolar cuff electrodes for the selective stimulation of nerve fascicles. *Journal of Neural Engineering, 8*(3), 036023.

Jun-Uk, C., Kang-Il, S., et al. (2012). Improvement of signal-to-interference ratio and signal-to-noise ratio in nerve cuff electrode systems. *Physiological Measurement, 33*(6), 943.

Kang, X., Liu, J. Q., et al. (2015). Self-closed parylene cuff electrode for peripheral nerve recording. *Journal of Microelectromechanical Systems, 24*(2), 319–332.

Kundu, A., Harreby, K. R., et al. (2014). Stimulation selectivity of the "thin-film longitudinal intrafascicular electrode" (tfLIFE) and the "transverse intrafascicular multi-channel electrode"; (TIME) in the large nerve animal model. *IEEE Transactions on Neural Systems and Rehabilitation Engineering, 22*(2), 400–410.

Kung, T. A. (2014). Regenerative peripheral nerve interface viability and signal transduction with an implanted electrode. *Plastic and Reconstructive Surgery (1963), 133*(6), 1380–1394.

Lago, N., Ceballos, D., et al. (2005). Long term assessment of axonal regeneration through polyimide regenerative electrodes to interface the peripheral nerve. *Biomaterials, 26*(14), 2021–2031.

Lago, N., Udina, E., et al. (2007a). Neurobiological assessment of regenerative electrodes for bidirectional interfacing injured peripheral nerves. *IEEE Transactions on Biomedical Engineering, 54*(6), 1129–1137.

Lago, N., Yoshida, K., et al. (2007b). Assessment of biocompatibility of chronically implanted polyimide and platinum intrafascicular electrodes. *IEEE Transactions on Biomedical Engineering, 54*(2), 281–290.

Langhals, N. B., Urbanchek, M. G., et al. (2014a). Update in facial nerve paralysis: Tissue engineering and new technologies. *Current Opinion in Otolaryngology & Head and Neck Surgery, 22*(4), 291–299.

Langhals, N. B., Woo, S. L., et al. (2014b). *Electrically stimulated signals from a long-term regenerative peripheral nerve interface.* 2014 36th annual international conference of the IEEE engineering in medicine and biology society.

Larsen, J. O., Thomsen, M., et al. (1998). Degeneration and regeneration in rabbit peripheral nerve with long-term nerve cuff electrode implant: A stereological study of myelinated and unmyelinated axons. *Acta Neuropathologica, 96*(4), 365–378.

Lawrence, S. M., Larsen, J. O., et al. (2002). Long-term biocompatibility of implanted polymer-based. *Journal of Biomedical Materials Research, 63*(5), 501–506.

Lawrence, S. M., Dhillon, G. S., et al. (2003). Fabrication and characteristics of an implantable, polymer-based, intrafascicular electrode. *Journal of Neuroscience Methods, 131*(1), 9–26.

Lee, Y. J., Kim, H.-J., et al. (2016). Characterization of nerve-cuff electrode interface for biocompatible and chronic stimulating application. *Sensors and Actuators B: Chemical, 237*(Supplement C), 924–934.

Lefurge, T., Goodall, E., et al. (1991). Chronically implanted intrafascicular recording electrodes. *Annals of Biomedical Engineering, 19*(2), 197–207.

Lertmanorat, Z., Montague, F. W., et al. (2009). A flat Interface nerve electrode with integrated multiplexer. *IEEE transactions on neural systems and rehabilitation engineering: A publication of the IEEE Engineering in Medicine and Biology Society, 17*(2), 176–182.

Leventhal, D. K., & Durand, D. M. (2003). Subfascicle stimulation selectivity with the flat interface nerve electrode. *Annals of Biomedical Engineering, 31*(6), 643–652.

Leventhal, D. K., & Durand, D. M. (2004). Chronic measurement of the stimulation selectivity of the flat interface nerve electrode. *IEEE Transactions on Biomedical Engineering, 51*(9), 1649–1658.

Lotfi, P., Garde, K., et al. (2011). Modality-specific axonal regeneration: Toward selective regenerative neural interfaces. *Front Neuroeng, 4*, 11.

Malagodi, M. S., Horch, K. W., et al. (1989). An intrafascicular electrode for recording of action potentials in peripheral nerves. *Annals of Biomedical Engineering, 17*(4), 397–410.

Malmstrom, J. A., McNaughton, T. G., et al. (1998). Recording properties and biocompatibility of chronically implanted polymer-based intrafascicular electrodes. *Annals of Biomedical Engineering, 26*(6), 1055–1064.

McNaughton, T. G., & Horch, K. W. (1996). Metallized polymer fibers as leadwires and intrafascicular microelectrodes. *Journal of Neuroscience Methods, 70*(1), 103–107.

McNeal, D. R., Baker, L. L., et al. (1989). Recruitment data for nerve cuff electrodes: Implications for design of implantable stimulators. *IEEE Transactions on Biomedical Engineering, 36*(3), 301–308.

Micera, S., Navarro, X., et al. (2008). On the use of longitudinal intrafascicular peripheral interfaces for the control of cybernetic hand prostheses in amputees. *IEEE Transactions on Neural Systems and Rehabilitation Engineering, 16*(5), 453–472.

Mirfakhraei, K., & Horch, K. (1994). Classification of action potentials in multi-unit intrafascicular recordings using neural network pattern-recognition techniques. *IEEE Transactions on Biomedical Engineering, 41*(1), 89–91.

Montgomery, K. L., Yeh, A. J., et al. (2015). Wirelessly powered, fully internal optogenetics for brain, spinal and peripheral circuits in mice. *Nature Methods, 12*(10), 969–974.

Musick, K. M., Rigosa, J., et al. (2015). Chronic multichannel neural recordings from soft regenerative microchannel electrodes during gait. *Scientific Reports, 5*, 14363.

Nannini, N., & Horch, K. (1991). Muscle recruitment with intrafascicular electrodes. *IEEE Transactions on Biomedical Engineering, 38*(8), 769–776.

Naples, G. G., Mortimer, J. T., et al. (1988). A spiral nerve cuff electrode for peripheral nerve stimulation. *IEEE Transactions on Biomedical Engineering, 35*(11), 905–916.

Navarro, X., Calvet, S., et al. (1996). Peripheral nerve regeneration through microelectrode arrays based on silicon technology. *Restorative Neurology and Neuroscience, 9*(3), 151–160.

Navarro, X., Calvet, S., et al. (1998). Stimulation and recording from regenerated peripheral nerves through polyimide sieve electrodes. *Journal of the Peripheral Nervous System, 3*(2), 91–101.

Navarro, X., Krueger, T. B., et al. (2005). A critical review of interfaces with the peripheral nervous system for the control of neuroprostheses and hybrid bionic systems. *Journal of the Peripheral Nervous System, 10*(3), 229–258.

Negredo, P., Castro, J., et al. (2004). Differential growth of axons from sensory and motor neurons through a regenerative electrode: A stereological, retrograde tracer, and functional study in the rat. *Neuroscience, 128*(3), 605–615.

Ortiz-Catalan, M., Brånemark, R., et al. (2012). On the viability of implantable electrodes for the natural control of artificial limbs: Review and discussion. *Biomedical Engineering Online, 11*, 33–33.

Ortiz-Catalan, M., Marin-Millan, J., et al. (2013). Effect on signal-to-noise ratio of splitting the continuous contacts of cuff electrodes into smaller recording areas. *Journal of Neuroengineering and Rehabilitation, 10*, 22–22.

Parisa, S., Milos, R. P., et al. (2017). Optimizing the design of bipolar nerve cuff electrodes for improved recording of peripheral nerve activity. *Journal of Neural Engineering, 14*(3), 036015.

Park, S. I., Brenner, D. S., et al. (2015). Soft, stretchable, fully implantable miniaturized optoelectronic systems for wireless optogenetics. *Nature Biotechnology, 33*(12), 1280–1286.

Parrini, S. (1997). *A modeling study to compare tripolar and monopolar cuff electrodes for selective activation of nerve fibers.* Conf. Neural Prosthesis: Motor Systems.

Patil, A. C., & Thakor, N. V. (2016). Implantable neurotechnologies: A review of micro- and nanoelectrodes for neural recording. *Medical & Biological Engineering & Computing, 54*(1), 23–44.

Perez-Orive, J., & Durund, D. M. (2000). Modeling study of peripheral nerve recording selectivity. *IEEE Transactions on Rehabilitation Engineering, 8*(3), 320–329.

Pflaum, C., Riso, R. R., et al. (1996). *Performance of alternative amplifier configurations for tripolar nerve cuff recorded ENG.* Proceedings of 18th annual international conference of the IEEE engineering in medicine and biology society.

Rabaey, J. M., M. Mark, et al. (2011). *Powering and communicating with mm-size implants.* 2011 design, automation & test in Europe.

Rahal, M., Taylor, J., et al. (2000a). The effect of nerve cuff geometry on interference reduction: A study by computer modeling. *IEEE Transactions on Biomedical Engineering, 47*(1), 136–138.

Rahal, M., Winter, J., et al. (2000b). An improved configuration for the reduction of EMG in electrode cuff recordings: A theoretical approach. *IEEE Transactions on Biomedical Engineering, 47*(9), 1281–1284.

Raspopovic, S., Capogrosso, M., et al. (2011). A computational model for the stimulation of rat sciatic nerve using a transverse intrafascicular multichannel electrode. *IEEE Transactions on Neural Systems and Rehabilitation Engineering, 19*(4), 333–344.

Rattay, F. (1986). Analysis of models for external stimulation of axons. *IEEE Transactions on Biomedical Engineering, BME-33*(10), 974–977.

Romero, E., Denef, J. F., et al. (2001). Neural morphological effects of long-term implantation of the self-sizing spiral cuff nerve electrode. *Medical and Biological Engineering and Computing, 39*(1), 90–100.

Rui, G., & Jing, L. (2017). Implantable liquid metal-based flexible neural microelectrode array and its application in recovering animal locomotion functions. *Journal of Micromechanics and Microengineering, 27*(10), 104002.

Rushton, W. A. H. (1927). The effect upon the threshold for nervous excitation of the length of nerve exposed, and the angle between current and nerve. *The Journal of Physiology, 63*(4), 357–377.

Sadeghlo, B. & Yoo, P. B. (2013). *Enhanced electrode design for peripheral nerve recording.* 2013 6th international IEEE/EMBS conference on neural engineering (NER).

Sahin, M., Haxhiu, M. A., et al. (1997). Spiral nerve cuff electrode for recordings of respiratory output. *Journal of Applied Physiology, 83*(1), 317–322.

Schiefer, M. A., Triolo, R. J., et al. (2005). *Modeling selective stimulation with A flat Interface nerve electrode for standing neuroprosthetic systems.* Conference proceedings. 2nd international IEEE EMBS conference on neural engineering, 2005.

Schiefer, M. A., Triolo, R. J., et al. (2008). A model of selective activation of the femoral nerve with a flat interface nerve electrode for a lower extremity neuroprosthesis. *IEEE Transactions on Neural Systems and Rehabilitation Engineering, 16*(2), 195–204.

Schiefer, M. A., Polasek, K. H., et al. (2010). Selective stimulation of the human femoral nerve with a flat interface nerve electrode. *Journal of Neural Engineering, 7*(2), 26006–26006.

Schiefer, M. A., Freeberg, M., et al. (2013). Selective activation of the human tibial and common peroneal nerves with a flat interface nerve electrode. *Journal of Neural Engineering, 10*(5), 056006.

Seo, D., Carmena, J. M., et al. (2015). Model validation of untethered, ultrasonic neural dust motes for cortical recording. *Journal of Neuroscience Methods, 244*(Supplement C), 114–122.

Seo, D., Neely, R. M., et al. (2016). Wireless recording in the peripheral nervous system with ultrasonic neural dust. *Neuron, 91*(3), 529–539.

Song, Y.-A., Melik, R., et al. (2011). Electrochemical activation and inhibition of neuromuscular systems through modulation of ion concentrations with ion-selective membranes. *Nature Materials, 10*, 980.

Srinivasan, A., Tahilramani, M., et al. (2015). Microchannel-based regenerative scaffold for chronic peripheral nerve interfacing in amputees. *Biomaterials, 41*(Supplement C), 151–165.

Stein, R. B., Charles, D., et al. (1975). Principles underlying new methods for chronic neural recording. *The Canadian Journal of Neurological Sciences, 2*(3), 235–244.

Stieglitz, T., Ruf, H. H., et al. (2002). A biohybrid system to interface peripheral nerves after traumatic lesions: Design of a high channel sieve electrod. *Biosensors and Bioelectronics, 17*(8), 685–696.

Struijk, J. J., Thomsen, M., et al. (1999). Cuff electrodes for long-term recording of natural sensory information. *IEEE Engineering in Medicine and Biology Magazine, 18*(3), 91–98.

Sweeney, J. D., Ksienski, D. A., et al. (1990). A nerve cuff technique for selective excitation of peripheral nerve trunk regions. *IEEE Transactions on Biomedical Engineering, 37*(7), 706–715.

Tanabe, Y., Ho, J. S., et al. (2017). High-performance wireless powering for peripheral nerve neuromodulation systems. *PLoS One, 12*(10), e0186698.

Tarler, M. D., & Mortimer, J. T. (2003). Comparison of joint torque evoked with monopolar and tripolar-cuff electrodes. *IEEE Transactions on Neural Systems and Rehabilitation Engineering, 11*(3), 227–235.

Tarler, M. D., & Mortimer, J. T. (2004). Selective and independent activation of four motor fascicles using a four contact nerve-cuff electrode. *IEEE Transactions on Neural Systems and Rehabilitation Engineering, 12*(2), 251–257.

Thompson, C. H., Zoratti, M. J., et al. (2016). Regenerative electrode interfaces for neural prostheses. *Tissue Engineering. Part B, Reviews, 22*(2), 125–135.

Triantis, I. F., & Demosthenous, A. (2008). Tripolar-cuff deviation from ideal model: Assessment by bioelectric field simulations and saline-bath experiments. *Medical Engineering & Physics, 30*(5), 550–562.

Tyler, D. J., & Durand, D. M. (2002). Functionally selective peripheral nerve stimulation with a flat interface nerve electrode. *IEEE Transactions on Neural Systems and Rehabilitation Engineering, 10*(4), 294–303.

Urbanchek, M. G., Wei, B., et al. (2011). Long-term stability of regenerative peripheral nerve interfaces (RPNI). *Plastic and Reconstructive Surgery Plastic and Reconstructive Surgery, 128*, 88–89.

Urbanchek, M. G., Kung, T. A., et al. (2016). Development of a regenerative peripheral nerve interface for control of a neuroprosthetic limb. *BioMed Research International, 2016*, 5726730.

Veltink, P. H., van Alste, J. A., et al. (1988). Influences of stimulation conditions on recruitment of myelinated nerve fibres: A model study. *IEEE Transactions on Biomedical Engineering, 35*(11), 917–924.

Veltink, P. H., van Alsté, J. A., et al. (1989a). Multielectrode intrafascicular and extraneural stimulation. *Medical and Biological Engineering and Computing, 27*(1), 19–24.

Veltink, P. H., Veen, B. K. v., et al. (1989b). A modeling study of nerve fascicle stimulation. *IEEE Transactions on Biomedical Engineering, 36*(7), 683–692.

Veraart, C., Grill, W. M., et al. (1993). Selective control of muscle activation with a multipolar nerve cuff electrode. *IEEE Transactions on Biomedical Engineering, 40*(7), 640–653.

Wallman, L., Rosengren, A., et al. (2000). *Geometric design and surface morphology of sieve electrodes-nerve regeneration and biocompatibility studies*. 1st annual international IEEE-EMBS special topic conference on microtechnologies in medicine and biology. Proceedings (Cat. No.00EX451).

Yoo, P. B., Sahin, M., et al. (2004). Selective stimulation of the canine hypoglossal nerve using a multi-contact cuff electrode. *Annals of Biomedical Engineering, 32*(4), 511–519.

Yoshida, K., & Horch, K. (1993). Selective stimulation of peripheral nerve fibers using dual intrafascicular electrodes. *IEEE Transactions on Biomedical Engineering, 40*(5), 492–494.

Yoshida, K., & Stein, R. B. (1999). Characterization of signals and noise rejection with bipolar longitudinal intrafascicular electrodes. *IEEE Transactions on Biomedical Engineering, 46*(2), 226–234.

Yoshida, K., Pellinen, D., et al. (2000). *Development of the thin-film longitudinal intra-fascicular electrode*. 5th annual conference of the International Functional Electrical Stimulation Society, Aalborg, Denmark, Center for Sensory-Motor Interaction (SMI), Department of Health Science and Technology, Aalborg University.

Yoshida, K., Farina, D., et al. (2010). Multichannel intraneural and intramuscular techniques for multiunit recording and use in active prostheses. *Proceedings of the IEEE, 98*(3), 432–449.

Chapter 6
Failure Modes of Implanted Neural Interfaces

Jean Delbeke, Sebastian Haesler, and Dimiter Prodanov

6.1 Introduction

6.1.1 Neural Implants: A Broad Spectrum of Devices and Applications

Implantable devices, capable of modulating electrical activity in the nervous system, promise to address major hitherto unmet clinical needs. A number of therapies already exploit bioelectronic medicine and provide clinical benefits. They can interface to the cortex (Schalk 2010), deep brain nuclei (Arle et al. 2008; Shils et al. 2008), retina (Barriga-Rivera et al. 2017), spinal cord (Nagel et al. 2017), cranial nerves (Giordano et al. 2017), spinal nerve roots (Donaldson et al. 1997), peripheral nerves (Ortiz-Catalan et al. 2012), and even intramuscular nerve endings (Nizard et al. 2012; Ordonez et al. 2012b; Peterson et al. 1994). The corresponding sensors or electrodes are shaped to fit the targeted anatomical structure. For example, epidural or subdural grids can directly contact the cortical surface or the spinal cord (Barriga-Rivera et al. 2017), while multi-contact slender electrodes have been developed to target the basal ganglia for deep brain stimulation (DBS) and cuff electrodes wrap around peripheral nerves. Slender wire electrodes can provide intra-neural as well as intra-cerebral interfaces, while spiral electrodes are needed to fit the helical shape of the cochlea (Gaylor et al. 2013).

J. Delbeke
4Brain, Neurology, Ghent University, Ghent, Belgium

S. Haesler
Neuroelectronics Research Flanders, Leuven, Belgium

D. Prodanov (✉)
Neuroelectronics Research Flanders, Leuven, Belgium

Environment, Health and Safety, Imec, Leuven, Belgium
e-mail: Dimiter.Prodanov@imec.be

© Springer Nature Switzerland AG 2020
L. Guo (ed.), *Neural Interface Engineering*,
https://doi.org/10.1007/978-3-030-41854-0_6

Multi-channel microelectrodes have been developed for interfacing the brain, as well as the peripheral nerves (Normann et al. 1997; Branner and Normann 2000). These assorted devices are used to target a diverse set of diseases. Neural interfaces have proven successful for the treatment of pain (Nizard et al. 2012; Rushton 2002), deafness (Gaylor et al. 2013), incontinence (Patton et al. 2013), Parkinson's disease (Weaver et al. 2017), essential tremor (Borretzen et al. 2014), epilepsy (Krishna et al. 2016), treatment-resistant depression (Bewernick and Schlaepfer 2015; Williams et al. 2016), obesity (Val-Laillet et al. 2015), phrenic nerve stimulation for diaphragm pacing (Sharkey et al. 1989; Peterson et al. 1994), sleep apnea (Fleury Curado et al. 2018), blindness (Lewis et al. 2015), hypertension (Ng et al. 2016), stroke (Kimberley et al. 2018), spasticity (Nagel et al. 2017), drop foot (Martin et al. 2016; Pečlin et al. 2016), and inflammatory diseases (Pavlov and Tracey 2015). Many more applications are being submitted for clinical trials. Invasive brain-computer interface (BCI) devices use brain electrodes to control prosthetic or robotic devices in tetraplegic patients (Hochberg et al. 2006; Gunasekera et al. 2015).

Vagus nerve stimulation implants treated more than 30,000 patients for various applications (Stieglitz 2019). Since the medical device approval in 2004, the number of implanted patients is continuously growing. Indications include epilepsy, severe forms of schizophrenia, and obsessive-compulsive disorder. It was even proposed that heart failure patients might benefit from vagus nerve stimulation, but this is still being disputed by the latest clinical studies (Gold et al. 2016; De Ferrari et al. 2017).

Electrodes on peripheral nerves can provide sensory feedback to control limb prostheses (Ortiz-Catalan et al. 2012). Stimulation can also be used to transmit feedback signals generated in external sensors (Schiefer et al. 2016). Recently, interest has grown for neural recording providing afferent control loop or allowing symptom-triggered therapies to be developed. The feedback loop can involve peripheral nerves but also the brain with BCI to support communication with or control of robotic prostheses in quadriplegic patients, for example (Wright et al. 2016). While the promise of innovative technologies has great potential, caution is necessary to ensure clinical benefits outweigh the risks associated with implanting devices into the human body. A better understanding of the biological mechanisms of action is required for many clinical applications, device performance needs to be optimized, and possible side effects must be considered.

6.1.2 The Timetable of Post-implantation Events

The interactions between the body and the implant are best known from animal studies of brain implantations. Features of brain responses appear relatively stereo-typic considering different microelectrode types, sterilization procedure, or implantations surgeries, which have been investigated in over 100 studies involving many species (Jorfi et al. 2015; Prodanov and Delbeke 2016).

Conventionally, the post-implantation period is divided into acute, sub-chronic, and chronic periods but no exact time-related definition has been agreed upon. On the other hand, evidence toward finer temporal granularity in the effects of implantation has been recently published (Potter et al. 2012; Xie et al. 2014). In our previous review, we have proposed such a more detailed classification of the post-implantation events. The following stages can indeed be discriminated with times given for rodents (Prodanov and Delbeke 2016):

- The **surgical stage** is characterized by possible hemorrhage and other consequences of vascular damage (Grand et al. 2010). Electrode breakages are also more frequent in this period, and other surgical accidents are reported as well.
- The **acute stage** corresponds to the first 24 hours. Three causal factors typify this period, namely, (i) insertion-related tissue dimpling, (ii) local surgical tissue damage, and (iii) implant volume-related pressure effects. The acute biologic reactions most visibly expressed as edema are amplified by vascular issues such as hemorrhages and/or ischemia. These dynamic complications are already detected within 6 hours post-implant showing axonal damage and loss of capillary perfusion on multiphoton imaging (Kozai et al. 2012; Michelson et al. 2018).
- The **progressive stage** runs between 24 hours and 3 weeks, and it corresponds to a whole spectrum of possible effects, including infections related to the lack of sterility. More typically, an acute brain inflammation with blood-brain barrier breach is reported eventually with hematoma resorption or damaging mechanical micro-motion. These factors can lead to the formation of a cell gap between the electrode and the target tissue.
- The **sub-chronic stage** spans between the 4 and 6 weeks after implantation. Tissue remodeling is the main feature of this period. Signs of chronic vascular damage and inflammation predominate. Any neuronal cell gap between the electrode and target tissue can further develop during this period.
- The **chronic stage** spans from the 6 to the 12 weeks after implantation. During this phase, restoration through neuronal migration becomes quite significant. Functional neurons can re-enter the neural cell gap. Neural migration is under the strict control of cytokines, and perturbations of diffusion or any chemical or mechanical force acting on the local cellular surroundings will play a major role. Traumatic insertion injury and a prolonged foreign body reaction have often been considered as the major causes of the functional failure of electrodes (Turner et al. 1999; Holecko et al. 2005; Ward et al. 2009). However, micro-motion and other mechanical forces chronically activate tissue reactions and significantly contribute to the mechanisms maintaining a distance between electrodes and the target neurons (Du et al. 2017).
- The **steady** or **persistent stage** is a modified but chronically stable state reached after 12 weeks. At this stage, a glial scar has formed through the coordinated action of several cell types, among which are astrocytes and activated microglial cells exchanging various messenger molecules and cytokines. An overview of cytokines secreted in this gliosis process can be found in Balasingam et al. (1994). Other cells participate as well, such as cells of mesodermal origin, fibroblasts, and endothelial and hematogenous cells.

The events described above can be applied as well to human studies. However, the times given above apply to rodents, while rates of metabolic mechanisms and the immune responses vary among species so that the timing of the various stages must be adapted. Unfortunately, there is no simple allometric rule to extrapolate the given milestones observed in rats to the human situation where similar data are still lacking (Mahmood 2018).

6.1.3   The Configuration of a Generic Neural Implant

Across the numerous devices listed above, a general three-component generic or "model" implant blueprint can be identified whose description will later be complemented with application-specific features. Conceptually, this generic neural implant can be divided into three sub-systems: (i) the electrodes, (ii) the leads, and (iii) the controller itself containing, for example, an implantable battery, signal processing unit, a logic unit, RF-links, etc. (see Fig. 6.1). This configuration of the neural implants can already be recognized in the cardiac pacemakers—historically the first successful neuro-muscular prosthesis. The electrodes consist of an insulating supporting structure holding sensors and/or neural electrical contacts in close contact with the targeted neural tissue. This is thus the proper neural interface, often described in the context of electrical stimulation or recording. However, in a general framework, other forms of energy exchange (photic stimulation or transmission), mechanical sensors, etc., must be considered as well.

From the perspective of the bottom-up assembly, the highly parallel organization of the nervous system suggests an ideal specification for a neural interface. This includes a high level of specificity and multi-channel input-output capacity. However, such blueprint would imply fractal branching of the unit design repeating simple geometrical structures on many scales. This favors the self-assembly mode of manufacturing. On the other hand, established mass manufacturing practices favor top-down approaches, where every electrode configuration must be specified in advance. Addressing simultaneously thousands of individual neurons in such a top-down manner is still science fiction. Contemporary electrodes are several orders of magnitude away from such a goal, and the modest successes in that direction are usually limited in their lifetime and not compatible with chronic clinical applications. Despite these limitations, existing systems have already proven beneficial in the treatment of the diseases listed above. However, neural electrodes clearly represent the main limiting factor in neural interfaces. Consequently, the literature about failure mechanisms has often concentrated on the performance and shortcomings of electrodes.

The second part of the model device, which we suggest to be called the implanted transmission unit (IXU), is enclosed in a hermetically sealed box, containing typically a battery and a simple stimulator. Historically, this was the configuration

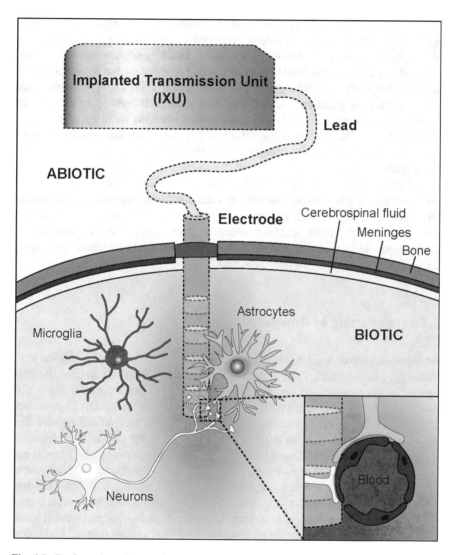

Fig. 6.1 Configuration of a generic neural implant. The implant components include Implanted Transmission Unit IXU, lead(s), and electrodes. The biotic components include the skull, meninges, the cerebrospinal fluid, and the brain tissue. The cellular components are drawn out of scale and include neurons, microglia, and astrocytes. The neurovascular unit is schematically drawn as an insert. The interactions between the implant and the brain span several spatial scales with prominent involvement of inertial, compression, shear, and stress force components, which have pleiotropic physiological effects

of the cardiac pacemakers—the first successful active implants.[1] In other applications, the electronic stimulator is more sophisticated and is called implantable stimulus generator. Nowadays, recording facilities become essential; some neural interfaces are even limited to the recording function so that the word "stimulator" no longer fits all applications. Today, the hermetically encapsulated box can also contain transcutaneous bidirectional communication links in addition to the power supply, a programmable processor with embedded software or firmware, memory, timers, stimulation current sources, and data recording and conditioning channels. The implant has become a transmission hub, hence the suggested name IXU.

The third component of the model device are the leads, which transmit signals and electrical power between the IXU and the electrodes. Although a lead can be integrated with the electrode and/or the IXU, it comes more often with connectors allowing in situ assemblies and later separate replacement, if needed. Therefore, the lead component includes the connectors to the IXU and/or to the electrodes if these are present.

6.1.4   Reporting of Adverse Events

As explained above, the key component of implanted neural interfaces is the electrodes (i.e., the sensors). Therefore, in order to gain insight about all clinical applications in use, a survey in the MAUDE[2] database was performed for the period between 2007 and 2018 using the term "electrode(s)" with contact OR degeneration OR failure OR fracture OR migration as qualifiers. This yielded 322 reports. The number of recorded "electrode failures" is dominated by the cochlear implants reported by audiologists. The cardiac pacemakers occupy second place. Clearly, and quite logically, the relative number of devices implanted influences the number of reported events. Most events are reported as "failures" usually with little information about the causal factors. Confusion between problem description and cause-seeking effort is common in the reports. In many cases, devices are not made available to the manufacturer, or no description of the cause of failure is provided. There is a clear lack of diagnostic effort. Similar findings were reported previously in a search performed about cochlear implants (Tambyraja et al. 2005). Clearly, auto-diagnostic features of implants and a systematic return of failed devices to the manufacturer could improve the present state of affairs and would allow for better mitigation measures to be implemented.

---

[1] The lay public recognized the importance of the power supply and often referred to the pacemaker as an implanted "battery." Later, and with the appearance of applications, such as the cochlear implant, power was supplied by transcutaneous means. The emphasis was put on the anatomically shaped electrode, hence the cochlear prosthesis.

[2] https://www.accessdata.fda.gov/scripts/cdrh/cfdocs/cfMAUDE/search.CFM

The available medical literature is only partially informative because most reports seem only interested in the relative risk of failures and overlook the preventive value of a detailed causal analysis. It is also possible, of course, that some events are underreported. Therefore, although based on a thorough literature search, and supported by the results of this survey, the first part of this chapter will attempt to systematically organize the possible events rather than simply summarize reported facts. This is, in such way, an attempt to provide a complete overview and to minimize the risk of overlooking issues. The second part will be devoted to a more in-depth exploration of the main causal mechanisms. The final goal of the chapter will be to place in perspective the development of preventive actions. Because of the rapid development of the field of neural interfaces and the broad spectrum of potential applications, this work will not only exploit data about currently approved clinical applications but also include preclinical and experimental findings coming from research projects.

6.2 Causal Failure Classification

6.2.1 Types of Neural Implant Failures

Neural implant failures can be broadly subdivided into two types, based on the mechanisms of action underlying the failure. Technical or abiotic failures are linked to the device itself or its mode of action (Prasad et al. 2014). Biologic or biotic failures result from the tissue and foreign body reaction to the implanted material and its actions on the tissues. These subdivisions are necessary to allow a systematic survey of the possible failures, but the various factors interfere with each other across categories.

Whether due to tissue or device damage, the malfunction can have an abrupt onset or be characterized by a progressive decline in performances. Observation of this temporal characteristic can help identify the cause of failure. Not considering the expected battery end-of-life, the lead and/or connector issues are probably the most frequent cause of failure (Blomstedt and Hariz 2005) requiring device replacement (Almassi et al. 1993). Unfortunately, in the literature, such failures are often only qualified as "high impedance" (Agarwal et al. 2011) or electrode breakage (Blomstedt and Hariz 2005), while any further systematic investigation of possible root causes, such as insulation breaks, corrosion, or metal fatigue breaks, would allow improving the system design. In other instances, the missing clinical details about the malfunction could have been of great help in diagnosing an IXU malfunction. Fortunately, altogether, these represent only about 10% of hardware-related complications (Blomstedt and Hariz 2005). More progressive declines in performance could also point to a subset of causes, such as contact displacement or electrode encapsulation. It should be stressed that the progressive failure of an implanted system does not necessarily mean system failure. Neural pathology or further progression of the disease can provide alternative explanations for decreased efficacy after one or several years of treatment (Ghika et al. 1998).

A further distinction must be made between side effects of the therapy and failures or complications resulting from a failure of the implanted system (Ben-Menachem et al. 2015; Günter et al. 2019). Side effects are, in principle, unavoidably linked to the therapy and often specific to a single clinical application. The present chapter focuses on possible corrective actions (see Sect. 6.6). Failures can lead to an abrupt loss of function or a progressive decline in performances limiting the value of therapies, such as DBS (Koller et al. 2001). The consequences are not limited to a deprivation of the corresponding therapy but also include the need for additional revision and replacement surgery or secondary side effects, such as pain or infection. While a failing device can be life-threatening for cardiac pacemakers and similar devices, the situation is less dramatic for neural interfaces (Borretzen et al. 2014; Gold et al. 2016). It should nevertheless be pointed out that, as an extreme consequence, the surgery itself, as well as severe infections, still confronts patients with a life-threatening risk (Révész et al. 2016). Failures can result from design errors, manufacturing problems, body reaction to the implant, surgical errors, inadequate therapy application, wrong handling or storage of the implant, and interference from other implants or external devices including magnetic resonance imaging (MRI). Failures can result from errors as well as inherent technological limitations, such as the battery lifetime.

Biotic failures can result from a whole range of biological effects, the major ones being described in Sect. 6.4. Extensive in vitro testing can ensure material "harmlessness." However, potential neural damage induced by an implanted active (i.e., with energy output) device is more difficult to predict and test, in part because every patient is unique. Moreover, surgical implantation protocols are being constantly developed and refined. The effect of implanted devices on tissue, which is commonly described as "foreign body reaction," is influenced by device properties, including shape and size, and material properties including surface chemistry (Lotti et al. 2017; Boehler et al. 2017; Skousen et al. 2011). Other major determinants of biotic failures are the cleanliness and the sterility of devices.

As an example, earlier publications about the cochlear implants suggest that device failure represents the most frequent and single most important complication (Luetje and Jackson 1997). Quite unlike what is seen with pharmaceuticals, the numbers of clinical indications for a specific implant are often relatively limited with the result that available data remain poorly representative and are often collected over a comparatively short period of time. On the other hand, the common features of these devices most often lead to generic causes of failure for which data from several applications (including pacemaker literature) can be pertinent and thus be pooled in order to increase and strengthen their statistical significance. A broad survey of the literature should thus combine all neural interface implementations when justified while still considering specific situations when appropriate. Implant failures are typically resulting from chronic multi-factorial biotic as well as abiotic processes from which simple mechanisms are difficult to sort out (Prasad et al. 2012). Of course, each clinical application might lead to a set of unique characteristic failures. For example, cerebrospinal fluid fistulas, cochlear ossifications, cholesteatomas, and mastoid fibrosis are specific complications cited in the frame of cochlear implants (Arnoldner et al. 2005; Balkany et al. 1999).

To partially overcome this shortcoming, the present review will be completed by some theoretical considerations to develop a more exhaustive failure analysis framework. The results are intended to lead to preventive actions or design decisions for overall device improvements.

In the following classification scheme, neural implant failures are considered from two very different points of view. First, events that might affect the patient-centered risk/benefit balance of clinically available therapies will be explored. The second point of view is that of researchers aiming at improving existing devices or developing new systems. Of course, with time, the same problems will be found at the root of the limitations identified by both groups, and this is the reason for combining these issues hereafter.

A simplified casual failure classification is schematized in Fig. 6.2. This will be the leading conceptual framework for the subsequent failure analysis. Conceptually, it will be useful to distinguish four major causes for failures: patient-related, device-operation-related interference with other devices or medical technologies (i.e., MRI scans), design errors, physician-related (iatrogenic) effects, and tissue remodeling. The main simplification depicted in Fig. 6.2 is the representation of the various factors as isolated elements. Clearly, interactions between the various causes is the rule. For example, a hemorrhage has mechanical consequences and produces some

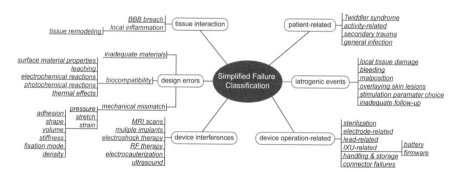

Fig. 6.2 Simplified casual failure classification

toxic substances that can affect the blood-brain barrier. For the purpose of clarity, the figure puts emphasis on the mechanical interactions leading to blood-brain barrier breach, as well as on the lack of biocompatibility without, however, excluding other interactions. As we will see below (Sect. 6.4.2), the tissue reaction to an implant is controlled by many signaling molecule exchange cascades involving several cell types and physiological structures. Therefore, control of the neuroinflammatory response could improve electrode chronic stability (Jorfi et al. 2015). Also, the blood-brain barrier (BBB) damage participates in the complex diffusion pattern. An important recent finding is the confirmation of the presence of bi-phasic reactive tissue response having a highly variable acute stage that appears to respond well to various interventions (Purcell et al. 2009; He et al. 2006) and relatively more stereotypic chronic phase. The importance of the diffusion of different factors might draw attention to the electrode and tissue geometrical structure, as well.

6.2.2 Patient-Related Effects

Because of the possible temporal changes, a regular follow-up and appropriate counseling are mandatory for failure prevention. Patients should be carefully instructed in such matters as the appropriate use of the battery charger or other external components. Similarly, attention should be given to a possible "Twiddler syndrome" leading to damage due to the implanted IXU mobilization as found in 1.3% of the patients of a cohort of 226. In all cases, securing the IXU in the chest pocket resolved the problem (Burdick et al. 2010). Twiddler syndromes can present as a lead fracture or lead retraction (Trout et al. 2013). Finally, accidental trauma with a direct impact on one of the implanted components is always possible. Prevention should take each patient's professional activities, age and habits into account. The mechanical and traumatic causes of failure are more frequent in children than in adults (Arnoldner et al. 2005).

6.2.3 Device Operation-Related Effects

We can highlight two important aspects: notably, the battery life and the handling and storage errors.

Handling and Storage Errors

Between the end-of-fabrication tests and their implantation, devices can be exposed to possibly damaging conditions, including the effects of time. Avoiding mechanical or thermal damage, preservation of sterile conditions, and protection from electrostatic discharges (ESD) during transport, storage, and pre-operative handling require appropriate packaging, clear instructions, and a validity limit date. In some instances, monitoring devices and a pre-implantation test can be justified.

Battery Life

Battery life has a large impact on the risk/benefit ratio and thus on the economic viability of implants. Therefore, it is important that an implanted battery performance be monitored. For many active implants, a theoretical end of life can be estimated based on the device battery capacity and the power consumption which is influenced by the applied stimulation regimen. Battery end-of-life is perhaps not a real failure, but it is nevertheless perceived by patients as a limiting factor requiring replacement surgery. For example, in vagus nerve stimulation (VNS) therapy, more than half of the replacement or revision surgeries are motivated by the battery depletion (Couch et al. 2016). It should be also noted that if patients complain about a loss of therapy efficiency, this might be due to a discharged battery.

Over the years, however, battery performances improve, and the power consumption of the devices decreases so that implants can remain functional for many years. At the same time, alternatives have appeared. Non-life-sustaining implants, such as cochlear implants, can use transcutaneous power transmission when the device is being used. Another alternative is the transcutaneously rechargeable batteries which become more and more popular. There is abundant research literature about energy-scavenging systems; however, these methods have not yet reached clinical applications.

6.2.4 Shortcomings in Design

The shortcomings in design can be a root cause, but they are not typically direct causes of failure. Therefore, they are very important but often hidden from view, and proper consideration must be given to them.

Good implant design should be perfectly patient-compatible. In practice, some initially unexpected failures are observed that could have been avoided with a more appropriate design of the implant. These can be minimized by design processes that include properly documented cycles of improvement of the application-oriented specifications. The systematic documentation guarantees that any shortcoming discovered later can be corrected through additional design cycles. An actively maintained feedback loop from the stakeholders to the designer is essential to progress. As "stakeholders," we designate the patient, the physicians in charge, and the surgeon implanting the system as well as the designer and manufacturer. For practical reasons, it seems that the patient's view will best be conveyed by the physician in charge. A well-designed data anonymization procedure can and is necessary to ensure proper privacy protection.

6.2.5   Material Issues

The term "biocompatibility" of an implanted device commonly refers to "the ability of a material to perform with an appropriate host response in a specific situation" (Williams 2008). This term is ubiquitously and traditionally used in the neural prosthesis literature because of some important safety concerns and regulatory requirements. Unfortunately, the definition is not very specific and therefore claims about biocompatibility in a functional sense are difficult to compare across studies. It can be used in different context often confusing the issues. Biocompatibility also refers to two different realities, both deserving specific attention. The term can indeed include both "biostability," that is, the indefinite preservation of the implanted device functionality, and "harmlessness," that is, the absence of damage to the body tissues and functions (Günter et al. 2019). In the section below, we discuss harmlessness as a notion including biocompatibility in the strict regulatory sense. The contrast between mere material biocompatibility and harmlessness is schematically presented in Fig. 6.3.

Medical device approval procedures and post-market surveillance methods appear to be very complex, and despite obvious efforts toward international standardization, details remain specific to each jurisdiction. In Europe, for instance, it is regulated at the EU level by the Medical Device Regulation 2017/475/EC supported by the European Medicines Agency and national conformity notified bodies.

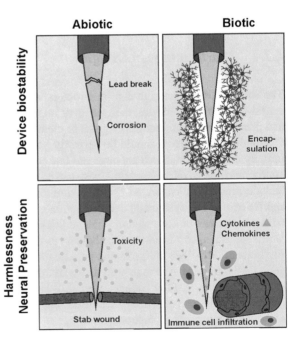

Fig. 6.3 From top to bottom and left to right: electrochemical corrosion, fibrotic electrode encapsulation, a mechanical stab wound and neural inflammation combine in four schematics representing the two failure causal factor categories (abiotic and biotic) times the two failure types (device breakdown and harm to neural tissue)

Material Biocompatibility as Harmlessness

In the narrowest sense, biocompatibility is a property of implant materials. All chemical compounds making up an implant and coming in direct contact with tissues and body fluids have to be proven biocompatible (in the sense of harmlessness), as defined by adopted standards, such as the ISO 10993 "Biological evaluation of medical devices."[3] Standards and regulatory requirements largely cover topics, such as cytotoxicity, sensitization, intracutaneous reactivity, hemotoxicity, systemic toxicity, chronic toxicity, pyrogenicity, genotoxicity, and carcinogenicity. The concept of biocompatibility can also be broadened from a material perspective toward entire devices. However, in such cases, it loses its regulatory context.

Most of the tests specified in the biocompatibility standards can be performed in vitro. In contrast, rare allergic reactions can occur only in vivo. Such reactions are difficult to predict since they are subject-dependent. Nevertheless, given the long history of existing materials being used in common applications, material biocompatibility is rarely an issue at present. However, the fact that the reaction to the implant is very sensitive to the fabrication method and not only to the nature of the material itself should not be overlooked. In that respect, novel fabrication techniques, such as additive manufacturing, bring about new challenges, in particular. Given their potential for personalizing the geometry of implants in a patient-specific manner, these challenges merit further studies.

Surface coatings, which draw a lot of attention in the literature, can also be discussed in the context of improving biocompatibility. Neural implants often undergo contact surface treatment and specific coating. As an example, the effect of Parylene-C coating of silicon microelectrodes suggests that a progressive failure can potentially be associated with CD68 immunoreactivity (Winslow et al. 2010). The resulting local damage includes demyelination and BBB breakdown near the implanted devices. In this study, the surface chemistry of the microelectrodes did not clearly affect the cyto-architectural component of the foreign body response. Various other types of coatings have been investigated for potential improvement of the tissue interface, such as hydrogels, bio-hybrid interfaces, cell-attachment polymers, etc. (reviewed in Aregueta-Robles et al. (2014)). Reported applications include improving the mechanical interface between the implant and tissues (Heo et al. 2016), protecting the tissues from the device surface material (Rao et al. 2012), improving the electrochemical exchange at the electric contact interface (Ludwig et al. 2011; Panic et al. 2010; Pierce et al. 2009), and allowing local release of preventive drugs (Winter et al. 2008; Taub et al. 2012; Mercanzini et al. 2010).

---

[3] Renewed in 2018 – https://www.fda.gov/downloads/medicaldevices/deviceregulationandguidance/guidancedocuments/ucm348890.pdf

Material Biocompatibility in the Sense of Biostability

On the other hand, biocompatibility in the sense of "biostability" is part of the patient compatibility of any medical device design. Selecting appropriate materials goes beyond the regulatory biocompatibility concept explained above. Survival in the body and chronic preservation of selected functional features are essential. Also, not only the bulk material must be considered but also fabrication contaminants and chemicals leaching from the devices especially from polymers and porous materials. Hermeticity failures could be responsible for the release of many toxic substances. Ensuring the systematic reproducibility and traceability is essential for a manufacturing process to identify and correct fabrication problems. Such quality management is only possible, however, if the entire production/use cycle is closed by accurate feedback from clinical issues to the manufacturer.

Indisputably, progress is linked to the discovery of new functional materials offering interesting but still very challenging perspectives (Wellman et al. 2018). However, the introduction of new materials reduces the support of past experience, and some authors have found less frequent technical failures in serially fabricated devices than in single unit prototypes (Baer et al. 1990). Noteworthy, the device failure rate seems to be particularly important in less common applications, such as sacral neuromodulation for lower urinary tract dysfunction (Shih et al. 2013).

Mechanical Mismatch

Some electrode designs themselves are sometimes near the limits of technical implantability. Such is the case with the mechanical buckling phenomenon occurring during insertion of some needle electrodes (Yoshida et al. 2007). Also, the brain tissue viscoelastic strain limit of about 0.1–0.3% is very low. Thus, acutely stretching of neural tissue during surgery must be limited as suggested by the exquisite sensitivity of the brain (Rashid et al. 2014) and the small elongation tolerance of peripheral nerves (Rickett et al. 2010). Damage induced by such acute mechanical action might not be immediately perceptible, while it triggers delayed cellular processes ending in apoptosis. On the other hand, functional losses due to physiological reactions, such as demyelination and axonal degeneration, can be temporary and subject to spontaneous recovery over weeks or months. Electrode displacement and migration are a common failure mode in epidural spinal cord stimulation for chronic pain (Kumar et al. 1991; Kumar et al. 1998). Such displacements can mimic lead fractures. Improved implantation techniques have been proposed to take care of this issue (Renard and North 2006).

6.2.6   Iatrogenic Events

From the preceding discussions, it is apparent that the physician in charge should be well trained not only to select the appropriate indication for the implantation but also to establish the most efficient stimulation regimen. Proper information exchange could prevent iatrogenic complications, such as trying, in an emergency, to access the jugular vein and accidentally puncturing the lead. Surgical clipping of the wire during surgery has also been described (Pearl et al. 2008).

Surgical Events

Lead failures can sometimes be traced back to inadequate implantation technique, resulting, for example, in cable kinking (Czerny et al. 2000; Holubec et al. 2015). Many surgical adverse events are more or less application-specific. Insertional trauma, for example, is typically observed in the delicate anatomical structures of the cochlea (Welling et al. 1993; Zrunek and Burian 1985). Also, the IXU of cochlear implants inserted between the skull and scalp represents a risk of disturbance of the overlying skin blood vessels, which is not the case when the device can be implanted deeper in the subclavian fossa, for example. Flap necrosis (Arnoldner et al. 2005) is described in 1.5–2% of patients with cochlear implants (Cervera-Paz et al. 1999). Skin pressure (e.g., in patients wearing helmets) and poor vascularization due to previous surgery are important factors (Ishida et al. 1997).

An estimated 16.8% of so-called drug-resistant epileptic patients receiving VNS suffer from complications related to surgery (Kahlow and Olivecrona 2013). However, improved surgical techniques seem to be able to significantly reduce this incidence (Lotan and Vaiman 2015). Anatomical variations in peripheral nerve anatomy can be a real challenge, which requires experienced surgeons. This might explain side effects or a lack of benefit in some cases (Hammer et al. 2015). Specific risks of VNS include rare instances of intraoperative bradycardia or even asystolia that can be observed during the implantation surgery (Giordano et al. 2017).

Electrode positioning is challenging in the brain due to blood vessel variations, and electrode malposition is also a possible cause of failure. Edema, blood clots, and bleeding represent typical risks for DBS. The surgical complications for this procedure are well described in the literature (Hariz 2002; Beric et al. 2001). Even subdural electrodes can lead to surgical complications (Fountas and Smith 2007). Device displacement represents 25% of the causes of failure in DBS, while the migration of the electrode is twice as frequent as the migration of the IXU (Blomstedt and Hariz 2005). Electrode migration (dislocation) is most frequently reported while cochlear implants (Balkany et al. 1999) and often linked to cochlear ossification (Connell et al. 2008). Mood disorders including depression, as well as voice alterations, dizziness, vomiting, heart rhythm problems, and even cognitive changes, can result from wrong stimulation parameters or electrode dislocation (Stieglitz 2019).

Placing an electrode around or on a peripheral nerve can induce local hemorrhage, a complication that can be avoided using endovascular leads as in pacemakers and other cardiac stimulation devices. However, these, in turn, bring the risk of long-term health issues, including lead failure, infections, and vein thrombosis (Daubert et al. 2014). The cochlear implant exploits the specific local anatomy, but this does not exclude the possibility of intraoperative bleeding (Kempf et al. 1999). The problem is, of course, much worse with invasive brain electrodes, which can drag the tissue along the implantation channel (Bjornsson et al. 2006). Signs of intra-parenchymal bleeding such as (ferritin immunoreactivity in microglia macrophages) do correlate with poor electrode performance. This is in sharp contrast with the absence of a similar correlation for neuro-inflammation (microglial activation) (Prasad et al. 2014).

Inadequate Device Use

Artificial neural stimulation allows for driving the activity of physiological systems at a level far above their natural functional level. Such an "overstimulation" can lead to a reversible loss of function or apoptosis thus resulting in cell loss. Such damaging effects have been described in the retina (Cohen 2009). The destructive influence of high-level noise on hearing is well known (Basta et al. 2018). Similarly, stimulation-induced depression of neuronal excitability (SIDNE) has been described by McCreery et al. (2000) and McCreery et al. (1997). Reduced sensitivity is not always linked to damage as neural adaptation can take place (Graczyk et al. 2018). Overall, however, very little work has been devoted to this important question. In addition, different physiological systems obviously can sustain a completely different stimulation regimen. Findings are thus difficult to extrapolate, and much more specific information is needed (Günter et al. 2019). One should also consider the full extent of the effects of neural stimulation, including effectors, such as muscles and bones that could be damaged by being driven above their normal working range. Finally, recruiting neighboring structures can induce avoidable side effects.

The stimulation regiment only does not necessarily lead to the restoration of a healthy state. Neural activation interferes with spontaneous activity and regulated systems typically including feedback loops. A good example is the importance of sensory integrity when stimulating spinal motor structures (Formento et al. 2018).

6.3   Abiotic Failure Mechanisms

6.3.1   Electromagnetic Interferences

Any exchange of energy (and signal transmission) creates the possibility for the implant to interfere with other devices or have its own function jeopardized by external equipment including other implants. The rapid development of new applications for active implants seriously increases the likelihood of several devices from different manufacturers to be implanted in the same patient. There is also a good chance that these will have been prescribed by different medical specialists. Technical coordination of such situations still needs to be organized. In addition, a large variety of medicotechnical procedures, such as magnetic resonance imaging (MRI), cardiac electroshock, transcranial magnetic stimulation (TMS), electroconvulsive therapy (ECT), and surgical electrocautery, are in common use or being developed. These could have serious disturbing or damaging effects on implanted material. Conversely, these procedures could be affected by implanted devices (see Sect. 6.3.4). Despite being less typically applied to a patient's body, non-medical devices must be considered as well. Finally, in our world of WiFi, Bluetooth, and other wireless RF-links, such as 5G, the transcutaneous communication facility of implants could open the possibility for interference with the proper implant function or for "hacking" by ill-motivated persons.

6.3.2   Hardware Failures

Despite the many identified causal parameters (Sankar et al. 2013), published studies tend to gather all device failures in a single category without further precision (Joint et al. 2002; Lyons et al. 2001). Root failures stemming from IXU hardware and software issues require a technical understanding of the device operation, which is usually only accessible to the manufacturers who are thus responsible for the necessary preventive actions. The pertinent technical details are mostly specific and often proprietary. They cannot be discussed in the frame of this chapter.

Leads and Connectors

Abnormal tethering of the leads can lead to "bowstringing," a complication of DBS associated with contracture of the patient's neck over the cable (Janson et al. 2010; Pearl et al. 2008).

A more common issue with leads is the conduction failures well-known from the cardiac pacemaker implants (Udo et al. 2012). Lead fracture and disconnection have also been frequently reported with neural implants such as the VNS system used in drug-resistant epilepsy patients (Kahlow and Olivecrona 2013), but lower failure rates (2.8%) seem to be found in more recent work (Révész et al. 2016). More optimistic figures are reported in the case of upper limb neuroprostheses (Kilgore et al. 2003). Lead failure does not necessarily result from an inadequate surgical technique such as exaggerated strain or cable kinking (Holubec et al. 2015). Conductor breakage could indeed result from metal fatigue on frequent bending or as the final stage of a corrosion process itself resulting from a lead insulation failure (see 6.3.2.2). A proper diagnosis of the primary cause requires clinical information about the timing of the failure and a description of clinical symptoms such as the patient feeling electric shocks at an abnormal location. Surgical findings during the revision/explantation surgery should also be reported, and finally, a detailed material analysis should be carried out on the breakage point.

A special and rare cause of lead break mimicking failure is the possibility of gas accumulation over the electrode contact, thus resulting in very high electrical impedances (Lasala et al. 1979).

Encapsulations, Passivation Layers, and Insulations

Hermetic encapsulation of electronic circuits is absolutely necessary to avoid moisture-induced failure. Encapsulation also protects tissues from the possible leakage of toxic substances making up electronic components. Insulation is also essential to preventing undesired current spreading to neighboring tissues and corrosion. Circuit passivation, lead insulation, and device encapsulation are thus critical fabrication steps that represent a significant challenge for very small device volumes (Ordonez et al. 2012b).

Delamination is a common problem that calls for improved adhesion layers (Ordonez et al. 2012a). With time, polymers lose their elasticity, and cracks can happen. Chronic movements are an important co-factor in the degradation of insulating material (Josset et al. 1984). Crossings between spare lead loops or pressure exerted by a hard IXU box on the polymer insulation can lead to damage.

Mechanical Forces

Implant failure can result from a direct mechanical impact. The mechanical forces deserve special attention since they are intimately linked to the biotic reactions described further (see Fig. 6.4). Well-protected intra-cranial electrodes are subject to inertial forces during head rotation, and blood and CSF flow create a mechanical load. Also, the blood pulsation causes variations in the pressure on the implant wall. Relative micro-motion of brain tissue generates mechanical stress near the less mobile brain implant. The importance of this mechanism and the resulting functional damage are often overlooked in the literature. However, as shown by Rennaker et al., the electrode longevity could be increased using mechanical

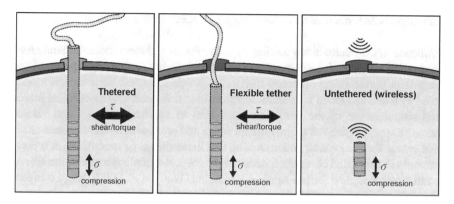

Fig. 6.4 Schematic drawings of different mechanical forces acting on the brain tissue. In the tethered configurations, the shear and torque components are expected to dominate. In the untethered configuration with wireless power supply, mostly pressure is present

insertion devices that prevent cortical compression. On the day of implantation, tissue micro-motion was found to be responsible for 12–55% of the steady-state stress on the microelectrode ($n = 4$). This decreased to a contribution of 2–21% after 4 weeks ($n = 4$) and 4–10% after 6–8 weeks post-implantation ($n = 7$) (Sridharan et al. 2013).

Muthuswamy and coworkers have measured brain micro-motion in animals (Muthuswamy et al. 2003; Gilletti and Muthuswamy 2006). In anesthetized rats, micro-motion evaluated as surface pulsations due to respiration was 10–30 μm larger when the dura mater was removed compared to the intact dura mater cases. Vascular pulsatility appeared to generate 2–4 μm displacements.

In monkeys and humans, brain motion has an even more impressive amplitude than in rats, because of the proportionally larger brain and subdural space. The periodic axial force of 5 mN has been measured as 0.22 Hz motions in monkey brains (Hosseini et al. 2007).

Tethering and Micro-Motion

While enclosed in the rigid cranial box, the brain floats in a cushion of cerebrospinal fluid (CSF) with the result that it becomes isolated from the external mechanical forces impacting the head during the normal active life. The tethering to the skull will increase the shear and tensile forces due to the implant torque during head rotation as well as due to the vascular pulsations. Because of the electrode tethering to the skull, all brain displacements will cause repeated stress on the interface. Such view is supported by the significantly reduced scars observed in untethered compared to tethered implants (Kim et al. 2004; Biran et al. 2007; Thelin et al. 2011; Thelin et al. 2011).

Uncompensated Mechanical Forces Trigger Deleterious Mechanisms

Evidence accumulates from animal experiments that chronic neuroinflammatory events and BBB breakdown carry a major responsibility for the degradation of the long-term performance of neural interfaces. A self-sustained BBB breach modulated by stress and strain is a central mechanism in this issue. The mechanical forces and displacement of the tissue can also lead to capillary damage and microhemorrhages. The impact of the implant shape and cross-section has not been clarified except for the general understanding that these should be minimized. A typical example is a reported correlation between the clinical complications and the size of brain electrodes used for pre-surgical evaluation (Wong et al. 2009). From this perspective, the inertia tensor of some implanted parts should be reduced in order to decrease the resulting torque forces. The design of specific implants must be optimized to minimize mechanical interaction given the constraints of the desired functionality (Lecomte et al. 2018). With the advent of computer-aided design (CAD) tools, such an optimization would be relatively straightforward. However, only very few studies have measured kinetic or dynamic parameters of the brain-implant interaction. Most of the studies focused on the insertion forces only covering the surgical and acute phases (Welkenhuysen et al. 2011; Andrei et al. 2012).

Mechanical Damage to the Electrode

Last but not least, mechanical forces can also be deleterious to the implant itself. Indeed, with the most recent miniature neural probes, the chronic strain between subcomponents of the electrode can be a cause of mechanical failure, especially during insertion (Rennaker et al. 2005). Failure of invasive microelectrodes can often be traced back to the mechanical interaction between the brain and implanted electrodes.

6.3.3 Material and Energy Exchange

A systematic review of energy exchanged between an implanted system and its surroundings is a simple method to ensure proper consideration of all possible abiotic interactions.

Ultrasound

Among the mechanical forms of pulsations, ultrasounds deserve particular attention, because of the development of various medical procedures whereby damaging ultrasound powers can be concentrated in a limited internal body volume. Lithotripsy and high-intensity, high-focus applications (Kirkhorn et al. 1997) represent the most extreme examples. However, less controlled ultrasound applications, such as in physical therapy or in teeth cleaning devices, should not be overlooked. The hazard of ultrasound for implanted material users is not always considered thoroughly. Typically, however, diagnostic ultrasounds work at safe low intensities, though image distortion is possible.

Electricity

The first source of electricity to consider is the stimulator. Contact corrosion and local production of toxic chemicals are the most immediate risk factors, but these can effectively be mitigated by the use of charge-balanced biphasic stimulation pulses (for a review on safety aspects, see Günter et al. (2019)). Low-level DC current, on the other hand, will rapidly be responsible for electrochemical changes resulting in local toxicity and severe corrosion. Excluding the possibility of DC current leakage requires careful design of the system itself and selected transmission schemes (Liu et al. 2007). Single failures should never lead to DC current leak. Electrochemical mechanisms characterize the electrode interface are discussed below. These electrode-tissue interactions are dependent on many parameters including the electrode size. It is important to stress that electrode contact size is not necessarily favorable, and it has even been suggested that small dimension might lead to lower rates of corrosion and nerve damage (Thoma et al. 1989).

Synkinetic (i.e., stimulus-related) activation of neighboring nerves can result from too strong stimuli, the presence of anatomical lesions, electrode malposition, or leakage from defective cable or connector insulation (Graham et al. 1989; Muckle and Levine 1994). In such cases, unexpected muscle contractions, discomfort, or pain can be observed.

Of course, the possibility of electrical damage to the target neural structure should also be considered. Typically, living tissue tolerates electricity very well if the injected power does not lead to unacceptable heating. A typical neural implant is unable to reach electrical field intensities that could induce electroporation. On the other hand, the stimulus can drive nerves and their target organs into quite damaging non-physiologic activity regimen (see Sect. 6.2.6).

Electricity generated outside the body can also reach an implant and damage it. Cardiac resuscitation electroshocks are typical examples. Depending on the position of the implant, such a treatment is likely to destroy the implant but save the patient, which is the preferred choice, of course. The surgical use of electrocauterization in the vicinity of an implant can be equally destructive and should be avoided. Careful use of electro-diagnostic examination procedures has also been recommended (Pease and Grove 2013). The Electro-convulsive Therapy ECT used in psychiatry can represent a significant risk of interference for implants such as a VNS system in a depressive patient, for example.

The exceptional failure of a cochlear implant found in a case of electric shock from a domestic electrical appliance has been traced back to the failure of a capacitor in the stimulator circuit (Woolford et al. 1995). At least in one instance, the functional failure of a VNS was attributed to a lightning strike in the vicinity of the patient (Terry et al. 2011). Electrostatic discharges (ESD) are potentially destructive for most electronic devices, but the body conductivity seems to offer enough protection when those are fully implanted. Note that this is not the case when an open surgical wound gives access to the device.

Electromagnetic Interference

Magnetic Resonance Imaging MRI is presently an indispensable medical diagnostic tool. Although noninvasive from the patient's point of view, this technique uses extremely strong magnets and electromagnetic fields that could lead to major issues for implanted devices. Image distortion is probably one of the most negligible consequences because, although it can impair a diagnostic evaluation, it is also easily recognized and taken into account by the experienced practitioner. The mechanical force on ferromagnetic material is known to induce "flying path lesions," that is, lesions induced by small objects freely moving in the surroundings. Some implants, such as clips or orthopedic devices made of ferromagnetic material, can be subject to dangerously strong mechanical forces. Implanted magnets have been used in some designs (cochlear implant) in order to facilitate proper alignment of a transcutaneous power supply. The magnetic field of an MRI system can then exert very significant forces on these, although not enough to induce a fracture of the skull (Sonnenburg et al. 2002).

Also, all implanted metallic parts (even remnants of partially explanted systems) will work as antennae subject to the induction of potentially hazardous voltages and high-frequency pulsations leading to heating (Muranaka et al. 2011). It should be remembered that these effects can be quite different for different MRI systems because they work at different frequencies (64 MHz for a 1.5 Tesla system and 128 MHz for 3 Tesla). Considering the different antennas being used for different types of investigations, it is difficult to generalize the findings of safety tests. New techniques, such as Transcranial Magnetic Stimulation TMS or magnetic evoked potentials (MEP) must also be considered for possible interference with implanted systems, especially, but not only, with intracranial devices.

Finally, through their communication systems, devices are potentially sensitive to electromagnetic sources. This includes commonly used safety gates but also GSM, WIFI, Bluetooth, RFID, 5G, etc. Power levels remain low, and appropriate encoding is applied in order to be safe, but false alarms are expected to be triggered, and "mala fide" or inadvertent interference with the software is a theoretical possibility.

Light

The steady growth of interest in optogenetics and the development of many medical laser applications have recently called attention for the consequences of photonic irradiation. Relatively high-power lasers are sometimes used (Rungta et al. 2017) resulting in neural tissue heating, sufficient to cause nuclear magnetic resonance (NMR) frequency shifts (Christie et al. 2013). Thermal effects could thus not only be damaging but also lead to misinterpretation of opto-fMRI experiments.

Ionizing Radiation

Although this has only been poorly studied, strong irradiations such as proton and X-ray beam therapy can clearly affect active implant (Gomez et al. 2013; Gossman et al. 2012). However, DBS treatment is not necessarily incompatible with brain radiotherapy (Kotecha et al. 2016). Risks of palliative cranial irradiation seem moderate, but changes in output currents have required reprogramming the speech map of a cochlear implant (Ralston et al. 1999).

Thermal Effects

All locally applied forms of energy will ultimately be converted into heat and could, therefore, damage neural tissue if the temperature rises above some critical value. The level of energy is usually well known, and the caloric capacity (relating heat energy to temperature) of the various tissues is roughly similar to water, that is, about 4.184 J/K/g (Stujenske et al. 2015). The temperature increase can thus easily be estimated for acute exposure. However, the maximal allowable temperature increase is unknown, is probably very different for various structures, and is dependent on the exposure duration. Usually, a temperature increase of below 2 °C is considered safe (Chou et al. 1997). The risks of chronic applications are much more difficult to evaluate. Heat transfer is dependent on the blood supply, which in turn changes with temperature via physiological reflexes (Rungta et al. 2017), thus leading to non-linear behavior. Hotspots will typically appear because of the tissue and device heterogeneities, but these are rarely considered. Safe thermal limits can thus only be established from empirical data specific to the application considered.

The high-frequency electric fields created by MRI scanners or by electrosurgical units can induce burnt lesions by interference with implants (Patterson et al. 2007). A reversible pain episode with the failure of DBS treatment with edema surrounding the electrode was observed after diathermy (Roark et al. 2008). Diathermy is contraindicated, of course.

Charge and Mass Transfer

From a chemical point of view, four issues should be considered. The first one is about the biocompatibility of the materials in contact with body tissues (see Sect. 6.2.5), the second issue is about the electrochemical consequences of the electric current exchanges. As a third point, we have the device surface triggering biotic reactions, and finally, we will consider the various preventive coatings that have been recently proposed.

Electrochemical electrode corrosion due to the stimulation current has been described under Sect. 6.3.3.2 above. Electrochemical exchanges become particularly critical with multichannel stimulating devices. Specific circuits have been described to maintain the small electrode contact within electrochemical water potentials and ensure that the electrode voltage remains zero (Kelly et al. 2014). Another consequence of electrochemical changes is the potential toxicity of products accumulating during the current flow. There is scarce literature about this aspect, probably because with electrode contact metals such as platinum, the toxic threshold is relatively high, thus rarely observed and difficult to measure (Kovach et al. 2016).

6.4 Biotic Failure Mechanisms

6.4.1 Bio-Mechanical Properties of the Brain

Biological tissue's viscoelastic properties can be determined from the stress-strain relationships. There are three types of moduli that characterize the linear response: Young modulus E, which describes the axial elasticity, when opposing forces are applied along an axis; shear modulus G, which describes the deformation of shape at constant volume, when the object is acted upon by opposing non-axial forces; and finally the bulk modulus K, which describes volumetric elasticity, or the object volumetric deformation, when uniformly compressed in all directions. Since biological tissue is inhomogeneous and anisotropic, all three moduli are independent of each other.

There have been many efforts to develop specific mathematical models of the various aspects of the brain mechanical response, but few have been validated (overviews in Miller (2011), Ch. 4, and van Dommelen et al. (2010)). The strong strain-rate sensitivity and non-linear viscoelasticity of brain tissue cannot yet be translated in widely used equations. The viscous transient behavior is often represented as the sum of a finite series of Maxwell elements (dashpots and springs). As a result, the relaxation modulus is described by the different time constants of a Prony series of exponentials. As an additional approximation, Fung has proposed a quasi-linear viscoelastic (QLV) theory, based on separation of different temporal and spatial scales of a continuous material body model (Fung 1993). The stress function can be separated in an elastic and a time-dependent portion, as introduced by the same author. The stress tensor is assumed to have the Boltzmann form:

$$\sigma(t) = \int_{-\infty}^{t} G(t-\tau)\frac{d\sigma_e}{d\tau}\,d\tau$$

While the kernel G is given by Prony series approximation normalized at $t = 0$:

$$G(t) = G_\infty + \sum_{k=0}^{N} G_k e^{-\frac{t}{\lambda_k}}, \qquad G(0) = 1$$

This is a flexible framework amenable to experimental parameter estimation. A starting point could be a single-exponent model (see Table 6.1).

The huge difference in Young's moduli between brain tissue and neural probes (0.1–10 MPa) remains a major issue. Materials typically used for neural probe fabrication have a high Young modulus contrasting with the brain's softness (see Tables 6.1 and 6.2).

However, in order to approach the brain characteristics, ultra-soft microelectrodes have been developed with Young's modulus lower than 1 MPa (Du et al. 2017). Investigation of the interface brain implant indeed confirms that (1) normal morphology is better preserved with definitely less deformation from mechanical stress; (2) the inflammatory gliosis seen with stiff electrodes is clearly reduced near soft electrodes; (3) this is linked to a clearly better long-term preservation of the BBB in the case of soft electrodes; and (4) tightly bonded neurons and other cells found on the surface of explanted soft wires also indicate a better integration with soft tissue.

In another study, after a 26–96 weeks' trial in the rabbit cortex, Sohal et al. found a reduced foreign body reaction to their new flexible microelectrode when comparing to microwires (Sohal et al. 2016). Also, the local neural density was most often increased alongside the probe. Over the relatively long implantation period, the soft wires also induced less gliosis than the stiff wires. The combined effect of an overall reduced micro-gliosis and a higher neuronal density with soft probes did contribute to this result. Microelectrode flexibility thus appears to reduce the trauma induced by micro-motions. The expected result is a prolonged electrode functional life. Isodense electrodes, that is, electrodes with the same stiffness characteristics as the brain, clearly represent an ideal situation which, unfortunately, is still quite challenging because of the conflicting requirements of implantability.

Table 6.1 Brain and hydrogel viscoelastic material constants

	Density (kg/mm³)	Bulk modulus, K (GPa)	Short-term shear modulus, G_o	Long-term shear modulus, G_∞	Decay constant, λ (ms^{-1})
Brain	1.05×10^{-6}	2.1	10 kPa	2 kPa	0.08
PVA			460 kPa		
Alginate			40–200 kPa[a]		

PVA Polyvinylalcohol
[a]Depending on the composition; references (Polanco et al. 2016; Argueta-Robles et al. 2014; Mancini et al. 1999)

Table 6.2 Young moduli of some materials used in implants compared to natural polymers

Material	Young modulus, E
Tungsten	400 GPa
Silicon	200 GPa
Polyimide	3 GPa
Parylene C	2–5 GPa
Collagen	0.5–12 kPa[a]
Agarose	1.5–2580 Pa[a]

[a]In the function of composition (Aregueta-Robles et al. 2014)

6.4.2 Biologic Reaction Mechanisms

Despite the insights provided by electrophysiology, material science, and histology, the picture of the biologic reaction mechanisms developed in the literature is far from complete. However, the functional state of the BBB or blood-nerve barrier (BNB) emerges as the main feature. Also, the brain is no longer considered as an "immune privileged" organ, and immune reactions should actively be studied.

The BBB and BNB

As already indicated above, the microenvironment of an implanted electrode can be subject to mechanical strain, BBB leakage, glial cell activation, deficient blood perfusion, secondary metabolic changes, and neuronal degeneration (Michelson et al. 2018).

These mechanisms are not independent, but they do strongly interact with each other. For example, neuronal health and activity level are very sensitive to the state of the neurovascular network (Kozai et al. 2012). In a 16-week experiment combining noninvasive imaging, electrophysiology, genomics, and histology in rats, Saxena et al. studied the functional consequences of a BBB breach on implanted electrodes (Saxena et al. 2013). The performances of intracortical electrodes (Michigan-style planar probes and microwire electrodes) inversely correlated with the BBB breach, while an enhanced wound healing response was demonstrated by the performed genomic analysis. The BBB permeability and the transcript levels of neuroinflammation-activating cytokines were lower in the cases with functionally stable electrodes. On the other hand, the accumulation of neurotoxic factors and the invasion by pro-inflammatory myeloid cells consecutive to the chronic BBB breach characterized the poorly functioning electrodes. The disruption of the BBB explains the observed deposition of albumin, globulins, fibrin/fibrinogen, thrombin, plasmin, complement components, and all plasma proteins normally not present in the CNS. The neuronal health is, of course, affected by these surroundings.

Michelson et al. reported that the widely used NeuN, Iba1, GFAP, and IgG histologic markers were not adequate to grasp the complexity of the biological mechanisms at the neural interface. A broader perspective should include both an evaluation of neuronal health and the ability to record functional electrophysiological activity from intact microelectrodes. Already 6 hours after implantation, multiphoton imaging in vivo could demonstrate a loss of capillary perfusion and signs of axonal injury (Michelson et al. 2018). It is still unclear whether this is a transient phenomenon or it can carry over time to degrade the tissue-implant interface. In the same study, molecular expression level analysis indicated a shift from excitatory transporters and ion channels to inhibitory transporters and ion channels in the course of 4 weeks. This can be expected if the neural interface stabilizes over time, as the neurofilament levels around the implants increase.

Saxena et al. hypothesized that a chronic breach of the BBB, resulting from the intracortical electrode insertion, is responsible for the chronic inflammation maintaining the BBB breach in a positive feedback loop. The long-term consequences are neurodegeneration and electrode functional failure. This view is in line with the differences in neuronal survival profile found by Potter et al. (2012) when comparing the effects of chronically implanted devices (non-functional planar single-shank arrays) with matched cortical stab wounds in animals. These authors found an enhanced IgG (a marker for BBB leakage) present, around devices implanted for 16 weeks, contrasting with the baseline levels observed in the stab injury region. Potter et al. concluded that an initial phase of neuronal loss resulted from the BBB leakage, while endogenous tissue events were responsible for the chronic neurodegeneration. Karumbaiah et al. also reported a comparatively higher permeable BBB and abundant active inflammatory cells and neurotoxic factors in the surroundings of chronically failing electrodes (Karumbaiah et al. 2013). Potter et al. performed a detailed assessment of the neuroinflammatory events and BBB integrity following implantation of non-functional planar single-shank arrays vs an identical cortical stab injury (Potter et al. 2012).

In a study of soft polymer probes compared to silicon devices, a reduction of microglia/macrophages and BBB leak could be demonstrated at 4 and 8 weeks after implantation as evidenced by fluorescent biomarkers (Lee et al. 2017).

Bedell et al. demonstrated that 2 weeks after implantation of Michigan type of neural probes, inhibiting innate immunity pathways associated with CD14 results in higher neuronal density and decreased glial scar, whereas implantation of a larger thiolene type of probes resulted in more microglia/macrophage activation and greater BBB leakage (Bedell et al. 2018). It should be noted that the probes' dimensions used in the study were not very well comparable. The differences reduced by 16th week post-implantation, which indicates that the persistent factors, were not affected by the genetic modification.

In a similar study, Hermann et al. implanted silicon planar non-recording neural probes into knockout mice lacking Toll-like receptor 2 (Tlr2−/−), knockout mice lacking Toll-like receptor 4 (Tlr4−/−), and wildtype control mice and evaluated endpoint histology at 2 and 16 weeks after implantation (Hermann et al. 2018). It was shown that Tlr4−/− mice exhibited significantly lower BBB permeability at acute and chronic time points but also demonstrated significantly lower neuronal survival at the chronic time point. Inhibition of the Tlr2 pathway had no significant effect compared to control animals. When investigating the neuroinflammatory response from 2 to 16 weeks, transgenic knockout mice exhibited similar histological trends to controls. Together, reported results indicate that complete genetic removal of TLR4 was detrimental to the recovery after, while inhibition of TLR2 had no impact.

Inflammatory Reaction

A distinction should be made here between the sterile inflammation, caused by the presence of a foreign object, and infection, causing reactive inflammation against brought in pathogens. It is clear that understanding and interfering with the neuro-inflammatory response could lead to progress in the long-term electrode stability. The chronic sterile inflammation and local neuronal degeneration and loss are still topics of active research (Mols et al. 2017; Michelson et al. 2018). It was hypothesized that the chronic immune reaction is a continuous response primarily sustained by the local mechanical stress created by the implanted device and the surface-released molecular factors (Prodanov and Delbeke 2016).

In the long term, the implanted neural electrodes get encapsulated by a fibrous tissue containing microglial cells, macrophages, meningeal fibroblasts, and reactive astrocytes. Numerous messenger molecule exchanges regulate the interactions between these cells. A full understanding of the temporal organization and the implant-triggered tissue modifications will require focusing more attention to often overlooked actors, such as oligodendrocytes and their precursor cells, as well as the pericytes associated with the BBB. This is clearly demonstrated in a very recent study using new histological markers (Wellman et al. 2019). Using these histological markers, Wellman et al. demonstrated apoptosis-induced oligodendrocyte cell death during the acute phase. Activation and proliferation of microglia, astrocytes, and NG2 glia were observed preferentially around inserted devices at distances about 150 μm. Furthermore, a novel subtype of reactive astrocytes was revealed around the site of implantation, potentially derived from a resident oligodendrocyte precursor population. Finally, chronic pericyte deficiency was noted alongside increased vascular dysfunction near inserted devices.

Using mechanically adaptative materials, it has been demonstrated that the cortical implant stiffness contributes to the neuroinflammatory response (Nguyen et al. 2014). In the brain, inflammation can induce vasogenic edema not to be confused with complications, such as a hemorrhage or infection (Englot et al. 2011). In peripheral nerves, functional electrical stimulation-associated trauma induces thickening of the connective tissue around the nerve and slightly compromises its conduction velocity (Pečlin et al. 2016).

Pharmacological means to mitigate the persistent inflammation are concisely reviewed in Capadona (2018). Briefly, these include steroids, for instance, dexamethasone, and antibiotics, for instance, minocycline, ILR-1a-modulating compounds, and caspase-1-modulating compounds. While these compounds alone cannot reverse the chronic inflammation, it is possible that some of them could contribute to the beneficial effect of optimal mechanical designs.

The causal link between micromotion and persistent local neuroinflammation is not a simple hypothesis to be demonstrated methodologically. There are many interactions and interdependencies, which are difficult to isolate. Use of the right biomarkers is particularly important. For example, NeuN, Iba1, GFAP, and IgG are traditional histological markers for tissue viability. However, some recent studies suggest that these are not adequate to predict long-term electrode performance (Kozai et al. 2014; Michelson et al. 2018). On the other hand, there is negative evidence coming from very stable preparations (Mols et al. 2017). Mols et al. have performed chronic multiunit recordings of cortical neurons using a tilted silicon probe tightly fitting an enclosure fixed to the skull. The electrophysiological signals remained stable for up to 10 weeks after implantation. The authors reported an increase in astrocyte fluorescence that remained unchanged between the second and the tenth week, which indicated a very stable glial scar and possibly reduced amplitude of micromotion.

The Reactive Oxygen Species (ROS)

A brief account of the contribution of ROS is given in Prodanov and Delbeke (2016). Peroxides, superoxide, hydroxyl radical, and singlet oxygen, functionally referred to as ROS, are present in every living cell. These molecules control redox homeostasis, thus playing a key role in many physiological processes and in pathology. In conjunction with reactive nitrogen species (RNS), they contribute to several cellular mechanisms. The corresponding regulatory processes use these compounds as mediators in redox signaling (review in Hsieh and Yang (2013)). Oxidative/nitrosative stress, corresponding to a deficient ROS elimination, will result in an oxidative imbalance causing irreversible tissue damage. ROS and oxidative stress have been well covered in the scientific literature. Interested readers are referred to the reviews in Abbott (2000), Hsieh and Yang (2013), and Uttara et al. (2009).

Oxidative stress effects on local neuronal cells are not the only consequence of ROS accumulation. The cerebral vasculature will be affected as well and in particular endothelial cells involved in the maintenance of BBB. As a consequence, multiple pathways lead to the participation of ROS in the observed BBB disruption:

- Direct damage to proteins, lipids, and DNA, that is, essential cellular building blocks
- Matrix metalloproteinase (MMP) activation
- Cytoskeleton reorganization
- Interference with tight junction proteins and inflammatory mediator release

Oxidative stress causing BBB breakdown is an often-met pathophysiologic mechanism in neurological diseases, such as stroke, amyotrophic lateral sclerosis, and multiple sclerosis. In these diseases, the neuronal malfunction can be traced back to ionic imbalances and abnormal interstitial fluid contents affecting neuro-transmission and metabolism. These changes easily explained by the BBB leakage are responsible for increased firing rates as observed during a few weeks after implantation. Finally, the catalytic function of Fe^{3+} in the blood clot also appears to be linked to the ROS activity.

In a recent study, the membrane integral protein named caveolin-1 (located at caveolae) was shown to reduce RNS and matrix metalloproteinase (MMP) activity (Gu et al. 2011). The result was the protection of tight junctions and BBB integrity. Caveolin-1 and RNS interact in a positive feedback loop considered as important components of brain-damaging mechanisms in ischemia-reperfusion.

Likewise, ROS cause secondary BBB leakage that maintain the neuroinflamma-tory response and cellular dysfunction. In a 4-week study of neurodegeneration around intracortical microelectrodes, Potter et al. indeed identified targets able to improve the local neuronal survival through a short-term inhibition of ROS accumu-lation and stabilization of BBB (Potter et al. 2013). The resulting therapeutic per-spectives suggest that the local BBB integrity and neuronal health could be improved by controlling the ROS production near the implant.

Diffuse Reactions

Implanted electrodes can be responsible for subtle brain damage requiring refined behavioral tests to be detected. Hence, the example of rats with chronic intracortical electrodes was found able to perform fine motor tasks but at the cost of a time-to-completion increase of 527% (Goss-Varley et al. 2018). In humans, gross move-ments can be affected by DBS electrodes before the onset of electrical stimulation. Several mechanisms could explain this microlesion effect (MLE) of the implanted device. Acute edema and hemorrhage are often considered, but a reduction in glu-cose metabolism could be a causal factor as well (Goss-Varley et al. 2018). On the positive side, MLE could provide evidence for correct electrode placement. Optical coherence tomography (OCT) of the surrounding tissue can be an interesting diag-nostic tool (Xie et al. 2014). In their 12-week study, these authors found an initial monotonic increase of the OCT signal reaching a plateau after 6 weeks post-implantation. This observation might be explained by the backscattering caused by the astrocytes that accumulate around the glass fiber. Impedance spectroscopy find-ings are in keeping with these changes up to 100 μm from the electrode.

Reactions to less-invasive electrodes have been also studied: for instance, subdu-ral electrodes in micro-ECoG. In a histopathology study in rats, moderate tissue reactivity is present at 25 weeks post-implantation, while new blood vessels grow through the perforated electrode substrate (Henle et al. 2011; Schendel et al. 2013). Nearly 2 years after subdural ECoG array implantation, in a rhesus monkey, the inflammatory response was minimal, and the implant was encapsulated in fibrous tissue (Degenhart et al. 2016). A comparison with the contralateral side did not reveal cortical damage under the implanted grid.

Resulting Tissue Changes

Neural tissue is far from a homogeneous medium. Even peripheral nerves contain several well-organized structures, such as the nerve fascicles, endoneurium, perineurium, and epineurium, with blood vessels and their BNB. Neural cells and axons are interacting heavily with the myelin-producing Schwann cells. In the brain, astrocytes, oligodendrocytes, and microglia far outnumber neurons. Fibroblasts and endothelial cells are major contributors to the foreign body reaction. Immunologically competent and vascular cells are by far not the only participants to the many cellular interactions. The large variety of cellular actors is also embedded in an extracellular matrix, where molecule diffusion mechanisms interfere with the various exchanges and signaling mechanisms.

Fibrosis and electrode encapsulation are often considered as a major source of functional degradation because of the resulting distance increase and the shielding effect between electrodes and their target. The post-implantation signal to noise ratio (SNR) showed more stability than the action potential (AP) amplitude, but, after a number of months, the SNR increasingly correlated with the neuronal density which corresponded to the neuronal loss near the microelectrodes (McCreery et al. 2016). This correlation was found at distances up to at least 140 μm from the microelectrodes, while the neuron density and glial fibrillary acidic protein (GFAP) density correlated up to about 80 μm. Immediately after implantation, the AP amplitude strongly reflects the histology marker density (GFAP and NeuN), while neuron density and SNR correlated better near the end of the study. On the other hand, GFAP density near the electrode did not significantly correlate with the SNR.

In order to improve cell attachment and differentiation, device surface coatings, such as hydrogel coating with collagen I or polylysine-laminin-1, have been proposed (Zhong et al. 2001). Also, the silicon surface immobilization of L1 biomolecules was shown to promote neuronal growth and, at the same time, prevent astrocyte attachment (Azemi et al. 2011).

Temporary demyelination/remyelination can explain some initial increases in stimulation threshold after implantation. Axonal degeneration is another cause of temporary failure not always easy to distinguish from neural cell migration. Neural apoptosis is another failure mechanism leading to cell death. It can contribute to the neural gap increasing the electrode to target distance.

Neural tissue is a living and adaptive structure. Activation through a stimulation implant will have a broad variety of consequences. Progressive functional loss in the frame of stimulation-induced depression of neuronal excitability (SIDNE) is an example. However, many physiological adaptive mechanisms exist, such as sensitization, habituation, and plasticity, whereby the response to a constant stimulus varies with time. Finally, one should not overlook the fact that neural interfaces are normally intended as a treatment, that is, to help a sick patient. The patient's disease can be degenerative (e.g., Parkinson's disease, among others), in which case a mere evolution of the pathology could also explain some failures despite initial success.

6.4.3 Infections

Basically, two infection mechanisms must be distinguished. Primary infections, are expected to appear locally, rapidly after surgery and could result, for example, from contamination, insufficient material sterilization, or a nonsterile implantation procedure. In the second instance, delayed or secondary infections are also possible because implanted foreign material is known to facilitate metastatic colonization by germs of distant origin. Surface adhesion (biofouling) mechanisms play a major role in such cases, and device micro-geometry might create spaces inaccessible to the immune cells. In cochlear implants, the prevalence of germs of the dental cavity suggests that preoperative dental care might have a preventive value (Kanaan et al. 2013).

Because the foreign material itself facilitates the infection, explantation of a properly working system is sometimes necessary. Infections are thus a relatively common cause of implant revision (Agarwal et al. 2011). In a 14-year follow-up of VNS, a revision incidence of 2% was reported (Couch et al. 2016). A somewhat higher rate of 3.5% necessitating device removal has been found in VNS-treated epileptic children (Smyth et al. 2003). Similar infection rates are reported for very different procedures. For example, a rate of 2.9% of infections has been found for phrenic nerve stimulator implantations (Weese et al. 1996).

Localization can be an important evaluation factor. The severe consequences of local brain infections represent an obvious example.

6.5 Phenomenology of Preclinical In Vivo Studies

Current-controlled stimulation electrodes can compensate for electrode impedance fluctuations. Stronger stimuli can compensate for activation threshold changes. Such an increase, however, typically will reduce the stimulus selectivity. Recording electrodes are even more exquisitely sensitive to the quality of their integration within the neural tissue. Their functional state (i.e., contact impedance, AP, and SNR) is often used as chronic monitoring of the interface evolution. Several types of highly parallel and/or selective neural interfacing (mostly cortical) electrodes have been developed that are under preclinical and clinical investigations. In animal studies using wire electrodes (Liu et al. 1999; Nicolelis et al. 2003) or silicon-based devices or multiwire arrays (Ward et al. 2009), recording longevity proved highly variable. However, the failure mode of neural prostheses still tends to be poorly explored. The very fragmentary data available limit potential translation to human applications.

6.5.1   Utah Array

A 3D microneedle electrode array, known as the Utah Array, can provide a dense parallel cortical access that has been exploited in visual prostheses (review in Schwartz (2004)). Recording from the motor cortex could lead to the development of neuromotor prostheses (Hochberg et al. 2006). However, short circuits in the electrode, cable, and/or connector ended in a failure after a trial of 6 months' duration. A second attempt by the same group in a 55-year-old subject was a little more successful with a 10-month recording period before an unexplained failure.

An eight-direction discrimination experiment exploiting a 100-electrode array in the primary motor cortex of three Macaque monkeys worked for 1.5 years (Suner et al. 2005). The electrode used was the Bionic silicon probe (Cyberkinetics, Inc., Foxboro, MA). Over time, that was, after 83, 179, and 569 days, there seemed to be a shift in the neurons contributing to the recorded waveform. Tuning to the reach direction was found in 66% of the recorded neurons.

With Utah arrays in rhesus macaque monkeys, inter-spike interval histograms and spike waveforms remained stable for 7, 10, and 15 days in 57, 43, and 39% of the initially recorded units, respectively (Dickey et al. 2009). In 56% of implantations in nonhuman primates, the intracortical Utah Array failed within a year of implantation (Barrese et al. 2013). Among mechanical causes (83% of the failures), connector issues explain 48% of the failures, in sharp contrast with only 24% of observable biological problems. Among these, shielding the tissue from the electrode array was due to a progressive meningeal reaction in 14.5% of the cases. In addition to the acute failures described above, a slow decline in the spike signal amplitude with increased noise was observed. A progressive loss of exploitable channels ended after 8 years in total recording failure. The recording quality did not parallel the impedance measurements, which after an initial increase progressively dropped over the duration of the experiment.

A detailed study of the tissue reaction to a 12-week Utah Array implantation has been performed in rats. Nolta et al. reported findings, such as plasma proteins point to BBB leakage, while the presence of activated macrophages and microglia and astrogliosis indicated persistent inflammation (Nolta et al. 2015). At the implantation site, a relatively large pyramidal-shaped cavity was often found that corresponds to a loss of brain tissue. This damage is likely to be caused by the surgical implantation itself, as similar lesions have been observed in stab wounds.

6.5.2   Wire Arrays

With only 8% successful cortical unit recording 5 months after implantation, the same slow decline has been observed with wire electrodes in the cat brain (Burns et al. 1974). Some implanted multiwire arrays are characterized by an early increase in the SNR and loss of recordable units (Williams et al. 1999). In other cases, however, the performance rapidly became stable after a limited initial deterioration. The experiments were terminated after 15–25 weeks, because of a skull cap loosening or medical issues. In another study with Ir wire electrodes (Liu et al. 1999), unstable recordings were observed during the tissue remodeling phase, while the following months were characterized by a slow and steady decrease in performances.

Minimal cell death was found at implantation of a tungsten wire microarray in the rat motor cortex (Freire et al. 2011). However, the number of recorded neurons progressively decreased over the 6-month study period. Immediate-early gene (IEG) expression was compatible with normal tissue physiology around the implant. However, the same exhaustive study identified local gliosis and detected signs of a mild inflammatory response.

Significant structural changes have been observed up to 6 months after implantation in rats of Pt/Ir electrode arrays (Prasad et al. 2014). This comprehensive abiotic-biotic study revealed delamination and cracking of insulation in almost all electrodes. However, irregular structural details including delamination and cracks of the insulation layer and recording surface were already clearly present before implantation. The long-term decline of the electrode impedance could thus be explained by the changes in the electrochemical interface area and leakage paths. Manufacturing variability and poor material quality are thus important contributors to electrode failure.

On the other hand, there was only a poor correspondence between the electrode performance and the impedance reduction. Intense ferritin immunoreactivity found in microglia and macrophages provided evidence for intra-parenchymal bleeding. This was correlated with a reduced electrode performance in sharp contrast with the unrelated microglial activation. Electrode failure could thus be caused by both biotic (intra-parenchymal bleeding) and abiotic (suboptimal electrode structure) factors.

6.5.3 Silicon Probes

For decades, multielectrode silicon probes have proven to be valuable research tools in animal experiments (Kuperstein and Eichenbaum 1985; Drake et al. 1988; Wise and Najafi 1991). On the other hand, the mechanical biocompatibility of many designs was rather limited compared to wire electrodes. It is also unclear if such electrodes will be introduced in the clinic or remain predominantly a research tool.

Electrode impedance, charge capacity SNR, recording stability, and immune response of implanted microelectrode arrays of different configurations have been compared in rats (Ward et al. 2009). The results were variable among the same electrodes as well as between electrode types, to such a degree that no "best" device could be identified.

At respectively 3 days and 12 weeks post-implantation, quantitative histology, transcriptomics, and electrophysiology have been exploited to compare the effect of the global shape (cylindrical or planar), size (15 μm, 50 μm, and 75 μm), and fixation of commercially available intracortical electrodes (Karumbaiah et al. 2013). Significantly more glial scarring was shown with tethered Michigan 50 μm electrodes comparing to the microwire and Michigan 15 μm design. Also, fewer neurons survived with the first two devices. No significant difference was found between the two observation time points. After 12 weeks, the microwire electrodes yielded significantly better electrophysiological results than all the other electrode types. These results show that the inflammatory response to intracortical electrodes can be minimized by selecting cylindrical shapes of small dimension and avoiding tethering.

6.6 Failure Mitigation

6.6.1 Severity Evaluation

Any design is confronted with choices between benefits, risks, and cost. For medical equipment, the severity of the risks of an adverse event or other consequences must be evaluated from the point of view of prospective patients. Failure evaluation must thus adopt a scale in line with that point of view. We could propose the following 6-grade severity scale:

1. The therapy requires *adjustment* because of temporary efficacy loss or side effects.
2. The therapy is a failure and the system must be *explanted*. The surgery can involve several regions (e.g., electrode and IXU).
3. The system requires explantation and *replacement* due to any cause including battery end-of-life.
4. Complications such as infection require delayed replacement surgery after explantation. Also, a specific treatment of the causal *infection* is necessary and can result in additional complications.
5. The therapy has generated *permanent but moderate* pain or side effect.
6. The therapy has ended in *severe permanent pain or disability or even death*.

Fortunately, active neural implant failure-induced risks remain mostly at levels 1–4 (Quigley et al. 2003). This is not the case for treatments of acute life-threatening conditions, such as heart pacing. Here, simple device failure can be deadly and is thus of severity level 6. It is thus obvious that the severity of the same technical failure can have a totally different impact in different therapeutic applications.

Secondary consequences or complications must also be considered. For example, any of the failures listed above will result in temporary or permanent, as well as partial or complete treatment deprivation, and/or discomfort. Often, secondary complications are application-specific, such as osteomyelitis, lateral sinus thrombosis, and temporal lobe infection seen after infection of a cochlear implant (Staecker et al. 1999).

Surgery is not only a relatively unpleasant experience, but it also exposes the patient to rare but major secondary complications. Replacement surgery justified by the end of battery life is not an adverse event but rather an expected part of the therapy (Lam et al. 2016). Lead revision is a relatively safe surgical procedure in patients with VNS lead fracture (Waseem et al. 2014).

Because of the relatively frequent need for replacement, it is important that the implanted device could be replaced safely, which seems to mostly be the case (Agarwal et al. 2011; Balkany et al. 1999; Dlouhy et al. 2012; Greenberg et al. 1992; Henson et al. 1999). Ultra-sharp monopolar coagulation can contribute to electrode revision safety (Ng et al. 2010). Only some authors suggest that the rate of adverse events can be slightly higher after replacement/revision than after first implantation (Giordano et al. 2017). In particular, a higher number of IXU replacements seems to correspond to a higher risk of infection (Kahlow and Olivecrona 2013).

6.6.2   Prevention

In general, prevention is better than cure. Professional quality work is characterized by an efficient exploiting of the knowledge acquired from previous experience (Welling et al. 1993). The analysis of failure modes throughout this chapter has come across possible preventive actions. For example, some researchers work on the development of soft implantable neuroprosthesis in order to minimize the mechanical mismatch between the devices and neural tissue (Lacour et al. 2016). Appropriate shapes can contribute to the compliance of neural implants (Sankar et al. 2013). Also, an appropriate failure reporting and analysis procedure is thus an essential part of any effort toward risk reduction and hazard prevention. Clearly, extending the information exchanges between users and manufacturers could lead to a significant acceleration of improvements.

6.6.3   Early Detection

Knowledge of possible failures would allow introducing prevention during the design phase by the inclusion of specific features, appropriate modeling, and bench testing. Similarly, fabrication quality includes exhaustive input and output testing as well as accurate process control. The premarket and clinical studies are pivotal in the preventive effort. In vitro and ex vivo testing on brain slices and cell cultures (Koeneman et al. 2004) can limit the need for animal studies. On the other hand, in vivo testing (in small and large animals) remains necessary for long-term studies (Vasudevan et al. 2016). However, evaluation of the expected chronic survival of an implant is far from obvious, and small animal experiments cannot always be considered as an accelerated model for the situation in humans. Existing allometric scaling methods are indeed probably not entirely justified (Glazier 2018).

Other aspects that should be considered are the clinical trials and post-market studies. Precise instructions should allow the surgeon to proceed to the last device check before implantation. At a later stage, the proper patient instructions and patient monitoring are essential preventive tools. The clinical efficiency of the therapy must be regularly evaluated because many forms of implant failures typically start with a re-emergence of the symptoms (Burdick et al. 2010). Pain and other new side effects or symptoms, stimulus-related or not, are also essential warning signs. Device maintenance tests must be planned. Auto-diagnostic features of the implanted device can also play a major role at this stage.

Steps in this direction have been recently undertaken by the European regulators with the new Regulation (EU) 2017/745 on medical devices. The high-risk devices are subject to stricter preventive controls with new pre-market scrutiny procedures and increased transparency based on a comprehensive EU database on medical devices. Also, the newly introduced unique device identification is expected to improve device traceability; a patient-linked "implant card" carrying implanted medical device information has been introduced; finally, the post-market surveillance requirements for manufacturers have been strengthened.

6.6.4   Diagnostic Means

Once placed in the body, an implanted device is by definition no longer accessible. Therefore, failures appearing after surgery place the caregivers before difficult decisions requiring an accurate diagnosis. This is when the availability of diagnostic means is essential.

The electrode impedance measurement is the most frequent test performed on a regular basis. High impedance values allow detecting electrode or connector breakages. Elevated impedance can also lead to failure because of the stimulator output voltage limit. Alternatively, low impedance can suggest insulation failures. Impedance signature could provide some information about the type of electrode encapsulation that has taken place (Cody et al. 2018).

When a battery is present, the battery voltage measurement is essential to predict the device end-of-life. However, the voltage is only an indirect predictor of charge capacity. Timely replacement of a battery is indeed desirable to avoid a resurgence of the pathology, abnormal stimulus delivery due to low power, and interrogation failure through the external system (Tatum et al. 2004).

Technical controls on the power supply transmission between implant and external components, checks on the user's input, and even software controls can be essential to detect partial failures.

Recording physiological parameters provide a direct functional evaluation. This can result in an improved follow-up of the efficiency of the therapy (Battmer et al. 1994). For example, physiologic potentials evoked by the therapeutic stimulus are an important confirmation of appropriate settings. With proper synchronization means, skin surface potentials generated by the stimulus can also be recorded (Grimonprez et al. 2014).

X-ray or computed tomography (CT) scan can demonstrate dislocation, twisting, and lead fractures (Burdick et al. 2010). It can be argued that conventional X-rays are preferable to CT scan in some instances (Czerny et al. 2000).

Attention for diagnostic or auto-diagnostic features seems relatively new but nonetheless very important to minimize the failure consequences for the patient. Implanted devices should include the necessary sensors and data logging facility to identify primary causes in case of failure.

6.7 Concluding Remarks

Current capabilities of neurotechnology are a limiting factor for the advancement of neuroscience and neuroprosthetics. Another hurdle toward further development is the inadequate understanding of the neural signal necessary to optimize designs and parameter trade-offs for meaning extraction (Kozai 2018). This can be considered as a grand challenge for cortical prostheses. At present, progress is being made about aspects, such as MRI compatibility, long-term power supply, and mechanical compatibility. Much work is devoted to the numerous interactions between the body and electrodes. The optimal integration of massively parallel multichannel, as well as selective recording and stimulation electrodes, remains a major challenge despite the ongoing very active research efforts.

Further development of high-quality implanted neural interfaces requires a proper understanding of the failure modes of existing applications. The present reporting system is insufficient, and more effort should be devoted to a complete causal analysis, which is clearly lacking at present. Each therapeutic use of a neural interface has its own specificities, but the main issues are common to most applications. Therefore, lessons should be learned not only from the limited experience with a single form of therapy, but also across the whole field of active implants.

References

Abbott, N. J. (2000). Inflammatory mediators and modulation of blood-brain barrier permeability. *Cellular and Molecular Neurobiology, 20*(2), 131–147.
Agarwal, G., Wilfong, A. A., & Edmonds, J. L. (2011). Surgical revision of vagus nerve stimulation electrodes in children. *Otolaryngology and Head and Neck Surgery, 144*(1), 123–124.
Almassi, G. H., Olinger, G. N., Wetherbee, J. N., & Fehl, G. (1993). Long-term complications of implantable cardioverter defibrillator lead systems. *The Annals of Thoracic Surgery, 55*(4), 888–892.
Andrei, A., Welkenhuysen, M., Nuttin, B., & Eberle, W. (2012). A response surface model predicting the in vivo insertion behavior of micromachined neural implants. *Journal of Neural Engineering, 9*(1), 016005.

Aregueta-Robles, U. A., Woolley, A. J., Poole-Warren, L. A., Lovell, N. H., & Green, R. A. (2014). Organic electrode coatings for next-generation neural interfaces. *Frontiers in Neuroengineering, 7*(15), 1–18.

Arle, J. E., Mei, L. Z., & Shils, J. L. (2008). Modeling parkinsonian circuitry and the DBS electrode. I. Biophysical background and software. *Stereotactic and Functional Neurosurgery, 86*(1), 1–15.

Arnoldner, C., Baumgartner, W. D., Gstoettner, W., & Hamzavi, J. (2005). Surgical considerations in cochlear implantation in children and adults: A review of 342 cases in Vienna. *Acta Oto-Laryngologica, 125*(3), 228–234.

Azemi, E., Lagenaur, C. F., & Cui, X. T. (2011). The surface immobilization of the neural adhesion molecule L1 on neural probes and its effect on neuronal density and gliosis at the probe/tissue interface. *Biomaterials, 32*(3), 681–692.

Baer, G. A., Talonen, P. P., Shneerson, J. M., Markkula, H., Exner, G., & Wells, F. C. (1990). Phrenic nerve stimulation for central ventilatory failure with bipolar and four-pole electrode systems. *Pacing and Clinical Electrophysiology, 13*(8), 1061–1072.

Balasingam, V., Tejada-Berges, T., Wright, E., Bouckova, R., & Yong, V. W. (1994). Reactive astrogliosis in the neonatal mouse brain and its modulation by cytokines. *The Journal of Neuroscience, 14*(2), 846–856.

Balkany, T. J., Hodges, A. V., Gomez-Marin, O., Bird, P. A., Dolan-Ash, S., Butts, S., et al. (1999). Cochlear reimplantation. *The Laryngoscope, 109*(3), 351–355.

Barrese, J. C., Rao, N., Paroo, K., Triebwasser, C., Vargas-Irwin, C., Franquemont, L., et al. (2013). Failure mode analysis of silicon-based intracortical microelectrode arrays in non-human primates. *Journal of Neural Engineering, 10*(6), 066014.

Barriga-Rivera, A., Bareket, L., Goding, J., Aregueta-Robles, U. A., & Suaning, G. J. (2017). Visual prosthesis: Interfacing stimulating electrodes with retinal neurons to restore vision. *Frontiers in Neuroscience [Internet], 11*. [cited 2017 Nov 29]. Available from: https://www.frontiersin.org/articles/10.3389/fnins.2017.00620/full?utm_source=F-AAE&utm_medium=EMLF&utm_campaign=MRK_468898_55_Neuros_20171128_arts_A.

Basta, D., Gröschel, M., & Ernst, A. (2018). Central and peripheral aspects of noise-induced hearing loss. *HNO, 66*(5), 342–349.

Battmer, R. D., Gnadeberg, D., Lehnhardt, E., & Lenarz, T. (1994). An integrity test battery for the nucleus mini 22 cochlear implant system. *European Archives of Oto-Rhino-Laryngology, 251*(4), 205–209.

Bedell, H. W., Song, S., Li, X., Molinich, E., Lin, S., Stiller, A., et al. (2018). Understanding the effects of both CD14-mediated innate immunity and device/tissue mechanical mismatch in the neuroinflammatory response to intracortical microelectrodes. *Frontiers in Neuroscience, 12*, 772.

Ben-Menachem, E., Revesz, D., Simon, B. J., & Silberstein, S. (2015). Surgically implanted and non-invasive vagus nerve stimulation: A review of efficacy, safety and tolerability. *European Journal of Neurology, 22*(9), 1260–1268.

Beric, A., Kelly, P. J., Rezai, A., Sterio, D., Mogilner, A., Zonenshayn, M., et al. (2001). Complications of deep brain stimulation surgery. *Stereotactic and Functional Neurosurgery, 77*(1–4), 73–78.

Bewernick, B., & Schlaepfer, T. E. (2015). Update on neuromodulation for treatment-resistant depression. *F1000Research, 4*, 1389.

Biran, R., Martin, D. C., & Tresco, P. A. (2007). The brain tissue response to implanted silicon microelectrode arrays is increased when the device is tethered to the skull. *Journal of Biomedical Materials Research. Part A, 82*(1), 169–178.

Bjornsson, C. S., Oh, S. J., Al Kofahi, Y. A., Lim, Y. J., Smith, K. L., Turner, J. N., et al. (2006). Effects of insertion conditions on tissue strain and vascular damage during neuroprosthetic device insertion. *Journal of Neural Engineering, 3*(3), 196–207.

Blomstedt, P., & Hariz, M. I. (2005). Hardware-related complications of deep brain stimulation: A ten year experience. *Acta Neurochirurgica (Wien), 147*(10), 1061–1064.

Boehler, C., Kleber, C., Martini, N., Xie, Y., Dryg, I., Stieglitz, T., et al. (2017). Actively controlled release of dexamethasone from neural microelectrodes in a chronic in vivo study. *Biomaterials, 129*, 176–187.

Borretzen, M. N., Bjerknes, S., Saehle, T., Skjelland, M., Skogseid, I. M., Toft, M., et al. (2014). Long-term follow-up of thalamic deep brain stimulation for essential tremor - patient satisfaction and mortality. *BMC Neurology, 14*(1), 120:1–13.

Branner, A., & Normann, R. A. (2000). A multielectrode array for intrafascicular recording and stimulation in sciatic nerve of cats. *Brain Research Bulletin, 51*(4), 293–306.

Burdick, A. P., Okun, M. S., Haq, I. U., Ward, H. E., Bova, F., Jacobson, C. E., et al. (2010). Prevalence of Twiddler's syndrome as a cause of deep brain stimulation hardware failure. *Stereotactic and Functional Neurosurgery, 88*(6), 353–359.

Burns, B. D., Stean, J. P., & Webb, A. C. (1974). Recording for several days from single cortical neurons in completely unrestrained cats. *Electroencephalography and Clinical Neurophysiology, 36*(3), 314–318.

Capadona, J. (2018). Anti-inflammatory approaches to mitigate the Neuroinflammatory response to brain-dwelling intracortical microelectrodes. *Journal of Immunological Sciences., 2*(4), 15–21.

Cervera-Paz, F. J., Manrique, M., Huarte, A., Garcia, F. J., Garcia-Tapia, R. (1999). Study of surgical complications and technical failures (correction of technical defects) of cochlear implants (published erratum appears in Acta Otorrinolaringol Esp 2000;51(1):96). *Acta Otorrinolaringologica Espanola, 50*(7), 519–24.

Chou, C. K., McDougall, J. A., & Chan, K. W. (1997). RF heating of implanted spinal fusion stimulator during magnetic resonance imaging. *IEEE Transactions on Biomedical Engineering, 44*(5), 367–373.

Christie, I. N., Wells, J. A., Southern, P., Marina, N., Kasparov, S., Gourine, A. V., et al. (2013). fMRI response to blue light delivery in the naïve brain: Implications for combined optogenetic fMRI studies. *NeuroImage, 66*, 634–641.

Cody, P. A., Eles, J. R., Lagenaur, C. F., Kozai, T. D. Y., & Cui, X. T. (2018). Unique electrophysiological and impedance signatures between encapsulation types: An analysis of biological Utah array failure and benefit of a biomimetic coating in a rat model. *Biomaterials [Internet], 161*, 117–128. [cited 2018 Oct 23]. Available from: http://www.sciencedirect.com/science/article/pii/S0142961218300310.

Cohen, E. D. (2009). Effects of high-level pulse train stimulation on retinal function. *Journal of Neural Engineering, 6*(3), 035005.

Connell, S. S., Balkany, T. J., Hodges, A. V., Telischi, F. F., Angeli, S. I., & Eshraghi, A. A. (2008). Electrode migration after cochlear implantation. *Otology and Neurotology., 29*(2), 156–159.

Couch, J. D., Gilman, A. M., & Doyle, W. K. (2016). Long-term expectations of vagus nerve stimulation: A look at battery replacement and revision surgery. *Neurosurgery, 78*(1), 42–46.

Czerny, C., Gstoettner, W., Adunka, O., Hamzavi, J., & Baumgartner, W. D. (2000). Postoperative imaging and evaluation of the electrode position and depth of insertion of multichannel cochlear implants by means of high-resolution computed tomography and conventional X-rays. *Wiener klinische Wochenschrift, 112*(11), 509–511.

Daubert, J. C., Behaghel, A., Leclercq, C., & Mabo, P. (2014). Future of implantable electrical cardiac devices. *Bulletin de l'Académie Nationale de Médecine, 198*(3), 473–487.

de Donaldson, N. N., Perkins, T. A., & Worley, A. C. (1997). Lumbar root stimulation for restoring leg function: Stimulator and measurement of muscle actions. *Artificial Organs, 21*(3), 247–249.

De Ferrari, G. M., Stolen, C., Tuinenburg, A. E., Wright, D. J., Brugada, J., Butter, C., et al. (2017). Long-term vagal stimulation for heart failure: Eighteen month results from the NEural Cardiac TherApy foR Heart Failure (NECTAR-HF) trial. *International Journal of Cardiology [Internet]*. Available from: http://www.sciencedirect.com/science/article/pii/S0167527317312950.

Degenhart, A. D., Eles, J., Dum, R., Mischel, J. L., Smalianchuk, I., Endler, B., et al. (2016). Histological evaluation of a chronically-implanted electrocorticographic electrode grid in a non-human primate. *Journal of Neural Engineering., 13*(4), 046019.

Dickey, A. S., Suminski, A., Amit, Y., & Hatsopoulos, N. G. (2009). Single-unit stability using chronically implanted multielectrode arrays. *Journal of Neurophysiology, 102*(2), 1331–1339.

Dlouhy, B. J., Viljoen, S. V., Kung, D. K., Vogel, T. W., Granner, M. A., Howard, M. A., III, et al. (2012). Vagus nerve stimulation after lead revision. *Neurosurgical Focus, 32*(3), E11.

Drake, K. L., Wise, K. D., Farraye, J., Anderson, D. J., & BeMent, S. L. (1988). Performance of planar multisite microprobes in recording extracellular single-unit intracortical activity. *IEEE Transactions on Biomedical Engineering, 35*(9), 719–732.

Du, Z. J., Kolarcik, C. L., Kozai, T. D. Y., Luebben, S. D., Sapp, S. A., Zheng, X. S., et al. (2017). Ultrasoft microwire neural electrodes improve chronic tissue integration. *Acta Biomaterialia, 53*, 46–58.

Englot, D. J., Glastonbury, C. M., & Larson, P. S. (2011). Abnormal t(2)-weighted MRI signal surrounding leads in a subset of deep brain stimulation patients. *Stereotactic and Functional Neurosurgery, 89*(5), 311–317.

Fleury Curado, T., Oliven, A., Sennes, L. U., Polotsky, V. Y., Eisele, D., & Schwartz, A. R. (2018). Neurostimulation treatment of OSA. *Chest, 154*(6), 1435–1447.

Formento, E., Minassian, K., Wagner, F., Mignardot, J. B., Le Goff-Mignardot, C. G., Rowald, A., et al. (2018). Electrical spinal cord stimulation must preserve proprioception to enable locomotion in humans with spinal cord injury. *Nature Neuroscience, 21*(12), 1728–1741.

Fountas, K. N., & Smith, J. R. (2007). Subdural electrode-associated complications: A 20-year experience. *Stereotactic and Functional Neurosurgery, 85*(6), 264–272.

Freire, M. A. M., Morya, E., Faber, J., Santos, J. R., Guimaraes, J. S., Lemos, N. A. M., et al. (2011). Comprehensive analysis of tissue preservation and recording quality from chronic multielectrode implants. *PLoS One, 6*(11), e27554.

Fung, Y.-C. (1993). *Biomechanics: Mechanical properties of living tissues [Internet].* New York: Springer. [cited 2019 Jan 1]. Available from: http://public.eblib.com/choice/publicfullrecord.aspx?p=3084753.

Gaylor, J. M., Raman, G., Chung, M., Lee, J., Rao, M., Lau, J., et al. (2013). Cochlear implantation in adults: A systematic review and meta-analysis. *JAMA Otolaryngology. Head & Neck Surgery, 139*(3), 265–272.

Ghika, J., Villemure, J. G., Fankhauser, H., Favre, J., Assal, G., & Ghika, S. F. (1998). Efficiency and safety of bilateral contemporaneous pallidal stimulation (deep brain stimulation) in levodopa-responsive patients with Parkinson's disease with severe motor fluctuations: A 2-year follow-up review. *Journal of Neurosurgery, 89*(5), 713–718.

Gilletti, A., & Muthuswamy, J. (2006). Brain micromotion around implants in the rodent somatosensory cortex. *Journal of Neural Engineering, 3*(3), 189–195.

Giordano, F., Zicca, A., Barba, C., Guerrini, R., & Genitori, L. (2017). Vagus nerve stimulation: Surgical technique of implantation and revision and related morbidity. *Epilepsia, 58*(Suppl 1), 85–90.

Glazier, D. S. (2018). Rediscovering and reviving old observations and explanations of metabolic scaling in living systems. *Systems., 6*(4), 28.

Gold, M. R., Van Veldhuisen, D. J., Hauptman, P. J., Borggrefe, M., Kubo, S. H., Lieberman, R. A., et al. (2016). Vagus nerve stimulation for the treatment of heart failure: The INOVATE-HF trial. *Journal of the American College of Cardiology [Internet]., 68*(2), 149–158. Available from: http://www.sciencedirect.com/science/article/pii/S0735109716324044.

Gomez, D. R., Poenisch, F., Pinnix, C. C., Sheu, T., Chang, J. Y., Memon, N., et al. (2013). Malfunctions of implantable cardiac devices in patients receiving proton beam therapy: Incidence and predictors. *International Journal of Radiation Oncology, Biology, Physics, 87*(3), 570–575.

Gossman, M. S., Ketkar, A., Liu, A. K., & Olin, B. (2012). Vagus nerve stimulator stability and interference on radiation oncology x-ray beams. *Physics in Medicine & Biology, 57*(20), N365–N376.

Goss-Varley, M., Shoffstall, A. J., Dona, K. R., McMahon, J. A., Lindner, S. C., Ereifej, E. S., et al. (2018). Rodent behavioral testing to assess functional deficits caused by microelectrode implantation in the rat motor cortex. *Journal of Visualized Experiments, 18*(138).

Graczyk, E. L., Delhaye, B. P., Schiefer, M. A., Bensmaia, S. J., & Tyler, D. J. (2018). Sensory adaptation to electrical stimulation of the somatosensory nerves. *Journal of Neural Engineering, 15*(4), 046002.

Graham, J. M., East, C. A., & Fraser, J. G. (1989). UCH/RNID single channel cochlear implant: surgical technique. *The Journal of Laryngology and Otology. Supplement, 18*, 14–19.

Grand, L., Wittner, L., Herwik, S., Göthelid, E., Ruther, P., Oscarsson, S., et al. (2010). Short and long term biocompatibility of NeuroProbes silicon probes. *Journal of Neuroscience Methods, 189*(2), 216–229.

Greenberg, A. B., Myers, M. W., Hartshorn, D. O., Miller, J. M., & Altschuler, R. A. (1992). Cochlear electrode reimplantation in the guinea pig. *Hearing Research, 61*(1–2), 19–23.

Grimonprez A, Raedt R, De Taeye L, Larsen LE, Delbeke J, Boon P, et al. (2014). A Preclinical Study of Laryngeal Motor-Evoked Potentials as a Marker Vagus Nerve Activation. Int. J Neural Syst. 2015 Sep 14;25(8(1550034)):1–10.

Gu, Y., Dee, C. M., & Shen, J. (2011). Interaction of free radicals, matrix metalloproteinases and caveolin-1 impacts blood-brain barrier permeability. *Frontiers in Bioscience (School Edition), 3*, 1216–1231.

Gunasekera, B., Saxena, T., Bellamkonda, R., & Karumbaiah, L. (2015). Intracortical recording interfaces: Current challenges to chronic recording function. *ACS Chemical Neuroscience [Internet], 6*(1):68–83. Available from: https://doi.org/10.1021/cn5002864.

Günter, C., Delbeke, J., & Ortiz-Catalan, M. (2019). Safety of long-term electrical peripheral nerve stimulation: Review of the state of the art. *Journal of Neuroengineering and Rehabilitation, 16*(1), 13.

Hammer, N., Glatzner, J., Feja, C., Kuhne, C., Meixensberger, J., Planitzer, U., et al. (2015). Human vagus nerve branching in the cervical region. *PLoS One, 10*(2), e0118006. 1–13.

Hariz, M. I. (2002). Complications of deep brain stimulation surgery. *Movement Disorders, 17*(Suppl 3), S162–S166.

He, W., McConnell, G. C., & Bellamkonda, R. V. (2006). Nanoscale laminin coating modulates cortical scarring response around implanted silicon microelectrode arrays. *Journal of Neural Engineering, 3*(4), 316–326.

Henle, C., Raab, M., Cordeiro, J. G., Doostkam, S., Schulze-Bonhage, A., Stieglitz, T., et al. (2011). First long term in vivo study on subdurally implanted Micro-ECoG electrodes, manufactured with a novel laser technology. *Biomedical Microdevices., 13*(1), 59–68.

Henson, A. M., Slattery, W. H., III, Luxford, W. M., & Mills, D. M. (1999). Cochlear implant performance after reimplantation: A multicenter study. *The American Journal of Otology, 20*(1), 56–64.

Heo, D. N., Song, S.-J., Kim, H.-J., Lee, Y. J., Ko, W.-K., Lee, S. J., et al. (2016). Multifunctional hydrogel coatings on the surface of neural cuff electrode for improving electrode-nerve tissue interfaces. *Acta Biomaterialia, 39*, 25–33.

Hermann, J. K., Lin, S., Soffer, A., Wong, C., Srivastava, V., Chang, J., et al. (2018). The role of toll-like receptor 2 and 4 innate immunity pathways in intracortical microelectrode-induced neuroinflammation. *Frontiers in Bioengineering and Biotechnology, 6*, 113.

Hochberg, L. R., Serruya, M. D., Friehs, G. M., Mukand, J. A., Saleh, M., Caplan, A. H., et al. (2006). Neuronal ensemble control of prosthetic devices by a human with tetraplegia. *Nature, 442*(7099), 164–171.

Holecko, M. M., Williams, J. C., & Massia, S. P. (2005). Visualization of the intact interface between neural tissue and implanted microelectrode arrays. *Journal of Neural Engineering, 2*(4), 97–102.

Holubec, T., Ursprung, G., Schonrath, F., Caliskan, E., Steffel, J., Falk, V., et al. (2015). Does implantation technique influence lead failure? *Acta Cardiologica, 70*(5), 581–586.

Hosseini, N. H., Hoffmann, R., Kisban, S., Stieglitz, T., Paul, O., & Ruther, P. (2007). Comparative study on the insertion behavior of cerebral microprobes. *Conference Proceedings: Annual International Conference of the IEEE Engineering in Medicine and Biology Society, 2007,* 4711–4714.

Hsieh, H.-L., & Yang, C.-M. (2013). Role of redox signaling in neuroinflammation and neurodegenerative diseases. *BioMed Research International, 2013,* 484613.

Ishida, K., Shinkawa, A., Sakai, M., Tamura, Y., & Naito, A. (1997). Cause and repair of flap necrosis over cochlear implant. *The American Journal of Otology, 18*(4), 472–474.

Janson, C., Maxwell, R., Gupte, A. A., & Abosch, A. (2010). Bowstringing as a complication of deep brain stimulation: Case report. *Neurosurgery, 66*(6), E1205.

Joint, C., Nandi, D., Parkin, S., Gregory, R., & Aziz, T. (2002). Hardware-related problems of deep brain stimulation. *Movement Disorders, 17*(Suppl 3), S175–S180.

Jorfi, M., Skousen, J. L., Weder, C., & Capadona, J. R. (2015). Progress towards biocompatible intracortical microelectrodes for neural interfacing applications. *Journal of Neural Engineering, 12*(1), 1–46. https://doi.org/10.1088/1741-2560/12/1/011001.

Josset, P., Meyer, B., Gegu, D., & Chouard, C. H. (1984). Implant material tolerance. *Acta Oto-Laryngologica. Supplementum (Stockh), 411*(9002-84-0 (Polytetrafluoroethylene)), 45–52.

Kahlow, H., & Olivecrona, M. (2013). Complications of vagal nerve stimulation for drug-resistant epilepsy: A single center longitudinal study of 143 patients. *Seizure, 22*(10), 827–833.

Kanaan, N., Winkel, A., Stumpp, N., Stiesch, M., & Lenarz, T. (2013). Bacterial growth on cochlear implants as a potential origin of complications. *Otology & Neurotology, 34*(3), 539–543.

Karumbaiah, L., Saxena, T., Carlson, D., Patil, K., Patkar, R., Gaupp, E. A., et al. (2013). Relationship between intracortical electrode design and chronic recording function. *Biomaterials, 34*(33), 8061–8074.

Kelly, S. K., Ellersick, W. F., Krishnan, A., Doyle, P., Shire, D. B., Wyatt, J. L., et al. (2014). Redundant safety features in a high-channel-count retinal neurostimulator. *IEEE Biomedical Circuits and Systems Conference., 2014,* 216–219.

Kempf, H. G., Johann, K., & Lenarz, T. (1999). Complications in pediatric cochlear implant surgery. *European Archives of Otorhinolaryngology, 256*(3), 128–132.

Kilgore, K. L., Peckham, P. H., Keith, M. W., Montague, F. W., Hart, R. L., Gazdik, M. M., et al. (2003). Durability of implanted electrodes and leads in an upper-limb neuroprosthesis. *Journal of Rehabilitation Research and Development, 40*(6), 457–468.

Kim, Y.-T., Hitchcock, R. W., Bridge, M. J., & Tresco, P. A. (2004). Chronic response of adult rat brain tissue to implants anchored to the skull. *Biomaterials, 25*(12), 2229–2237.

Kimberley, T. J., Pierce, D., Prudente, C. N., Francisco, G. E., Yozbatiran, N., Smith, P., et al. (2018). Vagus nerve stimulation paired with upper limb rehabilitation after chronic stroke. *Stroke, 49*(11), 2789–2792.

Kirkhorn, T., Almquist, L. O., Persson, H. W., & Holmer, N. G. (1997). An experimental high energy therapeutic ultrasound equipment: Design and characterisation. *Medical & Biological Engineering & Computing, 35*(3), 295–299.

Koeneman, B. A., Lee, K. K., Singh, A., He, J., Raupp, G. B., Panitch, A., et al. (2004). An ex vivo method for evaluating the biocompatibility of neural electrodes in rat brain slice cultures. *Journal of Neuroscience Methods, 137*(2), 257–263.

Koller, W. C., Lyons, K. E., Wilkinson, S. B., Troster, A. I., & Pahwa, R. (2001). Long-term safety and efficacy of unilateral deep brain stimulation of the thalamus in essential tremor. *Movement Disorders, 16*(3), 464–468.

Kotecha, R., Berriochoa, C. A., Murphy, E. S., Machado, A. G., Chao, S. T., Suh, J. H., et al. (2016). Report of whole-brain radiation therapy in a patient with an implanted deep brain stimulator: Important neurosurgical considerations and radiotherapy practice principles. *Journal of Neurosurgery, 124*(4), 966–970.

Kovach, K. M., Kumsa, D. W., Srivastava, V., Hudak, E. M., Untereker, D. F., Kelley, S. C., et al. (2016). High-throughput in vitro assay to evaluate the cytotoxicity of liberated platinum compounds for stimulating neural electrodes. *Journal of Neuroscience Methods, 273,* 1–9.

Kozai, T. (2018). The history and horizons of microscale neural interfaces. *Micromachines, 9*(9), 445.

Kozai, T. D. Y., Vazquez, A. L., Weaver, C. L., Kim, S.-G., & Cui, X. T. (2012). In vivo two-photon microscopy reveals immediate microglial reaction to implantation of microelectrode through extension of processes. *Journal of Neural Engineering, 9*(6), 066001.

Kozai, T. D. Y., Li, X., Bodily, L. M., Caparosa, E. M., Zenonos, G. A., Carlisle, D. L., et al. (2014). Effects of caspase-1 knockout on chronic neural recording quality and longevity: Insight into cellular and molecular mechanisms of the reactive tissue response. *Biomaterials, 35*(36), 9620–9634.

Krishna, V., Sammartino, F., King, N. K. K., So, R. Q. Y., & Wennberg, R. (2016). Neuromodulation for epilepsy. *Neurosurgery Clinics of North America, 27*(1), 123–131.

Kumar, K., Nath, R., & Wyant, G. M. (1991). Treatment of chronic pain by epidural spinal cord stimulation: A 10-year experience. *Journal of Neurosurgery, 75*(3), 402–407.

Kumar, K., Toth, C., Nath, R. K., & Laing, P. (1998). Epidural spinal cord stimulation for treatment of chronic pain--some predictors of success a 15-year experience. *Surgical Neurology, 50*(2), 110–120.

Kuperstein, M., & Eichenbaum, H. (1985). Unit activity, evoked potentials and slow waves in the rat hippocampus and olfactory bulb recorded with a 24-channel microelectrode. *Neuroscience, 15*(3), 703–712.

Lacour, S. P., Courtine, G., & Guck, J. (2016). Materials and technologies for soft implantable neuroprostheses. *Nature Reviews Materials, 1*.

Lam, J., Lin, Y., Curry, D. J., Reddy, G. D., & Warnke, P. C. (2016). Revision surgeries following vagus nerve stimulator implantation. *Journal of Clinical Neuroscience, 30*, 83–87.

Lasala, A. F., Fieldman, A., Diana, D. J., & Humphrey, C. B. (1979). Gas pocket causing pacemaker malfunction. *Pacing and Clinical Electrophysiology, 2*(2), 183–185.

Lecomte, A., Descamps, E., & Bergaud, C. (2018). A review on mechanical considerations for chronically-implanted neural probes. *Journal of Neural Engineering, 15*(3), 031001.

Lee, H. C., Ejserholm, F., Gaire, J., Currlin, S., Schouenborg, J., Wallman, L., et al. (2017). Histological evaluation of flexible neural implants; flexibility limit for reducing the tissue response? *Journal of Neural Engineering., 14*(3), 036026.

Lewis, P. M., Ackland, H. M., Lowery, A. J., & Rosenfeld, J. V. (2015). Restoration of vision in blind individuals using bionic devices: A review with a focus on cortical visual prostheses. *Brain Research [Internet], 1595*, 51–73. [cited 2017 Jul 3]. Available from: http://www.sciencedirect.com/science/article/pii/S0006899314015674.

Liu, X., McCreery, D. B., Carter, R. R., Bullara, L. A., Yuen, T. G., & Agnew, W. F. (1999). Stability of the interface between neural tissue and chronically implanted intracortical microelectrodes. *IEEE Transactions on Rehabilitation Engineering, 7*(3), 315–326.

Liu, X., Demosthenous, A., & Donaldson, N. (2007). Implantable stimulator failures: Causes, outcomes, and solutions. *Conference Proceedings: Annual International Conference of the IEEE Engineering in Medicine and Biology Society*, 5786–5790.

Lotan, G., & Vaiman, M. (2015). Treatment of epilepsy by stimulation of the vagus nerve from head-and-neck surgical point of view. *The Laryngoscope, 125*(6), 1352–1355.

Lotti, F., Ranieri, F., Vadalà, G., Zollo, L., & Di Pino, G. (2017). Invasive intraneural interfaces: Foreign body reaction issues. *Frontiers in Neuroscience, 11*, 497.

Ludwig, K. A., Langhals, N. B., Joseph, M. D., Richardson-Burns, S. M., Hendricks, J. L., & Kipke, D. R. (2011). Poly(3,4-ethylenedioxythiophene) (PEDOT) polymer coatings facilitate smaller neural recording electrodes. *Journal of Neural Engineering, 8*(1), 014001:1–8.

Luetje, C. M., & Jackson, K. (1997). Cochlear implants in children: What constitutes a complication? *Otolaryngology Head and Neck Surgery, 117*(3 Pt 1), 243–247.

Lyons, K. E., Koller, W. C., Wilkinson, S. B., & Pahwa, R. (2001). Long term safety and efficacy of unilateral deep brain stimulation of the thalamus for parkinsonian tremor. *Journal of Neurology, Neurosurgery, and Psychiatry, 71*(5), 682–684.

Mahmood, I. (2018). Misconceptions and issues regarding allometric scaling during the drug development process. *Expert Opinion on Drug Metabolism & Toxicology, 14*(8), 843–854.

Mancini, M., Moresi, M., & Rancini, R. (1999). Mechanical properties of alginate gels: Empirical characterisation. *Journal of Food Engineering., 39*(4), 369–378.

Martin, K. D., Polanski, W. H., Schulz, A.-K., Jöbges, M., Hoff, H., Schackert, G., et al. (2016). Restoration of ankle movements with the ActiGait implantable drop foot stimulator: A safe and reliable treatment option for permanent central leg palsy. *Journal of Neurosurgery, 124*(1), 70–76.

McCreery, D. B., Yuen, T. G., Agnew, W. F., & Bullara, L. A. (1997). A characterization of the effects on neuronal excitability due to prolonged microstimulation with chronically implanted microelectrodes. *IEEE Transactions on Biomedical Engineering, 44*(10), 931–939.

McCreery, D. B., Yuen, T. G., & Bullara, L. A. (2000). Chronic microstimulation in the feline ventral cochlear nucleus: Physiologic and histologic effects. *Hearing Research, 149*(1–2), 223–238.

McCreery, D., Cogan, S., Kane, S., & Pikov, V. (2016). Correlations between histology and neuronal activity recorded by microelectrodes implanted chronically in the cerebral cortex. *Journal of Neural Engineering, 13*(3), 036012.

Mercanzini, A., Reddy, S. T., Velluto, D., Colin, P., Maillard, A., Bensadoun, J. C., et al. (2010). Controlled release nanoparticle-embedded coatings reduce the tissue reaction to neuroprostheses. *Journal of Controlled Release, 145*(3), 196–202.

Michelson, N. J., Vazquez, A. L., Eles, J. R., Salatino, J. W., Purcell, E. K., Williams, J. J., et al. (2018). Multi-scale, multi-modal analysis uncovers complex relationship at the brain tissue-implant neural interface: New emphasis on the biological interface. *Journal of Neural Engineering., 15*(3), 033001.

Miller, K. (Ed.). (2011). *Biomechanics of the brain*. New York: Springer.

Mols, K., Musa, S., Nuttin, B., Lagae, L., & Bonin, V. (2017). In vivo characterization of the electrophysiological and astrocytic responses to a silicon neuroprobe implanted in the mouse neocortex. *Scientific Reports [Internet], 7*(1). [cited 2019 Apr 6]. Available from: http://www.nature.com/articles/s41598-017-15121-1.

Muckle, R. P., & Levine, S. C. (1994). Facial nerve stimulation produced by cochlear implants in patients with cochlear otosclerosis. *The American Journal of Otology, 15*(3), 394–398.

Muranaka, H., Horiguchi, T., Ueda, Y., & Tanki, N. (2011). Evaluation of RF heating due to various implants during MR procedures. *Magnetic Resonance in Medical Sciences, 10*(1), 11–19.

Muthuswamy, J., Gilletti, A., Jain, T., & Okandan, M. (2003). Microactuated neural probes to compensate for brain micromotion. Engineering in Medicine and Biology Society, 2003. *Proceedings of the 25th Annual International Conference of the IEEE [Internet]*, 1941–1943. Available from: http://ieeexplore.ieee.org/stamp/stamp.jsp?arnumber=1279819.

Nagel, S. J., Wilson, S., Johnson, M. D., Machado, A., Frizon, L., Chardon, M. K., et al. (2017). Spinal cord stimulation for spasticity: Historical approaches, current status, and future directions. *Neuromodulation.*

Ng, W. H., Donner, E., Go, C., Abou-Hamden, A., & Rutka, J. T. (2010). Revision of vagal nerve stimulation (VNS) electrodes: Review and report on use of ultra-sharp monopolar tip. *Childs Nervous Systems, 26*(8), 1081–1084.

Ng, F. L., Saxena, M., Mahfoud, F., Pathak, A., & Lobo, M. D. (2016). Device-based therapy for hypertension. *Current Hypertension Reports [Internet], 18*(8), 61. [cited 2017 Jul 6]. Available from: https://link.springer.com/article/10.1007/s11906-016-0670-5.

Nguyen, J. K., Park, D. J., Skousen, J. L., Hess-Dunning, A. E., Tyler, D. J., Rowan, S. J., et al. (2014). Mechanically-compliant intracortical implants reduce the neuroinflammatory response. *Journal of Neural Engineering, 11*(5), 056014–2560/11.

Nicolelis, M. A. L., Dimitrov, D., Carmena, J. M., Crist, R., Lehew, G., Kralik, J. D., et al. (2003). Chronic, multisite, multielectrode recordings in macaque monkeys. *Proceedings of the National Academy of Sciences of the United States of America, 100*(19), 11041–11046.

Nizard, J., Raoul, S., Nguyen, J. P., & Lefaucheur, J. P. (2012). Invasive stimulation therapies for the treatment of refractory pain. *Discovery Medicine, 14*(77), 237–246.

Nolta, N. F., Christensen, M. B., Crane, P. D., Skousen, J. L., & Tresco, P. A. (2015). BBB leakage, astrogliosis, and tissue loss correlate with silicon microelectrode array recording performance. *Biomaterials, 53*, 753–762.

Normann, R. A., Maynard, E. M., Rousche, P. J., Nordhausen, C. T., Warren, D. J., & Guillory, K. S. (1997). The Utah 100 microelectrode array: An experimental platform for cortically based vision prosthesis. *Investigative Ophthalmology & Visual Science, 38*(4), S41.

Ordonez, J. S., Boehler, C., Schuettler, M., & Stieglitz, T. (2012a). Improved polyimide thin-film electrodes for neural implants. *Conference Proceeding: Annual International Conference of the IEEE Engineering in Medicine and Biology Society, 2012*(0 (Coated Materials, Biocompatible);0 (Membranes, Artificial);0 (Resins, Synthetic);39355–34-5 (polyimide resin)), 5134–5137.

Ordonez, J. S., Schuettler, M., Ortmanns, M., & Stieglitz, T. (2012b). A 232-channel retinal vision prosthesis with a miniaturized hermetic package. *Conference Proceeding: Annual International Conference of the IEEE Engineering in Medicine and Biology Society, 2012*(0 (Hermetic);0 (Zinc Oxide-Eugenol Cement)), 2796–2799.

Ortiz-Catalan, M., Branemark, R., Hakansson, B., & Delbeke, J. (2012). On the viability of implantable electrodes for the natural control of artificial limbs: Review and discussion. *Biomedical Engineering Online., 11*(33), 1–30.

Panic, V. V., Dekanski, A. B., Mitric, M., Milonjic, S. K., Miskovic-Stankovic, V. B., & Nikolic, B. Z. (2010). The effect of the addition of colloidal iridium oxide into sol-gel obtained titanium and ruthenium oxide coatings on titanium on their electrochemical properties. *Physical Chemistry Chemical Physics, 12*(27), 7521–7528.

Patterson, T., Stecker, M. M., & Netherton, B. L. (2007). Mechanisms of electrode induced injury. Part 2: Clinical experience. *American Journal of Electroneurodiagnostic Technology, 47*(2), 93–113.

Patton, V., Wiklendt, L., Arkwright, J. W., Lubowski, D. Z., & Dinning, P. G. (2013). The effect of sacral nerve stimulation on distal colonic motility in patients with faecal incontinence. *The British Journal of Surgery, 100*(7), 959–968.

Pavlov, V. A., & Tracey, K. J. (2015). Neural circuitry and immunity. *Immunologic Research [Internet], 63*(1–3), 38–57. [cited 2017 Jul 6]. Available from: https://link.springer.com/article/10.1007/s12026-015-8718-1.

Pearl, P. L., Conry, J. A., Yaun, A., Taylor, J. L., Heffron, A. M., Sigman, M., et al. (2008). Misidentification of vagus nerve stimulator for intravenous access and other major adverse events. *Pediatric Neurology, 38*(4), 248–251.

Pease, W. S., & Grove, S. L. (2013). Electrical safety in electrodiagnostic medicine. *PM & R: The Journal of Injury, Function, and Rehabilitation, 5*(5 Suppl), S8–S13.

Pečlin, P., Rozman, J., Krajnik, J., & Ribarič, S. (2016). Evaluation of the efficacy and robustness of a second generation implantable stimulator in a patient with hemiplegia during 20 years of functional electrical stimulation of the common peroneal nerve. *Artificial Organs, 40*(11), 1085–1091.

Peterson, D. K., Nochomovitz, M. L., Stellato, T. A., & Mortimer, J. T. (1994). Long-term intramuscular electrical activation of the phrenic nerve: Safety and reliability. *IEEE Transactions on Biomedical Engineering, 41*(12), 1115–1126.

Pierce, A. L., Sommakia, S., Rickus, J. L., & Otto, K. J. (2009). Thin-film silica sol-gel coatings for neural microelectrodes. *Journal of Neuroscience Methods, 180*(1), 106–110.

Polanco, M., Bawab, S., & Yoon, H. (2016). Computational assessment of neural probe and brain tissue interface under transient motion. *Biosensors (Basel)., 6*(2), 27.

Potter, K. A., Buck, A. C., Self, W. K., & Capadona, J. R. (2012). Stab injury and device implantation within the brain results in inversely multiphasic neuroinflammatory and neurodegenerative responses. *Journal of Neural Engineering, 9*(4)(046020–2560/9):1–15.

Potter, K. A., Buck, A. C., Self, W. K., Callanan, M. E., Sunil, S., & Capadona, J. R. (2013). The effect of resveratrol on neurodegeneration and blood brain barrier stability surrounding intracortical microelectrodes. *Biomaterials, 34*(29), 7001–7015.

Prasad, A., Xue, Q.-S., Sankar, V., Nishida, T., Shaw, G., Streit, W. J., et al. (2012). Comprehensive characterization and failure modes of tungsten microwire arrays in chronic neural implants. *Journal of Neural Engineering., 9*(5), 056015.

Prasad, A., Xue, Q. S., Dieme, R., Sankar, V., Mayrand, R. C., Nishida, T., et al. (2014). Abiotic-biotic characterization of Pt/Ir microelectrode arrays in chronic implants. *Frontiers in Neuroengineering, 7*(2), 1–15.

Prodanov, D., & Delbeke, J. (2016). Mechanical and biological interactions of implants with the brain and their impact on implant design. *Frontiers in Neuroscience, 10*, 11.

Purcell, E. K., Thompson, D. E., Ludwig, K. A., & Kipke, D. R. (2009). Flavopiridol reduces the impedance of neural prostheses in vivo without affecting recording quality. *Journal of Neuroscience Methods, 183*(2), 149–157.

Quigley, D. G., Arnold, J., Eldridge, P. R., Cameron, H., McIvor, K., Miles, J. B., et al. (2003). Long-term outcome of spinal cord stimulation and hardware complications. *Stereotactic and Functional Neurosurgery, 81*(1–4), 50–56.

Ralston, A., Stevens, G., Mahomudally, E., Ibrahim, I., & Leckie, E. (1999). Cochlear implants: Response to therapeutic irradiation. *International Journal of Radiation Oncology, Biology, Physics, 44*(1), 227–231.

Rao, L., Zhou, H., Li, T., Li, C., & Duan, Y. Y. (2012). Polyethylene glycol-containing polyurethane hydrogel coatings for improving the biocompatibility of neural electrodes. *Acta Biomaterialia, 8*(6), 2233–2242.

Rashid, B., Destrade, M., & Gilchrist, M. D. (2014). Mechanical characterization of brain tissue in tension at dynamic strain rates. *Journal of the Mechanical Behavior of Biomedical Materials, 33*, 43–54.

Renard, V. M., & North, R. B. (2006). Prevention of percutaneous electrode migration in spinal cord stimulation by a modification of the standard implantation technique. *Journal of Neurosurgery. Spine, 4*(4), 300–303.

Rennaker, R. L., Street, S., Ruyle, A. M., & Sloan, A. M. (2005). A comparison of chronic multichannel cortical implantation techniques: Manual versus mechanical insertion. *Journal of Neuroscience Methods, 142*(2), 169–176.

Révész, D., Rydenhag, B., & Ben-Menachem, E. (2016). Complications and safety of vagus nerve stimulation: 25 years of experience at a single center. *Journal of Neurosurgery. Pediatrics, 18*(1), 97–104.

Rickett, T., Connell, S., Bastijanic, J., Hegde, S., & Shi, R. (2010). Functional and mechanical evaluation of nerve stretch injury. *Journal of Medical Systems, 6*, 1–7.

Roark, C., Whicher, S., & Abosch, A. (2008). Reversible neurological symptoms caused by diathermy in a patient with deep brain stimulators: Case report. *Neurosurgery, 62*(1), E256.

Rungta, R. L., Osmanski, B.-F., Boido, D., Tanter, M., & Charpak, S. (2017). Light controls cerebral blood flow in naive animals. *Nature Communications, 8*, 14191.

Rushton, D. N. (2002). Electrical stimulation in the treatment of pain. *Disability and Rehabilitation, 24*(8), 407–415.

Sankar, V., Sanchez, J. C., McCumiskey, E., Brown, N., Taylor, C. R., Ehlert, G. J., et al. (2013). A highly compliant serpentine shaped polyimide interconnect for front-end strain relief in chronic neural implants. *Frontiers in Neurology, 4*(124), 1–11.

Saxena, T., Karumbaiah, L., Gaupp, E. A., Patkar, R., Patil, K., Betancur, M., et al. (2013). The impact of chronic blood-brain barrier breach on intracortical electrode function. *Biomaterials, 34*(20), 4703–4713.

Schalk, G. (2010). Can electrocorticography (ECoG) support robust and powerful brain-computer interfaces? *Frontiers in Neuroengineering., 3*(9), 1–2.

Schendel, A. A., Thongpang, S., Brodnick, S. K., Richner, T. J., Lindevig, B. D. B., Krugner-Higby, L., et al. (2013). A cranial window imaging method for monitoring vascular growth around chronically implanted micro-ECoG devices. *Journal of Neuroscience Methods, 218*(1), 121–130.

Schiefer, M., Tan, D., Sidek, S. M., & Tyler, D. J. (2016). Sensory feedback by peripheral nerve stimulation improves task performance in individuals with upper limb loss using a myoelectric prosthesis. *Journal of Neural Engineering, 13*(1), 016001.

Schwartz, A. B. (2004). Cortical neural prosthetics. *Annual Review of Neuroscience, 27*, 487–507.

Sharkey, P. C., Halter, J. A., & Nakajima, K. (1989). Electrophrenic respiration in patients with high quadriplegia. *Neurosurgery, 24*(4), 529–535.

Shih, C., Miller, J. L., Fialkow, M., Vicars, B. G., & Yang, C. C. (2013). Reoperation after sacral neuromodulation therapy: A single-institution experience. *Female Pelvic Medicine & Reconstructive Surgery, 19*(3), 175–178.

Shils, J. L., Mei, L. Z., & Arle, J. E. (2008). Modeling parkinsonian circuitry and the DBS electrode. II. Evaluation of a computer simulation model of the basal ganglia with and without subthalamic nucleus stimulation. *Stereotactic and Functional Neurosurgery, 86*(1), 16–29.

Skousen, J. L., Merriam, S. M. E., Srivannavit, O., Perlin, G., Wise, K. D., & Tresco, P. A. (2011). Reducing surface area while maintaining implant penetrating profile lowers the brain foreign body response to chronically implanted planar silicon microelectrode arrays. *Progress in Brain Research [Internet], 194*, 167–180. Available from: https://doi.org/10.1016/B978-0-444-53815-4.00009-1.

Smyth, M. D., Tubbs, R. S., Bebin, E. M., Grabb, P. A., & Blount, J. P. (2003). Complications of chronic vagus nerve stimulation for epilepsy in children. *Journal of Neurosurgery, 99*(3), 500–503.

Sohal, H. S., Clowry, G. J., Jackson, A., O'Neill, A., & Baker, S. N. (2016). Mechanical flexibility reduces the foreign body response to long-term implanted microelectrodes in rabbit cortex. *PLoS One, 11*(10), e0165606.

Sonnenburg, R. E., Wackym, P. A., Yoganandan, N., Firszt, J. B., Prost, R. W., & Pintar, F. A. (2002). Biophysics of cochlear implant/MRI interactions emphasizing bone biomechanical properties. *The Laryngoscope, 112*(10), 1720–1725.

Sridharan, A., Rajan, S. D., & Muthuswamy, J. (2013). Long-term changes in the material properties of brain tissue at the implant-tissue interface. *Journal of Neural Engineering, 10*(6), 066001.

Staecker, H., Chow, H., & Nadol, J. B., Jr. (1999). Osteomyelitis, lateral sinus thrombosis, and temporal lobe infarction caused by infection of a percutaneous cochlear implant. *The American Journal of Otology, 20*(6), 726–728.

Stieglitz, T. (2019). Why neurotechnologies? about the purposes, opportunities and limitations of neurotechnologies in clinical applications. *Neuroethics [Internet]*, [cited 2019 Apr 18]. Available from: http://link.springer.com/10.1007/s12152-019-09406-7.

Stujenske, J. M., Spellman, T., & Gordon, J. A. (2015). Modeling the spatiotemporal dynamics of light and heat propagation for in vivo optogenetics. *Cell Reports [Internet], 12*(3), 525–534. [cited 2018 Oct 28]. Available from: http://www.sciencedirect.com/science/article/pii/S2211124715006488.

Suner, S., Fellows, M. R., Vargas-Irwin, C., Nakata, G. K., & Donoghue, J. P. (2005). Reliability of signals from a chronically implanted, silicon-based electrode array in non-human primate primary motor cortex. *IEEE Transactions on Neural Systems and Rehabilitation Engineering, 13*(4), 524–541.

Tambyraja, R. R., Gutman, M. A., & Megerian, C. A. (2005). Cochlear implant complications: Utility of federal database in systematic analysis. *Archives of Otolaryngology – Head & Neck Surgery, 131*(3), 245–250.

Tatum, W. O., Ferreira, J. A., Benbadis, S. R., Heriaud, L. S., Gieron, M., Rodgers-Neame, N. T., et al. (2004). Vagus nerve stimulation for pharmacoresistant epilepsy: Clinical symptoms with end of service. *Epilepsy & Behavior, 5*(1), 128–132.

Taub, A. H., Hogri, R., Magal, A., Mintz, M., & Shacham-Diamand, Y. (2012). Bioactive anti-inflammatory coating for chronic neural electrodes. *Journal of Biomedical Materials Research. Part A, 100*(7), 1854–1858.

Terry, G. E., Conry, J. A., Taranto, E., & Yaun, A. (2011). Failure of a vagus nerve stimulator following a nearby lightning strike. *Pediatric Neurosurgery, 47*(1), 72–73.

Thelin, J., Jörntell, H., Psouni, E., Garwicz, M., Schouenborg, J., Danielsen, N., et al. (2011). Implant size and fixation mode strongly influence tissue reactions in the CNS. *PLoS One, 6*(1), e16267.

Thoma, H., Girsch, W., Holle, J., & Mayr, W. (1989). Technology and long-term application of an epineural electrode. *ASAIO Transactions, 35*(3), 490–494.

Trout, A. T., Larson, D. B., Mangano, F. T., & Gonsalves, C. H. (2013). Twiddler syndrome with a twist: A cause of vagal nerve stimulator lead fracture. *Pediatric Radiology, 43*(12), 1647–1651.

Turner, J. N., Shain, W., Szarowski, D. H., Andersen, M., Martins, S., Isaacson, M., et al. (1999). Cerebral astrocyte response to micromachined silicon implants. *Experimental Neurology, 156*(1), 33–49.

Udo, E. O., Zuithoff, N. P. A., van Hemel, N. M., de Cock, C. C., Hendriks, T., Doevendans, P. A., et al. (2012). Incidence and predictors of short- and long-term complications in pacemaker therapy: The FOLLOWPACE study. *Heart Rhythm, 9*(5), 728–735.

Uttara, B., Singh, A. V., Zamboni, P., & Mahajan, R. T. (2009). Oxidative stress and neurodegenerative diseases: A review of upstream and downstream antioxidant therapeutic options. *Current Neuropharmacology, 7*(1), 65–74.

Val-Laillet, D., Aarts, E., Weber, B., Ferrari, M., Quaresima, V., Stoeckel, L. E., et al. (2015). Neuroimaging and neuromodulation approaches to study eating behavior and prevent and treat eating disorders and obesity. *Neuroimage Clinical., 8*, 1–31.

van Dommelen, J. A. W., Hrapko, M., & Peters, G. W. M. (2010). Constitutive modelling of brain tissue for prediction of traumatic brain injury. In L. E. Bilston (Ed.), *Neural tissue biomechanics [Internet]* (pp. 41–67). Berlin/Heidelberg: Springer Berlin Heidelberg. [cited 2019 Jan 1]. Available from: http://link.springer.com/10.1007/8415_2010_16.

Vasudevan, S., Patel, K., & Welle, C. (2016). Rodent model for assessing the long term safety and performance of peripheral nerve recording electrodes. *Journal of Neural Engineering [Internet], 14*(1), 016008. [cited 2017 Jul 3]. Available from: http://iopscience.iop.org/article/10.1088/1741-2552/14/1/016008/meta.

Ward, M. P., Rajdev, P., Ellison, C., & Irazoqui, P. P. (2009). Toward a comparison of microelectrodes for acute and chronic recordings. *Brain Research, 1282*, 183–200.

Waseem, H., Raffa, S. J., Benbadis, S. R., & Vale, F. L. (2014). Lead revision surgery for vagus nerve stimulation in epilepsy: Outcomes and efficacy. *Epilepsy & Behavior, 31*, 110–113.

Weaver, F. M., Stroupe, K. T., Smith, B., Gonzalez, B., Huo, Z., Cao, L., et al. (2017). Survival in patients with Parkinson's disease after deep brain stimulation or medical management. *Movement Disorders, 32*(12), 1756–1763.

Weese, M., Silvestri, J. M., Kenny, A. S., Ilbawi, M. N., Hauptman, S. A., Lipton, J. W., et al. (1996). Diaphragm pacing with a quadripolar phrenic nerve electrode: An international study. *Pacing and Clinical Electrophysiology, 19*(9), 1311–1319.

Welkenhuysen, M., Andrei, A., Ameye, L., Eberle, W., & Nuttin, B. (2011). Effect of insertion speed on tissue response and insertion mechanics of a chronically implanted silicon-based neural probe. *IEEE Transactions on Biomedical Engineering, 58*(11), 3250–3259.

Welling, D. B., Hinojosa, R., Gantz, B. J., & Lee, J. T. (1993). Insertional trauma of multichannel cochlear implants. *The Laryngoscope, 103*(9), 995–1001.

Wellman, S. M., Eles, J. R., Ludwig, K. A., Seymour, J. P., Michelson, N. J., McFadden, W. E., et al. (2018). A materials roadmap to functional neural interface design. *Advanced Functional Materials [Internet], 28*(12), 1–38.[cited 2018 Oct 23]. Available from: https://onlinelibrary.wiley.com/doi/abs/10.1002/adfm.201701269.

Wellman, S. M., Li, L., Yaxiaer, Y., McNamara, I., & Kozai, T. D. Y. (2019). Revealing spatial and temporal patterns of cell death, glial proliferation, and blood-brain barrier dysfunction around implanted intracortical neural interfaces. *Frontiers in Neuroscience, 13*, 493.

Williams, D. F. (2008). On the mechanisms of biocompatibility. *Biomaterials, 29*(20), 2941–2953.

Williams, J. C., Rennaker, R. L., & Kipke, D. R. (1999). Long-term neural recording characteristics of wire microelectrode arrays implanted in cerebral cortex. *Brain Research. Brain Research Protocols, 4*(3), 303–313.

Williams, N. R., Short, E. B., Hopkins, T., Bentzley, B. S., Sahlem, G. L., Pannu, J., et al. (2016). Five-year follow-up of bilateral epidural prefrontal cortical stimulation for treatment-resistant depression. *Brain Stimulation* [Internet], *9*(6), 897–904. [cited 2016 Dec 1]. Available from: http://www.sciencedirect.com/science/article/pii/S1935861X16301917.

Winslow, B. D., Christensen, M. B., Yang, W. K., Solzbacher, F., & Tresco, P. A. (2010). A comparison of the tissue response to chronically implanted Parylene-C-coated and uncoated planar silicon microelectrode arrays in rat cortex. *Biomaterials, 31*(35), 9163–9172.

Winter, J. O., Gokhale, M., Jensen, R. J., Cogan, S. F., & Rizzo, J. F., III. (2008). Tissue engineering applied to the retinal prosthesis: Neurotrophin-eluting polymeric hydrogel coatings. *Materials Science & Engineering, C: Materials for Biological Applications, 28*(3), 448–453.

Wise, K. D., & Najafi, K. (1991). Microfabrication techniques for integrated sensors and microsystems. *Science, 254*(5036), 1335–1342.

Wong, C. H., Birkett, J., Byth, K., Dexter, M., Somerville, E., Gill, D., et al. (2009). Risk factors for complications during intracranial electrode recording in presurgical evaluation of drug resistant partial epilepsy. *Acta Neurochirurgica, 151*(1), 37–50.

Woolford, T. J., Saeed, S. R., Boyd, P., Hartley, C., & Ramsden, R. T. (1995). Cochlear reimplantation. *Annals of Otology, Rhinology, and Laryngology Supplement, 166*, 449–453.

Wright, J., Macefield, V. G., van Schaik, A., & Tapson, J. C. (2016). A review of control strategies in closed-loop neuroprosthetic systems. *Frontiers in Neuroscience, 10*(312), 1–13.

Xie, Y., Martini, N., Hassler, C., Kirch, R. D., Stieglitz, T., Seifert, A., et al. (2014). In vivo monitoring of glial scar proliferation on chronically implanted neural electrodes by fiber optical coherence tomography. *Frontiers in Neuroengineering., 7*(34), 1–10.

Yoshida, K., Lewinsky, I., Nielsen, M., & Hylleberg, M. (2007). Implantation mechanics of tungsten microneedles into peripheral nerve trunks. *Medical & Biological Engineering & Computing, 45*(4), 413–420.

Zhong, Y., Yu, X., Gilbert, R., & Bellamkonda, R. V. (2001). Stabilizing electrode-host interfaces: A tissue engineering approach. *Journal of Rehabilitation Research and Development, 38*(6), 627–632.

Zrunek, M., & Burian, K. (1985). Risk of basilar membrane perforation by intracochlear electrodes. *Archives of Oto-Rhino-Laryngology, 242*(3), 295–299.

Chapter 7
Strategies to Improve Neural Electrode Performance

Katrina Guido, Ana Clavijo, Keren Zhu, Xinqian Ding, and Kaimin Ma

7.1 Introduction

The electrode is one of the most crucial components in neural interfacing for direct neural signal recording and stimulating. Neural interface technology has emerged as one of the main tools for treating various nervous system disorders and conditions in both research-related and clinical applications. Many individuals with conditions such as Parkinson's disease (Kim et al. 2018) and motor cortex injuries (Murphy et al. 2016) directly benefit from the use of implanted electrodes to help support and improve quality of life. Despite the importance of this technology, many challenges, which can broadly be categorized as those pertaining to electrode design/engineering and those pertaining to biological response, remain.

The traditional approach for neural interfacing requires the insertion of a metal-based electrode near the neuron(s) of interest to record or mimic/stimulate a neural signal. However, this method cannot completely replicate the original neural signal because the electrodes are seen as intruders to the body's immune system. This identification initiates the foreign body response (FBR). Ultimately, FBR results in loss of signal due to the formation of a glial sheath around the electrodes. As a result, the traditional recording electrode can only function properly for a few months up to 2 years (Seymour et al. 2017). However, FBR is only one of the many problems that are associated with classic electrodes. Aside from biological issues, technological failures (e.g., electrical failure of wiring, exposure of delicate electronics to the biological environment from the dissolution of protective coatings) also contribute to the short lifespan of current neural electrodes (Chen et al. 2017).

K. Guido (✉) · K. Zhu · X. Ding · K. Ma
Department of Electrical and Computer Engineering, The Ohio State University,
Columbus, OH, USA
e-mail: guido.26@osu.edu

A. Clavijo
Department of Neuroscience, The Ohio State University, Columbus, OH, USA

© Springer Nature Switzerland AG 2020
L. Guo (ed.), *Neural Interface Engineering*,
https://doi.org/10.1007/978-3-030-41854-0_7

New techniques from a variety of research fields such as material science, electrical engineering, and biology need to be integrated to overcome these challenges. These new techniques aim to create more stable and better performing chronically implanted electrodes for neural interfacing applications. Strategies currently under investigation include, but are not limited to, altering the electrode material, coating, size, shape, etc. (Guo 2016; Rousche and Normann 1998; Schwartz et al. 2006). The purpose of this chapter is to explore strategies to improve the performance of the classic neural electrode model.

7.2 Improvement Strategies

The following sections discuss insertion methods, physical design parameters, material choices, and novel technologies designed to improve electrode performance.

7.2.1 Implantation Methods

In order to record or stimulate with a neural electrode system, a brain-computer interface can be established by directly accessing the brain. This invasive procedure is referred to as a craniotomy. During this procedure, the many protective layers covering the brain are removed or reflected, revealing the target location for electrical recording or stimulation by the electrode system. First, the topmost layers of hair, skin, and connective tissue are removed. Then, the visible bone of the skull is cleaned and dried, and holes are drilled into the skull to attach the stereotactic frame with several screws (see Fig. 7.1). These screws are very important because if they are too loose, the skull could slip and prevent localization of the target, but if they are too tight, skull damage could occur, particularly at the cranial sutures (Zrinzo 2012).

The stereotactic frame is designed to immobilize the skull during the operation and to serve as a guide for the insertion of the electrode (Schuhmann Jr et al. 2018). In order to guide the electrode to the target structure, preoperative MRI imaging data is used in combination with the frame to localize the target and plan the trajectory of the probe (Zrinzo 2012). There are multiple types of stereotactic systems. The most common are arc frame stereotactic systems (see Fig. 7.2), mini-frame stereotactic systems, and frameless stereotactic systems. Arc frame systems are the most accurate at the target location, while frameless and mini-frame systems are most accurate at the point of insertion. A study conducted on deep brain stimulation (DBS) implants in the thalamus found that arc frame systems were more accurate than the other two at hitting the target location; however, patient functional outcomes were similar for all groups (Anon 2018; Zrinzo 2012).

Fig. 7.1 A craniotomy in progress with reflected meninges and exposed skull. (Reproduced with permission from Gordon et al. (2018))

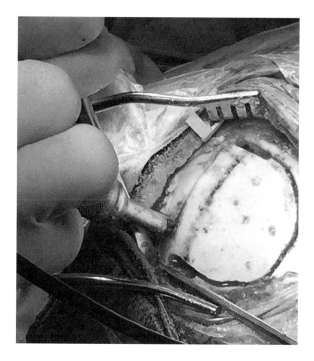

Following the placement of the frame, the bone is then drilled and removed. In some surgical protocols, the dura mater is also removed, while in other insertion protocols, the dura mater is left intact, and the electrode system pierces it to access the region of interest. The electrode is inserted into the stereotactic frame and calibrated to reach the target depth of the region of interest. Finally, once the device is inserted into the target area, the skull fragment is replaced, and the wound is sutured and closed (Gage et al. 2012a).

Due to the highly invasive nature of these operations and insertion procedures, there are several possible complications that may occur. Complications include damage to brain tissue or blood vessels when drilling or cutting through the bone and contaminating or damaging the electrode with excessive bleeding (Kook et al. 2016). Potential complications are also associated with the process of insertion. The trauma caused by insertion and surgery forms the primary wound, which can have consequences including vascular damage, increased intracranial pressure, and dimpling of the dura mater. Additionally, vascular damage may occur in two forms. The first is edema or hemorrhage caused by bleeding due to tears in the small vessels of the brain, and the second is loss of perfusion, or a lack of blood flow, resulting from the severing of upstream blood supply to a cell or small area. Age and hypertension are two important risk factors for vascular complications (Zrinzo 2012).

Another concern is the partial breakdown of the blood-brain barrier, which protects and isolates the central nervous system from the periphery of the body. For example, one study found that, after the electrode-insertion process, the biological molecule carbidopa, which normally is present only in the periphery, was present in

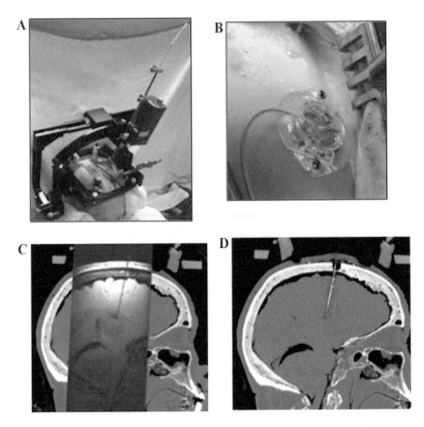

Fig. 7.2 Attachment of stereotactic frame from external (top) and internal (bottom) view. (Reproduced with permission from Edwards et al. (2018))

the central nervous system, which indicates a breakdown of the blood-brain barrier (Kozai et al. 2015). Finally, due to trauma and injury to the neural tissue, the neural inflammation process, FBR, is initiated through the activation of microglia. This immune response continues to be an obstacle for implanted electrodes throughout their lifetime (Kook et al. 2016; Kozai et al. 2015).

The severity of these complications and extent of damage are determined by the specific electrode system and insertion protocol used (Kozai et al. 2015). Several important factors must be considered when a surgical protocol is designed. The first is the force necessary to penetrate the brain tissue. The amount of force depends on variables such as the species or age of the subject, brain region, cell type and density, and the shape of the electrode tip. For example, a mature rat cortex requires approximately 1.5 mN of penetrating force to insert an electrode (Weltman et al. 2016). Speed is the second factor to consider. Previous studies have found that slower insertions require less force because brain stiffness increases at high insertion speeds (Weltman et al. 2016). However, previous research has also indicated slow insertion speeds as a potential cause of acidosis and increased vascular damage and bleeding after the insertion (Weltman et al. 2016). In classic neural electrode implantation surgeries, the speed of the insertion has ranged from approximately 10 μm/s to 1 mm/s (Weltman et al. 2016). Additionally, elastic tissue compression occurs when the electrode system pushes down and compresses the tissue instead of piercing the surface and entering the region of interest. This compression is associated with increased primary injury to the neural tissue (Rousche and Normann 1992). Finally, the specific anatomy and vasculature of the subject must be considered. Mapping of the major blood vessels on the surface of the brain is important to ensure the insertion process will not damage them (see Fig. 7.3). Hitting these

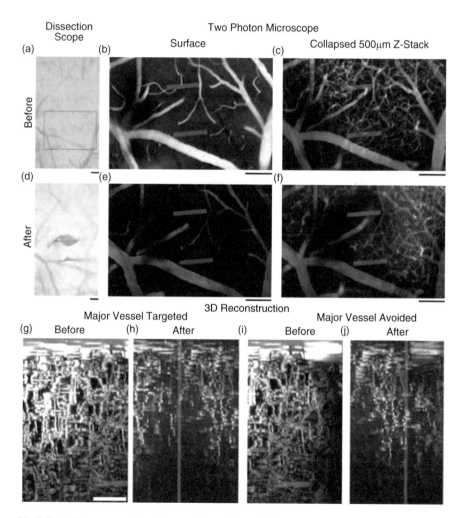

Fig. 7.3 A 3D reconstruction of the neural vasculature before and after when a major blood vessel is targeted (**g**), (**h**) or avoided (**i**), (**j**). (**a**)-(**c**), (**g**), (**i**) show cortical vasculature prior to insertion of the implant, (**d**) is following insertion, and (**e**), (**f**), (**h**), (**j**) is explantation of the implant. (Reproduced with permission from Kozai et al. (2015))

vessels results in increased bleeding and vascular damage. This mapping can be done with a variety of techniques, such as two-photon endoscopy (Kook et al. 2016). Additionally, with the use of MRI imaging, an approach to insertion through the crest of a gyrus is preferred to reduce the risk of damaging a blood vessel or allowing escape of cerebrospinal fluid (Zrinzo 2012). Due to the number of factors and variance between electrode systems, designing an insertion protocol optimal in every scenario is not realizable (Kook et al. 2016).

As novel electrode designs are developed, insertion protocols must evolve as well. One example of a new electrode design that has a dramatically different surgical insertion protocol is the flexible electrode. These electrodes have a distinct advantage during chronic recording or stimulation, because their flexibility causes less damage and a decreased immune response. These electrodes follow the same craniotomy procedure as the classic electrodes; however, during insertion, the flexible electrodes are unable to penetrate the neural tissue with enough force to pierce the surface and enter the region of interest. Instead, the flexible electrodes compress the tissue and result in elastic tissue compression leading to increased primary trauma (Weltman et al. 2016). Several design solutions can be implemented to temporarily increase stiffness or sharpness of the electrode during the insertion process to overcome this problem. Soluble, biocompatible coatings such as silk, gelatin, or biologically degrading polymers can coat the electrode and provide stiffness during insertion and then safely dissolve in the physiological environment once the system is in place (Weltman et al. 2016). Another strategy is the attachment of a shuttle system to deliver the electrode to the target location, which then safely disengages and can be removed after insertion. These probes can be attached with dissolvable adhesives such as polyethylene glycol (PEG) (Felix et al. 2013; Weltman et al. 2016) or simply hooked to the shuttle device (Luan et al. 2017). Additionally, the electrode system can be frozen with liquid nitrogen immediately prior to insertion. The liquid nitrogen increases electrode stiffness during insertion, but when the system warms up to physiological temperature, the electrode becomes flexible and soft again (Kook et al. 2016).

Another promising electrode design is the mesh electrode system. These electrodes can be delivered by syringe, so, though a craniotomy is still necessary to access the brain, the size of the hole is smaller, even though the electrode can cover a greater area compared with a classic electrode system. This smaller hole has many advantages, including less disruption of normal neurological function, shorter recovery time, and lower risk of surgical complications, though insertion is still a concern as with any invasive procedure (Schuhmann Jr et al. 2018). The syringe used to insert the electrodes can be mounted in the stereotactic frame to help locate the target area and guide the injection. As the electrode is injected with a minimal amount of fluid, the syringe is slowly removed from the brain. The injection must occur with the mesh in a specific orientation: the recording or stimulating region of the electrode exits the syringe first, followed by the input and output region (Schuhmann Jr et al. 2018). The insertion of these electrodes leads to a much lower immune response in the tissue after surgery. A recent study found that levels of GFAP, a protein that indicates the presence of glial cells and is found in elevated amounts during inflammation and immune response, are nearly normal after insertion of the mesh electrode system (Liu et al. 2015). In less dense areas of neural tissue, such as the ventricles, these electrodes spread out over a very wide area, while in areas of more densely packed cells, such as the hippocampus, they spread slightly following insertion but remain more compact (Wozny and Richardson 2015). The syringes used to insert the electrode depend on the individual electrode system specifications. For example, an inner diameter of a typical needle used ranges from 150 μm to 1.17 mm and an outer diameter from 250 μm to 1.5 mm (Schuhmann Jr et al. 2018). This illustrates that mesh electrodes must be very compressible and cannot adhere to the sides of the syringe, which is an important consideration for the design.

7.2.2   Electrode Fixation

Any probe or electrode system must be connected and able to communicate with the outside world. Commonly, the electrode probe is anchored or tethered to the skull, and then wires pass through this system to allow for communication with an external processing system (Wise et al. 2004). These probes can be attached to a cable that is fixed with PEG or dental cement to a stainless steel or titanium jewelry screw in the skull (Gage et al. 2012b; Kook et al. 2016). These systems exert force on the tissue in two places, at the insertion site on the surface of the brain and at the tip of the probe at the interior of the tissue (Kook et al. 2016). Other systems have attempted to more firmly fix the probe in the neural tissue with small barbs on the shaft of the probe known as anchors. These anchors are attached to the probe at the end closest to the tip and extend backward from the shaft of the probe at a 60-degree angle. This allows for easy insertion, but once placed, the anchors hook into the tissue and prevent movement. However, the anchor system can cause extensive damage to neural tissue in cases where the electrode needs to be removed at a later date for any reason, such as surgical complications or revisions and replacements of parts (Kook et al. 2016). Cable attachment has also been shown to be a major source of strain and tissue damage for electrodes (Kook et al. 2016).

A major flaw in this approach is that while the electrode is firmly held in place by the tethering system, the brain is not a static organ. It is free floating in cerebrospinal fluid (CSF) and is in constant motion due to normal physiological functions such as respiration and circulation. During respiration, brain tissue exerts between 80 and 130 μN of force on the probe system, which causes movement between 2 and 25 μm. Circulation occurs at a frequency of approximately 5–6 Hz, and these vascular forces exerted on the probe are between 14 and 25 μN. These forces cause movement between 1 and 3 μm (Weltman et al. 2016). This constant movement against the probe system causes extensive, chronic tissue damage, especially if the probe is rigid or has sharp edges (Weltman et al. 2016).

Several strategies have emerged to reduce movement and minimize the forces that cause damage to the tissue. Flexible probes allow the electrode itself to move

Fig. 7.4 Mouse subjects after mesh electrode is injected, with complete system of mesh electrode, ZIF connector, and PCB, wired to an external processing system. (Reproduced with permission from Schuhmann Jr et al. (2018))

with the brain, which reduces damage by absorbing the forces of respiration and circulation, reducing the strain by up to 94% (Weltman et al. 2016). Other strategies to reduce strain include a Teflon barrier between the probe and dura mater and a Goretex barrier between the dura and skull, S-shaped cables, sinusoidal-shaped shanks, and a dissolvable PEG matrix to attach traces of an electrode system to allow free, independent movement of individual traces once the system is implanted (Kook et al. 2016; Weltman et al. 2016). Mesh probes are essentially free floating once injected. On the edges of the polymer, there are I/O pads that attach to a zero insertion force (ZIF) connector and then to a custom printed circuit board (PCB) (see Fig. 7.4) (Fu et al. 2017; Zhou et al. 2017).

Studies have shown that untethered systems perform better than the classic tethered systems. In a direct comparison, untethered systems caused less immune response in the body, including less astrocyte and microglia response and higher neuronal density after 4 weeks (Thelin et al. 2011). The foreign body response is the cause of much of the signal degradation in classic systems as the body attacks and forms a scar around the electrode (Thelin et al. 2011). Additionally, tethered systems have a large cavity between the electrode and the neurons in the region of interest which is not present or is much smaller in untethered systems (Thelin et al. 2011). The mesh probes that are currently being developed have shown even less immune response and in fact have been shown in some studies of NeuN and DAPI stainings to be neurotrophic (Wozny and Richardson 2015).

Wireless systems are being developed as well. While wires are necessary for an electrode system's power supply and data transfer, they are also the primary source of infections, manufacturing cost, discomfort to patients, and electrode system failure. These systems are quite complex and have many challenges to maintain functionality and practicality. These requirements include a sufficiently high power supply for the system and neural stimulation, if necessary. This power system is usually based on electromagnetic radio frequency (RF) between two closely coupled coils, though infrared or acoustic energy can also be used (Wise et al. 2004). Additionally, a large enough range, such as several centimeters, and a high bandwidth, usually greater than 1 or 2 Mb/s, to allow fast speeds of data transfer to an external computing system are necessary. These devices must also be applicable for multiple types of neural electrode systems and biocompatible to withstand chronic implantation in the physiological environment (Wise et al. 2004). However, they show great promise to overcome some of the issues that are present in wired systems, both tethered and untethered.

7.2.3 Physical Design Parameters of the Electrode

Since most neural interfaces utilize invasive techniques to stimulate and record electrical signals, one of the biggest concerns is tissue response. Despite the mechanical trauma caused by the insertion process, the acute tissue response heals after a given amount of time (Schwartz et al. 2006). However, the chronic tissue response caused by FBR remains for the life of the implant and thus influences the performance of the whole interface (Schwartz et al. 2006; Stensaas and Stensaas 1967; Szarowski et al. 2003). One method to combat FBR is focusing on the physical design of the interface (e.g., the shape, size, and stiffness).

As mentioned above, FBR is the covering of the electrodes by activated microglia and reactive astrocytes (Schwartz et al. 2006; Marin 2010), so the design of the electrode tip is extremely important. Upon insertion into neural tissue, a structure will compress/stretch the surrounding tissue or tear the cells surrounding the structure. Stretching and compression, in physical terms, only cause minimal harm to the tissue and cells due to the abnormal leakage of certain ions. If the insertion of the structure breaks neuronal cells (changing their action potential capability), these cells die because of an imbalanced ionic state and thus the consumption of large amounts of energy due to a large influx of calcium (Ermak and Davies 2002). For example, a blunt electrode tip can break the blood vessel walls during the insertion process, causing bleeding within the brain. Microhemorrhages within the brain are very dangerous and can cause brain dysfunction (Walton 1978). However, inserting a sharp and smooth electrode probe can reduce harm to vessel walls caused by piercing or severing (Edell et al. 1992). At the same time, the surrounding tissue can still provide enough tension to prevent microhemorrhages.

Aside from sharpness, the shape of the tip is another engineering aspect to consider. The shape of electrode tips can be generally divided into two categories: conical and blade. A conical tip probe first opens a tiny hole, and upon insertion, the surrounding tissue is gently pushed away. However, due to the fibrous structure of neural tissue, the tip tears the immediate tissue area but leaves the adjacent tissue uncut (Walton 1978). As the probe is pushed further into the neural tissue, the adjacent tissue becomes tighter, requiring additional force to finish the insertion. This additional force spreads out into the adjacent tissue causing the tissue to compact. A

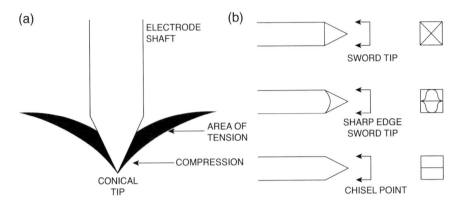

Fig. 7.5 (**a**) What may happen when a conical tip is inserted. (**b**) Three possible designs of blade-type probe tip. (Reproduced with permission from Edell et al. (1992))

microhemorrhage may occur as a result of this increased tension (Edell et al. 1992). Figure 7.5a shows what may happen when a conical tip is inserted.

Another basic type is the blade which includes microwires and the chisel-point shape (Mccreery et al. 1990). To create a sword/blade tip from a conical tip, a small point is added to the tip of the conical geometry. That small point is used to first cut tissue and then separate tissue under a triangular-shaped edge. For minimizing damage, the tip edges are designed to be sharp enough so that the least possible amount of force is necessary to cut the tissue. Other parts of tissue will provide enough tension to reduce the possibility of microhemorrhages (Edell et al. 1992). This design is theoretically preferred, since it requires cutting only a small amount of tissue with minimal applied force, but the shape is difficult to build (Edell et al. 1992). Comparing the separation of the tissue around the tip and the difficulty of implementation, the best compromise is a chisel-point blade shape. This geometry also cuts the minimum possible amount of tissue and retains enough tension within the residual tissue (Edell et al. 1992). Figure 7.5b shows three possible designs of the blade-type probe tip.

At the same time, concerning FBR, the electrode must also maintain the quality of the recorded signal. The signal can be affected by the scalp, skin, body noise, and other factors (Nicolas-Alonso and Gomez-Gil 2012). If the target neural area is small and the probe is placed with high precision, the recording results are more reliable with less noise (Schmidt et al. 1976; Kipke et al. 2008). One possible design is the Utah Intracortical Electrode Array (UIEA), shown in Fig. 7.6a, which is composed of 100 isolated silicon 1.2 mm long needles placed on a single thin substrate (Maynard et al. 1997; Normann et al. 1998). An integrated wireless electrode, shown in Fig. 7.7, based on the Utah Array, was also implemented, containing a signal processer, power coil, and capacitors (Kim et al. 2008a). Considering the compatibility and combination of each component's materials and integration, the technology offers a platform for the implantable microelectrode (Kim et al. 2008a). Another unique design that is based on the microelectrode and that allows the

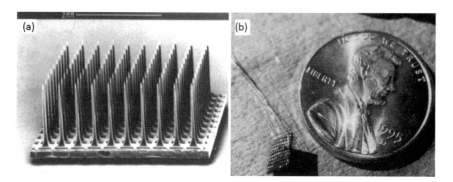

Fig. 7.6 (**a**) Utah Intracortical Electrode Array (**b**) UIEA next to a US penny (diameter of 1.9 cm). (Reproduced with permission from Normann et al. (1998))

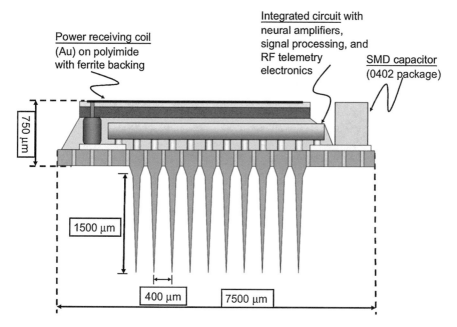

Fig. 7.7 Schematic of the integration and packaging concept of an integrated wireless neural interface. (Reproduced with permission from Kim et al. (2008a))

realization of a fully implantable system is called the floating multi-electrode array (FMA), shown in Fig. 7.8 (Musallam et al. 2007). The FMA uses anchors to make sure the electrode stays attached to the intended area of the brain.

The stiffness is also one of the factors influencing electrode performance and is a consideration in material choice, as discussed in the following section. Probe material can significantly affect the balance between recorded signal quality and probe biocompatibility (Schendel et al. 2014). Additionally, the focus of designing the electrode has moved from how to improve rigid and needle-like probes to

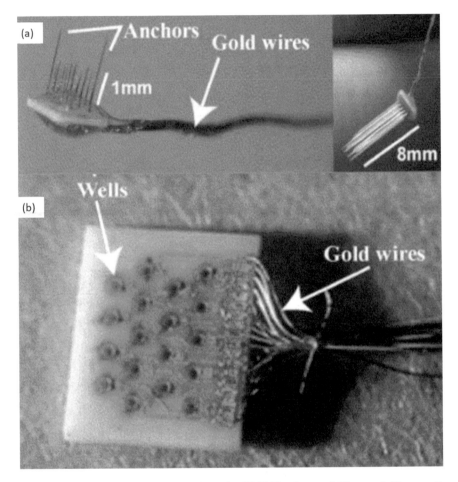

Fig. 7.8 (**a**) FMAs with short and long electrodes (**b**) FMA substrate, 1.95 mm × 2.45 mm, with attached 0.001 in. Parylene-C-insulated gold wires (Musallam et al. 2007)

considering the use of soft and polymeric material to design novel and ultrathin electrodes (Rivnay et al. 2017).

7.2.4   Material Composition of the Electrode

One of the key aspects/challenges for any neural electrode design is material choice. Classic electrodes are made with metallic conductors that enable the transition from electron flow in the electrode to ion flow in the tissue (Cogan 2008). However, conventional metal-based electrodes are rigid, stiff, and inefficient in stimulation mode. Introducing these components into the target subject can result in problems such as deformation of the existing cells, inaccurate signal recording, and poor long-term performance (Negi et al. 2012). For this reason, electrode material composition is an area with the potential to greatly improve neural interfacing.

From an electrical perspective, materials can be selected to achieve a low electrode impedance and high charge-injection capacity (CIC) (Cogan 2008). Low electrode impedance ensures a high signal-to-noise ratio (SNR) for electrodes, meaning better signal quality. Typically, the electrode impedance is measured at 1 kHz, and ideal SNR is around 5:1 (Cogan 2008). However, the challenge is to achieve high signal selectivity or the ability to activate a small population of neurons. Increased selectivity requires the electrode to be smaller in size, and this size decrease in turn increases the impedance. To overcome this trade-off, high-CIC materials are desired (Negi et al. 2012). High CIC, on one hand, lowers the potential required for stimulation and, on the other hand, allows a smaller electrode to achieve high current density within stimulation safety limits (Negi et al. 2012).

In literature, many materials have been studied to improve the electrode's electrical performance (i.e., increasing the SNR, lowering the impedance, and increasing the CIC). The most popular materials used for electrodes include but are not limited to Pt (Negi et al. 2010), PtIr alloys (Robblee et al. 1983), activated iridium oxide (Cogan 2008), thermal iridium oxide (Robblee et al. 1985), sputtered iridium oxide (Negi et al. 2010), titanium nitride (Weiland et al. 2002), tantalum oxide (Cogan 2008), poly(3,4-ethylenedioxythiophene) polystyrene sulfonate (PEDOT) (Cogan 2008), and carbon nanotubes (CNT) (Cogan 2008).

One material characteristic that reduces electrode impedance is texture (i.e., dimples in the surface resulting in increased electrochemical surface area) (Green et al. 2012). Among the aforementioned materials, CNT are well-known for having low electrical impedance and increased SNR (Shoval 2009). PEDOT-coated electrodes have been shown to have the highest CIC (15 mC/cm^2) (Cogan 2008). The good electrical performance (i.e., CIC) of PEDOT is primarily due to its intrinsically rough morphology (Aregueta-Robles et al. 2014). To further increase CIC, introducing additional dopants to the PEDOT increases the intrinsic surface roughness. As is seen in Fig. 7.9, molecularly large dopants such as poly(styrene sulfonate) (PSS) generally make the surface smoother, while small dopants such as paratoluene sulfonate (pTS) generally create a more nodular surface, in turn increasing the surface area and ultimately improving the electrical performance (Green et al. 2012).

From a biological perspective, materials can be selected to reduce tissue response and improve biocompatibility and chronic performance. Typically, two methods are used to achieve this effect: (1) applying a biomaterial organic coating to the electrode outer layer or integrating the electrode on a biocompatible substrate and (2) directly using surface topography modifications on the electrode itself to improve the electrode neuron cell adhesion.

Biomaterials such as biocompatible conductive polymers (CPs) can easily camouflage as part of the target environment, improving chronic performance and reducing inflammation (Seidlits et al. 2008). The biomaterial coating aims to mimic the cell environment while supporting the growth and survival of tissues near the

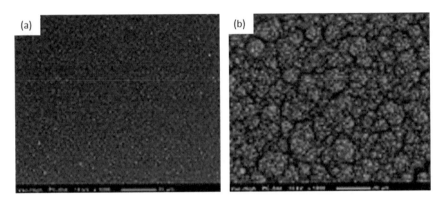

Fig. 7.9 (**a**) PEDOT doped with PSS (**b**) PEDOT doped with pTS. (Reproduced with permission from Aregueta-Robles et al. (2014))

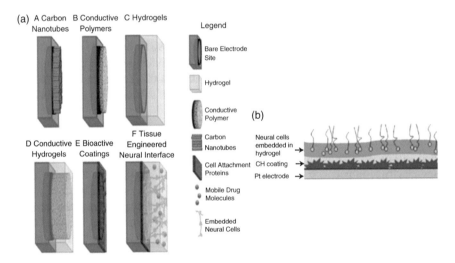

Fig. 7.10 (**a**) Schematic of example coating material for traditional electrode (Aregueta-Robles et al. 2014) (**b**) One example structure of the bio-material coating using conductive hydrogels (CH). (Reproduced with permission from Green et al. (2013))

implanted electrodes. Some of the commonly used coating materials and their structures are summarized in Fig. 7.10a. The CNT, as their name suggests, are physically small; thus, when infiltrating the tissue, CNT are less likely to be identified as a foreign body by the immune system (Bolin et al. 2009). Consequently, when coating the electrode with CNT, less scar tissue is formed around the electrode, improving the long-term performance of the electrode.

All of the rest of the coating materials shown in Fig. 7.10a have been shown to be cytocompatible with various neural cells. For example, various CPs have been shown to be compatible with cells including but not limited to neuroblastomas (Bolin et al. 2009), spiral ganglion cells (Evans et al. 2009), and neural cell lines

(Schmidt et al. 1997). Conductive hydrogels (CH) have shown the ability to support neural cells (i.e., PC12) for 12 days with no visible neuronal death using the layer schematic shown in Fig. 7.10b (Green et al. 2013). However, even though the CP coating can easily support the target neuron cells, those cells cannot directly attach to the electrode surface itself, resulting in increased background noise and undesirable electrical coupling of the neuron to the electrode. As a result, the electrode sensitivity and signal accuracy are reduced (Waser 2012).

To overcome adhesion issues, modifications can be directly applied to the conductive electrode material via surface topography modifications to increase neuronal adhesion on the bare electrode, further improving electrode sensitivity and signal accuracy (Hai et al. 2009). One of the most promising electrode materials that have shown a high level of bioacceptance is graphene (Schmidt 2012). The study in (Veliev et al. 2016) has shown healthy neuronal adhesion and growth on pristine graphene (uncoated) through the main stages of neuron growth. A 5-day in vitro (DIV) experiment in Ref. (Veliev et al. 2016) assessed cell growth from adhesion kinetics (DIV1, DIV2) to neuritogenesis (DIV4) and finally to first synaptic contacts formation (DIV5). Figure 7.11 shows the experimental results comparing

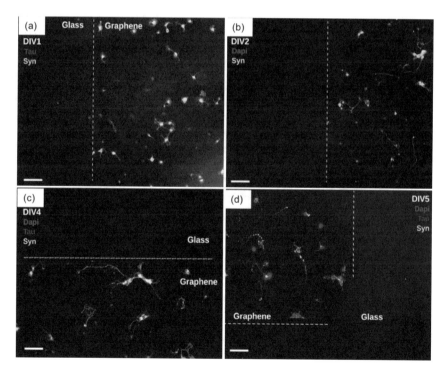

Fig. 7.11 Mouse embryonic hippocampal neurons' growth on bare single-layer graphene and uncoated glass. The dashed line separates the glass with bare graphene. Nucleus is stained with DAPI in blue, pre-synaptic vesicles are stained with synapsin in green, and actin and axon are stained with phalloïdin and tau, respectively, in red. (Reproduced with permission from Veliev et al. (2016))

neuronal growth on uncoated glass with high-quality, single-layer bare graphene through DIV1 to DIV5. It is seen that single-layer graphene has the ability to support healthy neuronal growth with a strong confinement to the material. In turn, this confirms graphene not only has a strong adhesive nature to the neuron but also is capable of supporting neurite branching, pre-synaptic vesicles agglomerating, and dendrites outgrowing for neurons to successfully connect with its neighboring cells (Veliev et al. 2016).

Moreover, combining the coating and surface topography modification techniques, coated electrodes have also shown promising results (Veliev et al. 2016; Li et al. 2011). Research in Ref. (Veliev et al. 2016) compares neuronal growth among poly-L-lysine (PLL)-coated glass, PLL-coated graphene, and bare graphene at different stages (Fig. 7.12). Results have shown that although all three materials demonstrated the ability to support healthy neuronal growth (Fig. 7.12a) with similar neuronal density at DIV5 (Fig. 7.12b), graphene coated with PLL had a significantly higher neuronal growth with respect to surface area (Fig. 7.12c, d). Along these lines, the graphene layer and PLL coating complement each other, creating a

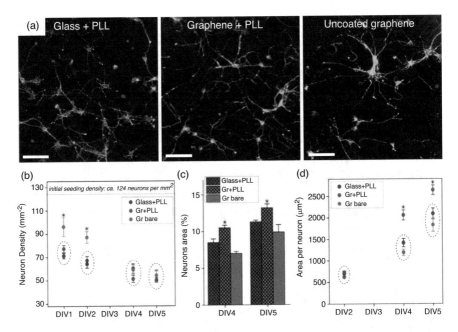

Fig. 7.12 (**a**) Mouse embryonic hippocampal neurons' growth at DIV5 on (left to right): PLL-coated glass (Glass+PLL), PLL-coated graphene (Graphene/Gr+PLL), and bare graphene (Uncoated graphene/Gr). Nucleus is stained with DAPI in blue, pre-synaptic vesicles are stained with synapsin in green, and actin and axon are stained with phalloïdin and tau, respectively, in red. (**b**) Neuron density vs. days of in vitro experiment for Glass+PLL, Gr+PLL, and Gr bare. (**c**) Comparison for percentage of area covered by neurons among Glass+PLL, Gr+PLL, and Gr bare at DIV4 and DIV5. (**d**) Area per neuron vs. days of in vitro experiment for Glass+PLL, Gr+PLL, and Gr bare. (Reproduced with permission from Veliev et al. (2016))

higher neuronal growth with respect to surface area after a few days of neuronal growth (Veliev et al. 2016).

From a mechanical perspective, choosing soft and flexible electrode materials can reduce cell deformation in the implantation area. Flexible electrodes using various substrate materials include but are not limited to silicone, polyimide, and parylene and can provide extreme flexibility (Seymour et al. 2017; Ghane-Motlagh and Sawan 2013). This flexibility minimizes cell deformation at the implantation site in the long run (Capogrosso et al. 2016). The elastic modulus was introduced to quantitatively characterize the stiffness of the electrode. For a classic Pt electrode, the elastic modulus is about 164 GPa which is significantly stiffer than that of the neural tissue (Weiland et al. 2006). As a comparison, most neural tissue has a modulus of less than 100 kPa (Lacour et al. 2010). Some of the most discussed materials that claim to have low stiffness can be summarized into two primary categories: CPs and hydrogels. One well-known example of using CPs as stated in (Minev et al. 2015) uses an elastomeric substrate made with transparent silicone integrated with stretchable gold interconnects and a compliant fluidic microchannel to achieve long-term implantation without significant cell deformation. A more detailed description of this system is discussed in the next section. Another example within the category of CPs uses PEDOT/PSS exhibiting an elastic modulus as low as 40 MPa (Green et al. 2012). Though still significantly stiffer than neural tissue, when compared to the conventional Pt electrode, this material reduces the modulus by 10^6 times.

Another example is using hydrogels to coat the electrode. Using a polyvinyl alcohol-heparin (PVA-Hep) hydrogel-coated Pt electrode, the modulus is reduced to around 100 kPa (Green et al. 2012). Combining the CPs and hydrogels, CH exhibit an elastic modulus generally between the two.

7.2.5 Novel Interface Technologies

Due to the recent development in materials science and nanotechnology, electronics are able to be fabricated on stretchable, flexible, and three-dimensional materials to form a non-planer and soft electrical functional unit (Minev et al. 2015; Kaltenbrunner et al. 2013; Kim et al. 2008b; Timko et al. 2009; Cherenack and Pieterson 2012; Martirosyan and Kalani 2011; Rousche et al. 2011). Three representative neural interfaces in this category are presented below.

E-dura The elastic and shear moduli of most electrodes are around GPa. These electrodes are considered "rigid" when compared to neural tissues with elastic and shear moduli in the range of 100–1500 kPa (Minev et al. 2015; Shanks and Konrad 2010). Rigid electrodes cause the acute and long-term tissue response following surgery. In order to resolve such a mismatch between the neural tissues and electrodes, a soft neural interface imitating the physical shape and mechanical behavior of the dura mater, called *E-dura*, was designed in 2015 (Kaltenbrunner et al. 2013; Hong et al. 2018).

E-dura is made up of one 120-μm-thick elastic silicone substrate, several 35-nm-thick stretchable gold interconnects, platinum-silicone-coated soft electrodes with a diameter of 300 μm, and one microfluidic channel of which the cross section is 100 μm by 50 μm. These are shown in Fig. 7.13a. Electrical excitation and collected bio-signals are transmitted through the interconnects and soft electrodes. Drugs are delivered through the fluidic microchannel. Due to the percolating network of microcracks in the gold interconnects and soft electrodes, e-dura has excellent stretchability and flexibility (Minev et al. 2015; Xie et al. 2015).

In e-dura control experiments, the long-term performance of e-dura was tested compared with stiff implants. Both e-dura and stiff implants were implanted into the subdural space of lumbosacral segments in healthy rats. In order to insert e-dura into the subdural environment, Ivan et al. developed a vertebral orthosis to support e-dura which is shown in Fig. 7.13b. A 25-mm-thick polyimide film was chosen as the comparing stiff implants. The sham-operated rats were used as the control group. After 6 weeks of implantation, three groups were tested together to give the motor performance. From the high-resolution kinematic recordings of the rats' movements when they were tested to cross a horizontal ladder, Fig. 7.13c shows that the motion behavior of rats with soft implants was indistinguishable from that of the sham-operated rats. Meanwhile, Fig. 7.13d illustrates that those rats inserted with stiff implants presented significant motor deficits and their spinal cord structure was damaged due to tissue response.

Another experiment was conducted on a model of a spinal cord to compare the physical damage of the rigid implant and e-dura to the spinal tissues. They used a hydrogel core to simulate spinal tissue and a silicone tube to mimic the dura mater. From Fig. 7.13e, the rigid implant flattened the raised circular structure. In contrast, e-dura kept the original shape of the model. When they bent the model with the implants, the rigid implant formed wrinkles and caused compression along the "spinal tissues." However, e-dura did not change the smoothness of the model. Compared to stiff interfaces, e-dura can be embedded into dura mater without pressing the tissue and damaging the neurons. A large number of experiments have shown that e-dura can be functional and biologically friendly during a long-term implantation in rats.

Syringe injectable mesh electronics Conventional implantable probes do not exhibit long-term compatibility due to the chronic immune response, FBR (Liu et al. 2015). Even flexible probes made of polyimide with a thickness of 10–20 μm are a few orders of magnitude stiffer than brain tissue (Hong et al. 2018). To improve the long-term performance, Harvard University's research team developed the "tissue-like" neural probes called mesh electronics in 2015 (Hong et al. 2018; Xie et al. 2015; Liu et al. 2015). Specifically, mesh electronics have a bending stiffness similar to that of brain tissues that is 4–6 orders of magnitude smaller than conventional implantable probes. One mesh width is about 10 μm, and the whole structure can be several millimeters wide according to the corresponding brain region, which

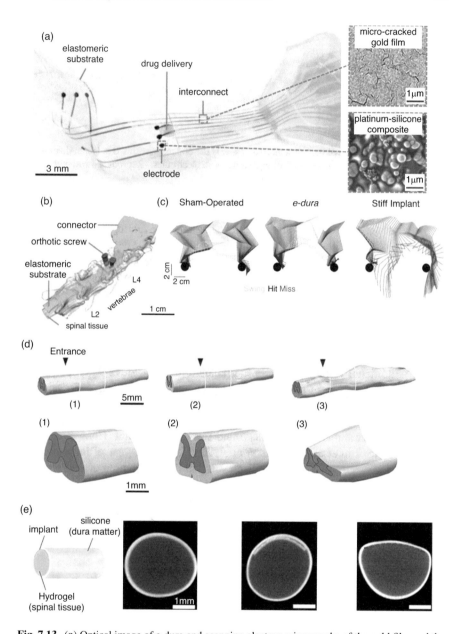

Fig. 7.13 (**a**) Optical image of e-dura and scanning electron micrographs of the gold film and the platinum-silicone composite. (**b**) 3D microcomputed tomography scans of the e-dura implanted into the rats' spinal cord. (**c**) Hindlimb kinematics during horizontal ladder walking 6 weeks after implantation. (**d**) 3D spinal cord reconstructions 6 weeks after implantation. The first picture in (**d**) shows the spinal cord of the sham-operated rat. The second and third ones belong to rat inserted with e-dura and stiff implant. (**e**) The scanning picture of the spinal cord model. The red line shown in the picture is the stiff implant (25 mm thick), not visualized because of scanner resolution. (Reproduced with permission from (Minev et al. 2015))

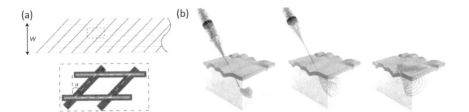

Fig. 7.14 (**a**) Schematic of the mesh electronics, w is the one mesh width, α is the angle of deviation from rectangular. (**b**) Schematics of injectable mesh electronics. (Reproduced with permission from Liu et al. (2015))

is shown in Fig. 7.14a (Zhou et al. 2017). Different from previous surgical implants, mesh electrodes can be injected directly through a needle. Giving positive pressure, the mesh electrodes that were drawn into the needle could be delivered to the target region (Hong et al. 2018). Mesh electronics are able to handle the issues from the mismatches between brain tissue and implants in size, mechanism, and topology due to the following reasons: Mesh electronics (1) has a nanometer to micrometer scale structure which is the same as that of the neural tissue, (2) features similar mechanical properties to those of brain tissues, and (3) can be accurately delivered to the target brain region through one syringe, which means acute damage caused during insertion is decreased as shown in Fig. 7.14b and discussed in a previous section. Additionally, the topology of mesh electronics provides a 90% open space, which allows diffusion of molecules through membranes and interpenetration by neurons. Also, previous experiments illustrated that neuronal soma and axons can interpenetrate the open space of the 3D mesh electronics several weeks after injection, so that no immune response was found around the implant (Hong et al. 2018; Xie et al. 2015; Liu et al. 2015; Zhou et al. 2017). Up to 1 year of immunohistochemical studies after implantation displayed that the distribution of neuronal cell bodies and axons at the reticular interface was nearly the same as the baseline of natural tissues for 4–6 weeks and kept this natural distribution for at least a year (Hong et al. 2018). Compared with stiff electrodes, mesh electronics show great flexibility and minimal immune response which provides better long-term performance after implantation.

Living electrodes Commercial implants are usually fabricated by biologically inert conductive materials (Geddes and Roeder 2003; Ward et al. 2009; Rutten 2002). The stimulating process involves the flow of electrons from the device to produce depolarization and action potentials in neural tissues. The electric field generated around the implant affects the local biological environment, which may worsen the scar tissue response that wraps and electrically segregates the implant. As a result, the efficiency of the implant is reduced over time (Onnela et al. 2019).

To improve the long-term performance of implants, "living electrodes" was first proposed in the 1980s by Ochiai et al. (1980). In this technique, living neurons were integrated with the electrode to produce a natural physiological tissue activation

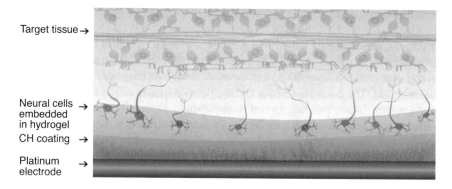

Fig. 7.15 Schematic of layered "living electrode" design. (Reproduced with permission from Goding et al. (2017))

pattern. Ochiai et al. used live blue-green algae embedded in alginate gel to form a photo-converter (Mahoney and Anseth 2006). In subsequent studies, the biological electrode tested was stable; however, the power conversion efficiency was as low as 0.1%. After that, many experiments had been carried out, but due to the inappropriate environment produced by electrodes and exposure to DC voltages, neural tissues could not survive long after implantation (Mahoney and Anseth 2006). They all ended in failure.

Recently, Green et al. developed a hybrid material which combined PEDOT and PVA hydrogel called the CH (Green et al. 2012). In the CH, the CP was produced in a doping hydrogel mesh to exhibit electrical activity comparable to conventional CP materials. The CH had similar elastic and shear moduli to neural tissues and was physically stable compared to classic medical electrodes. Based on the CH, a layer approach was put forward to solve the problems of the living electrode technique which is shown in Fig. 7.15. When encapsulated in an overlying hydrogel layer above the CH coating, cells were isolated from high DC voltage and the solvents that were not suitable for neural tissues. As the overlying hydrogel was degradable, neurons were able to grow and differentiate after implantation. After a period of time, with the degradation of hydrogel, cells would generate their own structure and form an effectively integrated electrode interface that would work for a long time without the FBR immune response (Goding et al. 2017).

7.3 Recording Versus Stimulation Electrodes

Up to this point, this chapter has discussed strategies to improve the electrode performance with only a minor mention of the differences between stimulation and recording electrodes. As many of the strategies mentioned (i.e., insertion methods) apply to both, this does not detract from what has been presented. However, differences and motivations regarding the challenges to overcome do exist.

With regard to decreasing the impedance of stimulation electrodes, as briefly mentioned in Sect. 7.3.1, the motivation is to increase the CIC, whereas in recording electrodes, the decreased impedance is necessary for an increased SNR. For stimulation electrodes, the increased CIC in turn decreases the voltage required to elicit an action potential (Cogan 2008) and therefore decreases the power consumption of the electrode. The decreased power consumption increases efficiency of the stimulation system. When building an end-to-end system, especially one leveraging batteries (e.g., deep brain stimulators), increased efficiency equates to fewer battery-replacement surgeries and therefore a decreased risk to the patient as well as a lower lifetime monetary cost. A decrease in the electrode surface voltage also decreases heating and subsequent irreversible faradaic reactions, which are harmful to the surrounding cells.

For recording electrodes, a decreased impedance at the electrode-tissue interface gives a decreased mismatch or resistance to the flow of charges between the electrode and its environment. A smaller mismatch leads to fewer ionic vibrations/collisions at the electrode surface, which leads to less thermal noise obscuring the intended signal.

Aside from electrode material impedance, the increased efficiency of stimulation electrodes is also achieved via field shaping (Howell and Grill 2015). Selectively placing electrode contacts to direct the electric field toward a specific target as opposed to placing contacts to radiate equally in all directions reduces the overall power supplied to an electrode required to elicit an action potential (Fig. 7.16). The increase in SNR for recording electrodes is also achieved via the use of differential recording (i.e., using multipolar recording) where the reference (negative input for the differential amplifier of the recording unit) for the measurement is moved closer to the recording site (Howell and Grill 2015).

Regarding tissue damage, FBR is an issue for both stimulation and recording. However, stimulation electrodes are able to overcome this challenge (albeit to a limited extent) by adjusting stimulation parameters (e.g., frequency, duration, and amplitude,) (Cogan 2008). The adjustment of stimulation parameters can also overcome issues with electrode migration, whereas recording electrodes have no corresponding adjustment methods (Howell and Grill 2015). Stimulation electrodes also have the added issue of stimulation-induced tissue damage. Issues include heat generation and irreversible charge transfer, which can be overcome by careful material choice, as discussed in the previous sections, and stimulation parameters, the details of which are situation dependent (Cogan 2008).

7.4 Conclusion

In spite of the frequent chronic failures of classic electrode technologies regarding engineering design and biological response, many strategies designed to ameliorate the various failure modes associated with the classic electrode model are in development. Strategies range from altering the geometrical properties of the electrode to

(a)   (b)   (c)

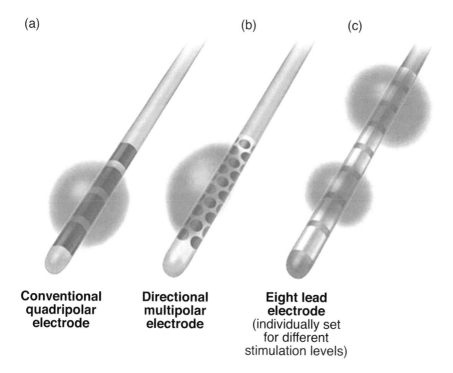

Conventional quadripolar electrode   **Directional multipolar electrode**   **Eight lead electrode** (individually set for different stimulation levels)

Fig. 7.16 By selectively locating the electrode contacts on a given lead, the electric field used to stimulate the surrounding neurons can be shaped so as to selectively stimulate a specific area or to overcome the formation of an irregular glial sheath. (Reproduced with permission from Hickey and Stacy (2016))

increasing electrode flexibility to using biocompatible coatings. This chapter leads the reader to a better understanding of how these various cutting-edge strategies can be employed to combat the classic electrode failure modes.

References

Anon. *Elekta AB* Leksell Stereotactic System®. 30 Dec 2018. https://www.elekta.com/neurosurgery/leksell-stereotactic-system/

Aregueta-Robles, U. A., Woolley, A. J., Poole-Warren, L. A., Lovell, N. H., & Green, R. A. (2014). Organic electrode coatings for next-generation neural interfaces. *Frontiers in Neuroengineering, 7*, 15.

Bolin, M. H., Svennersten, K., Wang, X., Chronakis, I. S., Richter-Dahlfors, A., Jager, E. W., & Berggren, M. (2009). Nano-fiber scaffold electrodes based on PEDOT for cell stimulation. *Sensors and Actuators B: Chemical, 142*(2), 451–456.

Capogrosso, M., Milekovic, T., Borton, D., Wagner, F., Moraud, E. M., Mignardot, J.-B., Buse, N., Gandar, J., Barraud, Q., Xing, D., Rey, E., Duis, S., Jianzhong, Y., Ko, W. K. D., Li, Q., Detemple, P., Denison, T., Micera, S., Bezard, E., Bloch, J., & Courtine, G. (2016). A brain–

spine interface alleviating gait deficits after spinal cord injury in primates. *Nature, 539*(7628), 284–288.

Chen, R., Canales, A., & Anikeeva, P. (2017). Neural recording and modulation technologies. *Nature Materials, 2*(2), 16093.

Cherenack, K., & Pieterson, L. V. (2012). Smart textiles: Challenges and opportunities. *Journal of Applied Physics, 112*(9). 091301.

Cogan, S. F. (2008). Neural stimulation and recording electrodes. *Annual Review of Biomedical Engineering, 10*(1), 275–309.

Edell, D., Toi, V., Mcneil, V., & Clark, L. (1992). Factors influencing the biocompatibility of insertable silicon microshafts in cerebral cortex. *IEEE Transactions on Biomedical Engineering, 39*(6), 635–643.

Edwards, C. A., Rusheen, A. E., Oh, Y., Paek, S. B., Jacobs, J., Lee, K. H., Dennis, K. D., Bennet, K. E., Kouzani, A. Z., Lee, K. H., & Goerss, S. J. (2018). A novel re-attachable stereotactic frame for MRI-guided neuronavigation and its validation in a large animal and human cadaver model. *Journal of neural engineering, 15*(6), 066003.

Ermak, G., & Davies, K. J. (2002). Calcium and oxidative stress: From cell signaling to cell death. *Molecular Immunology, 38*(10), 713–721.

Evans, A. J., Thompson, B. C., Wallace, G. G., Millard, R., Oleary, S. J., Clark, G. M., Shepherd, R. K., & Richardson, R. T. (2009). Promoting neurite outgrowth from spiral ganglion neuron explants using polypyrrole/BDNF-coated electrodes. *Journal of Biomedical Materials Research Part A, 91A*(1), 241–250.

Felix, S. H., Shah, K. G., Tolosa, V. M., Sheth, H. J., Tooker, A. C., Delima, T. L., Jadhav, S. P., Frank, L. M., & Pannu, S. S. (2013). Insertion of flexible neural probes using rigid stiffeners attached with biodissolvable adhesive. *Journal of Visualized Experiments*, (79), e50609.

Fu, T.-M., Hong, G., Viveros, R. D., Zhou, T., & Lieber, C. M. (2017). Highly scalable multichannel mesh electronics for stable chronic brain electrophysiology. *Proceedings of the National Academy of Sciences, 114*(47), 10046–10055.

Gage, G. J., Stoetzner, C. R., Richner, T., Brodnick, S. K., Williams, J. C., & Kipke, D. R. (2012a). Urgical implantation of chronic neural electrodes for recording single unit activity and electrocorticographic signals. *Journal of Visualized Experiments*, (60).

Gage, J., Stoetzner, C. R., Richner, T., Brodnick, S. K., Williams, J. C., & Kipke, D. R. (2012b). Surgical implantation of chronic neural electrodes for recording single unit activity and electrocorticographic signals. *Journal of Visualized Experiments*, (60), 3565.

Geddes, L. A., & Roeder, R. (2003). Criteria for the selection of materials for implanted electrodes. *Annals of Biomedical Engineering, 31*, 879–890.

Ghane-Motlagh, B., & Sawan, M. (2013). Design and implementation challenges of microelectrode arrays: A review. *Materials Sciences and Applications, 4*(8), 483–495.

Goding, J., Gilmour, A., Robles, U. A., Poole-Warren, L., Lovell, N., Martens, P., & Green, R. (2017). A living electrode construct for incorporation of cells into bionic devices. *MRS Communications, 7*(3), 487–495.

Gordon, W. E., Michael Ii, L. M. & VanLandingham, M. A. (2018). Exposure of Dural Venous Sinuses: A Review of Techniques and Description of a Single-piece Troughed Craniotomy. *Cureus, 10*(2), 2184.

Green, R. A., Hassarati, R. T., Bouchinet, L., Lee, C. S., Cheong, G. L., Yu, J. F., Dodds, C. W., Suaning, G. J., Poole-Warren, L. A., & Lovell, N. H. (2012). Substrate dependent stability of conducting polymer coatings on medical electrodes. *Biomaterials, 33*(25), 5875–5886.

Green, R. A., Lim, K. S., Henderson, W. C., Hassarati, R. T., Martens, P. J., Lovell, N. H., & Poole-Warren, L. A.. (2013). Living electrodes: Tissue engineering the neural interface. *35th Annual International Conference of the IEEE Engineering in Medicine and Biology Society (EMBC)*.

Guo, L. (2016). The pursuit of chronically reliable neural interfaces: A materials perspective. *Frontiers in Neuroscience, 10*, 599.

Hai, A., Dormann, A., Shappir, J., Yitzchaik, S., Bartic, C., Borghs, G., Langedijk, J. P. M., & Spira, M. E. (2009). Spine-shaped gold protrusions improve the adherence and electrical coupling of neurons with the surface of micro-electronic devices. *Journal of the Royal Society Interface, 6*(41), 1153–1165.

Hickey, P., & Stacy, M. (2016). Deep brain stimulation: A paradigm shifting approach to treat parkinson's disease. *Fronteirs in Neuroscience, 10,* 173.

Hong, G., Yang, X., Zhou, T., & Lieber, C. M. (2018). Mesh electronics: A new paradigm for tissue-like brain probes. *Current Opinion in Neurobiology, 50,* 33–41.

Howell, B., & Grill, W. (2015). Design of electrodes for stimulation and recording. In *Implantable neuroprostheses for restoring function* (pp. 59–93). Woodhead Publishing: Cambridge, UK.

Kaltenbrunner, M., Sekitani, T., Reeder, J., Yokota, T., Kuribara, K., Tokuhara, T., Drack, M., Schwödiauer, R., Graz, I., Bauer-Gogonea, S., Bauer, S., & Someya, T. (2013). An ultra-lightweight design for imperceptible plastic electronics. *Nature, 499*(7459), 458–463.

Kim, S., Bhandari, R., Klein, M., Negi, S., Rieth, L., Tathireddy, P., Toepper, M., Oppermann, H., & Solzbacher, F. (2008a). Integrated wireless neural interface based on the Utah electrode array. *Biomedical Microdevices, 11*(2), 453–466.

Kim, D.-H., Ahn, J.-H., Choi, W. M., Kim, H.-S., Kim, T.-H., Song, J., Huang, Y. Y., Liu, Z., Lu, C., & Rogers, J. A. (2008b). Stretchable and foldable silicon integrated circuits. *Science, 320*(5875), 507–511.

Kim, J. H., Lee, G. H., Kim, S., Chung, H. W., Lee, J. H., Lee, S. M., Kang, C. Y., & Lee, S. H. (2018). Flexible deep brain neural probe for localized stimulation and detection with metal guide. *Biosensors and Bioelectronics, 117,* 436–443.

Kipke, D. R., Shain, W., Buzsaki, G., Fetz, E., Henderson, J. M., Hetke, J. F., & Schalk, G. (2008). Advanced neurotechnologies for chronic neural interfaces: New horizons and clinical opportunities. *Journal of Neuroscience, 28*(46), 11830–11838.

Kook, G., Lee, S., Lee, H., Cho, I.-J., & Lee, H. (2016). Neural probes for chronic applications. *Micromachines, 7*(10), 179.

Kozai, T. D. Y., Jaquins-Gerstl, A. S., Vazquez, A. L., Michael, A. C., & Cui, X. T. (2015). Brain tissue responses to neural implants impact signal sensitivity and intervention strategies. *ACS Chemical Neuroscience, 6*(1), 48–67.

Lacour, S. P., Benmerah, S., Tarte, E., Fitzgerald, J., Serra, J., Mcmahon, S., Fawcett, J., Graudejus, O., Yu, Z., & Morrison, B. (2010). Flexible and stretchable micro-electrodes for in vitro and in vivo neural interfaces. *Medical & Biological Engineering & Computing, 48*(10), 945–954.

Li, N., Zhang, X., Song, Q., Su, R., Zhang, Q., Kong, T., Liu, L., Jin, G., Tang, M., & Cheng, G. (2011). The promotion of neurite sprouting and outgrowth of mouse hippocampal cells in culture by graphene substrates. *Biomaterials, 32*(35), 9374–9382.

Liu, J., Fu, T.-M., Cheng, Z., Hong, G., Zhou, T., Jin, L., Duvvuri, M., Jiang, Z., Kruskal, P., Xie, C., Suo, Z., Fang, Y., & Lieber, C. M. (2015). Syringe injectable electronics. *Nature Nanotechnology, 10*(7), 629–636.

Luan, L., et al. (2017). Ultraflexible nanoelectronic probes form reliable, glial scar-free neural integration. *Science Advances, 3*(2), e1601966.

Mahoney, M. J., & Anseth, K. S. (2006). Three-dimensional growth and function of neural tissue in degradable polyethylene glycol hydrogels. *Biomaterials, 27*(10), 2265–2274.

Marin, C. (2010). Biocompatibility of intracortical microelectrodes: Current status and future prospects. *Frontier in Neuroengineering, 3,* 8.

Martirosyan, N., & Kalani, M. Y. S. (2011). Epidermal electronics. *World Neurosurgery, 76*(6), 485–486.

Maynard, E. M., Nordhausen, C. T., & Normann, R. A. (1997). The Utah Intracortical Electrode Array: A recording structure for potential brain-computer interfaces. *Electroencephalography and Clinical Neurophysiology, 102*(3), 228–239.

Mccreery, D., Agnew, W., Yuen, T., & Bullara, L. (1990). Charge density and charge per phase as cofactors in neural injury induced by electrical stimulation. *IEEE Transactions on Biomedical Engineering, 37*(10), 996–1001.

Minev, I. R., Musienko, P., Hirsch, A., Barraud, Q., Wenger, N., Moraud, E. M., & Lacour, S. P. (2015). Electronic dura mater for long-term multimodal neural interfaces. *Science, 347*(6218), 159–163.

Murphy, M. D., Guggenmos, D. J., Bundy, D. T., & Nudo, R. J. (2016). Current challenges facing the translation of brain computer interfaces from preclinical trials to use in human patients. *Frontiers in Cellular Neuroscience, 9,* 497.

Musallam, S., Bak, M. J., Troyk, P. R., & Andersen, R. A. (2007). A floating metal microelectrode array for chronic implantation. *Journal of Neuroscience Methods, 160*(1), 122–127.

Negi, S., Bhandari, R., Rieth, L., & Solzbacher, F. (2010). In vitro comparison of sputtered iridium oxide and platinum-coated neural implantable microelectrodes arrays. *Biomedical Materials, 5*(1), 15007.

Negi, S., Bhandari, R., & Solzbacher, F.. (2012). A novel technique for increasing charge injection capacity of neural electrodes for efficacious and safe neural stimulation. *2012 Annual International Conference of the IEEE Engineering in Medicine and Biology Society.*

Nicolas-Alonso, L. F., & Gomez-Gil, J. (2012). Brain computer interfaces, a review. *Sensors, 12*(2), 1211–1279.

Normann, P., Rousche, J., & R. A. (1998). Chronic recording capability of the Utah Intracortical Electrode Array in cat sensory cortex. *Journal of Neuroscience Methods, 82*(1), 1–15.

Ochiai, H., Shibata, H., Sawa, Y., & Katoh, T. (1980). 'Living electrode' as a long-lived photo-converter for biophotolysis of water. *Proceedings of the National Academy of Sciences, 77*(5), 2442–2444.

Onnela, N., Takeshita, H., Kaiho, Y., Kojima, T., Kobayashi, R., Tanaka, T., & Hyttinen, J. (2019). Comparison of electrode materials for the use of retinal prosthesis. *Scientific Publication Data, 2*(2), 83–97.

Rivnay, J., Wang, H., Fenno, L., Deisseroth, K., & Malliaras, G. G. (2017). Next-generation probes, particles, and proteins for neural interfacing. *Science Advances, 3*(6), e1601649.

Robblee, L., Mchardy, J., Agnew, W., & Bullara, L. (1983). Electrical stimulation with Pt electrodes. VII. Dissolution of Pt electrodes during electrical stimulation of the cat cerebral cortex. *Journal of Neuroscience Methods, 9*(4), 301–308.

Robblee, L. S., Mangaudis, M. J., Lasinsky, E. D., Kimball, A. G., & Brummer, S. B. (1985). Charge injection properties of thermally-prepared iridium oxide films. *MRS Proceedings, 55.*

Rousche, P. J., & Normann, R. A. (1992). A method for pneumatically inserting an array of penetrating electrodes into cortical tissue. *Annals of Biomedical Engineering, 20*(4), 413–422.

Rousche, P. J., & Normann, R. A. (1998). Chronic recording capability of the Utah Intracortical Electrode Array in cat sensory cortex. *Journal of Neuroscience Methods, 82*(1), 1–15.

Rousche, P., Pellinen, D., Pivin, D., Williams, J., Vetter, R., & Kipke, D. (2011). Flexible polyimide-based intracortical electrode arrays with bioactive capability. *IEEE Transactions on Biomedical Engineering, 48*(3), 361–371.

Rutten, W. L. C. (2002). Selective electrical interfaces with the nervous system. *Annual Review of Biomedical Engineering, 4,* 407–452.

Schendel, A. A., Eliceiri, K. W., & Williams, J. C. (2014). Advanced materials for neural surface electrodes. *Current Opinion in Solid State and Materials Science, 18*(6), 301–307.

Schmidt, C. (2012). Bioelectronics: The bionic material. *Nature, 483*(7389), S37.

Schmidt, E., Bak, M., & Mcintosh, J. (1976). Long-term chronic recording from cortical neurons. *Experimental Neurology, 52*(3), 496–506.

Schmidt, C. E., Shastri, V. R., Vacanti, J. P., & Langer, R. (1997). Stimulation of neurite outgrowth using an electrically conducting polymer. *Proceedings of the National Academy of Sciences, 94*(17), 8948–8953.

Schuhmann, T. G., Jr., Zhou, T., Hong, G., Lee, J. M., Fu, T.-M., Park, H.-G., & Lieber, C. M. (2018). Syringe-injectable mesh electronics for stable chronic rodent electrophysiology. *Journal of Visualized Experiments*, (137), e58003.

Schwartz, A. B., Cui, X. T., Weber, D. J., & Moran, D. W. (2006). Brain-controlled interfaces: Movement restoration with neural prosthetics. *Neuron, 52*(1), 205–220.

Seidlits, S. K., Lee, J. Y., & Schmidt, C. E. (2008). Nanostructured scaffolds for neural applications. *Nanomedicine, 3*(2), 183–199.

Seymour, J. P., Wu, F., Wise, K. D., & Yoon, E. (2017). State-of-the-art MEMS and microsystem tools for brain research. *Microsystems & Nanoengineering, 3*, 16066.

Shanks, T., & Konrad, P. (2010). Implantable brain computer interface: Challenges to neurotechnology translation. *Neurobiology of Disease, 38*(3), 369–375.

Shoval, A. (2009). Carbon nanotube electrodes for effective interfacing with retinal tissue. *Frontiers in Neuroengineering, 2*, 4.

Stensaas, S. S., & Stensaas, L. J. (1967). The reaction of the cerebral cortex to chronically implanted plastic needles. *Acta Neuropathologica, 35*(3), 187–203.

Szarowski, D., Andersen, M., Retterer, S., Spence, A., Isaacson, M., Craighead, H., Turner, J., & Shain, W. (2003). Brain responses to micro-machined silicon devices. *Brain Research, 983*(1–2), 23–35.

Thelin, J., Jörntell, H., Psouni, E., Garwicz, M., Schouenborg, J., Danielsen, N., & Linsmeier, C. E. (2011). Implant size and fixation mode strongly influence tissue reactions in the CNS. *PLoS One, 6*(1), e16267.

Timko, B. P., Cohen-Karni, T., Yu, G., Qing, Q., Tian, B., & Lieber, C. M. (2009). Electrical recording from hearts with flexible nanowire device arrays. *Nano Letters, 9*(2), 914–918.

Veliev, F., Briançon-Marjollet, A., Bouchiat, V., & Delacour, C. (2016). Impact of crystalline quality on neuronal affinity of pristine graphene. *Biomaterials, 86*, 33–41.

Wozny, R. M., & Richardson, T. A. (2015). The future of neural recording devices. *Neurosurgery, 77*(6), N17–N19.

Walton, J. N. (1978). The fine structure of the nervous system — The neurons and supporting cells. *Journal of the Neurological Sciences, 35*(1), 168.

Ward, M. P., Rajdev, P., Ellison, C., & Irazoqui, P. P. (2009). Toward a comparison of microelectrodes for acute and chronic recordings. *Brain Research, 1282*, 183–200.

Waser, R. (2012). *Nanoelectronics and information technology: Advanced electronic materials and novel devices.* Weinheim: Wiley-VCH.

Weiland, J. D., Anderson, D. J., & Humanyun, M. S. (2002). In vitro electrical properties for iridium oxide versus titanium nitride stimulating electrodes. *IEEE Transactions on Biomedical Engineering, 49*(12), 1574–1579.

Weiland, R., Lupton, D. F., Fischer, B., Merker, J., Scheckenbach, C., & Witte, J. (2006). High-temperature mechanical properties of the platinum group metals. *Platinum Metals Review, 50*(4), 158–170.

Weltman, A., Yoo, J., & Meng, E. (2016). Flexible, penetrating brain probes enabled by advances in polymer microfabrication. *Micromachines, 7*(10), 180.

Wise, K., Anderson, D., Hetke, J., Kipke, D., & Najafi, K. (2004). Ireless implantable microsystems: High-density electronic interfaces to the nervous system. *Proceedings of the IEEE, 92*(1), 76–97.

Xie, J., Liu, T.-M., Fu, X., Dai, W. Z., & Lieber, C. M. (2015). Three-dimensional macroporous nanoelectronic networks as minimally invasive brain probes. *Nature Materials, 14*(12), 1286–1292.

Zhou, T., Hong, G., Fu, T.-M., Yang, X., Schuhmann, T. G., Viveros, R. D., & Lieber, C. M. (2017). Syringe-injectable mesh electronics integrate seamlessly with minimal chronic immune response in the brain. *Proceedings of the National Academy of Sciences, 114*(23), 5894–5899.

Zrinzo, L. (2012). Pitfalls in precision stereotactic surgery. *Surgical Neurology International, 3*(2), 53. https://www.elekta.com/neurosurgery/leksell-stereotactic-system/.

Chapter 8
3D Cell Culture Systems
for the Development of Neural Interfaces

Omaer Syed, Chris Chapman, Catalina Vallejo-Giraldo, Martina Genta,
Josef Goding, Emmanuel Kanelos, and Rylie Green

8.1 Introduction

8.1.1 The Role of 3D Culture Neuronal Interface Research

The basis of neural interface research is the study of connections between the
nervous system and devices or within the nervous system itself. Applications span
both the central nervous system (CNS) and the peripheral nervous system (PNS),
including recording and stimulation, therapeutic applications and neural prostheses.
Some more specific applications include brain-machine interfaces (BMIs), cochlear
implants and functional electrical stimulation (FES) (Hatsopoulos and
Donoghue 2009).

Animal models remain the gold standard for neural interface research; however,
with increasing weight of ethical considerations, the time scales required (usually
chronic, >30 days) and high cost, these are not always easily undertaken.
Furthermore, the ability of these models to truly represent the human nervous system with differences in the genetic, biochemical and metabolic functions can be
brought into question (Zhuang et al. 2018). Consequently, the need has arisen to
produce better in vitro research models.

Conventional 2D models possess several limitations, as they do not have the
capacity to replicate critical features of the in vivo environment, having inadequate
cell–cell and cell–ECM interactions (Ko and Frampton 2016). Cells grown in 2D
are usually adhered to a flat substrate and communicate only laterally with adjacent
cells, having minimal interactions across their cytoplasm, which is exposed to the
fluid media. This is starkly different to the endogenous 3D neural cell environment.
It is known that the simple process of placing a neural or neuroglial cell on a 2D

O. Syed · C. Chapman · C. Vallejo-Giraldo · M. Genta · J. Goding · E. Kanelos
· R. Green (✉)
Department of Bioengineering, Imperial College London, London, UK
e-mail: rylie.green@imperial.ac.uk

© Springer Nature Switzerland AG 2020   201
L. Guo (ed.), *Neural Interface Engineering*,
https://doi.org/10.1007/978-3-030-41854-0_8

substrate will alter its phenotypical behaviour. Cultures with 3D approaches have been developed with a view to overcome these limitations. As such, complex 3D environments are expected to act as better models and have been used to improve prediction for the in vivo behaviours of neural cells and tissues.

Several 3D culture systems have been developed to study different aspects of neural responses, electrophysiology, function and development. While models using 3D cell culture cannot replicate an entire neural interface, a number of important subjects can be studied and used to inform performance aspects critical to neural interfaces. Biocompatibility and cell interactions are important considerations that can be studied using such cell culture models. Tissue response, including the inflammatory response in both acute and chronic time scales, can be modelled to give insight on the in vivo response to an interface or device. The impact of the biological environment on biomaterials can also be studied through these cell culture models, including impacts on the electrical properties for conductive materials, the changes in stiffness for dynamic materials and the effect of surface properties. Application specific properties such as the delivery of a neuromodulating signal can be analysed in vitro prior to progression to in vivo. Some neural interface devices may contain cells within them, and in vitro studies can be used to investigate the efficacy of these cell components including their survival and function Finally, in vitro cell cultures enable useful insight into the controlled release of drugs and biomolecules at the neural interface (Fattahi et al. 2014; Green et al. 2013; Nam 2012).

The 3D cell models currently used to investigate this broad range of neural interfacing properties can be categorized into explant or ex vivo models, cell-only approaches and biomaterials and scaffold approaches. The ability of these three different approaches to model neural function and how they aid towards the broader goals of neural interfacing between synthetic constructs and neural tissue are discussed.

8.1.2   Cells and ECM Components of the Nervous System

To develop 3D cell culture models, there must be an adequate understanding of the tissues being modelled. The nervous system is one of the most complex systems in the human body and is composed of multiple different types of cells and ECM components. It contains approximately 171 billion cells, divided almost equally into neuronal and non-neuronal cells (Benam et al. 2015). Non-neuronal cells include microglia, oligodendrocytes, astrocytes, endothelial cells and pericytes (Zhuang et al. 2018). Each of these cell types is characterized by substantial variability in terms of function, morphology, size and gene expressions, further amplifying the diversity of the whole system (Benam et al. 2015).

The nervous system is divided into two parts: the central nervous system (CNS), which includes the brain, spinal cord and auditory, olfactory and vision systems, and the peripheral nervous system (PNS), which consists of all the nerves departing

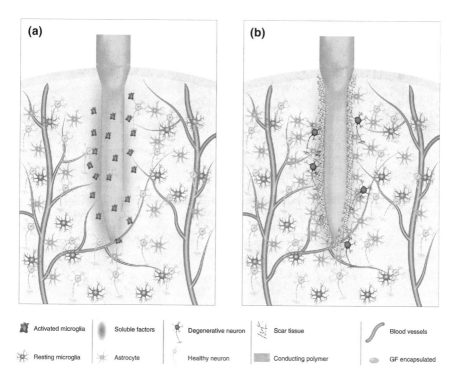

Activated microglia	Soluble factors	Degenerative neuron	Scar tissue	Blood vessels
Resting microglia	Astrocyte	Healthy neuron	Conducting polymer	GF encapsulated

Fig. 8.1 Onset of the gliosis response at the brain-electrode interface. (**a**) Representation of the acute inflammation after a device implantation. (**b**) Chronic inflammation leading to the formation of glial scar. (Figure adapted and reproduced with permission from Vallejo-Giraldo et al. (2014))

from the brain (with the exception of the optic nerve which is part of the CNS) and the spinal cord (Schmidt and Leach 2003). Each of these components is characterized by different structures as well as specialized cells. Therefore, when designing neuronal interfaces and modelling the nervous system's inflammatory response and subsequent gliosis (Fig. 8.1) (Moshayedi et al. 2014), it is essential to consider which part of the nervous system is being targeted and interfaced. Furthermore, the gliosis process in the CNS and PNS involves different cell types. In the CNS, this is driven by the astrocytes, whereas in the PNS Schwann cells represent the main cellular component in this reaction (Tian et al. 2015).

All cells are embedded in ECM, which not only gives support through a unique physical environment but also contains a multitude of biochemical cues, such as soluble factors, growth factors, chemokines and cytokines. These cues play an integral role in the regulation of cell behaviour and fate (Zhuang et al. 2018). In most human tissues, ECM structurally consists of multiple components including fibronectin, collagen, elastin and laminin. However, brain ECM is mainly composed of hyaluronic acid (HA) and proteoglycans (Mouw et al. 2014). In particular, glycosaminoglycans (GAGs), the core protein of proteoglycans, have been shown to play an essential role in neural tissue development and pathological states (Miller and

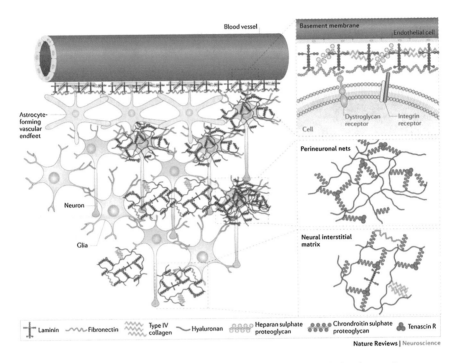

Fig. 8.2 Brain extracellular matrix heterogeneity and cell interaction in the three main compartments of the CNS. (Figure reproduced with permission from Lau et al. 2013)

Hsieh-Wilson 2015; Yu et al. 2017). Furthermore, amongst the family of GAGs, chondroitin sulphate and heparan sulphate are the most important in the CNS, specifically in the peri-neuronal nets, neural interstitial matrix and the basement membrane (Fig. 8.2) (Lau et al. 2013).

The interactions between different brain cells and ECM are of great importance, and it has become clear that leveraging these components is essential to the successful design of neural interfaces. For this reason, the incorporation of ECM biomolecules as a functionalization approach for neural interfaces has been adopted to better mimic the native neural tissue and elucidate functionality of the neural system (Chen et al. 2018). The simplest approach to maintaining this interaction is to use the intact tissue as a model in the form of explants for ex vivo models.

8.2 Explant and Ex Vivo Approaches

Historically, the most common approach to studying or modelling tissue responses has been to use whole tissues extracted from animal subjects. In the neural system, this has included whole brain, brain slices, whole nerve and in some cases organs, such as the ocular orbits (Famm 2013; Humpel 2015a; Ogilvie et al. 1999). These tissue preparations are intrinsically 3D and therefore recapitulate a myriad of relevant cell processes. The main advantage to using explanted or ex vivo tissues is that the cellular environment is preserved, and with techniques that enable preservation of these tissues in vitro, there is capacity for increased throughput compared to typical in vivo models. Although these preparations provide the most realistic replication of the 3D cellular environment in vitro, many of these tissue preparations can suffer from lack of longevity, degradation of their 3D microenvironment during the culture period as well as challenging preparation steps. The use of these preparations is strongly recommended when cellular connectivity and replication of the in vivo environment are extremely important parameters for assessing neural interfaces.

8.2.1   Explant Culture Types

The most well-known and characterized method of studying the nervous system through explanted tissue is the brain slice, which can be taken and sustained in vitro for chronic time periods from weeks to months (Humpel 2015b). Many different techniques have been developed for slicing and maintaining regions of the brain, the most notable of which are the cortex (Eugene et al. 2014; Humpel 2015b) and hippocampus (De Simoni and My Yu 2006; Gogolla et al. 2006; Hsiao et al. 2015), as shown in Fig. 8.3a. Other explant systems include cultures of explanted retina (Fig. 8.3d) (Caffe et al. 2001; Feigenspan et al. 2009; Germer et al. 1997; Ogilvie et al. 1999; Rzeczinski et al. 2006), dorsal root ganglion (DRG) (Fig. 8.3b) (Fornaro et al. 2018; Livni et al. 2019; Melli and Hoke 2009) and spinal cord sections (Fig. 8.3c) (Gerardo-Nava et al. 2014; Glazova et al. 2015; Heidemann et al. 2014; Ravikumar et al. 2012). There has also been some work carried out on the spiral ganglion of the cochlea (Evans et al. 2009; Hahnewald et al. 2016; Mullen et al. 2012; Zheng and Gao 1996). With the recent increase in interest surrounding peripheral nerve stimulation (Famm 2013; Waltz 2016), explanted whole nerve section has seen use as an acute ex vivo model.

Slice cultures of the brain are typically taken from rodents in the early post-natal (p0–p7) development stages or in some cases from the embryonic stage for thalamic tissue (Humpel 2015b). The early post-natal development stages are ideal for the culturing of brain slices due to the larger size, more developed cytoarchitecture and improved survival of cells in the culture. Although the surgical techniques to separate the tissue of interest differ, the slices are typically prepared similarly after dissection. Hundreds of micrometre-thick slices are cut using a vibratome or tissue slicer and secured to a semi-porous membrane culture substrate. These cultures have been kept in stable in vitro conditions for weeks to months from preparation (Gogolla et al. 2006). Interestingly, it has been shown in hippocampal slices that while oligodendrocytes and microglia maintain normal activity, astrocytes lose their layer-specific distribution (Hailer et al. 1996). Since normal synaptic transmission and excitation is seen within these cultures, they are still appropriate models for

a) Hippocampal slice   b) DRG explant   c) Spinal cord slice

d) Retinal explant

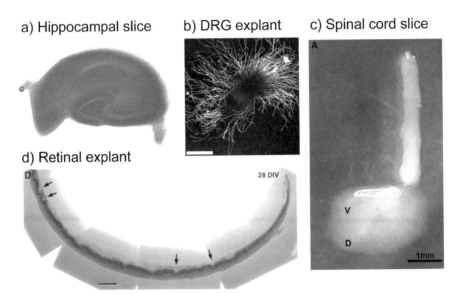

Fig. 8.3 Light microscopy images of explanted cultures. (**a**) Side view of an optimal hippocampal slice with clearly visible cell layers. (Reproduced with permission from Gogolla et al. 2006). (**b**) Anti-β-III tubulin immunostaining showing axonal outgrowth from DRG explant in vitro. (Reproduced with permission from Livni et al. 2019). (**c**) A combination of spinal cord slice (bottom) interfacing with an explanted nerve (top) cultured 7 days in culture. (Reproduced with permission from Gerardo-Nava et al. 2014). (**d**) Retinal explants grown to maturity in organ culture. (Reproduced with permission from Ogilvie et al. 1999)

electrophysiological studies for both extra- and intracellular recordings (Gähwiler et al. 1997; Hsiao et al. 2015).

Similar to the brain slice cultures, retinal slice cultures are typically taken from early post-natal rodents (p0–p2). The cells that make up the retina are morphologically driven, and therefore the preserved environment in the slice cultures is the main benefit of this system. Mouse retinal slices have been successfully cultured for over 4 weeks in vitro using a method published by Ogilvie et al. (1999). Retinal explant cultures are essential for developing neural interfaces for stimulating the tissue. However, the preserved morphology of retinal slices has been reported to limit the efficacy of fluorescent dyes that are frequently used to assess electrophysiological activity (Briggman and Euler 2011). Additionally, as the preserved environment is particularly crucial for retinal explants, these cultures suffer more from the in vitro degradation.

The DRG is a major collection of afferent nerve fibres in the PNS. Ex vivo culturing of the DRG can be done using tissue from embryonic, postnatal and adult animals. Typically, embryonic DRG are used for ex vivo preparations owing to the high content of neurons that can be isolated (Melli and Hoke 2009). Many rodent DRG explant protocols have been published (Fornaro et al. 2018; Livni et al. 2019).

These preparations are particularly useful when investigating pharmaceutical inter-actions due to the known afferent fibre composition. Additionally, the morphology of the explant enables control of the sprouted neurites, leading to easy compartmen-talization and use with microfluidic systems (Mobini et al. 2018). One major limita-tion of DRG explant cultures is that although the 3D microenvironment is preserved for the neuron bodies in the ganglion, the new sprouted neurites are typically grown in 2D. The preservation of the ganglion is important to maintaining relevance to the in vivo system; however, for applications that require 3D neurite outgrowth, this culture system can be limiting.

Another common ganglionic explant culture system is the spiral ganglion from the cochlea. These explant cultures are typically used from either mature (Hahnewald et al. 2016) or post-natal (Evans et al. 2009) rats. Similar to the DRG cultures, these explant cultures maintain a 3D microenvironment for the neuron bodies in the gan-glion, but the neurite outgrowth is in 2D. Explant cultures from the spiral ganglion have been used to investigate both electrical (Evans et al. 2009; Hahnewald et al. 2016) and chemical cues (Mullen et al. 2012; Zheng and Gao 1996) on neurite out-growth from the ganglion.

Slice cultures from the whole spinal cord are frequently used to investigate spinal cord regeneration (Gerardo-Nava et al. 2014; Heidemann et al. 2014), degeneration (Ravikumar et al. 2012) and neurogenesis (Glazova et al. 2015). These cultures are prepared similarly to the brain slice cultures. The spinal cord sections are typically taken from post-natal (p7–p9) rodents to ensure the cellular organization in the spi-nal cord is well developed; however, methods for embryonic slices have also been presented (Pakan and McDermott 2014). The preservation of cellular arrangement in the spinal cord slices is particularly important when investigating neural regen-eration and connectivity from specific branches in the spinal cord. However, a major limitation of spinal cord slices is their lack of preserved vasculature as well as end targets for regenerating axons (Gerardo-Nava et al. 2014).

While several tissue slice models have been used for CNS interrogation, the most common approach in the PNS is the use of whole nerve explants. These explant models have historically been used from non-mammalian animals such as frogs, cephalopods, crustaceans and nematodes (Hodgkin 1948). With the recent interest in PNS stimulation, these whole nerve explant models have been adapted to sustain activity from mammalian nerves of interest, such as the sciatic (Park et al. 2015) or vagus nerve (Peclin and Rozman 2014). In these cultures, the nerve of interest is dissected from an adult animal and kept alive in a heated bath of physiological solu-tion. A major limitation of these preparations is that they are only effective as acute models kept for approximately 1 day at a time. This is due to the cell bodies being transected during dissection. Although the lack of culture time is a major limitation, a benefit is that with the system consisting only of peripheral nerve fibres, it is much easier to decouple changes in activity from cellular changes. These culture systems are typically used for the development of cuff electrodes for electrical neuromodu-lation devices (Grill et al. 2009; Navarro et al. 2005) prior to in vivo studies.

8.2.2   Use of Ex Vivo Systems in Neural Interface Development

Ex vivo culture systems have been applied to study a wide variety of neural interfaces in which their preserved cytoarchitecture is of a direct benefit to the parameters being studied, specifically, for electrophysiological studies involving extra- or intracellular stimulation and recording, studies involving investigation of changes in the cellular environment and studies on long-term neural regeneration. Specific use of explanted tissue cultures for the development of neural interfaces is discussed, detailing investigation of electrical, chemical and mechanical characteristics of interfaces.

Electrical Stimulation and Recording Interfaces

One of the most common uses of explanted tissue slices is to investigate the efficacy of neuro-electronic interfaces. These studies focus on understanding neural connectivity through stimulation and recording paradigms, where it is paramount that the cellular architecture is preserved. Owing to the importance of preserving the cellular connections, there have been many electrical neural interfaces developed in the form of multiple electrode arrays (MEAs) for the integrated culturing and recording of explanted brain slice cultures (Fig. 8.4a) (Berdichevsky et al. 2009, 2010; Egert et al. 1998; van Bergen et al. 2003). In the case of devices being developed for implantation into the nervous system, the long-term cellular coupling becomes another factor of interest, particularly when new electrode surfaces are being developed. Typical implants tested this way are planar MEAs; however, acute ex vivo nerve cultures are increasingly being used to test efficacy of peripheral nerve cuffs (Fig. 8.4b). Planar MEAs are particularly amenable to development with explanted tissues due to their flat surfaces upon which the explanted tissue can be mounted (Ou et al. 2012; Raz-Prag et al. 2017). Devices with new electrode materials such as carbon nanotubes (Kuzum et al. 2014; Raz-Prag et al. 2017), graphene (Kuzum et al. 2014) and conductive polymers (Aregueta-Robles et al. 2014; Khodagholy et al. 2013) have all been tested using explanted neural tissues. For example, Kuzum et al. (2014) demonstrated the efficacy of transparent flat graphene electrodes to enable both electrical and optical recording and stimulation from a hippocampal slice culture model.

Understanding Mechanical, Chemical and Physical Interactions

When developing neural interfaces, the cellular reactions to the synthetic device or interface at both acute and chronic time points are important. Explanted slice cultures enable these long-term cellular reactions and interactions to be studied. These explanted tissue slices are most often used in the development of pharmaceuticals and devices designed to form chemical interfaces with the nervous system (Daviaud

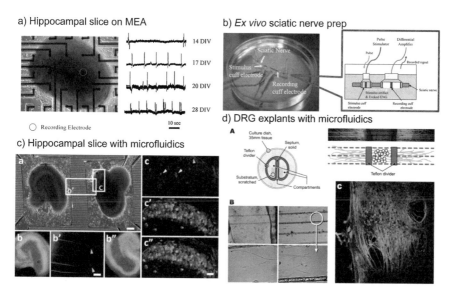

a) Hippocampal slice on MEA

b) *Ex vivo* sciatic nerve prep

c) Hippocampal slice with microfluidics

d) DRG explants with microfluidics

Fig. 8.4 Examples of uses for ex vivo and explanted tissue cultures ranging from electrical inter-faces to chemical and mechanical. (**a**) Hippocampal slice culture at DIV 20 on a planar multiple electrode array (MEA) and the associated signals recorded. (Reproduced with permission from Berdichevsky et al. 2009). (**b**) Image and schematic of a stimulation and recording set-up for ex vivo rat sciatic nerve. (Reproduced with permission from Park et al. 2015). (**c**) Organotypic hippocampal slices in PDMS compartments to investigate neurite outgrowth and connectivity between slices. (Reproduced with permission from Berdichevsky et al. 2010). (d) Compartmentalized DRG cultures in microfluidic channels to investigate drug effects on neurite outgrowth. (Reproduced with permission from Melli and Hoke 2009)

et al. 2013; Kim et al. 2013; Yi et al. 2015). When studying the effects of newly developed pharmaceutical agents, it is important that the cellular connections and the surrounding architecture are preserved as much as possible to mimic the envi-ronment in which the cells are processing the drug. These cultures can also enable the long-term imaging of changes in the cells and their environment as a result of a disease state (Fig. 8.4c, d) (Le Duigou et al. 2018). However, the difficulty in prepa-ration and the reduced throughput in comparison to a purely in vitro model can make these systems less attractive for doing bulk chemical screening. These culture preparations are particularly useful when a pharmaceutical of interest has been identified and a more realistic model system needs to be used before performing in vivo work.

Explanted slice cultures have also been used with devices to study the effect of mechanical trauma on neural tissue. One area of specific interest is to study how mechanical trauma to the CNS (such as traumatic brain injury) alters the neural con-nections and architecture. Devices using stretchable materials have been developed to study the reaction of explanted slice cultures to the application of a known force (Morrison 3rd et al. 2006). In addition to providing a relevant platform to examine the effects of chemical and/or mechanical cues, explanted cultures are also often used for the development of devices involved in neural regeneration. Explanted cultures of spinal cord slices are predominantly used in studies in conjunction with engineered scaffolds to measure their potential in aiding either neurogenesis (Glazova et al. 2015) or the regeneration of severed connections (Gerardo-Nava et al. 2014; Heidemann et al. 2014).

While explant models have advantages, particularly for certain applications, they do suffer from a degradation of their 3D microenvironment during the culture period and, in many cases, a lack of longevity. They can also be challenging to prepare, and high-throughput studies may not be possible. There remains the issue of requiring animal tissues, which has practical and ethical concerns. Understandably, the vast majority of research has turned to cell culture to model neural tissues.

8.3 Cell Biology–Based Models

Characteristic features of the native neural tissue rely on the large assortment of biomolecules and cell types within the nervous system (Hopkins et al. 2015; Zhuang et al. 2018). Therefore, successful in vitro 3D models must embrace the native cell diversity to provide cell–cell interactions and enhanced intrinsic organization, a fundamental requirement to the assembly of key signals for neural networking (Hopkins et al. 2015). The move to 3D does mean that the support of a matrix becomes unavoidable. While biomaterial approaches are discussed later, cell approaches which use only basic ECM components to create 3D cell models are detailed in this section.

The simplest form of a cell culture model for the neural interface is a pure neuronal model. These provide important neuron to neuron interactions, allowing for the assessment of cell attachment for neural implants and micro-devices, as well as neurite outgrowth and synapse formation (Gao et al. 2002). Systems involving neuronal cell cultures have been developed using 3D engineering approaches. A study conducted by Frega et al. (2014) showed the conformation of a 3D structure by using layers of glass microbeads. The confined microbeads, together with the dissociated hippocampal neurons, were maintained on a porous membrane forming a structure with hexagonal geometry (Fig. 8.5a). The suspension of microbeads and hippocampal neurons were then moved from the porous membrane to the active areas of the MEA for direct recording and stimulation (Fig. 8.6b). Their results showed that the 3D-engineered hippocampal assemblies presented with cell morphology and neural network connectivity similar to those observed in an in vivo environment.

Another type of pure neuronal model can be made with ECM components, such as the development of a 3D culture using collagen gel constructs with hippocampal neurons (Bourke et al. 2018). This technique allows unrestricted cell positioning and neurite outgrowth within the gel. The 3D cell constructs were transferred to MEAs for electrophysiological recordings and showed extended neuronal burst

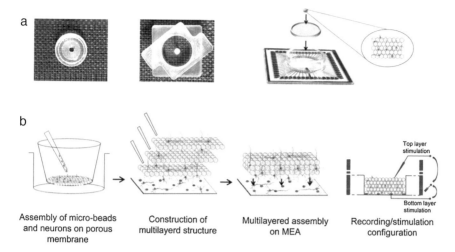

Fig. 8.5 Graphical representation of the structure of the 3D neural cell construct. (**a**) Representation of the PDMS mould used to layer microbeads and neurons in the final configuration. (**b**) Main steps for the creation of a 3D neural network. (Figure adapted and reproduced with permission from Frega et al. 2014)

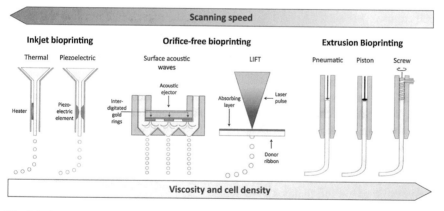

Fig. 8.6 Overview of the most widespread bioprinting approaches and critical parameters governing the bioprinting process for each. (Figure reproduced with permission from Holzl et al. (2016))

patterns suggesting a fully interconnected neural network formation when compared to the 2D controls (Bourke et al. 2018).

Although neuronal cell models promote the study of neural networking and neurite outgrowth, they may not be the most useful for assessing device/material interactions, as they do not capture the important inflammatory response. This is because most of the significant cellular components involved in this reaction are the non-neuronal glial cells (Gilmour et al. 2016; Kawano et al. 2012; Sofroniew 2015).

8.3.1 Non-neuronal/Mixed Cell Type Culturing Approaches

Models with both neuronal and glial cell types provide a more complete approach in determining interface reactions. Mechanical mismatch between an electrode probe and tissue (brain cell components) and micromotion are key drivers of the inflammatory response. Glial cell mixed population cultures produce important morphological hallmarks of inflammation and the glial scar enabled by the presence of microglia and astrocytes.

Mixed glial cells can be used to model device interactions and understand the role of glia on the neural system. Jeffery et al. (2014) have described an in vitro 3D culture model of a glial scar formed in response to a microelectrode. A methacrylated HA hydrogel was formed into an insoluble network scaffold. Mixed glial cells (microglia and astrocytes) taken at postnatal day 1 were encapsulated in the HA hydrogel, and the inflammatory response to the microelectrode inserted into the hydrogel was assessed over a period of 2 weeks (Jeffery et al. 2014). The pro-inflammatory phenotype of microglia was first increased at the peri-electrode region followed by astrocytes at the electrode. These results demonstrated the potential of the 3D in vitro model system to assess glial scarring.

3D collagen gels have been used to show the role of electrode micromotion in driving glial scar formation around neural implants using embryonic mixed neural cell cultures (Spencer et al. 2017). In this study, the presence of reactive astrocytes was highlighted in cell areas and perimeters in response to micromotion when compared to a static control (Spencer et al. 2017). However, the neuronal component of the culture was not fully assessed. Due to the complexity of cell–cell and cell–ECM interactions, there remains limited knowledge of neural network functionality in 3D cell culture–based systems. This may be due to the difficulties in elucidating electrophysiological mechanisms and the lack of more comprehensive tools for cell data analysis (Hopkins et al. 2015). Despite this limitation, when compared to 2D culture, the 3D models have been shown to better mimic the in vivo CNS environment (Griffith and Humphrey 2006; Hopkins et al. 2015; Szarowski et al. 2003). However, creating more biomimetic neural interfaces capable of interconnecting with the native brain tissue still remains a challenge.

Current methods are still limited for modelling acute and chronic inflammatory responses at the electrode-brain interface. Moreover, the 3D cell-based models are restricted in size as well as in supporting heterogeneous cell architectures which do not accurately reflect the nervous system organization and functionality (Hopkins et al. 2015). The incorporation of microfluidics into 3D cell culture scaffolds (discussed in Sect. 8.4.3) has been proposed for the development of pseudo-vascularized models (Cullen et al. 2007; van Duinen et al. 2015) which could improve translation of the culture responses to in vivo responses. Increasing the complexity and hence accuracy of a 3D model can be achieved through manipulating the properties of the matrix in which the cells are being cultured. A large number of biomaterials and synthesizing techniques exist and have been applied to neuronal models.

8.4 Materials and Engineering-Based Models

The use of biomaterials for creating 3D cell culture models covers a very wide range of techniques and approaches. Creating scaffolds for cell culture can be as varied as advanced 3D printing to as simple as bulk polymer synthesis. This section focuses on material approaches and considerations related to either adding cells after scaffold synthesis or materials into which cells must be incorporated during the synthesis process.

8.4.1 Biomaterial Approaches

To more accurately recreate a neuronal environment, cells are required to be suspended in a nutrient-permeable matrix. Using biomaterials allows for the creation of a biochemically, spatially and mechanically relevant cellular environment. In producing a suitable 3D environment, there is greater capacity for representing the complex cell–cell and cell–biomaterial interactions that occur in vivo. This gives rise to complex patterning, highly controlled differentiation and morphogenic characteristics not possible in 2D. There is a plethora of biomaterial scaffolds available, both synthetic and natural, with a variable and often tailored array of properties. As a result, care must be taken when selecting a biomaterial to ensure the desired outcome.

Commonly used biomaterials for neural cell culture include fibrin (Willerth et al. 2006), alginate (Gu et al. 2016) and polyethylene glycol (PEG) (Schwartz et al. 2015). Furthermore, proteins and peptides such as laminins and collagens have been used to promote the adhesion properties of otherwise synthetic materials (Schwartz et al. 2015). Fibrin gels have been widely employed as they represent a low-cost scaffold with satisfactory cell adhesion properties and biomechanics. Willerth et al. were able to incorporate embryonic stem cells into fibrin gels and differentiated them into neurons, astrocytes and oligodendrocytes (Willerth et al. 2006, 2008). Similar results were attained by Johnson et al. (2010), and this was further developed to demonstrate an interface with an in vivo model of spinal cord injury. Despite the encouraging results, fibrin possesses slow gelation mechanics and thus is rarely used on its own. It is often mixed with other biomaterials that have rapid gelation rates, such as alginate (Gu et al. 2016). Despite exhibiting adequate biorecognition and fostering cell adhesion, fibrin gels support higher neurite outgrowth when proteins such as laminin are incorporated within the hydrogel carrier (Pittier et al. 2005). There remains some concern over immunogenicity of biologically sourced materials such as fibrin, including fear of disease transfer and batch-to-batch variation. The uncontrollable biofunctionalization of biological hydrogel carriers has led to researchers pursuing synthetic biomaterials as an alternative or adjunct.

Synthetic hydrogels have several attractive properties. These include defined pore size and mechanics, lower immunogenicity, ease of biofunctionalization, possessing extracellular matrix components or growth factors and controllable degradation (Shao et al. 2015). For example, peptide-functionalized PEG gels have been shown to sustain diverse neuronal and glial populations when seeded with neural precursor cells (NPCs), microglial precursors, mesenchymal stem cells and endothelial cells and even exhibit some vascular networks (Schwartz et al. 2015).

The range of biomaterials and their capacity for modification are extensive; however, a more limited number of approaches have been used in the context of neural cell culture. Some examples of synthetic and natural biomaterials used in 3D cultures are listed in Table 8.1. As evidenced, a range of material properties will impact on the cell survival and function, including biological activity, immunogenicity, biomechanics and the degree of functionalization or modification. In addition, degradation kinetics must be closely evaluated to ensure that the cells are properly supported. Overall, most scaffolds used in neural cultures degrade over 2–8 weeks via hydrolysis, ionic exchange or enzymatic degradation (Thomas and Willerth 2017). Critically, degradation creates by-products (e.g. salts, acids, macromolecules and peptides), which can affect cell growth or initiate immune responses and thus should be considered carefully.

Table 8.1 highlights the wide range of materials which have been used to culture 3D neural tissues, and this list is by no means exhaustive. While many studies have been used to investigate the effects of specific material variables on neurons for modelling the neural interface, there are also many more studies which use 3D cell culture as a means for neural tissue engineering. This extensive subject has been reviewed elsewhere (Ko and Frampton 2016; Zhuang et al. 2018).

A major obstacle in 3D culture is controlling the spatial arrangement of cells in order to generate relevant tissues (e.g. specific brain areas) which are able survive and function as desired. An example of this is organoids being used to generate 3D brain-like structures (Lancaster et al. 2013). Here, the organoids cultured in Matrigel droplets were able to form a neuroepithelium and possessed structures reminiscent of a cerebral cortex while surviving up to 10 months in a bioreactor. However, cultures could only be grown to a maximal size of 4 mm due to the lack of vascular network and eventually began to die.

Biomaterial approaches, both biological and synthetic, usually involve scaffolds created by 3D printing, electrospinning or bulk material synthesis (Deb et al. 2018). Incorporating cells into these 3D scaffolds can be a challenge. Cells can be incorporated either during synthesis or after synthesis, depending on the material fabrication approach. When a scaffold is porous, the cells are usually introduced post-fabrication. When cells are encapsulated within a hydrogel volume, they must be incorporated during synthesis. This latter approach may limit the type of shapes and structures which can be made, as conditions for gelation of the hydrogel can impact on cell survival and hence the encapsulation process must be designed to be cytocompatible and rapid. To overcome this challenge and provide additional advantages, alternative approaches have been developed, such as bioprinting and microfluidics, which incorporate cells during the material synthesis process.

Table 8.1 Comparison between biomaterial-based 3D in vitro models for neural tissues

Biomaterial	Cell type	Investigation	Culture time	Result	Reference
PEG	Rat E14 forebrain cells	Stiffness/mechanical	7 days	Lower percentage macromer resulted in higher metabolic activity (measured by ATP content). Large reduction observed in >10% macromer gels. Increased proliferation in 7.5 wt% PEG hydrogel	Lampe et al. (2010)
Gelatin methacrylate	Murine iPSCs differentiated into neural stem cells (NSC)	Spinal cord regeneration in mice	7 days	Higher level of differentiation markers such as Tuj-1 in soft and medium hydrogels. Neuronal extension was higher in both these groups. Stiff hydrogels resulted in lower synaptic marker synaptophysin and increase in GFAP	Fan et al. (2018)
Gelatin-hydroxyphenyl propionic acid	Human mesenchymal stem cells (MSC)	Mechanical properties	21 days	In the lower stiffness material proliferation increased (increase in total DNA content). More neuronal markers were found on the less stiff variants	Wang et al. (2010)
Sodium alginate	Rat hippocampal NSCs	Mechanical properties	7 days	Differentiation markers and proliferation increased in softest hydrogels	Banerjee et al. (2009)
Elastin-like proteins with elastin-like domains and bioactive domains (RGD) Also PEG/RGD	Adult murine NPCs from micro-dissected dentate gyrus	Scaffold properties (remodelling)	14 days	Initial stiffness did not correlate with neural stem cell phenotype. Instead degradation did correlate and was necessary for maintaining the phenotype. Effect was independent of the maintenance of cytoskeletal tension and presentation of adhesive ligands	Madl et al. (2019)
PEG with matrix metalloproteinase degradable linker	U87-MG cells (glioblastoma cell line)	Scaffold properties (remodelling)	14 days	Degradable linkers allowed for increased cell proliferation and increased cell spreading	Wang et al. (2017)
Hydrolytically degradable PEG	Pheochromocytoma cells (PC12)	Scaffold properties (remodelling)	7 days	Cross-linking densities above 15% (and thus slower degradation) resulted in lower proliferation	Zustiak et al. (2013)

(continued)

Table 8.1 (continued)

Biomaterial	Cell type	Investigation	Culture time	Result	Reference
PEGylated fibrinogen	Dorsal root ganglion cells (E8-E11 chicken embryos)	Cell capacity to remodel scaffolds	3 days	When treated with MMP inhibitors, cell growth was inhibited	Sarig-Nadir and Seliktar (2010)
PEG-RGD/tenascin gel	Rat neural stem cells (NSC) induced from bone marrow stem cells	Cell growth and survivability within a scaffold	7 days	Enhanced survival proliferation and differentiation of NSCs in gel (evaluated by nestin, neurofil 68, beta tubulin III, GFAP, MBP and oct4 expression)	Naghdi et al. (2016)
Collagen hydrogel with epidermal growth factor fused to a collagen binding protein	NSCs (E16 EGFP-transgenic rats)	Cell growth and survivability within a scaffold	7 days	Both collagen with free epidermal growth factor and collagen with chimeric epidermal growth factor increased proliferation though the latter did so to a larger degree. Cells in gels with bound epidermal growth factor also expressed higher beta tubulin	Egawa et al. (2011)
Self-assembled fmoc-DIKVAV peptide loaded with GDNF	Mice embryonic ventral midbrain cells (VM)	Drug delivery	Minimal as the hydrogel was used as a carrier	Cells within gel showed higher proliferation than cells alone. Cells in GDNF loaded gels showed highest proliferation and showed significant difference compared to cells alone	Rodriguez et al. (2018)
Self-assembled peptide system: RADA16 and RADA16-IKVAV	Rat neural stem cells (HCN-A94-2)	Cell growth and survivability within a scaffold	14 days	IKVAV gels resulted in higher rates of neuronal differentiation	Cheng et al. (2013)
Hyaluronic acid gel loaded with brain-derived neurotrophic factor	Rat striatum cells embryonic day 16	Cell growth and survivability within a scaffold	3 days	Brain-derived neurotropic factor-loaded gels showed higher viability	Nakaji-Hirabayashi et al. (2009)
Hyaluronic acid gel (peptide modified)	Neutralized H9 human embryonic stem cells	Cell growth and survivability within a scaffold	10 weeks	Demonstrated long-term viability in gel (70 days) and RGD modified gels showed modest increase in viability. 3D cultures showed decreased reactive astrocytes	Seidlits et al. (2019)

Biomaterial	Cell type	Investigation	Culture time	Result	Reference
Poly(ε-caprolactone) (PCL) and gelatin aligned fibres with retinoic acid	Human mesenchymal stem cells	Topography and controlled drug release	14 days	Plain topography resulted in increased differentiation markers. (Combined with drug release, morphological changes took place.) Topography was required for the expression of synaptophysin	Jiang et al. (2012)
Mixture of collagen and magnetic particles	Pheochromocytoma cells (PC12)	Physical properties (alignment)	7 days	Cells showed good viability and elongated in the orientation of the fibres	Antman-Passig and Shefi (2016)
Different ratios of hyaluronic acid and PCL	SH–SY5Y human neuroblastoma cell	Cell growth and survivability within a scaffold	14 days	Improved attachment and proliferation on PCL/hyaluronic acid	Entekhabi et al. (2016)
Methacrylamide chitosan	Rat NSCs from forebrain	Physical properties (porosity)	14 days	Improved survivability of cells in porous material vs non-porous material	Li et al. (2012)

8.4.2 Bioprinting

Bioprinting is computer-aided precise deposition of biomaterials within a defined space (Tarassoli et al. 2018). This process differs from conventional 3D printing in that it typically incorporates cells into the biomaterial matrix or bioink. Bioprinting allows for high geometric control over cell and scaffold placement. This includes control over the ECM feature size, mechanical properties/gradients and generation of chemical gradients. Ultimately, this allows the user to generate exceedingly complex constructs (Chia and Wu 2015).

The bioink acts as a cell supporting environment which can mimic some aspects of the ECM, promoting natural processes such as cell adhesion, differentiation and proliferation. They differ from other materials used in 3D printing in that they are printed at much lower temperatures, are commonly naturally derived and have cross-linking conditions that have been designed to preserve cell function and prevent unwanted matrix degradation (Malda et al. 2013). The design criteria for bioinks and the various systems used within the literature have been extensively reviewed (Chen et al. 2018; Holzl et al. 2016; Hospodiuk et al. 2017; Tarassoli et al. 2018).

Bioprinting Methods

Bioprinting allows for both cells and supporting materials to be positioned at the same time. This is beneficial in that fabrication is usually done in a single step, enabling rapid production of constructs that minimize impact on cell survival. The process covers a number of different approaches broadly defined as extrusion, droplet printing and laser. Figure 8.6 outlines some of the technologies covered by each approach.

Extrusion-based methods have been employed in a wide number of applications. The advantage and popularity of this technique are in part that it relies on a very simple and clear method which has led to diversity and predictability. The viscosity of the bioinks is in the range of 30 mPa.s to 6×10^7 mPa.s. At the upper end of this range extrusion is the only way in which such a viscous material can be formed (Derakhshanfar et al. 2018). This capacity for high viscosity supports printing of high cell densities including cell spheroids (Holzl et al. 2016). The drawbacks of this approach are relatively low resolution (200–1000 μm), potential nozzle clogging and decreased cell viability post-printing due to shear stress (Chang et al. 2008; Malda et al. 2013).

Extrusion printing can also be employed to produce gel-in-gel structures. This is where a secondary gel is printed within a primary supporting gel. Specifically, structures are made where one material encapsulates the other. Photocrosslinking can be used as a secondary step to stabilize one or both components of the structure. When one hydrogel component is photosensitive, the uncrosslinked component can be washed away to produce voids or complex 3D shapes (Holzl et al. 2016; Wu et al.

2011). Highley et al. (2015) printed modified shear thinning HA inks with self-healing properties into the shape of a bifurcating channel. The construct was then exposed to UV light to stabilize the supporting gel component. The printing ink was then removed by flow applied through needles inserted into the gel in the inlet and outlet of the channel structure, leaving behind an open channel with the printed pattern and dimensions within a supporting structure.

Droplet printing involves the use of an inkjet bioprinter to deliver small droplets of bioink (1–100 picolitres; 10–50 µm diameter) (Nakamura et al. 2005; Pereira and Bártolo 2015) which can then be deposited sequentially onto a substrate. Inkjet printing has the advantage of being a rapid process and able to simultaneously print multiple cell types (Thomas and Willerth 2017). It also has the advantages of low cost due to similar components with commercial printers (Bourke et al. 2018) and relatively high cell viability (usually from 80% to 90%), as described in a number of studies (Cui et al. 2012a, b, 2013). The disadvantage of this technique is that the bioinks have to be relatively low in viscosity, typically lower than 10 mPa•s (Derakhshanfar et al. 2018). With low bioink viscosity, the resulting constructs often possess poor mechanical properties (Thomas and Willerth 2017). The cell densities possible are also lower when compared to extrusion-based techniques, as increased cell number within the bioink will increase the average viscosity, resulting in clogging of the printing head (Bourke et al. 2018). This technique also produces small droplets potentially limiting the number of cells per unit volume (Holzl et al. 2016). Within inkjet printing, there are two main technologies used to achieve bioprinting: piezoelectric and thermal bioprinting (Holzl et al. 2016; Saunders and Derby 2014) (see Fig. 8.6). These two methods along with electrostatic inkjet bioprinting form the most commonly used subdivision of bioprinting, which includes drop-on demand inkjet bioprinting, continuous-inkjet bioprinting and electro-hydrodynamic jet bioprinting (Benam et al. 2015).

Laser-induced forward transfer (LIFT) is a technique which has been applied to bioprinting and involves a pulsed laser beam being focussed and scanned over a donor substrate coated with a laser absorbing layer (e.g. gold or titanium), below which is a layer of bioink. The result of the laser pulsing is the localized evaporation at the absorbing layer creating a high-pressure bubble that pushes small volumes of bioink onto a collector. The advantage of this technique is that it is nozzle-free. Both the risk of clogging and the impacts from shear stress on the cells are eliminated (Holzl et al. 2016). The resolution of LIFT is quite high, being in the range of 10–100 µm (Duocastella et al. 2007; Guillotin et al. 2010). Laser bioprinting is considered suitable for bioinks with viscosity ranging between 1 and 300 mPa.s and medium cell densities of approximately 108 cells/ml (Barron et al. 2004; Guillotin et al. 2010; Holzl et al. 2016). The disadvantage of the system is that the cell survival is often below 85% (Hopp et al. 2012). This is due to thermal damage by the nanosecond laser pulses (Derakhshanfar et al. 2018); however, this problem might be alleviated by newer picosecond- and femtosecond-based techniques (Petit et al. 2017). A variation on the method which potentially avoids this issue uses surface acoustic waves (Demirci and Montesano 2007; Tasoglu and Demirci 2013).

The different techniques highlight the number of technical issues which need to be taken into consideration when using bioprinting. Design of system and fabrication approach must take into account the application, cell type and matrix material. Bioprinting has been applied to numerous tissues and applications (Benam et al. 2015; Bourke et al. 2018; Cullen et al. 2007; Derakhshanfar et al. 2018; Murphy and Atala 2014). The technique remains a versatile one which can be tailored to a wider variety of applications. While the number of 3D neural culture applications remains relatively small, these have demonstrated the versatility of bioprinting for neural applications.

3D Printed Neural Cell Culture Systems

Through the use of 3D printed neural culture systems, it has been shown that cell survival, function and growth of neural networks depend on a wide range of factors. The growth of the cells can be influenced by the scaffold or carrier base material used, the incorporation of mobile bioactive factors, tailoring of systems with inclusions and topographies and the length of culture time. A number of different approaches can be taken to bioprint 3D neural culture systems.

The simplest form of bioprinting involves homogenous cells being printed in a bioink to produce tissue-like constructs. Gu et al. showed that small tissue-like constructs could be produced using bioprinting (Gu et al. 2016). In this instance, frontal cortical human neural stem cells were printed in a matrix made of alginate, carboxymethyl-chitosan and agarose. The constructs were made using an extrusion method and built up through a series of criss-cross patterned extruded lengths with the matrix forming a gel by chemical cross-linking following extrusion. The method produced a homogenous cell distribution with high viability throughout the printed construct. Immunophenotyping and reverse transcription quantitative PCR (RT-qPCR) analysis demonstrated successful differentiation into functional neurons and supporting neuroglia. A similar concept was applied to human-induced pluripotent stem cells (iPSCs), and while they were shown to differentiate into cell aggregates demonstrating different cell linages, with the use of neural induction medium they were able to obtain a tissue construct consisting of neural cell types (Gu et al. 2017). The constructs, also made of a combination of alginate, carboxymethyl-chitosan and agarose, were found to have negligible cell death over a period of 7 days, and after 10 days, pluripotency markers were also confirmed with flow cytometry. The bioprinted iPSCs were shown to be differentiated into phenotypes that represented neuronal subtypes and microglia. This demonstrated not only that iPSCs could be printed but also that tissue-like constructs could be created and differentiated into desired cell types. Both studies also demonstrated the suitability of alginate, carboxymethyl-chitosan and agarose as a bioink which can be used with neural cell types.

This approach to producing tissue-like constructs can also be performed using other methods such as inkjet printing. Xu et al. (Miller and Hsieh-Wilson 2015) were able to inkjet-print 3D cell sheets using NT2 neurons and within a fibril gel matrix. This was created through a layer-by-layer process using fibrinogen, followed by thrombin to produce the fibrin gel. The cells were printed onto the layer, and the process was repeated. Each compound layer was 50–70 μm with a total 3D sheet dimension being 5 mm × 5 mm × 1 mm. In this culture, the cells were able to spread and exhibited neurite outgrowth after 12 days. The fibril gel was found to have a loose microporous structure which provided an efficient means of supplying nutrients and oxygen to the cells.

Previous examples have described bioprinting approaches using single cell types; however, this is not a limitation of the technique. In a study by Lee et al., a layered 3D approach was used to make constructs with mixed rat astrocytes and embryonic neurons, which were printed in a multi-layered collagen gel (Lee et al. 2009). The viabilities of both the neurons and astrocytes were between 75% and 80% and not statistically different from the controls. A natural extension of this would be bioprinting of neural spheroids. The use of neural spheroids in neural research is increasing (Dingle et al. 2015; Kraus et al. 2015), but only limited work on bioprinting neural spheroids has been carried out (Han and Hsu 2017).

While the approaches used have largely consisted of cells and matrices, additional components can also be printed along with the cells. Lee et al. used a collagen and fibrin matrix with the fibrin having been loaded with VEGF (Lee et al. 2010). It was observed that mouse neural progenitor cells (C17.2 cells) exhibited significantly higher degrees of morphological changes after 3 days, compared to the control. The cells also had greater than 90% viability and were observed migrating towards the fibrin. It was concluded that the VEGF was effectively bound to the printed fibrin gel and, through its release, was able to support cell function. This illustrates the potential for adding bioactive molecules to bioinks.

In addition to the bioactive molecules, the properties of the matrix can also be modified. Neurons are electrically excitable and may benefit from materials that can transport charge. This has been investigated using graphene which, along with its derivatives, is considered a potential material component for imparting electrical conductivity to biomaterials (Kawano et al. 2012). Wei et al. bioprinted murine NSCs within gelatin methacrylate (GelMA) modified with graphene nanoplatelets (Wei et al. 2016). The cells showed high viability and were also observed to differentiate normally. However, the addition of the graphene nanoplatelets resulted in no significant difference when compared to the control. It is important to note that in this study the electrical properties of the scaffold were not characterized. While not successful in this instance, it is now known that there is a clear lack of understanding of the long-term effects of electrical inclusions, with potential cytotoxicity also being a concern for nanoscale inclusions (Sofroniew 2015). In contrast, entirely electroactive scaffold materials have been shown to impact on cell differentiation and promote the formation of neural networks (Inal et al. 2017; Wan et al. 2015).

While simple tissue constructs can be used for a variety of study applications, bioprinting can be used to create more complex 3D shapes. Lozerno et al. demonstrated the creation of a novel 3D layered construct with a more complex structure containing discrete layers of primary cortical neurons (murine cells harvested from E18 embryos) with the ambition to mimic the layered structure of the human cortex

(Lozano et al. 2015). The structure was stable for 7 days, and viability of encapsulated cells after printing remained stable and over 70% for 5 days. The base material was an RGD peptide-modified gellan gum hydrogel, and each layer can be produced with a different composition, an advantage unique to printing processes, resulting in a free-standing construct containing six layers with different composition. The layered structure was tested with cells in three layers with the top and bottom containing cells and the middle containing no cells. Axon penetration was observed from cell-containing layers to the middle layer which lacked cells. Survival and function past this time point were not investigated; however, the principle of making a more complex tissue structure was demonstrated.

While 3D bioprinted neural cultures can be used to study a variety of topics, they do not appear to have been used in any great capacity to study electrophysiology, a topic relevant to neural interfaces. Xu et al. bioprinted 2D cultures using primary embryonic hippocampal and cortical neurons onto collagen-based biopapers (Xu et al. 2006). Amongst other analysis, the electrophysiological characteristics of the printed neurons were also evaluated by the patch-clamp technique. After 15 days in culture, both printed hippocampal and cortical cells exhibited normal recordings of voltage-gated outward K^+ currents and inward Na^+ currents, suggesting that these cells expressed normal voltage-gated ion channels. Despite being in 2D, this study answered the important question of whether printed neurons could retain normal electrophysiological functions.

Most bioprinting studies in the literature are reported in culture for relatively short time scales. While this may be suitable for these particular experiments, investigation into a longer time scale would be important for creating more informative neural interface models. In a study by de la Vega et al. (2018), human-induced pluripotent stem cells (hiPSCs) were bioprinted in a layered fibrin-based bioink for 30 days. The bioprinting produced good cell viability of greater than 81%. The cells in the neural tissue were found to differentiate normally and after 15 days express the neuronal marker, T-III and markers associated with spinal cord motor neurons (MNs), such as Olig2 and HB9. While the cells at 30 days did appear to be stable and differentiated as indicated by the mature MN marker (ChaT), it was not clear how long these cells could be cultured. There is a clear need and drive for the development of long-term 3D neural cultures. It has been argued that CNS models which can span beyond the time of the initial neural wound healing and remodelling phases would allow for a more meaningful study output (Gilmour et al. 2016).

The technologies and the scope for 3D bioprinting of neurons have the potential to impart insight into neural interfaces for both modelling and implant design. Neurons have been shown to survive the printing processes at high viability, with capacity for tailoring the environment through growth factors, cell types, matrix components and geometries. Crucially, a 30-day culture time (de la Vega et al. 2018) and electrophysiology of bioprinted cells may provide improved translation of models through understanding of the neural network development within these structures (Xu et al. 2006). In the future, bioprinting will undoubtedly provide opportunities to create complex geometries critical to creating tissue models (such as layered tissues). There do remain some challenges to overcome, as not all

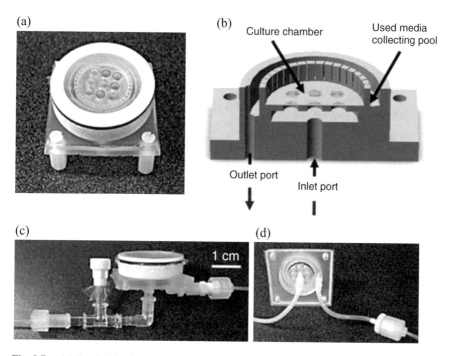

Fig. 8.7 (**a**) Microfluidic chamber designed for 3D neural cell cultures by Cullen et al. (**b**) Cross section showing PDMS culture chambers and polytetrafluoroethylene (PTFE) housing. (**c, d**) Microfluidic device incorporated in bioreactor/perfusion platforms. (Reproduced from with permission from Cullen et al. 2007)

materials can be easily printed with some requiring specialized conditions, such as photopolymerization in the presence of an initiator, which can complicate the printing process and reduce cell viability (Carrow et al. 2015). Despite the numerous advantages of 3D cultures, the viability of cells over time does show some degree of variability depending on the approach (Thomas and Willerth 2017).

8.4.3 Microfluidics in 3D Culture

Microfluidic Approaches

Microfluidic devices utilize precise control and manipulation of fluid flow at sub-millimetre scales based on capillary action, micropumps and microvalves. Microfluidic devices for cell culture have traditionally been fabricated via photolithography and injection moulding processes; however, 3D printing techniques have been a boon to laboratory-scale production of microfluidic devices. The devices are typically made from soft, elastic materials such as polydimethylsiloxane (PDMS) (Fig. 8.7) or hydrogels such as Matrigel or GelMA, which enable them to be

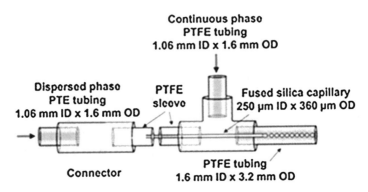

Fig. 8.8 Microfluidic device for the fabrication of microspheres. (Adapted with permission from Young et al. 2013)

compatible with cell culturing and optical imaging (Gupta et al. 2016; Millet and Gillette 2012). An important advantage of these systems that they enable continuous, high-resolution spatiotemporal control of the extracellular microenvironment via controlled perfusion of media. This active perfusion of media overcomes some limitations inherent to traditional 3D cell culture such as mass transport limitations, hence avoiding the development of necrotic cores (Ayuso et al. 2016; Barisam et al. 2018). They also allow for a large degree of flexibility, in both device design and the types of experiments which can be conducted. Other advantages include a high degree of possible control, the potential for automation, low reagent volumes and low cell numbers.

Another application of microfluidics in 3D cell culture is in the fabrication of hydrogel microspheres for cellular encapsulation. Microfluidic-based devices are capable of fabricating monodisperse, core-shell microspheres by delivering a continuous outer phase from an aperture to shear off droplets of an immiscible, cell-loaded hydrogel-phase from an in-line capillary (Fig. 8.8) (Young et al. 2013). The creation of microspheres is due to the coaxial laminar flow interaction between the two immiscible phases. Furthermore, it is possible to create microspheres with dual-shell layers by using coaxial needles. This can confer functionality to the microspheres, such as mimicking the blood-brain barrier through the use of a tightly cross-linked outer shell, while the internal phase is designed to support neuronal growth. The main advantage of the microfluidic systems here is that they overcome issue of the polydispersity found amongst microspheres produced using droplet extrusion techniques (Young et al. 2013).

Application of Microfluidics in 3D Neural Culture

The flexibility of microfluidic systems means that they can be designed and tailored to specific research applications and hence can create custom-made 3D neural cell cultures for probing specific research questions. An example of this is the use of microfluidic systems in attempting to replicate the high-density cell–cell and cell–matrix interactions that occur in native neural tissue. Cullen et al. demonstrated the use of a microfluidic system (Fig. 8.7) for the co-culture of primary murine neurons (E17–E18) and astrocytes (P0–P1) at high cellular densities (greater than 10^4 cells/mm^3) with viabilities over 90% when medium perfusion was 10–11 µL/min (Cullen et al. 2007). Lower perfusion rates resulted in lower viabilities with survival being dependent upon proximity to the perfusion source. While low cell density can be seen as an advantage of microfluidic cultures, here a high cell density demonstrates the flexibility of the approach.

The flexibility of these systems can also be used to exert a high degree of control over the culture environment. Kunze et al. describe a microfluidic system which utilizes micropatterned, laminated agarose hydrogel layers with neurotrophic chemogradient channels to spatially control the synaptic density throughout the construct (Kunze et al. 2011). In addition to trying to replicate the complex in vivo neural environment, microfluidic systems can also be used to probe specific cell–cell interactions, by controlling the introduction and geographical location of individual cell types. Additionally, microfluidic systems can be designed to investigate specific pathologies. Osaki et al. developed a microfluidic-based 3D model of stem cell-derived motor neuron spheroids and endothelial cells to investigate neurovascular coupling and its role in neurodegenerative diseases such as motor neuron disease (Osaki et al. 2018).

The other method in which microfluidic culture can be used is in the production of microspheres. Kim et al. encapsulated mouse embryonic carcinoma (EC) cells within alginate hydrogel microspheres and cultured them within a microfluidic device for 10 days to study the effect of retinoic acid on the neuronal differentiation of EC cells (Kim et al. 2012). Such an experimental set-up can be used as a 3D neural model for high-throughput screening of bioactive agents. Allesandri et al. designed a dual-shell microsphere system comprising of an outer alginate hydrogel layer and an internal layer of reconstituted extracellular matrix (ECM) (Alessandri et al. 2016). The ECM layer is only a few microns thick and designed to mimic the basal membrane of a cellular niche. The microspheres were used to encapsulate human neural stem cells derived from iPSCs and further differentiate them into neurons with approximately 98% viability.

Both methods demonstrate a large degree of flexibly and control and confer some important other advantages. The low cell numbers and small reagent volumes have the potential for high-throughput experiments. The systems also benefit from the ability to be automated. However, there are a number of disadvantages with using these approaches. With the microfluidic system, there may be technical challenges to consider such as the requirement of complex operational control and design. The use of non-standard culture protocols and novel culture surface (such as PDMS) also makes comparative analysis difficult which can be limiting from research perspective. Furthermore, with the use of small volumes, there might be challenges to subsequent analytical chemistry (Halldorsson et al. 2015).

8.5 Conclusions

Culture systems, being inherently outside of the natural biological environment, are limited in what they can model. However, they provide valuable insights into cell behaviours and potential interventions that are not easily explored in vivo.

Explant/ex vivo models are able to provide intact 3D tissue to model electro-physiological properties of cells allowing extra- or intracellular stimulation and recording. In the case of explant slices, their intact structure also means that they can be used to study cellular reactions to a synthetic device or interface. However, while certain tissues can be cultured for weeks or even months, most tissues last for short time periods. All explanted tissues gradually deteriorate over time, with changes occurring in the cellular structure, such as the loss of astrocyte layer-specific distribution in hippocampal explants. Explants and in vivo tissues are considered difficult to prepare, being both technique sensitive and time-consuming.

The use of cell only 3D cultures provides an obvious alternative. Cell cultures remain a viable model for interface work being cost-effective, well-established in research and with the capacity for high-throughput studies. They can be used to model a number of relevant subjects such as glial scar responses, drug delivery and biomaterial interactions. However, when used in isolation, without ECM support, their complexity in modelling the 3D microenvironment is limited.

The combination of cells with 3D biomaterial constructs provides the capacity to study the widest range of factors. This approach seems to represent the greatest potential to produce complex 3D cultures to model the different components of neural interfaces. Previously using scaffolds required incorporating cells post synthesis or into materials at the cost of a reduced viability. Newer techniques including bioprinting and microfluidics can incorporate cells into the synthesis process with improved viability. These techniques provide an increasing number of options for 3D culture models with the ability to tailor the properties of scaffolds allowing neural cultures to be created and shaped according to the requirements of specific studies.

While progress has been made over the last few decades on modelling neural tissue, there remain a number of important challenges. The length of time these models are generally cultured remains too short to accurately model the chronic implantation of an interface device. There remains the need for more chronic time scale work to produce better models. Additionally, viable tissue models may inherently be limited in size due to the lack of vasculature. Furthermore, no model can yet capture the complex interconnectivity of an intact nervous system. However, it can be envisioned that with ongoing developments in biomaterial technologies and with continued advancement of 3D printing on the microscale, models for neural interfaces will be produced which better replicate the physiological behaviour and functionality of neural and surrounding tissues.

References

Alessandri, K., Feyeux, M., Gurchenkov, B., Delgado, C., Trushko, A., Krause, K. H., Vignjevic, D., Nassoy, P., & Roux, A. (2016). A 3D printed microfluidic device for production of functionalized hydrogel microcapsules for culture and differentiation of human neuronal stem cells (hNSC). *Lab on a Chip, 16*(9), 1593–1604. https://doi.org/10.1039/c6lc00133e.

Antman-Passig, M., & Shefi, O. (2016). Remote magnetic orientation of 3D collagen hydrogels for directed neuronal regeneration. *Nano Letters, 16*(4), 2567–2573. https://doi.org/10.1021/acs.nanolett.6b00131.

Aregueta-Robles, U. A., Woolley, A. J., Poole-Warren, L. A., Lovell, N. H., & Green, R. A. (2014). Organic electrode coatings for next-generation neural interfaces. *Frontiers in Neuroengineering, 7*, 421. https://doi.org/10.3389/fneng.2014.00015.

Ayuso, J. M., Virumbrales-Muñoz, M., Lacueva, A., Lanuza, P. M., Checa-Chavarria, E., Botella, P., Fernández, E., Doblare, M., Allison, S. J., Phillips, R. M., Pardo, J., Fernandez, L. J., & Ochoa, I. (2016). Development and characterization of a microfluidic model of the tumour microenvironment. *Scientific Reports, 6*, 36086. https://doi.org/10.1038/srep36086, https://www.nature.com/articles/srep36086#supplementary-information.

Banerjee, A., Arha, M., Choudhary, S., Ashton, R. S., Bhatia, S. R., Schaffer, D. V., & Kane, R. S. (2009). The influence of hydrogel modulus on the proliferation and differentiation of encapsulated neural stem cells. *Biomaterials, 30*(27), 4695–4699. https://doi.org/10.1016/j.biomaterials.2009.05.050.

Barisam, M., Saidi, M. S., Kashaninejad, N., & Nguyen, N.-T. (2018). Prediction of necrotic core and hypoxic zone of multicellular spheroids in a microbioreactor with a U-shaped barrier. *Micromachines, 9*(3), 94. https://doi.org/10.3390/mi9030094.

Barron, J. A., Wu, P., Ladouceur, H. D., & Ringeisen, B. R. J. B. M. (2004). Biological laser printing: a novel technique for creating heterogeneous 3-dimensional cell patterns. *Biomedical Microdevices, 6*(2), 139–147. https://doi.org/10.1023/B:BMMD.0000031751.67267.9f.

Benam, K. H., Dauth, S., Hassell, B., Herland, A., Jain, A., Jang, K. J., Karalis, K., Kim, H. J., MacQueen, L., Mahmoodian, R., Musah, S., Torisawa, Y. S., van der Meer, A. D., Villenave, R., Yadid, M., Parker, K. K., & Ingber, D. E. (2015). Engineered in vitro disease models. *Annual Review of Pathology, 10*(1), 195–262. https://doi.org/10.1146/annurev-pathol-012414-040418.

Berdichevsky, Y., Sabolek, H., Levine, J. B., Staley, K. J., & Yarmush, M. L. (2009). Microfluidics and multielectrode array-compatible organotypic slice culture method. *Journal of Neuroscience Methods, 178*(1), 59–64. https://doi.org/10.1016/j.jneumeth.2008.11.016.

Berdichevsky, Y., Staley, K. J., & Yarmush, M. L. (2010). Building and manipulating neural pathways with microfluidics. *Lab on a Chip, 10*(8), 999–996. https://doi.org/10.1039/b922365g.

Bourke, J. L., Quigley, A. F., Duchi, S., O'Connell, C. D., Crook, J. M., Wallace, G. G., Cook, M. J., & Kapsa, R. M. I. (2018). Three-dimensional neural cultures produce networks that mimic native brain activity. *Journal of Tissue Engineering and Regenerative Medicine, 12*(2), 490–493. https://doi.org/10.1002/term.2508.

Briggman, K. L., & Euler, T. (2011). Bulk electroporation and population calcium imaging in the adult mammalian retina. *Journal of Neurophysiology, 105*(5), 2601–2609. https://doi.org/10.1152/jn.00722.2010.

Caffe, A. R., Ahuja, P., Holmqvist, B., Azadi, S., Forsell, J., Holmqvist, I., Soderpalm, A. K., & van Veen, T. (2001). Mouse retina explants after long-term culture in serum free medium. *Journal of Chemical Neuroanatomy, 22*(4), 263–273. https://doi.org/10.1016/S0891-0618(01)00140-5.

Carrow, J. K., Kerativitayanan, P., Jaiswal, M. K., Lokhande, G., & Gaharwar, A. K. (2015). Polymers for bioprinting. In *Essentials of 3D biofabrication and translation* (pp. 229–248). Elsevier. https://doi.org/10.1016/B978-0-12-800972-7.00013-X.

Chang, R., Nam, J., & Sun, W. (2008). Effects of dispensing pressure and nozzle diameter on cell survival from solid freeform fabrication–based direct cell writing. *Tissue Engineering. Part A, 14*(1), 41–48. https://doi.org/10.1089/ten.a.2007.0004.

Chen, N., Luo, B., Yang, I. H., Thakor, N. V., & Ramakrishna, S. (2018). Biofunctionalized platforms towards long-term neural interface. *Current Opinion in Biomedical Engineering,* *6*(March), 81–91. https://doi.org/10.1016/j.cobme.2018.03.002.

Cheng, T. Y., Chen, M. H., Chang, W. H., Huang, M. Y., & Wang, T. W. (2013). Neural stem cells encapsulated in a functionalized self-assembling peptide hydrogel for brain tissue engineering. *Biomaterials, 34*(8), 2005–2016. https://doi.org/10.1016/j.biomaterials.2012.11.043.

Chia, H. N., & Wu, B. M. (2015). Recent advances in 3D printing of biomaterials. *Journal of Biological Engineering, 9,* 4. https://doi.org/10.1186/s13036-015-0001-4.

Cui, X., Breitenkamp, K., Finn, M. G., Lotz, M., & D'Lima, D. D. (2012a). Direct human cartilage repair using three-dimensional bioprinting technology. *Tissue Engineering. Part A, 18*(11–12), 1304–1312. https://doi.org/10.1089/ten.TEA.2011.0543.

Cui, X., Breitenkamp, K., Lotz, M., & D'Lima, D. (2012b). Synergistic action of fibroblast growth factor-2 and transforming growth factor-beta1 enhances bioprinted human neocartilage formation. *Biotechnology and Bioengineering, 109*(9), 2357–2368. https://doi.org/10.1002/bit.24488.

Cui, X., Gao, G., & Qiu, Y. J. B. L. (2013). Accelerated myotube formation using bioprinting technology for biosensor applications. *Biotechnology Letters, 35*(3), 315–321. https://doi.org/10.1007/s10529-012-1087-0.

Cullen, D. K., Vukasinovic, J., Glezer, A., & Laplaca, M. C. (2007). Microfluidic engineered high cell density three-dimensional neural cultures. *Journal of Neural Engineering, 4*(2), 159–172. https://doi.org/10.1088/1741-2560/4/2/015.

Daviaud, N., Garbayo, E., Schiller, P. C., Perez-Pinzon, M., & Montero-Menei, C. N. (2013). Organotypic cultures as tools for optimizing central nervous system cell therapies. *Experimental Neurology, 248,* 429–440. https://doi.org/10.1016/j.expneurol.2013.07.012.

de la Vega, L., Gomez, D. A. R., Abelseth, E., Abelseth, L., da Silva, V. A., & Willerth, S. M. (2018). 3D bioprinting human induced pluripotent stem cell-derived neural tissues using a novel lab-on-a-printer technology. *Applied Sciences, 8*(12), 2414. https://doi.org/10.3390/app8122414.

De Simoni, A., & My Yu, L. (2006). Preparation of organotypic hippocampal slice cultures: Interface method. *Nature Protocols, 1*(3), 1439–1445. https://doi.org/10.1038/nprot.2006.228.

Deb, P., Deoghare, A. B., Borah, A., Barua, E., & Das Lala, S. (2018). Scaffold development using biomaterials: A review. *Materials Today: Proceedings, 5*(5, Part 2), 12909–12919. https://doi.org/10.1016/j.matpr.2018.02.276.

Demirci, U., & Montesano, G. (2007). Single cell epitaxy by acoustic picolitre droplets. *Lab on a Chip, 7*(9), 1139–1145. https://doi.org/10.1039/B704965J.

Derakhshanfar, S., Mbeleck, R., Xu, K., Zhang, X., Zhong, W., & Xing, M. (2018). 3D bioprinting for biomedical devices and tissue engineering: A review of recent trends and advances. *Bioactive Materials, 3*(2), 144–156. https://doi.org/10.1016/j.bioactmat.2017.11.008.

Dingle, Y. T., Boutin, M. E., Chirila, A. M., Livi, L. L., Labriola, N. R., Jakubek, L. M., Morgan, J. R., Darling, E. M., Kauer, J. A., & Hoffman-Kim, D. (2015). Three-dimensional neural spheroid culture: An in vitro model for cortical studies. *Tissue Engineering Part C, Methods, 21*(12), 1274–1283. https://doi.org/10.1089/ten.TEC.2015.0135.

Duocastella, M., Colina, M., Fernández-Pradas, J. M., Serra, P., & Morenza, J. L. (2007). Study of the laser-induced forward transfer of liquids for laser bioprinting. *Applied Surface Science, 253*(19), 7855–7859. https://doi.org/10.1016/j.apsusc.2007.02.097.

Egawa, E. Y., Kato, K., Hiraoka, M., Nakaji-Hirabayashi, T., & Iwata, H. (2011). Enhanced proliferation of neural stem cells in a collagen hydrogel incorporating engineered epidermal growth factor. *Biomaterials, 32*(21), 4737–4743. https://doi.org/10.1016/j.biomaterials.2011.03.033.

Egert, U., Schlosshauer, B., Fennrich, S., Nisch, W., Fejtl, M., Knott, T., Muller, T., & Hammerle, H. (1998). A novel organotypic long-term culture of the rat hippocampus on substrate-integrated multielectrode arrays. *Brain Research. Brain Research Protocols, 2*(4), 229–242. papers3://publication/uuid/7E2EF671-F4E1-4CB1-A26B-F327723D0BE4.

Entekhabi, E., Haghbin Nazarpak, M., Moztarzadeh, F., & Sadeghi, A. (2016). Design and manufacture of neural tissue engineering scaffolds using hyaluronic acid and polycaprolactone nanofibers with controlled porosity. *Materials Science & Engineering. C, Materials for Biological Applications, 69*, 380–387. https://doi.org/10.1016/j.msec.2016.06.078.

Eugene, E., Cluzeaud, F., Cifuentes-Diaz, C., Fricker, D., Le Duigou, C., Clemenceau, S., Baulac, M., Poncer, J. C., & Miles, R. (2014). An organotypic brain slice preparation from adult patients with temporal lobe epilepsy. *Journal of Neuroscience Methods, 235*, 234–244. https://doi.org/10.1016/j.jneumeth.2014.07.009.

Evans, A. J., Thompson, B. C., Wallace, G. G., Millard, R., O'Leary, S. J., Clark, G. M., Shepherd, R. K., & Richardson, R. T. (2009). Promoting neurite outgrowth from spiral ganglion neuron explants using polypyrrole/BDNF-coated electrodes. *Journal of Biomedial Materials Research Part A, 91A*(1), 241–250. https://doi.org/10.1002/jbm.a.32228.

Famm, K. (2013). Drug discovery: A jump-start for electroceuticals. *Nature, 496*(7445), 159–161. papers3://publication/uuid/6214CB3D-B1DF-45EE-AF58-F784E75A9495.

Fan, L., Liu, C., Chen, X., Zou, Y., Zhou, Z., Lin, C., Tan, G., Zhou, L., Ning, C., & Wang, Q. (2018). Directing induced pluripotent stem cell derived neural stem cell fate with a three-dimensional biomimetic hydrogel for spinal cord injury repair. *ACS Applied Materials & Interfaces, 10*(21), 17742–17755. https://doi.org/10.1021/acsami.8b05293.

Fattahi, P., Yang, G., Kim, G., & Abidian, M. R. (2014). A review of organic and inorganic biomaterials for neural interfaces. *Advanced Materials, 26*(12), 1846–1885. https://doi.org/10.1002/adma.201304496.

Feigenspan, A., Bormann, J., & Wässle, H. (2009). Organotypic slice culture of the mammalian retina. *Visual Neuroscience, 10*(02), 203–217. https://doi.org/10.1017/s0952523800003618.

Fornaro, M., Sharthiya, H., & Tiwari, V. (2018). Adult mouse DRG explant and dissociated cell models to investigate neuroplasticity and responses to environmental insults including viral infection. *Journal of Visualized Experiments*, (133). https://doi.org/10.3791/56757.

Frega, M., Tedesco, M., Massobrio, P., Pesce, M., & Martinoia, S. (2014). Network dynamics of 3D engineered neuronal cultures: A new experimental model for in-vitro electrophysiology. *Scientific Reports, 4*, 5489. https://doi.org/10.1038/srep05489.

Gähwiler, B. H., Capogna, M., Debanne, D., McKinney, R. A., & Thompson, S. M. (1997). Organotypic slice cultures: A technique has come of age. *Trends in Neurosciences, 20*(10), 471–477. https://doi.org/10.1016/S0166-2236(97)01122-3.

Gao, H. M., Jiang, J., Wilson, B., Zhang, W., Hong, J. S., & Liu, B. (2002). Microglial activation-mediated delayed and progressive degeneration of rat nigral dopaminergic neurons: Relevance to Parkinson's disease. *Journal of Neurochemistry, 81*(6), 1285–1297. https://doi.org/10.1046/j.1471-4159.2002.00928.x.

Gerardo-Nava, J., Hodde, D., Katona, I., Bozkurt, A., Grehl, T., Steinbusch, H. W., Weis, J., & Brook, G. A. (2014). Spinal cord organotypic slice cultures for the study of regenerating motor axon interactions with 3D scaffolds. *Biomaterials, 35*(14), 4288–4296. https://doi.org/10.1016/j.biomaterials.2014.02.007.

Germer, A., Kuhnel, K., Grosche, J., Friedrich, A., Wolburg, H., Price, J., Reichenbach, A., & Mack, A. F. (1997). Development of the neonatal rabbit retina in organ culture. 1. Comparison with histogenesis in vivo, and the effect of a gliotoxin (alpha-aminoadipic acid). *Anatomy and Embryology (Berl), 196*(1), 67–79. papers3://publication/uuid/A3725BB1-8323-46B5-B900-612B96200FB7.

Gilmour, A. D., Woolley, A. J., Poole-Warren, L. A., Thomson, C. E., & Green, R. A. (2016). A critical review of cell culture strategies for modelling intracortical brain implant material reactions. *Biomaterials, 91*, 23–43. https://doi.org/10.1016/j.biomaterials.2016.03.011.

Glazova, M. V., Pak, E. S., & Murashov, A. K. (2015). Neurogenic potential of spinal cord organotypic culture. *Neuroscience Letters, 594*, 60–65. https://doi.org/10.1016/j.neulet.2015.03.041.

Gogolla, N., Galimberti, I., DePaola, V., & Caroni, P. (2006). Preparation of organotypic hippocampal slice cultures for long-term live imaging. *Nature Protocols, 1*(3), 1165–1171. https://doi.org/10.1038/nprot.2006.168.

Green, R. A., Lim, K. S., Henderson, W. C., Hassarati, R. T., Martens, P. J., Lovell, N. H., & Poole-Warren, L. A. (2013). Living electrodes: Tissue engineering the neural interface. In *Conference proceedings: Annual international conference of the IEEE engineering in medicine and biology society 2013* (pp. 6957–6960). https://doi.org/10.1109/embc.2013.6611158.

Griffith, R. W., & Humphrey, D. R. (2006). Long-term gliosis around chronically implanted platinum electrodes in the Rhesus macaque motor cortex. *Neuroscience Letters, 406*(1–2), 81–86. https://doi.org/10.1016/j.neulet.2006.07.018.

Grill, W. M., Norman, S. E., & Bellamkonda, R. V. (2009). Implanted neural interfaces: Biochallenges and engineered solutions. *Annual Review of Biomedical Engineering, 11*, 1–24. https://doi.org/10.1146/annurev-bioeng-061008-124927.

Gu, Q., Tomaskovic-Crook, E., Lozano, R., Chen, Y., Kapsa, R. M., Zhou, Q., Wallace, G. G., & Crook, J. M. (2016). Functional 3D neural mini-tissues from printed gel-based bioink and human neural stem cells. *Advanced Healthcare Materials, 5*(12), 1429–1438. https://doi.org/10.1002/adhm.201600095.

Gu, Q., Tomaskovic-Crook, E., Wallace, G. G., & Crook, J. M. (2017). 3D bioprinting human induced pluripotent stem cell constructs for in situ cell proliferation and successive multi-lineage differentiation. *Advanced Healthcare Materials, 6*(17). https://doi.org/10.1002/adhm.201700175.

Guillotin, B., Souquet, A., Catros, S., Duocastella, M., Pippenger, B., Bellance, S., Bareille, R., Rémy, M., Bordenave, L., Amédée, J., & Guillemot, F. (2010). Laser assisted bioprinting of engineered tissue with high cell density and microscale organization. *Biomaterials, 31*(28), 7250–7256. https://doi.org/10.1016/j.biomaterials.2010.05.055.

Gupta, N., Liu, J. R., Patel, B., Solomon, D. E., Vaidya, B., & Gupta, V. (2016). Microfluidics-based 3D cell culture models: Utility in novel drug discovery and delivery research. *Bioengineering and Translational Medicine, 1*(1), 63–81. https://doi.org/10.1002/btm2.10013.

Hahnewald, S., Roccio, M., Tscherter, A., Streit, J., Ambett, R., & Senn, P. (2016). Spiral ganglion neuron explant culture and electrophysiology on multi electrode arrays. *Journal of Visualized Experiments, 116*, 7. https://doi.org/10.3791/54538.

Hailer, N. P., Jarhult, J. D., & Nitsch, R. (1996). Resting microglial cells in vitro: Analysis of morphology and adhesion molecule expression in organotypic hippocampal slice cultures. *Glia, 18*(4), 319–331. https://doi.org/10.1002/(SICI)1098-1136(199612)18:43.0.CO;2-S.

Halldorsson, S., Lucumi, E., Gómez-Sjöberg, R., & Fleming, R. M. T. (2015). Advantages and challenges of microfluidic cell culture in polydimethylsiloxane devices. *Biosensors and Bioelectronics, 63*, 218–231. https://doi.org/10.1016/j.bios.2014.07.029.

Han, H.-W., & Hsu, S.-H. (2017). Using 3D bioprinting to produce mini-brain. *Neural Regeneration Research, 12*(10), 1595–1596. https://doi.org/10.4103/1673-5374.217325.

Hatsopoulos, N. G., & Donoghue, J. P. (2009). The science of neural interface systems. *Annual Review of Neuroscience, 32*, 249–266. https://doi.org/10.1146/annurev.neuro.051508.135241.

Heidemann, M., Streit, J., & Tscherter, A. (2014). Functional regeneration of intraspinal connections in a new in vitro model. *Neuroscience, 262*(C), 40–52. https://doi.org/10.1016/j.neuroscience.2013.12.051.

Highley, C. B., Rodell, C. B., & Burdick, J. A. (2015). Direct 3D printing of shear-thinning hydrogels into self-healing hydrogels. *Advanced Materials, 27*(34), 5075–5079. https://doi.org/10.1002/adma.201501234.

Hodgkin, A. L. (1948). The local electric changes associated with repetitive action in a non-medullated axon. *The Journal of Physiology, 107*(2), 165–181. papers3://publication/uuid/4DAD2DD3-9845-40B8-850A-2EEEAD275B5D.

Holzl, K., Lin, S., Tytgat, L., Van Vlierberghe, S., Gu, L., & Ovsianikov, A. (2016). Bioink properties before, during and after 3D bioprinting. *Biofabrication, 8*(3), 032002. https://doi.org/10.1088/1758-5090/8/3/032002.

Hopkins, A. M., DeSimone, E., Chwalek, K., & Kaplan, D. L. (2015). 3D in vitro modeling of the central nervous system. *Progress in Neurobiology, 125*, 1–25. https://doi.org/10.1016/j.pneurobio.2014.11.003.

Hopp, B., Smausz, T., Szab, G., Kolozsvri, L., Kafetzopoulos, D., Fotakis, C., & Ngrdi, A. (2012). Femtosecond laser printing of living cells using absorbing film-assisted laser-induced forward transfer. *Optical Engineering, 51*(1), 014302. https://doi.org/10.1117/1.oe.51.1.014302.

Hospodiuk, M., Dey, M., Sosnoski, D., & Ozbolat, I. T. (2017). The bioink: A comprehensive review on bioprintable materials. *Biotechnology Advances, 35*(2), 217–239. https://doi.org/10.1016/j.biotechadv.2016.12.006.

Hsiao, M. C., Yu, P. N., Song, D., Liu, C. Y., Heck, C. N., Millett, D., & Berger, T. W. (2015). An in vitro seizure model from human hippocampal slices using multi-electrode arrays. *Journal of Neuroscience Methods, 244*, 154–163. https://doi.org/10.1016/j.jneumeth.2014.09.010.

Humpel, C. (2015a). Neuroscience forefront review organotypic brain slice cultures: A review. *Neuroscience, 305*(C), 86–98. https://doi.org/10.1016/j.neuroscience.2015.07.086.

Humpel, C. (2015b). Organotypic brain slice cultures: A review. *Neuroscience, 305*(C), 86–98. https://doi.org/10.1016/j.neuroscience.2015.07.086.

Inal, S., Hama, A., Ferro, M., Pitsalidis, C., Oziat, J., Iandolo, D., Pappa, A.-M., Hadida, M., Huerta, M., Marchat, D., Mailley, P., & Owens, R. M. (2017). Conducting polymer scaffolds for hosting and monitoring 3D cell culture. *Advanced Biosystems, 1*(6), 1700052. https://doi.org/10.1002/adbi.201700052.

Jeffery, A. F., Churchward, M. A., Mushahwar, V. K., Todd, K. G., & Elias, A. L. (2014). Hyaluronic acid-based 3D culture model for in vitro testing of electrode biocompatibility. *Biomacromolecules, 15*(6), 2157–2165. https://doi.org/10.1021/bm500318d.

Jiang, X., Cao, H. Q., Shi, L. Y., Ng, S. Y., Stanton, L. W., & Chew, S. Y. (2012). Nanofiber topography and sustained biochemical signaling enhance human mesenchymal stem cell neural commitment. *Acta Biomaterialia, 8*(3), 1290–1302. https://doi.org/10.1016/j.actbio.2011.11.019.

Johnson, P. J., Tatara, A., Shiu, A., & Sakiyama-Elbert, S. E. (2010). Controlled release of neurotrophin-3 and platelet-derived growth factor from fibrin scaffolds containing neural progenitor cells enhances survival and differentiation into neurons in a subacute model of SCI. *Cell Transplantation, 19*(1), 89–101. https://doi.org/10.3727/096368909X477273.

Kawano, H., Kimura-Kuroda, J., Komuta, Y., Yoshioka, N., Li, H. P., Kawamura, K., Li, Y., & Raisman, G. (2012). Role of the lesion scar in the response to damage and repair of the central nervous system. *Cell and Tissue Research, 349*(1), 169–180. https://doi.org/10.1007/s00441-012-1336-5.

Khodagholy, D., Doublet, T., Quilichini, P., Gurfinkel, M., Leleux, P., Ghestem, A., Ismailova, E., Herve, T., Sanaur, S., Bernard, C., & Malliaras, G. G. (2013). In vivo recordings of brain activity using organic transistors. *Nature Communications, 4*(1), 1575. https://doi.org/10.1038/ncomms2573.

Kim, C., Bang, J. H., Kim, Y. E., Lee, J. H., & Kang, J. Y. (2012). Stable hydrodynamic trapping of hydrogel beads for on-chip differentiation analysis of encapsulated stem cells. *Sensors and Actuators B: Chemical, 166*, 859–869. https://doi.org/10.1016/j.snb.2012.02.008.

Kim, H., Kim, E., Park, M., Lee, E., & Namkoong, K. (2013). Organotypic hippocampal slice culture from the adult mouse brain: A versatile tool for translational neuropsychopharmacology. *Progress in Neuro-Psychopharmacology & Biological Psychiatry, 41*(C), 36–43. https://doi.org/10.1016/j.pnpbp.2012.11.004.

Ko, K. R., & Frampton, J. P. (2016). Developments in 3D neural cell culture models: The future of neurotherapeutics testing? *Expert Review of Neurotherapeutics, 16*(7), 739–741. https://doi.org/10.1586/14737175.2016.1166053.

Kraus, D., Boyle, V., Leibig, N., Stark, G. B., & Penna, V. (2015). The neuro-spheroid—A novel 3D in vitro model for peripheral nerve regeneration. *Journal of Neuroscience Methods, 246*, 97–105. https://doi.org/10.1016/j.jneumeth.2015.03.004.

Kunze, A., Valero, A., Zosso, D., & Renaud, P. (2011). Synergistic NGF/B27 gradients position synapses heterogeneously in 3D micropatterned neural cultures. *PLoS One, 6*(10), e26187. https://doi.org/10.1371/journal.pone.0026187.

Kuzum, D., Takano, H., Shim, E., Reed, J. C., Juul, H., Richardson, A. G., de Vries, J., Bink, H., Dichter, M. A., Lucas, T. H., Coulter, D. A., Cubukcu, E., & Litt, B. (2014). Transparent and flexible low noise graphene electrodes for simultaneous electrophysiology and neuroimaging. *Nature Communications, 5*(1), 5259. https://doi.org/10.1038/ncomms6259.

Lampe, K. J., Mooney, R. G., Bjugstad, K. B., & Mahoney, M. J. (2010). Effect of macromer weight percent on neural cell growth in 2D and 3D nondegradable PEG hydrogel culture. *Journal of Biomedical Materials Research. Part A, 94*(4), 1162–1171. https://doi.org/10.1002/jbm.a.32787.

Lancaster, M. A., Renner, M., Martin, C. A., Wenzel, D., Bicknell, L. S., Hurles, M. E., Homfray, T., Penninger, J. M., Jackson, A. P., & Knoblich, J. A. (2013). Cerebral organoids model human brain development and microcephaly. *Nature, 501*(7467), 373–379. https://doi.org/10.1038/nature12517.

Lau, L. W., Cua, R., Keough, M. B., Haylock-Jacobs, S., & Yong, V. W. (2013). Pathophysiology of the brain extracellular matrix: A new target for remyelination. *Nature Reviews Neuroscience, 14*(10), 722–729. https://doi.org/10.1038/nrn3550.

Le Duigou, C., Savary, E., Morin-Brureau, M., Gomez-Dominguez, D., Sobczyk, A., Chali, F., Milior, G., Kraus, L., Meier, J. C., Kullmann, D. M., Mathon, B., de la Prida, L. M., Dorfmuller, G., Pallud, J., Eugene, E., Clemenceau, S., & Miles, R. (2018). Imaging pathological activities of human brain tissue in organotypic culture. *Journal of Neuroscience Methods, 298*, 33–44. https://doi.org/10.1016/j.jneumeth.2018.02.001.

Lee, W., Pinckney, J., Lee, V., Lee, J.-H., Fischer, K., Polio, S., Park, J.-K., & Yoo, S.-S. (2009). Three-dimensional bioprinting of rat embryonic neural cells. *Neuroreport, 20*(8), 798–803. https://doi.org/10.1097/WNR.0b013e32832b8be4.

Lee, Y.-B., Polio, S., Lee, W., Dai, G., Menon, L., Carroll, R. S., & Yoo, S.-S. (2010). Bio-printing of collagen and VEGF-releasing fibrin gel scaffolds for neural stem cell culture. *Experimental Neurology, 223*(2), 645–652. https://doi.org/10.1016/j.expneurol.2010.02.014.

Li, H., Wijekoon, A., & Leipzig, N. D. (2012). 3D differentiation of neural stem cells in macroporous photopolymerizable hydrogel scaffolds. *PLoS One, 7*(11), e48824. https://doi.org/10.1371/journal.pone.0048824.

Livni, L., Lees, J. G., Barkl-Luke, M. E., Goldstein, D., & Moalem-Taylor, G. (2019). Dorsal root ganglion explants derived from chemotherapy-treated mice have reduced neurite outgrowth in culture. *Neuroscience Letters, 694*, 14–19. https://doi.org/10.1016/j.neulet.2018.11.016.

Lozano, R., Stevens, L., Thompson, B. C., Gilmore, K. J., Gorkin, R., Stewart, E. M., in het Panhuis, M., Romero-Ortega, M., & Wallace, G. G. (2015). 3D printing of layered brain-like structures using peptide modified gellan gum substrates. *Biomaterials, 67*, 264–273. https://doi.org/10.1016/j.biomaterials.2015.07.022.

Madl, C. M., LeSavage, B. L., Dewi, R. E., Lampe, K. J., & Heilshorn, S. C. (2019). Matrix remodeling enhances the differentiation capacity of neural progenitor cells in 3D hydrogels. *Advanced Science, 6*(4), 1801716. https://doi.org/10.1002/advs.201801716.

Malda, J., Visser, J., Melchels, F. P., Jungst, T., Hennink, W. E., Dhert, W. J., Groll, J., & Hutmacher, D. W. (2013). 25th anniversary article: Engineering hydrogels for biofabrication. *Advanced Materials, 25*(36), 5011–5028. https://doi.org/10.1002/adma.201302042.

Melli, G., & Hoke, A. (2009). Dorsal root ganglia sensory neuronal cultures: A tool for drug discovery for peripheral neuropathies. *Expert Opinion on Drug Discovery, 4*(10), 1035–1045. https://doi.org/10.1517/17460440903266829.

Miller, G. M., & Hsieh-Wilson, L. C. (2015). Sugar-dependent modulation of neuronal development, regeneration, and plasticity by chondroitin sulfate proteoglycans. *Experimental Neurology, 274*(Pt B), 115–125. https://doi.org/10.1016/j.expneurol.2015.08.015.

Millet, L. J., & Gillette, M. U. (2012). Over a century of neuron culture: From the hanging drop to microfluidic devices. *The Yale Journal of Biology and Medicine, 85*(4), 501–521.

Mobini, S., Song, Y. H., McCrary, M. W., & Schmidt, C. E. (2018). Advances in ex vivo models and lab-on-a-chip devices for neural tissue engineering. *Biomaterials, 198*, 146. https://doi.org/10.1016/j.biomaterials.2018.05.012.

Morrison, B., 3rd, Cater, H. L., Benham, C. D., & Sundstrom, L. E. (2006). An in vitro model of traumatic brain injury utilising two-dimensional stretch of organotypic hippocampal slice cultures. *Journal of Neuroscience Methods, 150*(2), 192–201. https://doi.org/10.1016/j. jneumeth.2005.06.014.

Moshayedi, P., Ng, G., Kwok, J. C., Yeo, G. S., Bryant, C. E., Fawcett, J. W., Franze, K., & Guck, J. (2014). The relationship between glial cell mechanosensitivity and foreign body reactions in the central nervous system. *Biomaterials, 35*(13), 3919–3925. https://doi.org/10.1016/j. biomaterials.2014.01.038.

Mouw, J. K., Ou, G., & Weaver, V. M. (2014). Extracellular matrix assembly: a multiscale deconstruction. *Nature Reviews. Molecular Cell Biology, 15*(12), 771–785. https://doi.org/10.1038/ nrm3902.

Mullen, L. M., Pak, K. K., Chavez, E., Kondo, K., Brand, Y., & Ryan, A. F. (2012). Ras/p38 and PI3K/Akt but not Mek/Erk signaling mediate BDNF-induced neurite formation on neonatal cochlear spiral ganglion explants. *Brain Research, 1430*, 25–34. https://doi.org/10.1016/j. brainres.2011.10.054.

Murphy, S. V., & Atala, A. (2014). 3D bioprinting of tissues and organs. *Nature Biotechnology, 32*, 773. https://doi.org/10.1038/nbt.2958.

Naghdi, P., Tiraihi, T., Ganji, F., Darabi, S., Taheri, T., & Kazemi, H. (2016). Survival, proliferation and differentiation enhancement of neural stem cells cultured in three-dimensional polyethylene glycol-RGD hydrogel with tenascin. *Journal of Tissue Engineering and Regenerative Medicine, 10*(3), 199–208. https://doi.org/10.1002/term.1958.

Nakaji-Hirabayashi, T., Kato, K., & Iwata, H. (2009). Hyaluronic acid hydrogel loaded with genetically-engineered brain-derived neurotrophic factor as a neural cell carrier. *Biomaterials, 30*(27), 4581–4589. https://doi.org/10.1016/j.biomaterials.2009.05.009.

Nakamura, M., Kobayashi, A., Takagi, F., Watanabe, A., Hiruma, Y., Ohuchi, K., Iwasaki, Y., Horie, M., Morita, I., & Takatani, S. (2005). Biocompatible inkjet printing technique for designed seeding of individual living cells. *Tissue Engineering, 11*(11–12), 1658–1666. https:// doi.org/10.1089/ten.2005.11.1658.

Nam, Y. (2012). Material considerations for in vitro neural interface technology. *MRS Bulletin, 37*(6), 566–572. https://doi.org/10.1557/mrs.2012.98.

Navarro, X., Krueger, T. B., Lago, N., Micera, S., Stieglitz, T., & Dario, P. (2005). A critical review of interfaces with the peripheral nervous system for the control of neuroprostheses and hybrid bionic systems. *Journal of the Peripheral Nervous System: JPNS, 10*(3), 229–258. https://doi. org/10.1111/j.1085-9489.2005.10303.x.

Ogilvie, J. M., Speck, J. D., Lett, J. M., & Fleming, T. T. (1999). A reliable method for organ culture of neonatal mouse retina with long-term survival. *Journal of Neuroscience Methods, 87*(1), 57–65. papers3://publication/uuid/FBF2051A-BFB3-4AC3-B997-F73F5058301D.

Osaki, T., Sivathanu, V., & Kamm, R. D. (2018). Engineered 3D vascular and neuronal networks in a microfluidic platform. *Scientific Reports, 8*(1), 5168. https://doi.org/10.1038/ s41598-018-23512-1.

Ou, Y. T., Lu, M. S., & Chiao, C. C. (2012). The effects of electrical stimulation on neurite outgrowth of goldfish retinal explants. *Brain Research, 1480*(C), 22–29. https://doi.org/10.1016/j. brainres.2012.08.041.

Pakan, J. M., & McDermott, K. W. (2014). A method to investigate radial glia cell behavior using two-photon time-lapse microscopy in an ex vivo model of spinal cord development. *Frontiers in Neuroanatomy, 8*, 22. https://doi.org/10.3389/fnana.2014.00022.

Park, S. J., Lee, Y. J., Heo, D. N., Kwon, I. K., Yun, K.-S., Kang, J. Y., & Lee, S. H. (2015). Functional nerve cuff electrode with controllable anti-inflammatory drug loading and release by biodegradable nanofibers and hydrogel deposition. *Sensors and Actuators B: Chemical, 215*, 133–141. https://doi.org/10.1016/j.snb.2015.03.036.

Peclin, P., & Rozman, J. (2014). Alternative paradigm of selective vagus nerve stimulation tested on an isolated porcine vagus nerve. *ScientificWorldJournal, 2014*(1), 310283. https://doi. org/10.1155/2014/310283.

Pereira, R. F., & Bártolo, P. J. (2015). 3D bioprinting of photocrosslinkable hydrogel constructs. *Journal of Applied Polymer Science, 132*(48). https://doi.org/10.1002/app.42458.

Petit, S., Kérourédan, O., Devillard, R., & Cormier, E. (2017). Femtosecond versus picosecond laser pulses for film-free laser bioprinting. *Applied Optics, 56*(31), 8648–8655. https://doi.org/10.1364/AO.56.008648.

Pittier, R., Sauthier, F., Hubbell, J. A., & Hall, H. (2005). Neurite extension and in vitro myelination within three-dimensional modified fibrin matrices. *Journal of Neurobiology, 63*(1), 1–14. https://doi.org/10.1002/neu.20116.

Ravikumar, M., Jain, S., Miller, R. H., Capadona, J. R., & Selkirk, S. M. (2012). An organotypic spinal cord slice culture model to quantify neurodegeneration. *Journal of Neuroscience Methods, 211*(2), 280–288. https://doi.org/10.1016/j.jneumeth.2012.09.004.

Raz-Prag, D., Beit-Yaakov, G., & Hanein, Y. (2017). Electrical stimulation of different retinal components and the effect of asymmetric pulses. *Journal of Neuroscience Methods, 291*, 20–27. https://doi.org/10.1016/j.jneumeth.2017.07.028.

Rodriguez, A. L., Bruggeman, K. F., Wang, Y., Wang, T. Y., Williams, R. J., Parish, C. L., & Nisbet, D. R. (2018). Using minimalist self-assembling peptides as hierarchical scaffolds to stabilise growth factors and promote stem cell integration in the injured brain. *Journal of Tissue Engineering and Regenerative Medicine, 12*(3), e1571–e1579. https://doi.org/10.1002/term.2582.

Rzeczinski, S., Victorov, I. V., Lyjin, A. A., Aleksandrova, O. P., Harms, C., Kronenberg, G., Freyer, D., Scheibe, F., Priller, J., Endres, M., & Dirnagl, U. (2006). Roller culture of free-floating retinal slices: A new system of organotypic cultures of adult rat retina. *Ophthalmic Research, 38*(5), 263–269. https://doi.org/10.1159/000095768.

Sarig-Nadir, O., & Seliktar, D. (2010). The role of matrix metalloproteinases in regulating neuronal and nonneuronal cell invasion into PEGylated fibrinogen hydrogels. *Biomaterials, 31*(25), 6411–6416. https://doi.org/10.1016/j.biomaterials.2010.04.052.

Saunders, R. E., & Derby, B. (2014). Inkjet printing biomaterials for tissue engineering: Bioprinting. *International Materials Reviews, 59*(8), 430–448. https://doi.org/10.1179/1743280414Y.0000000040.

Schmidt, C. E., & Leach, J. B. (2003). Neural tissue engineering: Strategies for repair and regeneration. *Annual Review of Biomedical Engineering, 5*(1), 293–347. https://doi.org/10.1146/annurev.bioeng.5.011303.120731.

Schwartz, M. P., Hou, Z., Propson, N. E., Zhang, J., Engstrom, C. J., Santos Costa, V., Jiang, P., Nguyen, B. K., Bolin, J. M., Daly, W., Wang, Y., Stewart, R., Page, C. D., Murphy, W. L., & Thomson, J. A. (2015). Human pluripotent stem cell-derived neural constructs for predicting neural toxicity. *Proceedings of the National Academy of Sciences of the United States of America, 112*(40), 12516–12521. https://doi.org/10.1073/pnas.1516645112.

Seidlits, S. K., Liang, J., Bierman, R. D., Sohrabi, A., Karam, J., Holley, S. M., Cepeda, C., & Walthers, C. M. (2019). Peptide-modified, hyaluronic acid-based hydrogels as a 3D culture platform for neural stem/progenitor cell engineering. *Journal of Biomedical Materials Research. Part A, 107*(4), 704–718. https://doi.org/10.1002/jbm.a.36603.

Shao, Y., Sang, J., & Fu, J. (2015). On human pluripotent stem cell control: The rise of 3D bioengineering and mechanobiology. *Biomaterials, 52*, 26–43. https://doi.org/10.1016/j.biomaterials.2015.01.078.

Sofroniew, M. V. (2015). Astrocyte barriers to neurotoxic inflammation. *Nature Reviews. Neuroscience, 16*(5), 249–263. https://doi.org/10.1038/nrn3898.

Spencer, K. C., Sy, J. C., Falcon-Banchs, R., & Cima, M. J. (2017). A three dimensional in vitro glial scar model to investigate the local strain effects from micromotion around neural implants. *Lab on a Chip, 17*(5), 795–804. https://doi.org/10.1039/c6lc01411a.

Szarowski, D. H., Andersen, M. D., Retterer, S., Spence, A. J., Isaacson, M., Craighead, H. G., Turner, J. N., & Shain, W. (2003). Brain responses to micro-machined silicon devices. *Brain Research, 983*(1–2), 23–35. https://doi.org/10.1016/S0006-8993(03)03023-3.

Tarassoli, S. P., Jessop, Z. M., Kyle, S., & Whitaker, I. S. (2018). 8 - Candidate bioinks for 3D bioprinting soft tissue. In D. J. Thomas, Z. M. Jessop, & I. S. Whitaker (Eds.), *3D bioprinting for reconstructive surgery* (pp. 145–172). Woodhead Publishing. https://doi.org/10.1016/B978-0-08-101103-4.00026-0.

Tasoglu, S., & Demirci, U. (2013). Bioprinting for stem cell research. *Trends in Biotechnology, 31*(1), 10–19. https://doi.org/10.1016/j.tibtech.2012.10.005.

Thomas, M., & Willerth, S. M. (2017). 3-D bioprinting of neural tissue for applications in cell therapy and drug screening. *Frontiers in Bioengineering and Biotechnology, 5*, 69–69. https://doi.org/10.3389/fbioe.2017.00069.

Tian, L., Prabhakaran, M. P., & Ramakrishna, S. (2015). Strategies for regeneration of components of nervous system: Scaffolds, cells and biomolecules. *Regenerative Biomaterials, 2*(1), 31–45. https://doi.org/10.1093/rb/rbu017.

Vallejo-Giraldo, C., Kelly, A., & Biggs, M. J. P. (2014). Biofunctionalisation of electrically conducting polymers. *Drug Discovery Today, 19*(1), 88–94. https://doi.org/10.1016/j.drudis.2013.07.022.

van Bergen, A., Papanikolaou, T., Schuker, A., Möller, A., & Schlosshauer, B. (2003). Long-term stimulation of mouse hippocampal slice culture on microelectrode array. *Brain Research Protocols, 11*(2), 123–133. https://doi.org/10.1016/S1385-299X(03)00024-2.

van Duinen, V., Trietsch, S. J., Joore, J., Vulto, P., & Hankemeier, T. (2015). Microfluidic 3D cell culture: From tools to tissue models. *Current Opinion in Biotechnology, 35*, 118–126. https://doi.org/10.1016/j.copbio.2015.05.002.

Waltz, E. (2016). A spark at the periphery. *Nature Biotechnology, 34*, 904–908.

Wan, A. M.-D., Inal, S., Williams, T., Wang, K., Leleux, P., Estevez, L., Giannelis, E. P., Fischbach, C., Malliaras, G. G., & Gourdon, D. (2015). 3D conducting polymer platforms for electrical control of protein conformation and cellular functions. *Journal of Materials Chemistry B, 3*(25), 5040–5048. https://doi.org/10.1039/C5TB00390C.

Wang, L. S., Boulaire, J., Chan, P. P., Chung, J. E., & Kurisawa, M. (2010). The role of stiffness of gelatin-hydroxyphenylpropionic acid hydrogels formed by enzyme-mediated crosslinking on the differentiation of human mesenchymal stem cell. *Biomaterials, 31*(33), 8608–8616. https://doi.org/10.1016/j.biomaterials.2010.07.075.

Wang, C., Tong, X., Jiang, X., & Yang, F. (2017). Effect of matrix metalloproteinase-mediated matrix degradation on glioblastoma cell behavior in 3D PEG-based hydrogels. *Journal of Biomedical Materials Research. Part A, 105*(3), 770–778. https://doi.org/10.1002/jbm.a.35947.

Wei, Z., Harris, B. T., & Zhang, L. G. (2016). Gelatin methacrylamide hydrogel with graphene nanoplatelets for neural cell-laden 3D bioprinting. In *Conference proceedings: Annual international conference of the IEEE Engineering in medicine and biology society 2016* (pp. 4185–4188). https://doi.org/10.1109/EMBC.2016.7591649.

Willerth, S. M., Arendas, K. J., Gottlieb, D. I., & Sakiyama-Elbert, S. E. (2006). Optimization of fibrin scaffolds for differentiation of murine embryonic stem cells into neural lineage cells. *Biomaterials, 27*(36), 5990–6003. https://doi.org/10.1016/j.biomaterials.2006.07.036.

Willerth, S. M., Rader, A., & Sakiyama-Elbert, S. E. (2008). The effect of controlled growth factor delivery on embryonic stem cell differentiation inside fibrin scaffolds. *Stem Cell Research, 1*(3), 205–218. https://doi.org/10.1016/j.scr.2008.05.006.

Wu, W., DeConinck, A., & Lewis, J. A. (2011). Omnidirectional printing of 3D microvascular networks. *Advanced Materials, 23*(24), H178–H183. https://doi.org/10.1002/adma.201004625.

Xu, T., Gregory, C. A., Molnar, P., Cui, X., Jalota, S., Bhaduri, S. B., & Boland, T. (2006). Viability and electrophysiology of neural cell structures generated by the inkjet printing method. *Biomaterials, 27*(19), 3580–3588. https://doi.org/10.1016/j.biomaterials.2006.01.048.

Yi, Y., Park, J., Lim, J., Lee, C. J., & Lee, S. H. (2015). Central nervous system and its disease models on a chip. *Trends in Biotechnology, 33*(12), 762–776. https://doi.org/10.1016/j.tibtech.2015.09.007.

Young, C., Rozario, K., Serra, C., Poole-Warren, L., & Martens, P. (2013). Poly(vinyl alcohol)-heparin biosynthetic microspheres produced by microfluidics and ultraviolet photopolymerisation. *Biomicrofluidics, 7*(4), 44109–44109. https://doi.org/10.1063/1.4816714.

Yu, C., Griffiths, L. R., & Haupt, L. M. (2017). Exploiting Heparan sulfate proteoglycans in human neurogenesis-controlling lineage specification and fate. *Frontiers in Integrative Neuroscience, 11*(October), 28. https://doi.org/10.3389/fnint.2017.00028.

Zheng, J. L., & Gao, W. Q. (1996). Differential damage to auditory neurons and hair cells by oto-toxins and neuroprotection by specific neurotrophins in rat cochlear organotypic cultures. *The European Journal of Neuroscience, 8*(9), 1897–1905. https://doi.org/10.1111/j.1460-9568.1996.tb01333.x.

Zhuang, P., Sun, A. X., An, J., Chua, C. K., & Chew, S. Y. (2018). 3D neural tissue models: From spheroids to bioprinting. *Biomaterials, 154*, 113–133. https://doi.org/10.1016/j.biomaterials.2017.10.002.

Zustiak, S. P., Pubill, S., Ribeiro, A., & Leach, J. B. (2013). Hydrolytically degradable poly(ethylene glycol) hydrogel scaffolds as a cell delivery vehicle: Characterization of PC12 cell response. *Biotechnology Progress, 29*(5), 1255–1264. https://doi.org/10.1002/btpr.1761.

Chapter 9
Conductive Hydrogels for Bioelectronic Interfaces

Teuku Fawzul Akbar, Christoph Tondera, and Ivan Minev

9.1 Introduction

Current neural interface technologies rely on metal electrodes such as platinum, iridium, stainless steel, and alloys for recording or stimulation (Merrill et al. 2005). The mechanical properties of these electrodes are significantly different in comparison to those of tissues. For example, stiffness of metal in comparison to neural tissue could differ for more than 7 orders of magnitude. Young's modulus of platinum or stainless steel is reported in the range of 100–200 GPa (Rho et al. 1993; Merker et al. 2001), while for the central nervous system tissue, the value is much lower, ranging between 100 Pa and 10 kPa (Lacour et al. 2016). Metal electrodes also have low elastic strain limit (<5%) (Jeong et al. 2015) in contrary to some types of neural tissue which can experience repeated 10–20% tensile strain (Lacour et al. 2016).

Mechanical mismatch between electrode and tissue could aggravate adverse tissue response to electrode especially in invasive and long-term implantation. This tissue response mainly arises as a trauma due to insertion of an electrode and chronic inflammation due to forces which appear during movement and micromotion (Goding et al. 2018a). The interaction between electrode and tissue further induces foreign body reaction that leads to the formation of a fibrous encapsulation layer wherein the case of the central nervous system is called glial scar (Polikov et al. 2005). The presence of this tissue increases impedance of the electrode and extends the distance between electrode and target neurons (Polikov et al. 2005) which consequently gradually reduces the quality of electrical signal after implantation

T. F. Akbar · C. Tondera
Biotechnology Center (BIOTEC) Technische Universität Dresden, Dresden, Germany

Leibniz-Institut für Polymerforschung Dresden, Dresden, Germany

I. Minev (✉)
Biotechnology Center (BIOTEC) Technische Universität Dresden, Dresden, Germany
e-mail: ivan.minev@tu-dresden.de

© Springer Nature Switzerland AG 2020
L. Guo (ed.), *Neural Interface Engineering*,
https://doi.org/10.1007/978-3-030-41854-0_9

(Leach et al. 2010). To minimize those unfavorable effects, several strategies have been attempted in producing soft and stretchable implantable electrodes. For example, Minev et al. integrated gold electrodes on polydimethylsiloxane (PDMS) substrate to create a soft implant for neuromodulation of the spinal cord (Minev et al. 2015). Rodger et al. used platinum and iridium electrodes on Parylene substrate as multielectrode array for stimulation of retina and spinal cord (Rodger et al. 2008). By employing serpentine shape Ti/Au electrodes as interconnect encapsulated by polyimide and low-modulus silicone, Park et al. created an implantable optoelectronic system (Park et al. 2015). Although in comparison to metals these electrodes are softer, the development of truly biomimetic electrode materials still remains a challenge.

Hydrogels are a class of crosslinked polymeric material with high water content which allows modifications of its mechanical and biochemical properties. The benefit of hydrogels for neural interface is not only limited by its tunable mechanical properties. Electrical functionality could also be added by incorporation of conductive materials such as metal nanoparticles, carbon nanoparticles, or conductive polymers. Conductive hydrogels are a promising material class for improving the biointegration of implanted electrode arrays.

There are two mechanisms in which flow of electrons in an electrode is converted into ionic flow in an electrolyte or vice versa to allow communications with cells (Merrill et al. 2005). The first mechanism is capacitive charge transfer. This mechanism involves charging and discharging of an electrical double layer that is formed between the electrode and electrolyte. The second mechanism is faradaic where reduction or oxidation reaction occurs. Faradaic mechanism in general delivers a higher amount of current, although in certain cases it could be undesirable due to the possibility of generation of potentially harmful ions (Rose et al. 1985).

By employing conductive hydrogel as an alternative to metal electrodes, the contact between conductive material and electrolyte occurs inside the bulk of the hydrogel. This significantly increases the electrochemical surface area of the electrode (Cogan et al. 2016a). The electrical double layer forms throughout the volume of the hydrogel resulting in volumetric capacitance as opposed to areal capacitance in metal electrode (Yuk et al. 2018). Expansion of electrochemical surface area results in increasing charge injection capacity that allows reduction of electrode polarization during stimulation (Cogan et al. 2016a). Lower polarization of the electrode reduces the likelihood of harmful reactions to occur (Cogan et al. 2016b).

Another benefit of hydrogel is the versatility in performing tissue engineering. As the porosity, mechanical properties, and biochemical composition of a hydrogel can be engineered to mimic extracellular matrix, cells could be embedded within or made to interact with the electrode (Goding et al. 2018b; Goding et al. 2017; Aregueta-Robles et al. 2014). One potential architecture for such an "electronic tissue" is presented in Fig. 9.1.

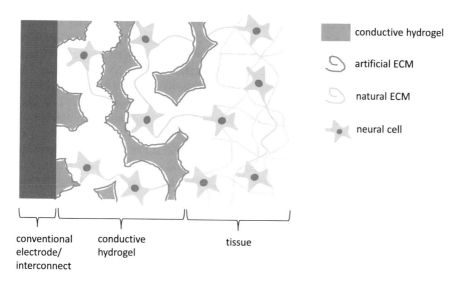

conductive hydrogel

artificial ECM

natural ECM

neural cell

conventional   conductive   tissue
electrode/   hydrogel
interconnect

Fig. 9.1 Concept for an electronic tissue architecture based on porous conductive hydrogel interacting with host tissue (cross-section view)

9.2   Classification of Conductive Hydrogels

9.2.1   Nanocomposite Hydrogels

Incorporation of electronically conductive materials into the hydrogel network could be done to impart electrical functionality to a hydrogel. The most common method of producing this nanocomposite hydrogel is by mixing nanomaterials into the hydrogel precursor before gel formation although other routes of incorporation are also possible. Nanomaterials could be physically added after hydrogel formation (Pardo-Yissar et al. 2001) or converted in situ within a hydrogel that contains nanomaterial precursors (Saravanan et al. 2007). Another method is employing nanomaterials as a crosslinker for the polymer network (Skardal et al. 2010). Examples of these strategies are outlined below.

Metallic nanoparticles Incorporating metal nanomaterials in hydrogels is a straightforward way to impart electrical conductivity. Metal nanomaterials are usually added together with hydrogel precursors before crosslinking. Noble metals such as gold, silver, and platinum were used in conductive nanocomposite hydrogels (Dvir et al. 2011; Lim et al. 2019; Zhai et al. 2013; Navaei et al. 2016). As for the use of hydrogels for neural interface, the metal must be stable in physiological environment and should not yield a toxic by-product that can harm cells under stimulation.

Incorporation of gold nanowires within alginate hydrogel was shown by Dvir et al. to improve electrical communication between cardiac cells even when the nanowires were not physically interconnected (Dvir et al. 2011). Besides, addition of the nanomaterials was also shown to improve the mechanical properties of the composite due to interaction between nanomaterials and the polymer matrix. The fabricated nanocomposite hydrogel had a compressive modulus of around 3.5 kPa which is significantly lower compared to that of native heart at around 425 kPa. Low compressive modulus hydrogel is beneficial as it does not inhibit contraction of cardiac tissue. With improved electrical and mechanical properties, the hydrogel could be used as a cardiac patch for treating damaged heart tissue. By applying electrical stimulation to the cells in contact with gold-embedded hydrogels cardiac patches, alignment of cells (Dvir et al. 2011), and uniformity of their beating behavior (Navaei et al. 2016) are improved.

Addition of some types of nanomaterials could increase the stiffness of the hydrogel significantly which is undesirable for some applications. For example, Lim et al. reported that incorporation of a 0.5% weight silver nanowire into alginate hydrogel increased Young's modulus of the material to around 15 MPa despite improvement of its electrical conductivity (Lim et al. 2019).

Carbon Graphene and carbon nanotubes (CNTs) are two allotropes of carbon that have been incorporated in carbon hydrogel nanocomposites. CNT has a cylindrical structure, while graphene has a two-dimensional layer structure. The use of carbon in conductive hydrogel is quite popular due to its high electrical conductivity, low mass density, and chemical stability in aqueous environment (Novoselov et al. 2012; Tasis et al. 2006). Moreover, carbon-based nanomaterial is a good candidate for neural stimulation as it provides high charge injection ability which operates mainly by capacitive charge transfer mechanism (Lu et al. 2016; Wang et al. 2006). Both materials have high surface to volume ratio that allows chemical modification to alter its surface properties.

Although the conductivity of graphene is very high, graphene is difficult to process as it lacks functional groups that interact with water or organic solvent for dispersion in the hydrogel phase. Graphene oxide (GO) on the other hand, which could be prepared from inexpensive graphite, has better dispersibility in water and organic solvents although in comparison to graphite, it has lower electrical conductivity which corresponds to its degree of oxidation (Ray 2015). However, GO can be further reduced into reduced graphene oxide (rGO) which improves its conductivity due to recovery of the conjugated structure. To form a hydrogel, graphene-based materials can be incorporated into the polymer matrix (Xiao et al. 2017; Han et al. 2017) or undergo self-assembly (Xu et al. 2010; Cong et al. 2012).

Incorporation of graphene oxide into polyvinyl alcohol and polyethylene glycol hydrogel was demonstrated for fabrication of electrocardiography (ECG) electrodes by Xiao et al. (2017). Addition of graphene oxide corresponds to generation of additional physical crosslinking points due to formation of hydrogen bonds between graphene oxide polar groups and the polymers. This interaction consequently improves stretchability of the hydrogel to more than 900% strain at around 1%

weight GO concentration. Graphene oxide was also shown by Han et al. for fabrication of a self-healable, tough, and skin adhesive electrode that could be implanted intramuscularly for electromyography (EMG) (Han et al. 2017). In that work, GO was incorporated into a polydopamine (PDA)-polyacrylamide (PAM) hydrogel network. PDA reduced some of GO to rGO, which provided electrical conductivity to the hydrogel, while the remaining unreduced GO improved the mechanical properties of the hydrogel by creating a network of noncovalent interaction with PDA and PAM via hydrogen bonding, π-π stacking, and electrostatic interactions. By fully reducing the GO, they could achieve a hydrogel with conductivity of 0.18 S/cm and extension ratio of 20.

CNT composite hydrogels have been shown to improve the viability and proliferation rate of some neuron-like cell lines such as PC12 (Shah et al. 2015) and RSC96 (Wu et al. 2017). A CNT nanocomposite hydrogel was also demonstrated for cardiac tissue engineering. Shin et al. incorporated CNTs into gelatin methacrylate hydrogel and used the composite hydrogel as cardiac patch to improve adhesion, viability, organization, and maturation of rat cardiomyocytes (Shin et al. 2013). By creating a freestanding structure, the gel cultured with cardiomyocytes could be used as an actuator as it showed beating behavior which could be modulated by the application of an external electric field.

Alignment of CNTs can be controlled to provide anisotropic electrical conduction by employing several methods. Ahadian et al. performed dielectrophoresis to align CNTs within gelatin methacryloyl hydrogel (Ahadian et al. 2016). Alignment of CNTs improved cardiac differentiation of mouse embryoid bodies which was observed from gene expression analysis and beating analysis. This improvement was further enhanced after stimulation of electric field. Alignment of CNTs prepared with the same method was also shown to improve generation of functional myofibers (Ahadian et al. 2014) and contractility of muscle cells (Ramón-Azcón et al. 2013). Shin et al. embedded aligned vertically grown CNTs on silicon substrate into a hydrogel (Shin et al. 2015). The hydrogel was then used to fabricate microelectrode array for cardiac tissue stimulation.

9.2.2 Ionogels

Ionogels are a type of gel characterized by charge transport mechanism that is dominated by moving ions instead of mobile electrons. Ionic liquid or salt is generally introduced into a hydrogel to provide mobile charge carriers. Ionogels can also be produced by polymerizing crosslinkable ionic liquid monomers although this method usually results in poor conductivity due to lack of mobile charge carriers (Yuan and Antonietti 2011). However, employing ionic liquid could restrict the applications of ionogel due to its cytotoxicity (Thuy Pham et al. 2010).

Liu et al. demonstrated the use of Fe^{3+} metal ion as a crosslinker to form a polyethylene glycol/poly(acrylamide-co-acrylic acid) double-network hydrogel (Liu et al. 2018). The gel exhibited self-healable behavior due to dynamic metal ion coordination bonds and hydrogen bonds. The resulting hydrogel achieved 1350% fracture strain, and due to free diffusion of Fe^{3+} and Cl^- ions, the conductivity could reach 0.0062 S/cm.

Another method to fabricate an ionogel is by soaking a hydrogel in sodium chloride solution. By embedding hydroxypropyl cellulose (HPC) biopolymer fibers in polyvinyl alcohol (PVA) hydrogel to provide space for ion migration, this method could achieve hydrogels with admittance as high as 0.034 S/cm at 1 MHz (Zhou et al. 2018a).

Ionic liquids could be immobilized in a polymeric network using electrostatic interaction. Ding et al. entrapped ionic liquid into negatively charged poly(2-acrylamido-2-methyl-1-propanesulfonic acid) (PAMPS)-based tough double-network hydrogel (Ding et al. 2017). The ionogel achieved conductivity of 0.019 S/cm and is optically transparent. This transparent behavior is not commonly observed in other types of conductive hydrogel. Optical transparency was also demonstrated by Odent et al. in an acrylamide, [2-(acryloyloxy)ethyl]trimethylammonium chloride (AETA)-based ionogel (Odent et al. 2017). The hydrogel showed admittance in the range of 10^{-2} S/cm at 1 MHz. Due to its transparency, the gel can be patterned using stereolithography by inducing photopolymerization of a hydrogel monomer.

9.2.3   Conductive Polymer-Based Hydrogels

Conductive polymers (CPs), in comparison to other conductive materials, could be advantageous due to the possibility to control their chemical, electrical, and physical properties (Balint et al. 2014). Incorporation of a conductive polymer in hydrogels usually results in materials with some common characteristics such as mixed electronic and ionic conductivity, flexibility, nontoxicity, and high specific surface area (Stejskal 2017).

Origin of CP conductivity   The conductivity of CPs stems from their conjugated structure which is indicated by alternating double and single bonds in the polymer backbone. This bond structure is formed by *sp2* hybridization, which results in existence of π-electrons attached by a π-bond. The weakly bound π-electrons compose delocalized electron orbitals throughout the polymer chain which is responsible for its conductivity.

An intrinsic conductive polymer in general has low conductivity. Hence, it is necessary to modify the material to improve its electrical properties by partial reduction (*n*-doping) or oxidation (*p*-doping) of the material, causing some electrons to be added or removed from the polymer chain. This could be achieved chemically by addition of oxidizing or reducing agent, or electrochemically by applying an electric field to the polymer in the presence of an electrolyte. The process of doping initiates geometric modification of the polymer chain and formation of charged defects in the form of soliton, polaron, or bipolaron (Bredas et al. 1984).

Addition of a certain chemical to a doped conductive polymer is often performed to further increase its conductivity. This process is called secondary doping (Elschner et al. 2011; MacDiarmid and Epstein 1994). MacDiarmid and Epstein noted that the difference between primary dopant and secondary dopant is that for secondary dopant, enhanced properties persist after complete removal of the dopant (MacDiarmid and Epstein 1994). The enhancement of conductivity could correspond to the change of crystal structure, reduction of anisotropy, change of proportion or interaction between the polymers and counterions, modification of morphology, or change of work function (Elschner et al. 2011).

Routes to Incorporating CPs in Hydrogels

Single-component hydrogels A conductive polymer network could be engineered to form a hydrogel without the need of a nonconductive polymer scaffold. This type of hydrogel could be advantageous in terms of its electrical properties as it does not contain an electrically nonconductive polymer network that hinders charge transport. Several methods such as self-assembly or modification of a conductive polymer to enable crosslinking could be performed to produce a conductive polymer hydrogel (Yuk et al. 2018; Mawad et al. 2016).

PEDOT/PSS dispersion in aqueous solution could form a hydrogel after addition of acid (Yao et al. 2017), salt (Leaf and Muthukumar 2016), or ionic liquid (Feig et al. 2018). The hydrogel formation could be caused by screening of electrostatic repulsions between PEDOT/PSS particles that leads to physical crosslinking due to π-π interactions (Feig et al. 2018). The PEDOT/PSS hydrogels produced by this method in general possess high conductivity which can still be enhanced after further treatment. For example, Yao et al. demonstrated a PEDOT/PSS hydrogel with conductivity of 0.46 S/cm after gelation which was increased to 8.8 S/cm after treatment with concentrated sulfuric acid (Yao et al. 2017). Phytic acid is a common crosslinker which also acts as a dopant for polyaniline. The crosslinking mechanism of this type of polyaniline hydrogel is likely based on the formation of hydrogen bonding (Stejskal 2017).

CP electropolymerization Electropolymerization usually employs a three-electrode setup where the scaffold hydrogel is in contact with a working electrode. The scaffold hydrogel as well as counter and reference electrodes are immersed in a solution that contains monomers of conductive polymer and counterions. Potential is applied between the working electrode and the reference electrode to start polymerization which occurs around the working electrode, filling the pores of the hydrogel with CP. One benefit of this method is the possibility to control the polymerization reaction by, for example, regulating the applied charge density. Higgins et al. showed that by increasing deposition charges during polymerization of pyrrole with gellan gum as a dopant, change of hydrogel morphology could be obtained (Higgins et al. 2011). They formed a polypyrrole/gellan gum hydrogel as a coating on gold electrode. The resulting hydrogel formed islands with fibrillar features at low deposition charges, while increasing the deposition charges resulted in a porous structure that corresponded to lower impedance.

Chemical polymerization of CP in hydrogel For comprehensive discussions about this topic, we refer the readers to a review by Stejskal (2017). Conductive hydrogels could be prepared by chemical polymerization of CP in a hydrogel matrix, causing the formation of an interpenetrating polymer network (IPN). This is usually achieved by immersing a monomer-containing hydrogel in the solution of an oxidizing agent or exposing a hydrogel which contains an oxidizing agent into the monomer solution. Diffusion causes the two solutions to mix, starting the polymerization reaction. In some cases where the oxidizing agent and the monomer solution are immiscible, interfacial polymerization occurs, which takes place only in the area close to the hydrogel surface. This was demonstrated by Wu et al. for polymerization of aniline in gelatin methacrylate (Wu et al. 2016). Gelatin methacrylate hydrogel was immersed in ammonium persulfate (APS) solution and then subsequently moved into hexane solution containing aniline monomers. A cross-sectional cut showed that polyaniline was formed in the outer layer of the hydrogel where the oxidizing agent and the monomers were in proximity.

Another method was demonstrated where polymerization of a hydrogel and CP was performed simultaneously using the same initiator and oxidizing agent to produce both hydrogel and CP. Tang et al. demonstrated polymerization of aniline and acrylamide with potassium peroxydisulfate (KPS) as oxidizing agent (Tang et al. 2008). However, the reaction was difficult to control as different concentrations of KPS were needed for polymerization of aniline and acrylamide. This could prevent the formation of one type of the polymers (Stejskal 2017).

Formation of a CP network in the hydrogel results in increasing stiffness which is sometimes accompanied by reduction of stretchability (Zhu et al. 2018; Kishi et al. 2014). Zhu et al. reported a decrease of maximum strain from more than 500% to less than 300% for polyion complex hydrogel without aniline in comparison to the hydrogel with polymerized 1.5 M aniline (Zhu et al. 2018). This change in mechanical properties could be a drawback for applying the hydrogel as a neural electrode.

Incorporation of CP dispersion Another approach to embed CP into a hydrogel is by incorporating CP dispersion into the hydrogel. The CP polymer dispersion is usually incorporated in the monomer mixture before scaffold polymerization. This method relies on homogeneous dispersion of the CP in water to make sure phase separation does not occur in the hydrogel composite. Poly(3,4-ethylenedioxythiophene) with polystyrene sulfonate (PEDOT/PSS), which has good dispersibility in water and widely commercially available, is often used to prepare this type of hydrogel composite. Xu et al. demonstrated a physical hydrogel system made of PEDOT/PSS and peptide-polyethylene glycol (PEG) (Xu et al. 2018). The hydrogel showed self-healing ability and injectability due to the formation of a dynamic network resulting from interaction between PEDOT/PSS and the peptide-PEG.

Examples of Hydrogels Containing CPs

Polyaniline (PANI) Depending on its oxidation states, polyaniline can be formed as leucoemeraldine (colorless), emeraldine (green/blue), and (per)nigraniline (blue/violet) (Stejskal et al. 1995). For each of its oxidation states, polyaniline could be protonated or deprotonated. The highest conductivity is observed for green protonated emeraldine which is also referred to as emeraldine salt. Due to the change of its protonation state, PANI loses its conductivity at neutral and high pH which significantly limits its applications. To overcome this problem, several methods were attempted such as introducing a negatively charged sulfonate group on polyaniline molecules (Yue et al. 1990) or trapping large-molecular-sized anions within the polyaniline matrix (Lukachova et al. 2003).

PANI/regenerated cellulose hydrogel was demonstrated by Fan et al. for peripheral nerve regeneration (Fig. 9.2) (Fan et al. 2017). The conductive hydrogel was produced as a conduit by employing the interfacial polymerization method of aniline in cellulose hydrogel. The hydrogel was then surgically implanted on an injured sciatic nerve of an adult male rat. The presence of polyaniline without external electrical stimulation increased expression of ciliary neutrophic factor (CNTF), brain-derived neurotrophic factor (BDNF), growth-associated protein-43, Tau, and α-tubulin. In addition, polyaniline was also shown to activate extracellular kinase (ERK) 1/2 signaling pathway which corresponded to nerve regeneration.

Polypyrrole (PPy) Polypyrrole (Fig. 9.3) is obtained by oxidation of pyrrole which usually results in the form of black powder (Vernitskaya and Efimov 1997). In the thin film form, as the oxidation takes place, the color changes from yellow to blue and finally black (Vernitskaya and Efimov 1997). Some forms of polypyrrole can retain its conductivity in highly alkaline condition (Stejskal et al. 2016).

The formation of a PPy hydrogel was demonstrated by using $FeCl_3$ as an oxidative polymerization initiator and tannic acid as a crosslinker and dopant (Zhou et al. 2018b). The conductivity could reach up to 0.18 S/cm with storage modulus between 0.3 and 2.2 kPa that can be controlled by changing tannic acid concentration. The conductive hydrogel was then implanted in the spinal cord of mice and was reported to promote endogenous neurogenesis in the lesion area and improved recovery of locomotion.

Poly(3,4-ethylenedioxythiophene) (PEDOT) PEDOT (Fig. 9.3) is one of the most widely investigated CPs due to its chemical stability (Yamato et al. 1995). As is the case with other CPs, one drawback of PEDOT is low solubility in water. To overcome this problem, polystyrene sulfonate (PSS) is usually added as counterion. PEDOT and PSS form a polyelectrolyte complex which improves its dispersibility in water.

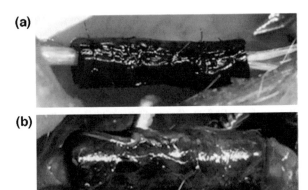

Fig. 9.2 Polyaniline/regenerated cellulose hydrogel on injured sciatic nerve of an adult male rat (**a**) just after operation and (**b**) 3 months after operation (Fan et al. 2017). (Reproduced with permission. Copyright 2017 Spandidos Publications UK Ltd.)

Fig. 9.3 Chemical structure of some common conductive polymers that have been incorporated in hydrogels: polypyrrole, poly(3,4-ethylenedioxythiophene), and polyaniline (Guimard et al. 2007). (Reproduced with permission. Copyright 2007 Elsevier)

Polypyrrole (PPy)

Poly(3,4-ethylene dioxythiophene) (PEDOT)

Polyaniline (PANI)

Other than PSS, biomolecules such as DNA, heparin, hyaluronic acid, and guar gum can be applied as counterions for PEDOT to create a stable dispersion (Sanchez-Sanchez et al. 2019). Wang et al. demonstrated a scaffold for neural stem cells made of PEDOT/hyaluronic acid nanoparticles in chitosan-gelatin matrix (Wang et al. 2017). Hyaluronic acid did not only function as counterions but also promoted attachment of PEDOT in the polymer matrix via chemical crosslinking of an amino and carboxyl group. Incorporation of PEDOT and hyaluronic acid was shown to improve proliferation and differentiation of neural stem cells into neurons and astrocytes. Recently, Tondera et al. reported a multinetwork hydrogel where PEDOT was polymerized in a scaffold formed by synthetic clay nanoparticles (Tondera et al. 2019). It was observed that negative charges of the nanoclay colloids contributed to doping of PEDOT as suggested by a dramatic increase in the conductivity of the hydrogel.

9.3 Electromechanical Properties of Conductive Hydrogels

9.3.1 Electrical Properties of Conductive Hydrogels

Electrochemical impedance spectroscopy Electrochemical impedance spectroscopy is a technique applied for the characterization of bioelectrodes. Measurement is done by applying a small sinusoidal voltage to an electrochemical cell while the current is being measured. Applied excitation signal is given in a broad frequency range, in which the magnitude should be small enough (<50 mV) to allow pseudo-linear current response. Hence, the response signal of the cell would have the same frequency as the applied signal with a certain phase shift. The impedance magnitude and phase angle could be extracted from the response signal in accordance to the applied excitation signal. Electrochemical impedance spectroscopy is also routinely applied for the characterization of conductive hydrogels.

Figure 9.4a shows the impedance spectra of a PPy hydrogel and film coating on a gold neural electrode as reported by Kim et al. (Kim et al. 2004). In comparison to the PPy film coating, impedance reduction of conductive hydrogel or conductive lyophilized hydrogel coating is more significant. This improvement, which shows the benefit of porous coating, could correspond to a larger contact area between the conductive material and electrolyte compared to film coating. Figure 9.4b shows the Nyquist plot of PPy hydrogel electrodes, demonstrated by Shi et al. (2014). The conductive hydrogel was prepared using the interfacial polymerization method by mixing an aqueous solution of oxidizing agent and dissolved pyrrole monomer in isopropyl alcohol. To make an electrode, the hydrogel was loaded to a carbon fiber cloth with varied mass loadings. The nearly vertical tail on the low-frequency range shows capacitive behavior of the electrodes. Increasing PPy concentration clearly reduces the real and imaginary components of the complex impedance.

Cyclic voltammetry Cyclic voltammetry (CV) is a method of characterization which employs a three-electrode setup consisting of the reference electrode, working electrode, and counter electrode. The potential of working electrode is swept cyclically in respect to a non-current-carrying reference electrode. The potential difference causes current to flow between the counter electrode and working electrode. This resulting current is measured and plotted as a function of the applied voltage.

The CV plot could be utilized to determine the charge storage capacity (CSC) of an electrode or hydrogel by calculating the integral of the current-voltage curve in a slow sweep rate. The integral of cathodic current and anodic current would show the cathodic (CSCc) and anodic (CSCa) charge storage capacity, respectively, although typically only CSCc is reported (Cogan 2008). This value is relevant to predict the number of charges that can be injected by an electrode during a tissue stimulation.

Other electrode properties could be derived from the measurement. The characteristics of the CV curve could hint the presence of capacitive charge transfer

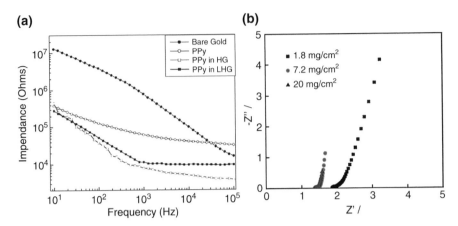

Fig. 9.4 (a) Impedance spectra of bare gold electrode, polypyrrole film (PPy), polypyrrole grown in intact hydrogel (PPy in HG), and polypyrrole grown in freeze-dried hydrogel (PPy in LHG) (Kim et al. 2004). (Reproduced with permission. Copyright 2004 John Wiley & Sons, Inc.) (b) Nyquist plot of polypyrrole hydrogel electrodes with different mass loadings (Shi et al. 2014). (Reproduced with permission. Copyright 2013 Royal Society of Chemistry)

mechanism, region of reduction and oxidation potential, reversibility of the reaction, chemical reaction rate, electrode stability, and concentration of the analytes.

As an example, CV was employed by Ismail et al. for characterization of an extrusion-formed microfiber made of chitosan PPy hydrogel (Ismail et al. 2011). The measurement employed a three-electrode setup with microfiber as working electrode, Ag/AgCl as reference electrode, and stainless steel plate as counter electrode in 1 M NaCl aqueous solution. At 10 mV/s scan rate, a broad anodic peak could be observed at 0.18 V showing the oxidation of polypyrrole, while a broad cathodic peak at −0.26 V showing the reduction potential (Fig. 9.5a).

The characterization was performed at varied scan rates from 2 to 200 mV/s (Fig. 9.5b). Lower scan rate resulted in decreased current as the diffusion layer surrounding the electrode grew thicker hence limiting the electrochemical reaction. The peak current was reported to be linearly proportional with square root of the scan rate until 50 mV/s (Fig. 9.5c). This behavior which followed Randles-Sevcik equation suggested electrochemical reversibility because the electron transfer process is fast in comparison to the voltage scan rate (Elgrishi et al. 2018). As the electron transfer reaction occurred fast, the reaction was controlled or limited by the diffusion of chemical species.

Figure 9.5d shows the effect of temperature to the potential where the electrochemical reaction occurred. The CV was performed in the temperature range from 10 °C to 40 °C. As the temperature increased, the anodic peak potential was reduced, and the cathodic peak potential was increased as the reaction occurred in a higher rate following Arrhenius equation (Ismail et al. 2011).

Conductivity The conductivity of a hydrogel is usually measured using a 4-probe setup. Four electrodes made of metal are brought into contact with the hydrogel.

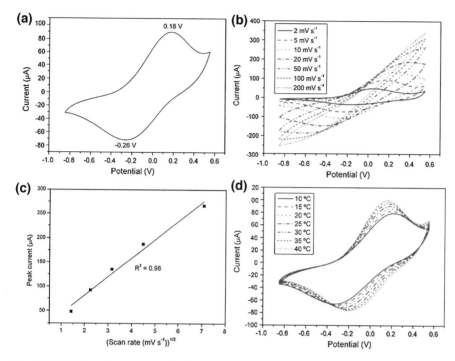

Fig. 9.5 Cyclic voltammogram of chitosan polypyrrole microfiber at (**a**) 10 mV/s and (**b**) 2–200 mV/s scan rate. (**c**) Linear relationship between the square root of the scan rate and the peak current suggesting electrochemical reversibility. (**d**) Temperature dependence of voltammetric response at 10 mV/s scan rate (Ismail et al. 2011). (Reproduced with permission. Copyright 2011 Elsevier)

Two probes are used as current-carrying electrodes and the other two are used for sensing the voltage. The purpose of using a separate pair of electrodes for feeding the current and sensing the potential difference is to minimize the effect of contact resistance between the probe and hydrogel. However, a 2-probe setup is also employed by various groups. Conductivity of a conductive hydrogel widely ranges from 10^{-4} S/cm to more than 1 S/cm (Yao et al. 2017; Tondera et al. 2019; Zhao et al. 2013). Table 9.1 shows the conductivity values of various types of conductive hydrogels. In the literature, there is a wide variation in the reported conductivities which depend on the composition and type of the hydrogel.

9.3.2 Mechanical Characterization

Several measurement techniques could be applied to assess the mechanical properties of a hydrogel. The measurements are based on applying force to cause deformation. Hydrogels behave as a viscoelastic material which exhibits time-dependent

Table 9.1 Examples of conductive hydrogels and reported conductivities

References	Hydrogel type	Conductivity (S/cm)	Measurement method	Concentration of conductive material
Carbon nanocomposite				
Han et al. (2017)	Polydopamine-rGO-polyacrylamide	0.18	Two-probe	4% rGO
Xu et al. (2010)	Hydrothermally reduced graphene	0.0049	Two-probe	2.6% rGO
Shah et al. (2015)	Polyethylene glycol-CNT	0.02	Two-probe	1.2% CNT
Ionogel				
Liu et al. (2018)	Polyethylene glycol-poly(acrylamide-co-acrylic acid)-Fe^{3+}	0.0062	Four-probe	0.057 M Fe^{3+}
Ding et al. (2017)	1-Ethyl-3-methylimidazolium dicyanamide ([EMIm][DCA])-poly(2-acrylamido-2-methyl-1-propanesulfonic acid) (PAMPS)	0.019	Four-probe	66.4% ionic liquid
Single-component conductive polymer				
Yao et al. (2017)	PEDOT/PSS (gelation with H_2SO_4)	0.46 to 8.8 after posttreatment	Four-probe	0.78% before posttreatment
Leaf et al. (2016)	PEDOT/PSS (gelation with ionic liquid EMIM BF4)	0.012	Four-probe	11 mg/ml PEDOT/PSS and 70 mM of EMIM BF4
Zhou et al. (2018b)	Polypyrrole-tannic acid	0.18	Two-probe	0.416 M aniline
Interpenetrating network of conductive polymer				
Feig et al. (2018)	PEDOT/PSS-polyacrylic acid	0.23	Four-probe	1.1% PEDOT/PSS
Zhu et al. (2018)	PANI-polyion complex	0.014	Four-probe	1.5 M aniline
Tondera et al. (2019)	PEDOT/laponite-PAAM	0.26	Four-probe	44% PEDOT (of the dry weight)

relation between applied stress and deformation due to molecular rearrangement of their polymeric chain. Hence, in addition to static measurements, frequency-based tests such as shear rheometry or dynamic mechanical analysis are usually performed.

To use conductive hydrogels as a neural electrode, the mechanical properties of the hydrogel must match those of neural tissue to minimize tissue damage. Young's modulus, a parameter which describes the stiffness of a material, is reported in the range of 100 Pa to 10 kPa for central nervous tissue (Lacour et al. 2016). Another parameter such as stretchability could be important considering some tissues are constantly in motion. For example, the human spinal cord undergoes 10–20%

tensile strain during normal postural movements (Lacour et al. 2016). These parameters are important to consider in designing the hydrogel to reduce mechanical mismatch between the hydrogel and tissue.

Compression and tensile   The elastic properties of hydrogels can be characterized by tensile or compression testing. For both methods, tensile or compression force is applied at constant rate usually on one end until the specimen reaches ultimate failure. The load is recorded in respect to displacement and presented as stress and strain by considering the geometry of the specimen. Young's modulus (E), maximum strain, yield strength, and ultimate tensile strength can be derived from the measurement.

The influence of monomer concentration and crosslinker concentration to the mechanical properties of a conductive hydrogel was demonstrated by Feig et al. (2018). They formed interpenetrating network conductive hydrogels by introducing ionic liquid in PEDOT/PSS aqueous solution to induce gelation followed by infiltrating the network with polyacrylic acid. Figure 9.6a and Table 9.2 show that reduction of monomer and crosslinker concentration could reduce the elastic

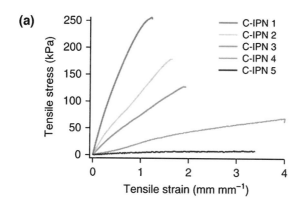

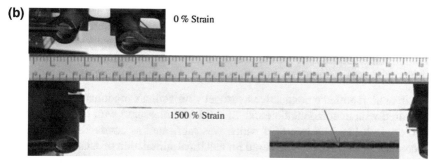

Fig. 9.6 (a) Stress-strain curve of PEDOT/PSS polyacrylic acid hydrogel with different concentrations of monomer and crosslinker (Feig et al. 2018). (Creative Commons BY). Elastic modulus and maximum strain can be tuned by modifying the concentration of the monomer and crosslinker. (b) Tensile testing of polypyrrole polyacrylic acid hydrogels. The hydrogel is stretched to 1500% strain (Darabi et al. 2017). (Reproduced with permission. Copyright 2017 John Wiley & Sons, Inc.)

Table 9.2 Mechanical properties of PEDOT/PSS polyacrylic acid hydrogel with varied concentrations of monomers and crosslinkers (Feig et al. 2018)

C-IPN formulation	AAc wt%	Bis/AAc wt ratio	Modulus (kPa)	Stain at break (%)
1	50	0.02	374	121
2	22	0.01	175	163
3	20	0.01	99	191
4	20	0.002	23	399
5	11	0.002	8	338

modulus of the hydrogel by almost two orders of magnitude, bringing it close to the stiffness of soft biological tissue in the range of 10 kPa. In this case, the reduction of elastic modulus is mostly accompanied by increasing stretchability. Although the hydrogel was formed by 2 polymer networks, the mechanical properties were mainly affected by polyacrylic acid as the percolation of PEDOT/PSS was achieved at very low concentration.

Darabi et al. demonstrated a highly stretchable self-healing conductive hydrogel based on covalently crosslinked polyacrylic acid, PPy, and ferric ions as physical crosslinker (Darabi et al. 2017). Compressive elastic modulus was in the range of 200–800 kPa depending on the concentration of PPy. The elastic modulus and stretchability of the hydrogel could also be tuned by modifying its covalent cross-linker concentration. Higher concentration of crosslinker corresponds to higher elastic modulus and lower stretchability. The hydrogel reached maximum strain of around 2000% without any covalent crosslinker (Fig. 9.6b).

Indentation Indentation is a method used to determine the mechanical properties of a material by locally compressing a specimen with a probe while the load and displacement depth are measured. Indentation can be done in static mode where the movement of the probe is held steady or in dynamic mode where the probe oscillates with certain frequency. Depending on the size of probe and specimen, this measurement could provide information about inhomogeneity of the material in nm and mm scale (Oyen 2014). The parameters reported are usually reduced elastic modulus and hardness (Guillonneau et al. 2012). Creep and relaxation behavior are also possible to be examined with this method (Ahearne et al. 2008).

Liu et al. reported a conductive hydrogel with Young's modulus of 24 ± 5.4 kPa measured with nanoindentation and 32 ± 5.1 kPa measured with compression test (Liu et al. 2019). The hydrogel which was fabricated as conductive material for microelectrode arrays was then used for electrical stimulation of sciatic nerve in live mice. The electrodes were shown to cause minimum damage and inflammatory tissue response in comparison to the plastic cuff electrode. The reduced adverse reaction was due to similarity of mechanical properties between the electrodes and surrounding tissue.

9.4 Fabrication Methods (Patterning) Compatible with Conductive Hydrogels

In order to be integrated in bioelectronic systems such as implants or artificial skins, conductive hydrogels have to be patterned into complex shapes on the meso- and microscale. Several techniques that have been investigated to date are described below.

Extrusion-based 3D printing Extrusion-based 3D printing employs regulated pressure inside a printing syringe to extrude an ink through a print-head nozzle while the movement of the nozzle is controlled. Hydrogel printing in the biomedical field is already demonstrated for various applications such as wound dressing (Murphy et al. 2013), mold for scaffold-free tissue spheroid growth (Tan et al. 2014), and different kinds of scaffolds such as aortic valve scaffold (Hockaday et al. 2012), scaffold for femur, branched coronary arteries, embryonic hearts and human brains (Hinton et al. 2015), and skeletal tissue scaffold (Fedorovich et al. 2007). Generally, preformed hydrogel is printed followed by polymer crosslinking to retain its structure. By controlling anisotropy of the printed hydrogel and taking advantage of its swelling behavior, Sydney Gladman et al. showed formation of complex structures after immersion of printed hydrogel in water (Sydney Gladman et al. 2016).

This method could also be used for creating a conductive hydrogel pattern. Lu et al. demonstrated printing of PEDOT/PSS aqueous solution with dimethyl sulfoxide (DMSO) as an additive (Fig. 9.7a) (Lu et al. 2019). After printing, the structure was subsequently dried. Due to the presence of DMSO, the PEDOT/PSS chain rearranged during drying which lead to formation of a pure PEDOT/PSS hydrogel after re-swelling. The anisotropy of swelling could be controlled by modifying the condition of drying. With this method, a resolution of 400 μm could be achieved.

One benefit of extrusion printing is its versatility in terms of cell handling. Huang et al. demonstrated a printable hydrogel composite consisting of polyurethane, graphene oxide, and pluronic-stabilized graphene (Huang et al. 2017). Neural stem cells were loaded into the gel before printing and have been shown to survive the

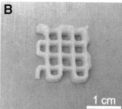
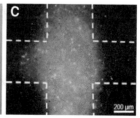

Fig. 9.7 Extrusion-based 3D printed construct of: (**a**) PEDOT/PSS hydrogel. (Modified from (Lu et al. 2019) (Creative Commons BY)) and (**b**) neural stem cells (NSC)-laden polyurethane-pluronic modified graphene hydrogel (Huang et al. 2017). (Reproduced with permission. Copyright 2013 Royal Society of Chemistry). (**c**) Fluorescence labelled NSCs embedded in the hydrogel (Huang et al. 2017). (Reproduced with permission. Copyright 2013 Royal Society of Chemistry)

printing process (Fig. 9.7b,c) as observed from gene expression analysis. Addition of 25-ppm graphene-based materials into the ink induced proliferation and differentiation of the cells due to improvement of cellular oxygen metabolism. The incorporation of those materials in this case was not meant to add electrical functionality to the hydrogel.

Stereolithography Stereolithography is another form of 3D printing that is based on spatially controlled layer-by-layer photopolymerization of monomers. Proper choice of hydrogel materials is necessary as the monomers must be able to photopolymerize at a rapid rate to fabricate the structure within a reasonable time. A polymer with an acrylate or methacrylate functional group is often used due to their photoreactivity (Wu et al. 2016; Arcaute et al. 2010).

Odent et al. printed a transparent and ionically conductive hydrogel with stereolithography (Fig. 9.8a) (Odent et al. 2017). The hydrogel was based on acrylamide and [2-(acryloyloxy)ethyl]trimethylammonium chloride (AETA) with riboflavin and triethanolamine as photoinitiator and co-initiator which enables free radical photopolymerization. Wu et al. polymerized polyaniline after printing a nonconductive gelatin methacrylate hydrogel (Fig. 9.8b) (Wu et al. 2016). Both routes resulted in conductive hydrogels with feature size less than 100 μm.

Photolithography Photolithography is the most common method used in the semiconductor industry to fabricate microscale devices. It uses light that passes through a patterned photomask to transfer the pattern by polymerizing the exposed photoresist. Liu et al. used this method to create a pattern on a PEDOT/PSS ionic liquid gel by etching exposed gel deposited under a patterned photoresist with oxygen plasma (Liu et al. 2019). The patterned gel was subsequently transformed into hydrogel by immersion in water to exchange the ionic liquid with water. The same technique was further employed to pattern fluorinated elastic photoresist on top of the conductive structure to create a passivation layer. The resulting structure could

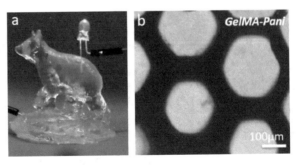

Fig. 9.8 (**a**) Transparent ionically conductive hydrogel (Odent et al. 2017). (Reproduced with permission. Copyright 2017 John Wiley & Sons, Inc.) (**b**) Gelatin methacrylate polyaniline hydrogel (Wu et al. 2016). (Reproduced with permission. Copyright 2016 Elsevier). Both structures were patterned using stereolithography

be patterned down to 5-μm resolution which was then used as microelectrode arrays (Fig. 9.9).

Patterned transfer via electropolymerization Electrochemical polymerization requires a conductive substrate/template. Hence, employing patterned electrodes as template for the polymerization of a conductive polymer could reproduce the pattern into the hydrogel. Ido et al. performed electropolymerization of PEDOT on platinum microelectrodes (Fig. 9.10a) (Ido et al. 2012). After attaching a hydrogel layer on the electrodes, a second PEDOT polymerization was performed to anchor the first layer of PEDOT to the hydrogel. The hydrogel with a patterned conductive polymer was then peeled off from the substrate. This method was shown to be versatile and can be used on a wide range of hydrogel substrates such as agarose, collagen, glucomannan, polyacrylamide, and polyHEMA. The resolution of the generated pattern greatly depends on the fabrication method of the template electrode. This method was employed by Sekine et al. to create an electrode array for stimulation and induction of muscle cell contraction (Sekine et al. 2010).

A highly conductive pattern made of PEDOT/polyurethane composite was produced using the method above (Sasaki et al. 2014). The resulting conductive hydrogel hybrid attained conductivity as high as 120 S/cm at 100% elongation and could support adhesion, proliferation, and differentiation of cultured neural and muscle

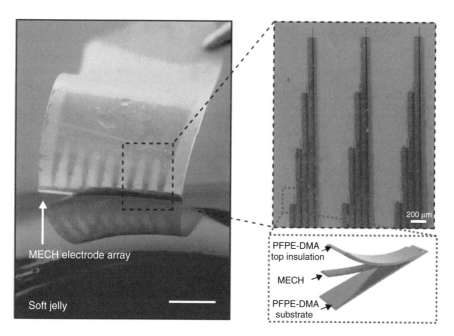

Fig. 9.9 Microelectrode array made of PEDOT/PSS hydrogel as material for electrodes and interconnects and dimethacrylate-functionalized perfluoropolyether for substrate and passivation layer (Liu et al. 2019). (Reproduced with permission. Copyright 2019 Springer Nature)

cells. Although the resolution of the fabricated pattern was not exceptionally high (0.5 mm), this was due to the limited resolution with which the template electrode arrays was patterned. An extension of this method has been used to create 3D conductive patterns by using a metal wire as a substrate for electropolymerization (Fig. 9.10b) (Sasaki et al. 2014).

Other routes Another method was used by Shin et al. to pattern carbon nanotubes inside of a hydrogel (Fig. 9.11) (Shin et al. 2015). They employed the chemical vapor deposition (CVD) method to vertically grow carbon nanotubes on thin patterned Fe film on silicon wafer. Fe was used as a catalyst for CNT growth. PEGDA hydrogel was then crosslinked on the structure followed by delamination and the second hydrogel network formation which was composed of CNT and modified

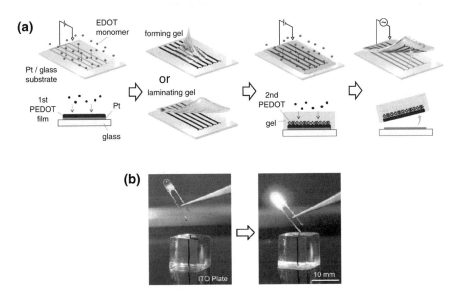

Fig. 9.10 (**a**) Electropolymerization of PEDOT to transfer a pattern from a platinum electrode template into the hydrogel (Ido et al. 2012). (Reproduced with permission. Copyright 2012 American Chemical Society). (**b**) Conductive line inside the bulk of a hydrogel produced by electropolymerization of PEDOT on a metal wire (Sasaki et al. 2014). The wire was removed subsequently. (Reproduced with permission. Copyright 2014 John Wiley & Sons, Inc.)

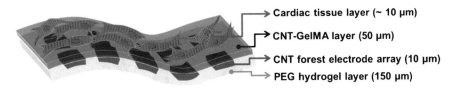

Fig. 9.11 Schematic of carbon nanotube hydrogel composite electrode array grown by the chemical vapor deposition method. The construct was used for cardiac tissue engineering (Shin et al. 2015). (Reproduced with permission. Copyright 2015 John Wiley & Sons, Inc.)

gelatin. The second hydrogel network acted as a bridge between CNT electrodes and the cardiac tissue layer. The whole hydrogel construct demonstrated beating activity after incorporation of cardiac tissue and was used as biohybrid actuator.

9.5   Applications

Electrode coating   Hydrogel coatings are often used to reduce the impedance of microelectrodes. They have also been considered as mechanical buffers to minimize the adverse effects associated with implantation of a rigid probe. Coating of cochlear electrode arrays (Fig. 9.12), for example, was reported by Hassarati et al. to increase the charge storage capacity and charge injection limit of the electrode for approximately one order of magnitude in comparison to a bare Pt electrode (Hassarati et al. 2014). Decrease of electrical impedance was also observed with the same coating which was stable for more than 1 billion stimulations. The coating was based on poly(vinyl alcohol) (PVA) and heparin methacrylate (Hep-MA) hydrogel with galvanostatically polymerized PEDOT as conductive material.

Besides coating cochlear implants, conductive hydrogel is also used for coating electrodes for implantation in the brain (Fig. 9.13a) (Kim et al. 2004, 2010). Alginate hydrogel with electrochemically grown PPy was used for coating a gold neural electrode for recording neural activity in the cerebellum of guinea pig (Kim et al. 2004). After coating, reduction of electrical impedance was more significantly observed for conductive-hydrogel-coated electrode in comparison to PPy-coated electrode without hydrogel. This shows the benefit of conductive hydrogel coating

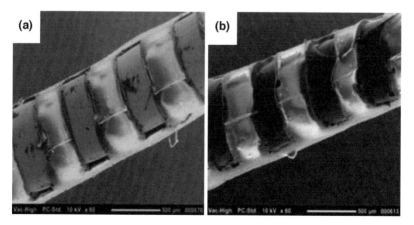

Fig. 9.12   (**a**) Uncoated and (**b**) PEDOT PVA Hep-MA hydrogel–coated cochlear electrode array at 60× magnification (Hassarati et al. 2014). Conductive hydrogel coating reduces electrical impedance and improves charge injection of the electrodes. (Reproduced with permission. Copyright 2014 IEEE)

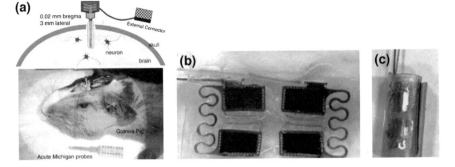

Fig. 9.13 (**a**) Conductive hydrogel coating on gold electrodes implanted into the auditory cortex of guinea pig (Kim et al. 2010). (Reproduced with permission. Copyright 2010 Elsevier). (**b**) Conductive hydrogel coating on PtIr cuff array for delivery of high-frequency pulses (Staples et al. 2018). (Creative Commons BY). (**c**) Curled PtIr cuff array (Staples et al. 2018). (Creative Commons BY)

in comparison to coating with conductive film in terms of improvement of electrical properties due to increase of conductive surface area.

Additional improvement could be introduced to the hydrogel coating such as formation of macropores by freeze drying. The constructed pores allowed growth of neurons inside the hydrogel which enabled better communication between electrodes and tissue (Kim et al. 2004).

Mechanical and electrical stability of an electrode is important to be properly designed especially for long-term implantation or stimulation. To improve adhesion stability, sometimes an additional layer is needed between the electrode and the hydrogel coating (Hassarati et al. 2014). In terms of electrical properties, Staples et al. showed that conductive hydrogel coating improved stability of metal electrodes especially for stimulation with long-term, high-frequency signal for nerve block (Fig. 9.13b, c) (Staples et al. 2018). High surface area introduced by conductive hydrogel coating increased charge storage capacity, which consequently lowered the voltage needed for stimulation thus improving the stability of an electrode (Hassarati et al. 2014). The authors demonstrated the stability of PEDOT-poly(vinyl alcohol)-coated stainless steel electrode array and platinum iridium (PtIr) for continuous delivery of 2-mA square pulse waveforms at 40 kHz for 42 days (Staples et al. 2018).

An additional benefit of hydrogel coating is the possibility of incorporation of bioactive molecules. Sericin and gelatin, for example, proteins with negatively charged groups, were covalently incorporated via methacrylate crosslinking within the hydrogel (Mario Cheong et al. 2014). The biomolecules improved electrical charge transfer due to their polarity while also promoted neural cell adhesion and proliferation. Furthermore, drug delivery capability was demonstrated by release of nerve growth factor to the target cells which promoted differentiation and neurite growth.

Cardiac tissue engineering Conductive hydrogels could be used to bridge electrical signaling between cardiac cells. This enables conductive hydrogels as potential materials for cardiac patch to treat myocardial infarction. As physical and electrical properties of the hydrogel can be tuned, the patch could be engineered to conformably interface with cardiac tissue. Liang et al. demonstrated a cardiac patch that can be painted directly onto the heart surface (Liang et al. 2018). Dopamine that was embedded in the hydrogel provided strong adhesion with cardiac tissue, while polypyrrole increased its electrical conductivity close to that of cardiac tissue ($\sim 10^{-4}$ S/cm). By employing the patch to treat infarcted rat heart, decrease of fibrosis area and infarct size was observed.

Several other conductive hydrogel composites were demonstrated as scaffolds for cardiac tissue growth. Incorporation of conductive materials such as carbon nanotube or gold nanorods in the hydrogel improved cell retention, viability (Navaei et al. 2016), adhesion and organization (Shin et al. 2015), tissue alignment (Dvir et al. 2011), and conduction velocity (Pok et al. 2014). Conductive materials could also affect beating behavior of the cells. Adding carbon nanotubes into gelatin methacrylate hydrogel scaffold, for example, increased spontaneous beating rates of embedded cells and lowered its stimulation excitation threshold (Shin et al. 2015). Low excitation threshold is beneficial to avoid damage of cardiac tissue due to unwanted electrochemical reaction.

Ahadian et al. further enhanced the electrical properties of scaffold by aligning carbon nanotube in gelatin methacryloyl hydrogel using dielectrophoresis (Ahadian et al. 2016). Applying electrical pulse stimulation in the scaffold increased cardiac protein and cardiac gene (*Tnnt2*, *Nkx2–5*, and *Actc1*) expression in comparison to unaligned scaffold, implying greater cardiac differentiation. This shows conductive hydrogels as promising materials for the treatment of cardiac tissue disorders and for in vitro studies.

Microelectrode arrays The use of conductive hydrogel as a material in microelectrode arrays for neural stimulation or recording in vivo without employing metal has been scarce due to current challenges in fabrication. In early 2019, Liu et al. published a work on the development of a microelectrode array made of PEDOT/PSS hydrogel for both electrodes and interconnect material (Liu et al. 2019). Dimethacrylate-functionalized perfluoropolyether was used as substrate and passivation layer. Using lithography, they patterned electrodes and insulation layer with 20-μm feature size. The conductive hydrogel possessed high charge storage capacity, more than 2 orders of magnitude higher than that of platinum. Hence, neuromodulation of mice sciatic nerve could be performed at a low voltage of 50 mV to observe leg movement. Low elastic modulus of the hydrogel, passivation layer, and substrate material allowed conformal wrapping of the array around the sciatic nerve with reduced immune response.

9.6   Conclusions and Future Outlook

Despite wide applications of long-established metal neural electrodes, conductive hydrogels hold a great potential as future alternative materials due to their superior mechanical, electrical, and biochemical properties. Various methods of hydrogel fabrication by using different polymers and crosslinkers have been demonstrated, resulting in diverse hydrogel mechanical properties such as tunable elastic modulus and stretchability, self-healing ability, and tissue adhesion. Hydrogels also allow modifications of their biochemical properties to improve adhesion, differentiation, and proliferation of cells. Methods to impart conductivity are shown by forming hydrogel composites, interpenetrating polymer network, conductive polymer hydrogel, or employing ions as charge carriers.

The resulting fabricated hydrogels are used for various applications such as in vitro tissue engineering and coating, where it is shown to improve the electrical properties of common metal electrodes. However, for some other applications, more complex structures are needed where patterning of the hydrogel is necessary. Here, the method to fabricate a device made of conductive hydrogels and elastomers as insulation is currently a challenge. To employ well-developed technologies such as additive manufacturing or photolithography for fabrication of complex structures, hydrogels must be adapted to fit the process. Solving this problem will bring hydrogels closer to more complex applications for in vivo recording and stimulations.

References

Ahadian, S., et al. (2014). Hybrid hydrogels containing vertically aligned carbon nanotubes with anisotropic electrical conductivity for muscle myofiber fabrication. *Scientific Reports, 4.* https://doi.org/10.1038/srep04271.

Ahadian, S., et al. (2016). Hybrid hydrogel-aligned carbon nanotube scaffolds to enhance cardiac differentiation of embryoid bodies. *Acta Biomaterialia, 31,* 134. https://doi.org/10.1016/j.actbio.2015.11.047.

Ahearne, M., Yang, Y., & Liu, K.-K. (2008). Mechanical characterisation of hydrogels for tissue engineering applications hydrogels for tissue engineering. In Ashammakhi, N., Reis, R., & Chiellini F. (Eds.), *Topics in Tissue Engineering, 4,* 4–13.

Arcaute, K., Mann, B., & Wicker, R. (2010). Stereolithography of spatially controlled multimaterial bioactive poly(ethylene glycol) scaffolds. *Acta Biomaterialia, 6,* 1047. https://doi.org/10.1016/j.actbio.2009.08.017.

Aregueta-Robles, U. A., et al. (2014). Organic electrode coatings for next-generation neural interfaces. *Frontiers in Neuroengineering, 7.* https://doi.org/10.3389/fneng.2014.00015.

Balint, R., Cassidy, N. J., & Cartmell, S. H. (2014). Conductive polymers: Towards a smart biomaterial for tissue engineering. *Acta Biomaterialia, 10,* 2341. https://doi.org/10.1016/j.actbio.2014.02.015.

Bredas, J. L., et al. (1984). The role of mobile organic radicals and ions (solitons, polarons and bipolarons) in the transport properties of doped conjugated polymers. *Synthetic Metals, 9,* 265–274.

Cogan, S. F. (2008). Neural stimulation and recording electrodes. *Annual Review of Biomedical Engineering, 10,* 275. https://doi.org/10.1146/annurev.bioeng.10.061807.160518.

Cogan, S. F., et al. (2016a). Tissue damage thresholds during therapeutic electrical stimulation. *Journal of Neural Engineering, 13*, 021001. https://doi.org/10.1088/1741-2560/13/2/021001.

Cogan, S. F., Garrett, D. J., & Green, R. A. (2016b). Electrochemical principles of safe charge injection. In *Neurobionics: The biomedical engineering of neural prostheses.* https://doi.org/10.1002/9781118816028.ch3.

Cong, H. P., et al. (2012). Macroscopic multifunctional graphene-based hydrogels and aerogels by a metal ion induced self-assembly process. *ACS Nano, 6*, 2693. https://doi.org/10.1021/nn300082k.

Darabi, M. A., et al. (2017). Skin-inspired multifunctional autonomic-intrinsic conductive self-healing hydrogels with pressure sensitivity, stretchability, and 3D printability. *Advanced Materials, 29.* https://doi.org/10.1002/adma.201700533.

Ding, Y., et al. (2017). Preparation of high-performance ionogels with excellent transparency, good mechanical strength, and high conductivity. *Advanced Materials, 29.* https://doi.org/10.1002/adma.201704253.

Dvir, T., et al. (2011). Nanowired three-dimensional cardiac patches. *Nature Nanotechnology, 6*, 720. https://doi.org/10.1038/nnano.2011.160.

Elgrishi, N., et al. (2018). A practical Beginner's guide to cyclic voltammetry. *Journal of Chemical Education, 95*, 197–206.

Elschner, A., et al. (2011). *PEDOT: Principles and applications of an intrinsically conductive polymer.* Boca Raton: CRC.

Fan, L., et al. (2017). Polyaniline promotes peripheral nerve regeneration by enhancement of the brain-derived neurotrophic factor and ciliary neurotrophic factor expression and activation of the ERK1/2/MAPK signaling pathway. *Molecular Medicine Reports, 16*, 7534–7540.

Fedorovich, N. E., et al. (2007). Hydrogels as extracellular matrices for skeletal tissue engineering: State-of-the-art and novel application in organ printing. *Tissue Engineering, 13*, 1905. https://doi.org/10.1089/ten.2006.0175.

Feig, V. R., et al. (2018). Mechanically tunable conductive interpenetrating network hydrogels that mimic the elastic moduli of biological tissue. *Nature Communications, 9*, 2740. https://doi.org/10.1038/s41467-018-05222-4.

Goding, J., et al. (2018a). Considerations for hydrogel applications to neural bioelectronics. *Journal of Materials Chemistry B, 7*, 1625. https://doi.org/10.1039/c8tb02763c.

Goding, J. A., et al. (2018b). Living bioelectronics: Strategies for developing an effective long-term implant with functional neural connections. *Advanced Functional Materials, 28.* https://doi.org/10.1002/adfm.201702969.

Goding, J., et al. (2017). A living electrode construct for incorporation of cells into bionic devices. *MRS Communications, 7*, 487. https://doi.org/10.1557/mrc.2017.44.

Guillonneau, G., et al. (2012). Determination of mechanical properties by nanoindentation independently of indentation depth measurement. *Journal of Materials Research, 27*, 2551. https://doi.org/10.1557/jmr.2012.261.

Guimard, N. K., Gomez, N., & Schmidt, C. E. (2007). Conducting polymers in biomedical engineering. *Progress in Polymer Science (Oxford), 32*, 876. https://doi.org/10.1016/j.progpolymsci.2007.05.012.

Han, L., et al. (2017). A mussel-inspired conductive, self-adhesive, and self-healable tough hydrogel as cell stimulators and implantable bioelectronics. *Small, 13.* https://doi.org/10.1002/smll.201601916.

Hassarati, R. T., et al. (2014). Improving Cochlear implant properties through conductive hydrogel coatings. *IEEE Transactions on Neural Systems and Rehabilitation Engineering, 22*, 411. https://doi.org/10.1109/tnsre.2014.2304559.

Higgins, T. M., et al. (2011). Gellan gum doped polypyrrole neural prosthetic electrode coatings. *Soft Matter, 7*, 4690. https://doi.org/10.1039/c1sm05063j.

Hinton, T. J., et al. (2015). Three-dimensional printing of complex biological structures by freeform reversible embedding of suspended hydrogels. *Science Advances, 1*, e1500758. https://doi.org/10.1126/sciadv.1500758.

Hockaday, L. A., et al. (2012). Rapid 3D printing of anatomically accurate and mechanically heterogeneous aortic valve hydrogel scaffolds. *Biofabrication, 4*, 035005. https://doi.org/10.1088/1758-5082/4/3/035005.

Huang, C. T., et al. (2017). A graphene-polyurethane composite hydrogel as a potential bioink for 3D bioprinting and differentiation of neural stem cells. *Journal of Materials Chemistry B, 5*, 8854–8864.

Ido, Y., et al. (2012). Conducting polymer microelectrodes anchored to hydrogel films. *ACS Macro Letters, 1*, 400–403.

Ismail, Y. A., et al. (2011). Sensing characteristics of a conducting polymer/hydrogel hybrid microfiber artificial muscle. *Sensors and Actuators B Chemical, 160*, 1180–1190.

Jeong, J. W., et al. (2015). Soft materials in neuroengineering for hard problems in neuroscience. *Neuron, 86*, 175. https://doi.org/10.1016/j.neuron.2014.12.035.

Kim, D. H., Abidian, M., & Martin, D. C. (2004). Conducting polymers grown in hydrogel scaffolds coated on neural prosthetic devices. *Journal of Biomedical Materials Research Part A, 71A*, 577. https://doi.org/10.1002/jbm.a.30124.

Kim, D. H., et al. (2010). Conducting polymers on hydrogel-coated neural electrode provide sensitive neural recordings in auditory cortex. *Acta Biomaterialia, 6*, 57. https://doi.org/10.1016/j.actbio.2009.07.034.

Kishi, R., et al. (2014). Mechanically tough double-network hydrogels with high electronic conductivity. *Journal of Materials Chemistry C, 2*, 736. https://doi.org/10.1039/c3tc31999g.

Lacour, S. P., Courtine, G., & Guck, J. (2016). Materials and technologies for soft implantable neuroprostheses. *Nature Reviews Materials, 1*. https://doi.org/10.1038/natrevmats.2016.63.

Leach, J., et al. (2010). Bridging the divide between neuroprosthetic design, tissue engineering and neurobiology. *Frontiers in Neuroengineering, 2*. https://doi.org/10.3389/neuro.16.018.2009.

Leaf, M. A., & Muthukumar, M. (2016). Electrostatic effect on the solution structure and dynamics of PEDOT:PSS. *Macromolecules, 49*, 4286. https://doi.org/10.1021/acs.macromol.6b00740.

Liang, S., et al. (2018). Paintable and rapidly bondable conductive hydrogels as therapeutic cardiac patches. *Advanced Materials, 30*. https://doi.org/10.1002/adma.201704235.

Lim, C., et al. (2019). Stretchable conductive nanocomposite based on alginate hydrogel and silver nanowires for wearable electronics. *APL Materials, 7*. https://doi.org/10.1063/1.5063657.

Liu, S., et al. (2018). Dual ionic cross-linked double network hydrogel with self-healing, conductive, and force sensitive properties. *Polymer (Guildf)., 144*, 111. https://doi.org/10.1016/j.polymer.2018.01.046.

Liu, Y., et al. (2019). Soft and elastic hydrogel-based microelectronics for localized low-voltage neuromodulation. *Nature Biomedical Engineering, 3*, 58–68.

Lu, Y., et al. (2016). Flexible neural electrode array based-on porous graphene for cortical microstimulation and sensing. *Scientific Reports, 6*. https://doi.org/10.1038/srep33526.

Lu, B., et al. (2019). Pure PEDOT: PSS hydrogels. *Nature Communications, 10*, 1043. https://doi.org/10.1038/s41467-019-09003-5.

Lukachova, L. V., et al. (2003). Electroactivity of chemically synthesized polyaniline in neutral and alkaline aqueous solutions. *Journal of Electroanalytical Chemistry, 544*, 59. https://doi.org/10.1016/s0022-0728(03)00065-2.

MacDiarmid, A. G., & Epstein, A. J. (1994). The concept of secondary doping as applied to polyaniline. *Synthetic Metals, 65*, 103. https://doi.org/10.1016/0379-6779(94)90171-6.

Mario Cheong, G. L., et al. (2014). Conductive hydrogels with tailored bioactivity for implantable electrode coatings. *Acta Biomaterialia, 10*, 1216. https://doi.org/10.1016/j.actbio.2013.12.032.

Mawad, D., Lauto, A., & Wallace, G. G. (2016). Conductive polymer hydrogels. In Kalia, S. (Ed.), *Polymeric hydrogels as smart biomaterials*. https://doi.org/10.1007/978-3-319-25322-0_2.

Merker, J., et al. (2001). High temperature mechanical properties of the platinum group metals: Elastic properties of platinum, rhodium and iridium and their alloys at high temperatures. *Platinum Metals Review, 45*, 74.

Merrill, D. R., Bikson, M., & Jefferys, J. G. R. (2005). Electrical stimulation of excitable tissue: Design of efficacious and safe protocols. *Journal of Neuroscience Methods, 141*, 171. https://doi.org/10.1016/j.jneumeth.2004.10.020.

Minev, I. R., et al. (2015). Electronic dura mater for long-term multimodal neural interfaces. *Science (80), 347*, 159–163.

Murphy, S. V., Skardal, A., & Atala, A. (2013). Evaluation of hydrogels for bio-printing applications. *Journal of Biomedical Materials Research Part A, 101A*, 272. https://doi.org/10.1002/jbm.a.34326.

Navaei, A., et al. (2016). Gold nanorod-incorporated gelatin-based conductive hydrogels for engineering cardiac tissue constructs. *Acta Biomaterialia, 41*, 133. https://doi.org/10.1016/j.actbio.2016.05.027.

Novoselov, K. S., et al. (2012). A roadmap for graphene. *Nature, 490*, 192–200.

Odent, J., et al. (2017). Highly elastic, transparent, and conductive 3D-printed ionic composite hydrogels. *Advanced Functional Materials, 27*, 1701807.

Oyen, M. L. (2014). Mechanical characterisation of hydrogel materials. *International Materials Review, 59*, 44. https://doi.org/10.1179/1743280413y.0000000022.

Pardo-Yissar, V., et al. (2001). Gold nanoparticle/hydrogel composites with solvent-switchable electronic properties. *Advanced Materials, 13*, 1320–1323.

Park, S. I., et al. (2015). Soft, stretchable, fully implantable miniaturized optoelectronic systems for wireless optogenetics. *Nature Biotechnology, 33*, 1280–1286.

Pok, S., et al. (2014). Biocompatible carbon nanotube–chitosan scaffold matching the electrical conductivity of the heart. *ACS Nano, 8*, 9822. https://doi.org/10.1021/nn503693h.

Polikov, V. S., Tresco, P. A., & Reichert, W. M. (2005). Response of brain tissue to chronically implanted neural electrodes. *Journal of Neuroscience Methods, 148*, 1. https://doi.org/10.1016/j.jneumeth.2005.08.015.

Ramón-Azcón, J., et al. (2013). Dielectrophoretically aligned carbon nanotubes to control electrical and mechanical properties of hydrogels to fabricate contractile muscle myofibers. *Advanced Materials, 25*, 4028. https://doi.org/10.1002/adma.201301300.

Ray, S. C. (2015). Application and uses of graphene oxide and reduced graphene oxide. In *Applications of graphene and graphene-oxide based nanomaterials*. https://doi.org/10.1016/b978-0-323-37521-4.00002-9.

Rho, J. Y., Ashman, R. B., & Turner, C. H. (1993). Young's modulus of trabecular and cortical bone material: Ultrasonic and microtensile measurements. *Journal of Biomechanics, 26*, 111–119.

Rodger, D. C., et al. (2008). Flexible parylene-based multielectrode array technology for high-density neural stimulation and recording. *Sensors and Actuators B Chemical, 132*, 449–460.

Rose, T. L., Kelliher, E. M., & Robblee, L. S. (1985). Assessment of capacitor electrodes for intracortical neural stimulation. *Journal of Neuroscience Methods, 12*, 181. https://doi.org/10.1016/0165-0270(85)90001-9.

Sanchez-Sanchez, A., Del Agua, I., Malliaras, G. G., & Mecerreyes, D. (2019). Conductive poly(3,4-Ethylenedioxythiophene) (PEDOT)-based polymers and their applications in bioelectronics. In *Smart Polymers and their Applications*. https://doi.org/10.1016/B978-0-08-102416-4.00006-5.

Saravanan, P., Padmanabha Raju, M., & Alam, S. (2007). A study on synthesis and properties of Ag nanoparticles immobilized polyacrylamide hydrogel composites. *Materials Chemistry and Physics, 103*, 278–282.

Sasaki, M., et al. (2014). Highly conductive stretchable and biocompatible electrode-hydrogel hybrids for advanced tissue engineering. *Advanced Healthcare Materials, 3*, 1919–1927.

Sekine, S., et al. (2010). Conducting polymer electrodes printed on hydrogel. *Journal of the American Chemical Society, 132*, 13174. https://doi.org/10.1021/ja1062357.

Shah, K., et al. (2015). Development and characterization of polyethylene glycol-carbon nanotube hydrogel composite. *Journal of Materials Chemistry B, 3*, 7950. https://doi.org/10.1039/c5tb01047k.

Shi, Y., et al. (2014). Nanostructured conductive polypyrrole hydrogels as high-performance, flexible supercapacitor electrodes. *Journal of Materials Chemistry A, 2*, 6086–6091.

Shin, S. R., et al. (2013). Carbon-nanotube-embedded hydrogel sheets for engineering cardiac constructs and bioactuators. *ACS Nano, 7*, 2369. https://doi.org/10.1021/nn305559j.

Shin, S. R., et al. (2015). Aligned carbon nanotube-based flexible gel substrates for engineering biohybrid tissue actuators. *Advanced Functional Materials, 25*, 4486. https://doi.org/10.1002/adfm.201501379.

Skardal, A., et al. (2010). Dynamically crosslinked gold nanoparticle-hyaluronan hydrogels. *Advanced Materials, 22*, 4736–4740.

Staples, N. A., et al. (2018). Conductive hydrogel electrodes for delivery of long-term high frequency pulses. *Frontiers in Neuroscience, 11*. https://doi.org/10.3389/fnins.2017.00748.

Stejskal, J. (2017). Conducting polymer hydrogels. *Chemical Papers, 71*, 269. https://doi.org/10.1007/s11696-016-0072-9.

Stejskal, J., Kratochvíl, P., & Jenkins, A. D. (1995). Polyaniline: Forms and formation. *Collection of Czechoslovak Chemical Communications, 60*, 1747. https://doi.org/10.1135/cccc19951747.

Stejskal, J., et al. (2016). Polypyrrole salts and bases: Superior conductivity of nanotubes and their stability towards the loss of conductivity by deprotonation. *RSC Advances, 6*, 88382. https://doi.org/10.1039/c6ra19461c.

Sydney Gladman, A., et al. (2016). Biomimetic 4D printing. *Nature Materials, 15*, 413. https://doi.org/10.1038/nmat4544.

Tang, Q., et al. (2008). Polyaniline/polyacrylamide conducting composite hydrogel with a porous structure. *Carbohydrate Polymers, 74*, 215. https://doi.org/10.1016/j.carbpol.2008.02.008.

Tan, Y., et al. (2014). 3D printing facilitated scaffold-free tissue unit fabrication. *Biofabrication, 6*, 024111. https://doi.org/10.1088/1758-5082/6/2/024111.

Tasis, D., et al. (2006). Chemistry of carbon nanotubes. *Chemical Reviews, 106*, 1105–1136.

Thuy Pham, T. P., Cho, C. W., & Yun, Y. S. (2010). Environmental fate and toxicity of ionic liquids: A review. *Water Research, 44*, 352–372.

Tondera, C., et al. (2019). Highly conductive, stretchable, and cell-adhesive hydrogel by nanoclay doping. *Small, 1901406*, 1–8.

Vernitskaya, T. V., & Efimov, O. N. (1997). Polypyrrole: a conducting polymer; its synthesis, properties and applications. *Russian Chemical Reviews, 66*, 443–457.

Wang, K., et al. (2006). Neural stimulation with a carbon nanotube microelectrode array. *Nano Letters, 6*, 2043. https://doi.org/10.1021/nl061241t.

Wang, S., et al. (2017). Neural stem cell proliferation and differentiation in the conductive PEDOT-HA/Cs/gel scaffold for neural tissue engineering. *Biomaterials Science, 5*, 2024. https://doi.org/10.1039/c7bm00633k.

Wu, Y., et al. (2016). Fabrication of conductive gelatin methacrylate-polyaniline hydrogels. *Acta Biomaterialia, 33*, 122. https://doi.org/10.1016/j.actbio.2016.01.036.

Wu, S., et al. (2017). Biocompatible chitin/carbon nanotubes composite hydrogels as neuronal growth substrates. *Carbohydrate Polymers, 174*, 830–840.

Xiao, X., et al. (2017). Preparation and property evaluation of conductive hydrogel using poly (vinyl alcohol)/polyethylene glycol/graphene oxide for human electrocardiogram acquisition. *Polymers (Basel), 9*. https://doi.org/10.3390/polym9070259.

Xu, Y., et al. (2010). Self-assembled graphene hydrogel via a one-step hydrothermal process. *ACS Nano, 4*, 4324. https://doi.org/10.1021/nn101187z.

Xu, Y., et al. (2018). Noncovalently assembled electroconductive hydrogel. *ACS Applied Materials & Interfaces, 10*, 14418. https://doi.org/10.1021/acsami.8b01029.

Yamato, H., Ohwa, M., & Wernet, W. (1995). Stability of polypyrrole and poly(3,4-ethylenedioxythiophene) for biosensor application. *Journal of Electroanalytical Chemistry, 397*, 163. https://doi.org/10.1016/0022-0728(95)04156-8.

Yao, B., et al. (2017). Ultrahigh-conductivity polymer hydrogels with arbitrary structures. *Advanced Materials, 29*. https://doi.org/10.1002/adma.201700974.

Yuan, J., & Antonietti, M. (2011). Poly(ionic liquid)s: Polymers expanding classical property profiles. *Polymer, 52*, 1469. https://doi.org/10.1016/j.polymer.2011.01.043.

Yue, J., Epstein, A. J., & Macdiarmid, A. G. (1990). Sulfonic acid ring-substituted polyaniline, a self-doped conducting polymer. *Molecular Crystals and Liquid Crystals Incorporating Nonlinear Optics, 189*, 255. https://doi.org/10.1080/00268949008037237.

Yuk, H., Lu, B., & Zhao, X. (2018). Hydrogel bioelectronics. *Chemical Society Reviews, 48*, 1642. https://doi.org/10.1039/C8CS00595H.

Zhai, D., et al. (2013). Highly sensitive glucose sensor based on Pt nanoparticle/polyaniline hydrogel heterostructures. *ACS Nano, 7*, 3540. https://doi.org/10.1021/nn400482d.

Zhao, Y., et al. (2013). 3D nanostructured conductive polymer hydrogels for high-performance electrochemical devices. *Energy and Environmental Science, 6*, 2856. https://doi.org/10.1039/c3ee40997j.

Zhou, Y., et al. (2018a). Highly stretchable, elastic, and ionic conductive hydrogel for artificial soft electronics. *Advanced Functional Materials, 29*. https://doi.org/10.1002/adfm.201806220.

Zhou, L., et al. (2018b). Soft conducting polymer hydrogels cross-linked and doped by tannic acid for spinal cord injury repair. *ACS Nano, 12*, 10957. https://doi.org/10.1021/acsnano.8b04609.

Zhu, F., et al. (2018). Tough and conductive hybrid hydrogels enabling facile patterning. *ACS Applied Materials & Interfaces, 10*, 13685. https://doi.org/10.1021/acsami.8b01873.

Chapter 10
Biofluid Barrier Materials and Encapsulation Strategies for Flexible, Chronically Stable Neural Interfaces

Jinghua Li

10.1 Introduction

Implantable electronic devices are essential for not only fundamental biomedical research but also the diagnostics and therapeutics of neurological diseases (Ahmad et al. 2013; Fu et al. 2017; Pei and Tian 2019; Rivnay et al. 2017; Won et al. 2018). Of particular interests are advanced technologies and platforms that can support real-time monitoring of brain activities with high spatiotemporal resolution across large areas for closed-loop neuromodulation. Ideally, the implantable platform should enable continuous sensing and stimulation over a long period of time during implantation with high fidelity, which can reach up to the whole lifespan of the patients based on the specific research target. To this end, an important research topic in this area is associated with improving the long-term stability of neural electrodes in biofluids for chronic operations. However, the permeation of biofluids, including water, ions, and other small molecules, can cause significant degradation in the electronic components of the devices, leading to device failures within a short time (e.g., from several hours to days) after implantation (Bazaka and Jacob 2013; Bowman and Meindl 1986; Liu et al. 1999; Swerdlow et al. 1999). This issue is even more critical for active electronic systems with on-chip field-effect transistors (FETs) for local signal amplification and multiplexing in the context of brain activity mapping with high spatiotemporal resolution (Viventi et al. 2010, 2011), as the leakage will result in not only catastrophic failures of transistors but also damages to surrounding tissues due to electrochemical reactions, additional potentials, and undesired stimulations. To address these challenges, one research direction focuses on the development of encapsulation materials to isolate the electronic devices from the surrounding environment. Thick and rigid bulk materials (e.g., metals and

J. Li (✉)
Department of Materials Science and Engineering, Center for Chronic Brain Injury, The Ohio State University, Columbus, OH, USA
e-mail: li.11017@osu.edu

© Springer Nature Switzerland AG 2020
L. Guo (ed.), *Neural Interface Engineering*,
https://doi.org/10.1007/978-3-030-41854-0_10

ceramics) can provide comparatively long lifetime, but at the price of compromising the compliance of bioelectronics to the biotissues for intimate coupling with minimal invasiveness and long-term biocompatibility (Mayberg et al. 2005; Wilson et al. 1991). As a result, the use of thin-film materials are of interests due to the form factors that can provide the mechanical flexibility to minimize the formation of scars and to enhance the chronic stability. In general, the ideal encapsulation materials should possess the following properties: (1) high structural integrity to minimize the penetration and permeation of small molecules such as water and ions, (2) low reactivity in biofluids in the presence of voltages/currents to avoid electrochemical reactions, (3) biocompatibility to yield a benign end product in solution environment, and (4) efficient electrical coupling at the biotic/abiotic interfaces for high signal-to-noise (SNR) ratio.

This chapter summarizes recent advances in the development of thin-film materials, structures, and integration strategies to build electronic platforms with necessary operational stability and biocompatibility for in vivo applications in neuroengineering. The chapter begins with an overview of different types of neural electrodes as persistent interfaces to the nervous system for chronic implantation with high fidelity. The following section discusses the capabilities of the most widely explored organic/inorganic materials as biofluid barriers with an emphasis on understanding the fundamental limitations associated with the extrinsic and intrinsic problems. The next part of this chapter presents a class of emerging materials derived from monocrystalline Si wafers for the encapsulation of flexible, actively multiplexed electronics that enable high spatiotemporal resolution in electrophysiological mapping and/or stimulation. Overall, this chapter aims to provide a framework for sophisticated neural electrodes with chronically stable performances to tackle grand challenges in human health. These advances suggest great potential in novel bio-integrated electronic devices as chronic implants for a broad scope of applications ranging from fundamental neuroscience research to clinical medicines.

10.2 Overview of Emerging Flexible Bioimplants for Neuroengineering

As stated in the *Introduction*, implantable bioelectronics are of growing interests for widespread applications in animal models and human subjects. In recent years, the development of these systems focuses on the design of materials, structures, and encapsulation strategies to provide the mechanics, operational stability, and biocompatibility necessary for chronic applications. Figure 10.1 shows a collection of examples with different geometries. Early works using techniques adapted from standard semiconductor procedures can produce interconnected microelectrodes on planar, rigid platforms such as "Utah arrays" (Kim et al. 2006) and/or "Michigan probes" (Fujishiro et al. 2014) for intracortical sensing of brain signals. Micro-optoelectrode array in these platforms based on monolithic microprobes of zinc

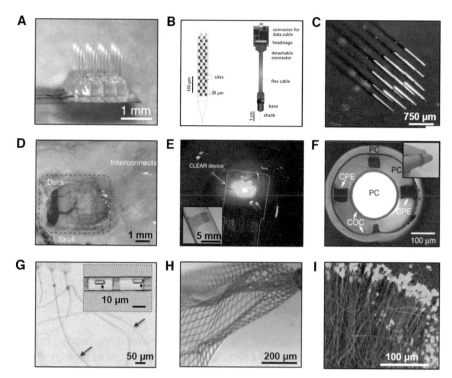

Fig. 10.1 (**a**) Photograph of a micro-optoelectrode array (MOA) device with 16 sensing nodes bonded on a polyimide electrical cable. (Reprinted with permission from Lee et al. (2015). Copyright 2015 Nature Publishing Group.) (**b**) Schematic illustration of Neuropixel probe highlighting electrode layouts and probe packaging. (Reprinted with permission from Jun et al. (2017). Copyright 2017 Nature Publishing Group.) (**c**) Photograph of four recording modules assembled into a 3D Si probe array device with 1024 electrodes. (Reprinted with permission from Rios et al. (2016). Copyright 2016 American Chemical Society.) (**d**) Optical image of ultrathin NeuroGrid implanted onto the surface of the somatosensory cortex of a rat. (Reprinted with permission from Khodagholy et al. (2015). Copyright 2014 Nature Publishing Group.) (**e**) Photography of a graphene-based, carbon-layered electrode array (CLEAR) device placed onto the cortex of a mouse. Inset: a CLEAR device in a bent configuration wrapping around a glass rod. (Reprinted with permission from Park et al. (2014). Copyright 2014 Nature Publishing Group.) (**f**) Optical image of the fiber probe tips for simultaneous electrical, optical, and chemical interrogation of neural circuits. Inset: a photograph showing a fiber wrapping around a finger. (Reprinted with permission from Canales et al. (2015). Copyright 2015 Nature Publishing Group.) (**g**) Optical image of multiple nanoelectronic thread (NET) electrodes suspended in water. Inset: optical image of two electrodes. (Reprinted with permission from Luan et al. (2017). Copyright 2017 AAAS.) (**h**) Optical image of macroporous flexible mesh electronics injected into aqueous solution using a syringe. (Reprinted with permission from Liu et al. (2015). Copyright 2015 Nature Publishing Group.) (**i**) 3D reconstructed interfaces between neuron-like electronics (NeuE, red) and neurons (green). (Reprinted with permission from Yang et al. (2019). Copyright 2019 Nature Publishing Group)

oxide (ZnO: height, 1.5 mm; width, 125 μm) can enable spatiotemporally selective light stimulation on the microscale through the ZnO-waveguide pillars and simultaneous electrophysiological recording through the indium tin oxide (ITO) coating (Fig. 10.1a) (Lee et al. 2015). The manufacturing processes developed in the integrated-chip industry allow the fabrication of Si neural probes with combined high spatiotemporal resolution and large volume coverage in a miniaturized format. Figure 10.1b shows a custom complementary metal–oxide–semiconductor (CMOS) fabrication process with micromachining on a planar silicon wafer to form a sophisticated electronic system with 960 sensing units on a rigid shank (length: 10 mm) with cross-sectional dimensions of 70 × 20 μm (known as "Neuropixel"), which offers multiplexed addressing and local amplification to enable chronic and low-noise electrophysiological mapping over a period of 6–8 weeks (Jun et al. 2017). Similarly, Fig. 10.1c shows another example of using three-dimensional (3D) electrode arrays for electrophysiology to enable neuronal activity monitoring with unprecedented spatial and temporal resolutions (Rios et al. 2016). These systems exploit mature microfabrication techniques and are readily scalable for large arrays and/or high densities that match to the needs of neuroscience and exploratory clinical studies. Challenges, however, are that the rigid construction and planar format interacting with soft biotissues may limit the sophistication at the bio-interface and lead to tissue/cellular damages.

To solve this problem, researchers pay attention to structural engineering to form systems with sufficient mechanical compliance to micromotions in biology due to their ability to establish intimate biotic–abiotic integration across curved, soft biotissues. Figure 10.1d shows an example of using ultrathin electrode arrays with sheet structures as a flexible micro-electrocorticographic (μ-ECoG) monitoring system. In this case, poly(3,4-ethylenedioxythiophene) doped with poly(styrene sulfonate) (PEDOT/PSS) serves as surface sensing electrodes with Au interconnects supported on an ultrathin membrane Parylene substrate (thickness: 4 μm). The sheetlike sensing electrodes conformally attach to the curved, dynamic, and wet surface of the rodent brain and record μ-ECoG signals from the surface of the cortex for a period of 10 days before device failure (Khodagholy et al. 2015). Aside from conventional organic/inorganic electronic materials, the use of emerging low-dimensional thin-film materials with broad-spectrum transparency could further enable simultaneous electrophysiology, optical imaging, and optogenetic activation of local brain tissues. Figure 10.1e shows an example of a graphene-based, carbon-layered electrode array (CLEAR) device for high-resolution neurophysiological recording (Park et al. 2014). The device demonstrates significant improvements in optical imaging capability compared to conventional metal-based electrodes, indicating broad utility for neuroengineering and other biomedical applications. Converting the geometries of electrodes from sheet to fiber and open-mesh structures can further enhance the flexibility for unique modes of integration. As illustrated in Fig. 10.1f, fibers fabricated from polymers by a thermal drawing method can enable the integration of multiple materials into neural interfaces for multimodal interrogation (Canales et al. 2015). In vivo operations, including recording, optical stimulation, and drug perturbation, confirm the stable performances for at least 2 months. Figure 10.1g shows another example of flexible brain probes with four passive electrodes embedded in a needlelike polymeric matrix (photo-definable epoxy, SU-8: thickness, 1.5 μm; width, 10 μm) that effectively decreases the bending stiffness to ensure high compliance to surrounding tissues. The design reduces probe–tissue interfacial forces to the level of cellular force (nanonewtons, nN), rendering it capable for chronic electrical recording in cortices for 4 months with low noise and stable impedance (Luan et al. 2017). Beyond probe geometries, complex 3D structures with tissue-like deformable mechanics and mesoscopic dimensions show advantages in device deployments and biotic/abiotic coupling. As shown in Fig. 10.1h, macroporous flexible mesh electronics consisting of Si nanowire transistors or Pt electrodes supported on photo-defined epoxy (SU-8) exhibits low bending stiffness (~ 0.087 nN m) and allows minimally invasive delivery through a syringe into tissue in a floating mesh structure (Liu et al. 2015). Another example, referred to as neuron-like electronics (NeuE), uses building blocks that mimic the subcellular structures and mechanical properties of neurons with polymer encapsulation layers. This design can yield bending stiffness comparable to those of individual neurons. A system with 16 channels demonstrates stable electrophysiological recordings for over 3 months with no evidence of loss of isolated neural signals during this period (Yang et al. 2019).

10.3 Conventional Encapsulation Strategies for Neural Electrodes

As outlined in Fig. 10.1, the considerations in form factors, geometries, and mechanics of bioimplants are essential to building systems for chronic in vivo applications. In the meantime, the choice of materials and their interactions with biofluids are of further interest to ensure the long-term stability of the electronic platforms, especially in the context of active electronic devices with advanced functionalities such as multiplexed addressing and local signal amplification. Figure 10.2 shows a collection of examples of material strategies to exploit organic and/or inorganic thin films for the encapsulation of implantable devices. A lot of works in this field focus on using thin-film materials deposited by spin coating, physical vapor deposition (PVD), chemical vapor deposition (CVD), and/or atomic layer deposition (ALD). Figure 10.2a provides an example of using organic electrochemical transistors (OECTs) for μ-ECoG recording (Khodagholy et al. 2013). Here, Parylene C, a Food and Drug Administration–approved polymer, is deposited by the CVD method as the substrate and the top-surface insulator. PEDOT/PSS patterned by photolithography serves as the transistor channels and the surface electrodes. The Au wires are isolated from the electrolyte with PEDOT/PSS or Parylene C. The total thickness of the system is 4 μm. In addition to polymers such as polyimide and Parylene C, there have been growing interests in liquid-crystal polymers (LCPs) for the packaging of bioimplants. The advantages of LCPs include their low moisture absorption rate,

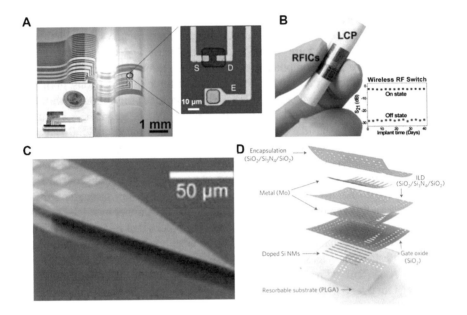

Fig. 10.2 (**a**) Photograph and optical image of ECoG probes that contain organic electrochemical transistors (OECTs) on a Parylene film (thickness: 2 μm) with PEDOT/PSS serving as the channel and surface electrodes. (Reprinted with permission from Khodagholy et al. (2013). Copyright 2013 Nature Publishing Group.) (**b**) Photograph of flexible RFICs on LCP substrate (thickness: 5 μm) with LCP encapsulation on the top (thickness: 25 μm). (Reprinted with permission from Hwang et al. (2013). Copyright 2013 American Chemical Society.) (**c**) SEM image of a Neuropixel probe tip for high-density recording of neural activity. (Reprinted with permission from Jun et al. (2017). Copyright 2017 Nature Publishing Group.) (**d**) Exploded-view schematic illustration of actively multiplexed Si transistor arrays with trilayers of SiO_2 (thickness: ~300 nm)/Si_3N_4 (thickness: ~400 nm thick)/SiO_2 (thickness: ~300 nm) as encapsulation. (Reprinted with permission from Yu et al. (2016). Copyright 2016 Nature Publishing Group)

low free volumes, and efficient chain packing within crystalline structures (Feng et al. 2004; Jeong et al. 2012), which lead to improved barrier properties compared to organic alternatives with similar thicknesses. Figure 10.2b shows an example of using LCPs for the encapsulation of flexible radio-frequency integrated circuits (RFICs) (Hwang et al. 2013). In this design, a LCP substrate (thickness: 50 μm) supports the flexible RFICs and is monolithically bonded with another layer of LCP as the cover (thickness: 25 μm). The LCPs exhibit excellent barrier properties as validated by in vitro soaking tests (projected to be ~2 years according to the Arrhenius equation) and in vivo tests (6 weeks). However, the relatively large thickness value limits the conformality of the platform to brain tissues for sensing and stimulation. Compared to organic materials, inorganic thin films adapted from CMOS technology can provide advantages in preparation and performance. An example of large-scale integration of Si probes appears in Fig. 10.2c, which contains 960 sensing units on a shank (length: 10 mm) with a $70 \times 20\ \mu m^2$ cross section (Jun et al. 2017). Here, plasma-enhanced chemical vapor deposition (PECVD) forms 800-nm SiO$_2$ which serves as the encapsulation with opening on the electrodes, and a 300-nm TiN layer deposited by reactive PVD serves as the electrical interfaces (Lopez et al. 2013). In vivo experiments by implanting a probe in the nervous system of a mouse suggest stable long-term recordings over 150 days. One commonly used method to enhance the barrier performance of the encapsulation is to use alternating multilayer stacks. Figure 10.2d shows an example of using a trilayer structure of SiO$_2$ (thickness: ~300 nm)/Si$_3$N$_4$ (thickness:~400 nm)/SiO$_2$ (thickness: ~400 nm) deposited by PECVD for the encapsulation of actively multiplexed Si transistor arrays (Yu et al. 2016). Failures of the device arise from the extrinsic imperfections (pinholes) associated with the deposition process due to the environment available to academic cleanroom facilities.

10.4 Thermally Grown SiO$_2$ as Capacitive Interface for Flexible Bio-integrated Electronics

As discussed in the previous session, the limitations in thin-film barrier materials prepared by conventional deposition approaches motivate studies to identify novel strategies for encapsulation. In general, disadvantages are sourced from both extrinsic (e.g., pinholes) and intrinsic (e.g., free volumes, reactivity with biofluids) problems. To address these issues, recent works report the use of ultrathin, thermally grown silicon dioxide on monocrystalline Si wafer (thermally grown SiO$_2$, t-SiO$_2$) as the encapsulation for various types of bioelectronics with unprecedented lifetime for applications as validated by in vitro and in vivo studies (Fang et al. 2016, 2017). Here, the monocrystalline structure of the growth substrate (Si wafer), together with the high oxidation temperature (typically above 1100 °C), helps to overcome the intrinsic and extrinsic deficiencies of deposited thin-film materials as stated above to yield a defect-free barrier layer that can effectively isolate electronic devices from water/ions in the surrounding environment. The well-established Si technology also enables the large-scale thin-film process with well-controlled spatial uniformity and thickness for targeted lifetimes. Figure 10.3a shows a schematic illustration of the integration scheme for transferring t-SiO$_2$ for encapsulation. The process begins with the fabrication of electronics on silicon-on-insulator (SOI) substrate with device-grade Si on the top. Bonding the top side of the device onto a temporal substrate followed by etching of the back-layer Si handle wafer exposes the buried oxide. Peeling off the device from the temporary substrate forms a piece of flexible electronics with t-SiO$_2$ to simultaneously serve as the chronically stable biofluid barrier and the capacitive sensing interface. Acceleration tests at elevated temperatures provide understanding on the chemistry and kinetics of the dissolution of t-SiO$_2$. As illustrated in Fig. 10.3b, the t-SiO$_2$ thin film shows a systematic reduction in thickness due to the hydrolysis (SiO$_2$ + 2H$_2$O → Si(OH)$_4$), and the lifetime linearly depends on the thickness. Temperature-dependent studies confirm that the dissolution rates depend exponentially on $1/T$, which is consistent with the

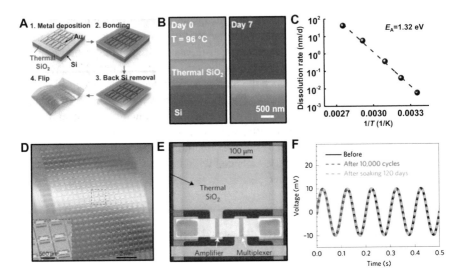

Fig. 10.3 (**a**) Schematic illustration of key fabrication steps exploiting thermal SiO_2 as biofluid barrier layer for implantable devices. (**b**) SEM images showing change in the thickness of a thermal SiO_2 (thickness: 1 μm) during soak test in PBS at 96 °C. (**c**) Dissolution speed of t-SiO_2 as a function of $1/T$ showing the linear relationship consistent with Arrhenius scaling. (Panel **a–c** reprinted with permission from Fang et al. (2016). Copyright 2016 National Academy of Sciences.) (**d**) Optical image of an actively multiplexed, capacitively coupled sensing system with t-SiO_2 encapsulation and 396 sensing nodes in a slightly bent configuration. (**e**) Optical image of a sensing node with two Si transistors serving as the multiplexer and the amplifier, respectively. (**f**) The output curves of a representative sensing unit to a sine wave input after 10,000 bending cycles and after soak test (120 days). (Panel **d–f** reprinted with permission from Fang et al. (2017). Copyright 2017 Nature Publishing Group)

Arrhenius scaling (Fig. 10.3c). The results indicate that the reaction proceeds exclusively at the surface without reactive diffusion into the film or permeation through defects. These findings enable the integration of t-SiO_2 with actively multiplexed Si transistor arrays for large-scale electrophysiology. Figure 10.3d shows a system consisting of 396 actively multiplexed sensing units with capacitive coupling. The design of a unit of this type appears in Fig. 10.3e, with a multiplexer connected to the back end to switch the unit and a source follower amplifier for signal processing. A layer of t-SiO_2 covers the entire interface, isolating all metal interconnects and electronic components from the biofluids. Bending (10,000 bending cycles) and soaking (120 days in phosphate-buffered saline [PBS], room temperature) tests highlight the mechanical flexibility and longevity of this platform, respectively (Fig. 10.3f). In vivo recording of the auditory cortex in rat models shows stable leakage current (~10^{-8} A) up to 20 days (Fang et al. 2017). The multiplexed transistor arrays can be readily scaled up to a very large coverage area and channel count, meeting the requirement for the study of brain activities in higher mammals such as primates.

10.5 Heavily Doped, Highly Conductive Monocrystalline Si Interface for Sensing and Stimulation

Despite the advantages of using t-SiO$_2$ as a novel type of relatively inert and defect-free biofluid barrier for various types of bioimplants, the intrinsic insulating property of SiO$_2$ sets fundamental limitations in its applications as the biotic/abiotic interface material in neuroengineering at high spatiotemporal resolution: in this design, t-SiO$_2$ is unpatterned, and the sensing takes place through capacitive coupling between the biotissues and transistors. Accordingly, the top gate dielectrics and the sensing interface form a voltage divider structure with two capacitors in series (Fang et al. 2017; Li et al. 2019, 2020). The voltage gain of the circuit can be calculated as

$$\text{Total gain} = V_{\text{out}}/V_{\text{In}} = V_{\text{TG}}/V_{\text{in}} = C_{\text{CAP}}/\left(C_{\text{TG}} + C_{\text{CAP}}\right) \qquad (10.1)$$

where V_{out} is the output voltage, V_{In} is the input voltage, V_{TG} is the voltage drop across the transistor, C_{CAP} is the interface capacitance, and C_{TG} is top gate capacitance. As a result, strategies to avoid signal attenuation involve maximizing the ratio of C_{CAP} to C_{TG}, which can be achieved through either reducing the thickness or increasing the lateral dimension of the t-SiO$_2$ capacitor according to the following equation:

$$C = \varepsilon_r \varepsilon_0 A/t \qquad (10.2)$$

where ε_r is the relative permittivity of the material of the capacitor, ε_0 is the vacuum permittivity ($\varepsilon_0 = 8.854 \times 10^{-12}$ F m^{-1}), and A and t are the area and thickness of the capacitor, respectively.

However, such types of scaling will either compromise the lifetime or the spatial resolution of the system (Feiner et al. 2016). To circumvent this problem, a recently reported study uses patterned p-type Si NMs (p^{++}-Si, concentration: 10^{20} cm^{-3}) intimately bonded to t-SiO$_2$ to form a conductive pathway that enables the direct electrical coupling to the top gate of the transistor. Although lightly doped Si dissolves in water at relatively high rate with transient performances in biofluids (~20 nm/day at 37 °C), p-type dopants can reduce the width of the depletion region at the water/Si interface by lowering the Fermi level in Si, thereby promoting the recombination of injected electrons with holes and impeding the reduction of water during the Si dissolution (Li et al. 2018; Seidel et al. 1990). As a result, p^{++}-Si demonstrates a much lower dissolution rate (~0.5 nm/day at 37 °C) and can serve as a chronically stable conductive sensing interface (sheet resistance: 32 Ω/sq) and biofluid barrier simultaneously. Figure 10.4a shows an illustration of a device of this type highlighting key functional layers (exploded view). The monolithic structure of p^{++}-Si and t-SiO$_2$ (denoted as p^{++}-Si//t-SiO$_2$) with chemically bonded interfaces formed at 1000 °C eliminates the capacitors at the surface of the electrode and enables direct electrical coupling between biotissues and transistors. Coating and patterning a layer of noble

276 J. Li

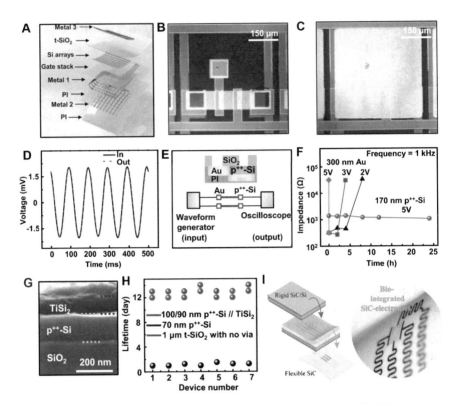

Fig. 10.4 (a) Schematic illustrations of active Si transistor arrays with p^{++}-Si//t-SiO$_2$ encapsulation and 64 sensing nodes. (**b, c**) Optical images of a sensing unit before (**b**) and after (**c**) the deposition of a metal coating. (**d**) Output characteristics of a sensing node with an AC input (voltage, 2.8 mV; frequency, 10 Hz). (**e**) Circuit diagram of a test system to evaluate efficiency and stability of p^{++}-Si stimulation electrodes. (**f**) Lifetimes of p^{++}-Si and Au electrode with different stimulation voltages. (Panel **a–f** reprinted with permission from Li et al. (2018). Copyright 2018 National Academy of Sciences.) (**g**) SEM images showing the side views of TiSi$_2$ during soak tests, suggesting no change in thickness at 96 °C for 10 days. (**h**) Statistics of lifetimes of transistors encapsulated by t-SiO$_2$, p^{++}-Si//t-SiO$_2$, and TiSi$_2$/p^{++}-Si//t-SiO$_2$. (Panel **g** and **h** reprinted with permission from Li et al. (2019, 2020). Copyright 2019 American Chemical Society.) (**i**) Schematic illustration of transferring crystalline SiC for use as flexible, long-lived, and multifunctional bioelectronics. (Panel **i** reprinted with permission from Phan et al. (2019). Copyright 2019 American Chemical Society)

metal on the side in contact with biofluids enables a 100% fill factor (defined as the ratio of sensing area to total unit area). Figure 10.4b, c present optical images of a sensing node before and after the coating of 300 nm Au, with p^{++}-Si only occupying a small portion (5%) of the whole area. This coupling strategy without signal attenuation provides a voltage gain of ~0.99 (design point: 1) (Fig. 10.4d), as expected based on Eq. (10.1). Systematic studies investigate key parameters (via opening size, fill factor of p^{++}-Si) on the scalability of active electronics with p^{++}-Si//t-SiO$_2$ interface for electrical recording with high spatial resolution (Li et al. 2019, 2020). Due to the direct electrical coupling, the devices show stable electrical performances with low variations in gain (~0.97–0.99) and noise (~30–80 µV) levels over a wide dynamic range (−100 Hz, input voltage: 2 mV). In contrast, devices with t-SiO$_2$ encapsulation demonstrate strong dependence on the lateral dimension of the interface capacitor (Fang et al. 2017). The structure of p^{++}-Si//t-SiO$_2$ provides a conductive interconnect to the top gate transistor and enables the capability for scaling to the resolution only limited by the dimension that can be achieved using semiconductor fabrication facilities. Such a monocrystalline, conductive interface can also deliver current for electrical stimulation in neuro- and cardiac science (Fig. 10.4e). Studies of electrochemical stability of the electrical interface during pulsed-mode stimulation evaluates the lifetime of p^{++}-Si (170 nm) and conventional thin-film materials (Cr/Au) (Li et al. 2018). While Au reacts with ions in PBS in the presence of a voltage (2–5 V) through the reaction (Au + 4Cl$^-$ → AuCl$_4^-$ + 3e$^-$) which lately causes thin-film dissolution and delamination from the substrate, p^{++}-Si remains stable under the same conditions, suggesting its excellent and reliable properties to simultaneously serve as electrical interface and encapsulation layer for stimulation electrodes.

In the meantime, the choice of conductive interface material is worth further attention, as p^{++}-Si has lower activation energy (0.82 eV) and, as a result, possesses a dissolution rate 1–2 orders higher than that of t-SiO$_2$ (1.32 eV) (Fang et al. 2016) and limits the lifetime of the system at body temperature to several years (with ~60-nm thickness). To address this issue, recent studies also focus on finding Si-based alternatives with higher conductivity and lower reactivity with water. One strategy is to use metal silicide alloys formed by the deposition of Ti on p^{++}-Si followed by thermal annealing (850 °C) (Li et al. 2019, 2020). The silicidation partially consumes top-layer Si and yields a TiSi$_2$/p^{++}-Si structure. The backside wafer etching steps described in Fig. 10.3 flip the stacks to form a p^{++}-Si/TiSi$_2$//t-SiO$_2$ structure with p^{++}-Si in contact with biofluids. Experimental studies and theoretical investigations using reactive diffusion modeling suggest the extrapolated lifetime of TiSi$_2$ to be approximately three orders of magnitude higher than that of a SiO$_2$ interface with a similar thickness. Consequently, the lifetime of the whole system with p^{++}-Si/TiSi$_2$//t-SiO$_2$ can be solely decided by the slow but finite dissolution of t-SiO$_2$ (Fig. 10.4g). However, the stability and suitability of TiSi$_2$ as chronic neural interface requires further studies. As one step further beyond this point, another study reports the use of transferred cubic silicon carbide (3C-SiC) nanomembranes (NMs) prepared by low-pressure chemical vapor deposition (LPCVD) as an alternative strategy for the encapsulation of bioimplants. The conductivity can be tuned by introducing dopants (e.g., aluminum or nitrogen) to form *p*- or *n*-type conductive interfaces (Phan et al. 2019). Systematic studies demonstrate that 230-nm SiC NMs show no hydrolysis behavior in PBS, with no detectable water permeability for 60 days at 96 °C, setting the record for conductive Si-based NMs as biofluid barriers among results reported so far. No Na$^+$ diffusions through SiC are detected at a thickness of 50 nm after soaking tests in PBS for 12 days at 96 °C. In addition to the encapsulation capability, SiC NMs are also well-suited as temperature and strain sensors due to the thermoresistive and piezoresistive phenomenon (Phan et al. 2019), respectively, suggesting a promising pathway towards multifunctional platforms for the applications in neural and cardiac electrophysiology.

10.6   Challenges and Perspectives

This chapter reviews recent progress in advanced materials and integration strategies for the development of chronically stable neural interfaces for long-term sensing and stimulation. While conventional encapsulation approaches show limitations due to intrinsic and extrinsic problems, the use of thin-film materials derived from monocrystalline Si appears to circumvent these challenges by providing ultrathin, flexible capacitive/conductive encapsulation layers with very low water vapor transmission rates. Incorporating these concepts with advanced neural circuits will likely provide further opportunities in developing bio-integrated electronics as the basis for closed-loop neuromodulations. While the studies reviewed here address challenges in the chronic stability of electronic devices, the biocompatibility of the system requires further attentions, as tissue reactions such as immunogenic fibrotic scar formation can isolate the devices from the surrounding environment to compromise sensing performances. This could reduce the effective lifetime of the implantable platform even if the sensor remains functional during chronic implantation. To avoid this issue, one solution is to use functionalized polymers (e.g., polymers with zwitterionic interfaces) and/or drug-eluting materials to suppress the immunogenic responses and the formation of insulating scars caused by nonspecific enzyme/cell binding on the neural electrodes. Additionally, although it would be highly desirable to build sensing platforms having the merits described above, the cost and morbidity due to additional procedures and surgeries to remove the bioimplants can lead to potential problems. To solve this, one solution is to substitute the conventional materials (e.g., Au, Cr, Ti, Cu) with more bioresorbable counterparts (e.g., Mo, Zn, Mg) (Yin et al. 2014, 2015). These materials together with Si electronics can go through complete bioresorption, producing benign end products that can be efficiently cleared via normal metabolic pathways (Bai et al. 2019). These bioresorbable electronic technologies could enable continuous, high-fidelity recording and data streaming over well-defined time periods matching the treatment process prior to complete dissolution, eliminating the danger and cost associated with nonresorbable metals. Overall, the future of the research field will bring together collaborative and multidisciplinary efforts in materials science, biomedical engineering, electrical engineering, and neuroscience. The continued interests in both fundamental and applied research will undoubtedly create a fertile research area for advanced healthcare.

References

Ahmad, J., Bazaka, K., Anderson, L. J., et al. (2013). Materials and methods for encapsulation of OPV: A review. *Renewable and Sustainable Energy Reviews, 27*, 104.

Bai, W. B., Shin, J., Fu, R. X., et al. (2019). Bioresorbable photonic devices for the spectroscopic characterization of physiological status and neural activity. *Nature Biomedical Engineering, 3*, 644.

Bazaka, K., & Jacob, M. V. (2013). Implantable devices: Issues and challenges. *Electronics, 2*, 1.

Bowman, L., & Meindl, J. D. (1986). The packaging of implantable integrated sensors. *IEEE Transactions on Biomedical Engineering, 33*, 248.

Canales, A., Jia, X. T., Froriep, U. P., et al. (2015). Multifunctional fibers for simultaneous optical, electrical and chemical interrogation of neural circuits in vivo. *Nature Biotechnology, 33*, 277.

Fang, H., Zhao, J. N., Yu, K. J., et al. (2016). Ultrathin, transferred layers of thermally grown silicon dioxide as biofluid barriers for biointegrated flexible electronic systems. *Proceedings of the National Academy of Sciences of the United States of America, 113*, 11682.

Fang, H., Yu, K. J., Gloschat, C., et al. (2017). Capacitively coupled arrays of multiplexed flexible silicon transistors for long-term cardiac electrophysiology. *Nature Biomedical Engineering, 1*, 0038.

Feiner, R., Engel, L., Fleischer, S., et al. (2016). Engineered hybrid cardiac patches with multifunctional electronics for online monitoring and regulation of tissue function. *Nature Materials, 15*, 679.

Feng, J., Berger, K., & Douglas, E. (2004). Water vapor transport in liquid crystalline and non-liquid crystalline epoxies. *Journal of Materials Science, 39*, 3413.

Fu, T.-M., Hong, G., Viveros, R. D., et al. (2017). Highly scalable multichannel mesh electronics for stable chronic brain electrophysiology. *Proceedings of the National Academy of Sciences of the United States of America, 114*, E10046.

Fujishiro, A., Kaneko, H., Kawashima, T., et al. (2014). In vivo neuronal action potential recordings via three-dimensional microscale needle-electrode arrays. *Scientific Reports, 4*, 4868.

Hwang, G. T., Im, D., Lee, S. E., et al. (2013). In vivo silicon-based flexible radio frequency integrated circuits monolithically encapsulated with biocompatible liquid crystal polymers. *ACS Nano, 7*, 4545.

Jeong, J., Lee, S. W., Min, K. S., et al. (2012). Liquid crystal polymer(LCP), an attractive substrate for retinal implant. *Sensors and Materials, 24*, 189.

Jun, J. J., Steinmetz, N. A., Siegle, J. H., et al. (2017). Fully integrated silicon probes for high-density recording of neural activity. *Nature, 551*, 232.

Khodagholy, D., Doublet, T., Quilichini, P., et al. (2013). In vivo recordings of brain activity using organic transistors. *Nature Communications, 4*, 1575.

Khodagholy, D., Gelinas, J. N., Thesen, T., et al. (2015). NeuroGrid: Recording action potentials from the surface of the brain. *Nature Neuroscience, 18*, 310.

Kim, S.-J., Manyam, S. C., Warren, D. J., et al. (2006). Electrophysiological mapping of cat primary auditory cortex with multielectrode arrays. *Annals of Biomedical Engineering, 34*, 300.

Lee, J., Ozden, I., Song, Y. K., et al. (2015). Transparent intracortical microprobe array for simultaneous spatiotemporal optical stimulation and multichannel electrical recording. *Nature Methods, 12*, 1157.

Li, J. H., Song, E. M., Chiang, C. H., et al. (2018). Conductively coupled flexible silicon electronic systems for chronic neural electrophysiology. *Proceedings of the National Academy of Sciences of the United States of America, 115*, E9542.

Li, J., Li, R., Du, H., et al. (2019). Ultrathin, transferred layers of metal silicide as faradaic electrical interfaces and biofluid barriers for flexible bioelectronic implants. ACS Nano, 13 , 660.

Li, J., Li, R., Chiang, C. H., et al. (2020). Ultrathin, high capacitance capping layers for silicon electronics with conductive interconnects in flexible, long-lived bioimplants. Advanced Materials Technologies.2020, 5(1):1900800

Liu, X., McCreery, D. B., Carter, R. R., et al. (1999). Stability of the interface between neural tissue and chronically implanted intracortical microelectrodes. *IEEE Transactions on Rehabilitation Engineering, 7*, 315.

Liu, J., Fu, T. M., Cheng, Z. G., et al. (2015). Syringe-injectable electronics. *Nature Nanotechnology, 10*, 629.

Lopez, C. M., Andrei, A., Mitra, S., et al. (2013). An implantable 455-active-electrode 52-channel CMOS neural probe. *IEEE Journal of Solid-State Circuits, 49*, 248.

Luan, L., Wei, X. L., Zhao, Z. T., et al. (2017). Ultraflexible nanoelectronic probes form reliable, glial scar-free neural integration. *Science Advances, 3*, e1601966.

Mayberg, H. S., Lozano, A. M., Voon, V., et al. (2005). Deep brain stimulation for treatment-resistant depression. *Neuron, 45,* 651.

Park, D. W., Schendel, A. A., Mikael, S., et al. (2014). Graphene-based carbon-layered electrode array technology for neural imaging and optogenetic applications. *Nature Communications, 5,* 5258.

Pei, F., & Tian, B. (2019). Nanoelectronics for minimally invasive cellular recordings. *Advanced Functional Materials,* Adv. Funct. Mater, 1906210.

Phan, H.-P., Zhong, Y., Nguyen, T.-K., et al. (2019). Long-lived, transferred crystalline silicon carbide nanomembranes for implantable flexible electronics. *ACS Nano, 13,* 11572.

Rios, G., Lubenov, E. V., Chi, D., et al. (2016). Nanofabricated neural probes for dense 3-D recordings of brain activity. *Nano Letters, 16,* 6857.

Rivnay, J., Wang, H., Fenno, L., et al. (2017). Next-generation probes, particles, and proteins for neural interfacing. *Science Advances, 3,* e1601649.

Seidel, H., Csepregi, L., Heuberger, A., et al. (1990). Anisotropic etching of crystalline silicon in alkaline solutions II. Influence of dopants. *Journal of The Electrochemical Society, 137,* 3626.

Swerdlow, C. D., Olson, W. H., O'Connor, M. E., et al. (1999). Cardiovascular collapse caused by electrocardiographically silent 60-Hz intracardiac leakage current: Implications for electrical safety. *Circulation, 99,* 2559.

Viventi, J., Kim, D.-H., Moss, J. D., et al. (2010). A conformal, bio-interfaced class of silicon electronics for mapping cardiac electrophysiology. *Science Translational Medicine, 2,* 24ra22.

Viventi, J., Kim, D.-H., Vigeland, L., et al. (2011). Flexible, foldable, actively multiplexed, high-density electrode array for mapping brain activity in vivo. *Nature Neuroscience, 14,* 1599.

Wilson, B. S., Finley, C. C., Lawson, D. T., et al. (1991). Better speech recognition with cochlear implants. *Nature, 352,* 236.

Won, S. M., Song, E., Zhao, J., et al. (2018). Recent advances in materials, devices, and systems for neural interfaces. *Advanced Materials, 30,* 1800534.

Yang, X., Zhou, T., Zwang, T. J., et al. (2019). Bioinspired neuron-like electronics. *Nature Materials, 18,* 510.

Yin, L., Cheng, H. Y., Mao, S. M., et al. (2014). Dissolvable metals for transient electronics. *Advanced Functional Materials, 24,* 645.

Yin, L., Farimani, A. B., Min, K., et al. (2015). Mechanisms for hydrolysis of silicon nanomembranes as used in bioresorbable electronics. *Advanced Materials, 27,* 1857.

Yu, K. J., Kuzum, D., Hwang, S. W., et al. (2016). Bioresorbable silicon electronics for transient spatiotemporal mapping of electrical activity from the cerebral cortex. *Nature Materials, 15,* 782.

Chapter 11
Regenerative Neural Electrodes

Gildardo Guzman, Muhammad Rafaqut, Sungreol Park, and Paul Y. Choi

11.1 Introduction

With nearly 2 million amputees in the United States and roughly 185,000 yearly amputations due to diabetes and peripheral arterial disease, prosthetic technology advancements have become an increasingly important topic (Kathryn Ziegler-Graham et al. 2008). Unfortunately, current prosthetic technology is far from perfect, as a review from 1980 to 2006 displayed a rejection rate of 45% for upper body prosthetics and 35% for body-powered prostheses and electric prostheses, respectively (Biddiss and Chau 2007). This points to current prostheses being inadequate replacements for lost limbs and suggests that there is much room for improvement. Over the past decades, huge leaps have been made in prosthetic technology and improved patient satisfaction. Advancements have led to the development of two major interfacing technologies: brain-machine interfaces which are implanted in the brain and translate signals from the cortex and peripheral nerve interfaces which are implanted directly into or around peripheral nerves to record signals. Both translate the recorded signals into movement of a prosthetic limb; however, these technologies each come with their own set of problems.

Brain-machine interfacing works with brain signals that may be too weak when recording outside of the cranium and require implantation within the cranium, making it a highly invasive procedure (Brown et al. 2004; Lee et al. 2008). Implant devices must be designed to effectively record brain signals without damage, but inserting an electrode array may potentially damage the brain (Edell et al. 1992). Brain signals are also harder to translate into movement since limb movements are not processed in an isolated area of the brain but rather multiple foci that represent any movement, making the recorded signals abstract and harder to interpret (Rao

G. Guzman · M. Rafaqut · S. Park · P. Y. Choi (✉)
Department of Electrical and Computer Engineering, The University of Texas Rio Grande Valley, Edinburg, TX, USA
e-mail: paul.choi@utrgv.edu

© Springer Nature Switzerland AG 2020
L. Guo (ed.), *Neural Interface Engineering*,
https://doi.org/10.1007/978-3-030-41854-0_11

et al. 1995; Sanes et al. 1995). In contrast, peripheral nerve interfaces (PNIs) are directly implanted on/in peripheral nerves, thus avoiding any risk of damage to the brain or central nervous system, and expose the peripheral nerves to minimal damage (Munger et al. 1992; Thanos et al. 1998; Branner et al. 2001; Stieglitz 2004; Navarro et al. 2005). Since signals from peripheral nerves have a much more obvious connection to movement than those found in the brain, a PNI allows for easier translation to prosthetic movement. A PNI uses electrodes to record neural signals and translates the information into specific functions such as movement. A limiting factor of PNIs is that extracellular action potentials are relatively weak and difficult to effectively detect and work with (Loeb and Peck 1996). Clinically viable PNIs must also be able to remain functional for years or even decades. These limitations have given rise to diverse PNIs, each seeking to address a certain limitation.

Myoelectric sensors paired with targeted muscle reinnervation can vastly improve the dexterity of a prosthetic device as compared to body-powered prosthetics (Kuiken et al. 2009). However, myoelectric technology is limited since the ability of myoelectric sensors to detect electrical muscle outputs becomes inhibited over extended periods of time by humidity and sweat which are quite common in prosthetic zones of attachment. The quality of information collected also suffers, because surface electrodes are limited to individual muscles as opposed to being nerve specific (Birbaumer et al. 2004). Kung et al. developed a PNI which used neurotized muscle units in a similar fashion to surface electrodes. The key difference was that the electrodes were subdermal and thus closer to the muscle and nerves, allowing for improved readings (Kung et al. 2014). Another subdermal alternative is cuff electrodes, which encircle nerves and offer higher selectivity than surface electrodes. Their cuff structure confines both reading and stimulation to what is inside the cuff's structure, increasing selectivity and decreasing cross-reading. However, due to their all-encompassing structure, they cannot record signals from distinct axonal fascicles (Loeb and Peck 1996). To address this problem, multichannel cuff electrodes (Rozman et al. 1993; Navarro et al. 2001) and flat-interface nerve electrodes have been developed, aiming to reshape nerves in order to increase selectivity (Durand et al. 2009).

In addition to extraneural interfaces, intraneural interfaces have also been developed to record neural signals. The main advantage of intraneural interfaces is that they can record stronger signals since intraneural action potential amplitudes are higher than extraneural potentials (Yoshida et al. 2000). They also have access to both motor and sensory axons allowing for modality-specific stimulation for sensation (Dhillon et al. 2004; Rossini et al. 2010). Longitudinal intrafascicular electrodes (LIFEs) are a type of electrode that is inserted through the ends of amputated nerves. Their elongated structure has allowed them to penetrate deep into neural fascicles for recording and stimulation purposes (Malagodi et al. 1989; McNaughton and Horch 1996; Yoshida et al. 2000; Lawrence et al. 2003). A few variations of LIFEs have been used to elicit graded sensation of touch, position, and joint movement (Dhillon et al. 2004; Lawrence et al. 2004). Penetrating interfaces are another form of intrafascicular interface, consisting of needlelike electrode arrays. They were designed for use in the central nervous system, but have been tested in the peripheral nervous system as well (Vallbo et al. 1979; Hagbarth 1993). After testing, however, only 10–20% of penetrating electrodes were shown to be capable of recording single-unit responses in mechanoreceptors (Branner and Normann 2000; Branner et al. 2001). Another example of a penetrating interface is the transverse intrafascicular multichannel electrode (TIME). In testing, this interface was implanted using a needle which penetrated longitudinally through the side of the nerve, pulling the electrode along behind it. The interface was held in place using fibrin glue and used to selectively stimulate and record different groups of axons in various fascicles in the nerve (Boretius et al. 2010). The drawback of penetrating interfaces is that they are relatively rigid. While this is not a disadvantage in the central nervous system where the implant is not subject to macro-motions, it is in the peripheral nervous system where implants are subject to macro-movements from the nerve itself and the surrounding muscles. This may result in premature failure of the interface.

All of these have led to the development of regenerative peripheral nerve interfaces (RPNIs). The idea of an RPNI was proposed in 1973 in a paper published by Llinas et al. (1973), but was first implemented in a design that allowed both recording and regeneration in 1991 (Kovacs et al. 1992). RPNIs work with transected nerve tissue by suturing the nerve stumps to a medium through which the nerve will regenerate (Edell 1986; Kovacs et al. 1992). An example of this is the sieve electrode interface, which is composed of a substrate structure (usually silicon, PDMS, or polyimide) with perforations known as via holes through which the axons grow (Llinas et al. 1973). These via holes may be surrounded by a ring electrode which is used to record and stimulate the axons (Rosen et al. 1990). RPNIs can offer great selectivity due to multiple electrodes reading individual axons. The electrode's close proximity to the nerve has also proven to be favorable for signal reading. However, effective readings may be dependent on the electrode's distance to a node of Ranvier (Hess and Young 1952; FitzGerald et al. 2009). Regenerative peripheral nerve microchannel interfaces can increase selectivity by isolating and recording from single axons, effectively eliminating cross-reading and increasing signal strength due to encompassment of the node of Ranvier by the microchannel. Figure 11.1 shows how the microchannel covers and records neural signals from individual axons through structural selectivity during nerve regeneration. Despite the current problems that RPNIs face, they are a great option for controlling a prosthetic device, since the residual nerve tissue may remain functional for years after amputation (Dhillon et al. 2004). They can provide a more intuitive form of control due to the peripheral nervous system's direct link to movements and avoid permanent brain damage, which is seen in some cases from brain-machine interfaces.

This chapter will focus on RPNIs and provide an overview of existing technologies, along with their application and viability. The first section will discuss conduits and how they are used for basic nerve regeneration but do not include electrodes. Then it will discuss conduits integrated with electrodes and how they work to record and stimulate nerves, but are limited in that unorganized nerve growth restricts the consistency and selectivity of recorded signals. Next is sieve electrode interfaces, which allow for nerves to regenerate through via holes

Fig. 11.1 Microchannel
combined with an
electrode for modulating a
single axon. The
microchannel is long
enough to record the neural
signal from the nodes
of Ranvier

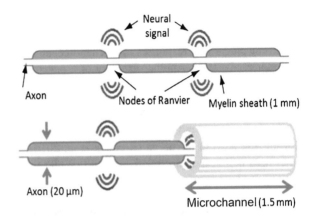

increasing the number of potential signals recorded, but are limited based on their proximity to a node of Ranvier. Microchannel interfaces, with their various structures and designs, such as rolled polyimide or PDMS interfaces or microwire fabricated channel interfaces, are then reviewed, focusing on their advantages, such as organized regeneration, high selectivity, and accurate recordings. The chapter will conclude with a discussion on how RPNIs can be improved and tested in the future, allowing for their use with human-controlled prosthetics.

11.2   Regenerative Conduit

Regenerative conduits guide along nerve regeneration and are essential for RPNIs in helping bridge the gaps between severed and damaged nerves. Several designs for regenerative conduits have been developed with their own purposes and approaches, and these conduits, and later RPNIs, have all been developed using biocompatible material used in neural gap bridging for decades (Lundborg et al. 1981; Suematsu et al. 1988; Wang et al. 1993). The conduit's synthetic biocompatible nature does not demonstrate any adverse effects such as an apparent inflammatory response or damage to nervous tissue, and the materials needed are widely available and relatively easy to fabricate (Pettersson et al. 2010; Srinivasan et al. 2015). For example, one method for conduit manufacture utilizes fibrin glue and a PDMS mold (Pettersson et al. 2010). Regeneration using the conduits occurs by suturing both the proximal and the distal nerve stumps to opposite ends of the conduit's tube as shown in Fig. 11.2. By altering the fabrication process, conduits may be customized, which can improve neural regeneration by avoiding dimensional mismatching and making the conduits extremely adaptable (Nichols et al. 2004).

Further research using these conduits and their application for regenerating nerves has also been studied. For example, axonal regeneration was successfully promoted over a 17-mm nerve gap in a rat model using aligned polymer fibers and demonstrated that conduits were functional in bridging long nerve gaps as well

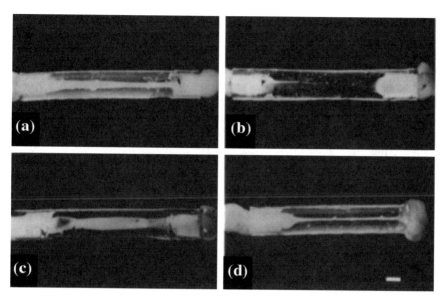

Fig. 11.2 Macroscopic view of 10-mm PDMS chambers with differing cell types at the distal ends. (**a**) Distal nerve at end (3 weeks). (**b**) Skin cells at end (1 week). (**c**) Tendon at end (1 week). (**d**) Open at end (3 weeks). Scale bar = 1 mml. (Adapted with permission from Williams et al. (1984))

(Kim et al. 2008). The regenerated neural tissue reinnervated the distal muscle and formed new neuromuscular junctions. Pairing conduits with neurotrophic factors was also studied. Neurotrophic factors such as brain-derived neurotrophic factor (BDNF), nerve growth factor (NGF), and neurotrophin-3 (NT-3) were shown to enhance cell survival, nerve regeneration, and functional recovery in peripheral nerves (Fischer et al. 1991; Nagahara et al. 2009; Brock et al. 2010; Rangasamy et al. 2010). Lotfi et al. used NGF and NT-3 to selectively entice growth of TrkA$^+$ nociceptive with NGF or TrkC$^+$ proprioceptive (Lotfi et al. 2011). The effects of either substance on growth were tested using "Y"-shaped tubing with one branch containing NGF and the other NT-3. This research showed that neurotrophic factors can be used to effectively promote specific axon types and could be used to segregate motor and sensory neurons in peripheral nerve interfaces. In addition to neurotrophic factors, the presence of certain cells has been shown to improve neural regeneration (Williams et al. 1984). For example, Schwann cell transplantations stimulated growth factor synthesis during nerve regeneration (Friedman et al. 1992), and Schwann cell proliferation in conduits may be improved by the presence of fibronectin (Evercooren et al. 1982). The effect of different distal ends for regenerating axons was also tested as shown in Fig. 11.2. When the distal end was replaced with skin or tendon or left open, regeneration was not successful. It was only successful when the distal end contained peripheral nerve tissue as shown in Fig. 11.2a.

Different conduit designs work to address different nerve regeneration problems. The standard design, a single tube, works to simply connect the two ends of the severed nerve and allow regeneration to occur. The problem with this method, however, was that axons regenerated in a disordered manner. This has led to the development of microchannel scaffolds which guide along and organize axon regeneration. The microchannel scaffolds differed from conduits in that they compartmentalized nerve growth into separate elongated channels to control axonal growth, as opposed to conduits which allowed relatively disorganized growth (Stokols et al. 2006). The regenerative microchannel scaffolds were developed using three different methods: pulling microwires out from casted PDMS, rolling a single microchannel layer, and a stacked multilayer microchannel scaffold (Lacour et al. 2009; Hossain et al. 2015). Fitzgerald et al. developed both the rolled microchannel scaffold and the pullout microchannel scaffold (FitzGerald et al. 2012). Srinivasan et al. improved on the rolled scaffold and developed their own version (Srinivasan et al. 2015), and Kim et al. advanced the pullout scaffold (Kim et al. 2015). Fitzgerald et al. manufactured the pullout implant by casting PDMS around bundles of nylon filament (Lacour et al. 2009; FitzGerald et al. 2012). Alternatively, the advanced pullout scaffold developed by Kim et al. was made by tightly packing microwires in PDMS tubes and then casting PDMS prepolymer to create the structure (Kim et al. 2015). The advantage of this technique was that no specialized equipment or cleanroom facilities were required, increasing accessibility for others attempting to fabricate a regenerative scaffold. Srinivasan et al. fabricated the rolled implant using PDMS, SU-8, and spin coating techniques (Srinivasan et al. 2015). Their technique for constructing the rolling scaffold differed from Fitzgerald et al. in that their microchannel scaffold contained a top and bottom layer, improving the design by more effectively sealing the scaffold and making a closed encompassing structure (Srinivasan et al. 2015). They also utilized PDMS to create the microchannel scaffold instead of polyimide (Lacour et al. 2008).

These regenerative conduits and microchannel scaffolds were manufactured using different methods, but all of them worked to organize nerve regeneration through conduits or microchannels. The next step as described in the following sections was integrating microelectrodes into the conduits and microchannel scaffolds for neural signal recording and stimulation.

11.3   Conduit Interface

There have been several different methods of integrating electrodes into conduit designs, each with their own advantages. All of the methods serve the basic function of neural recording and/or stimulation. One method combines a conduit with a penetrating electrode interface. It combines an area of pin electrodes with a nerve conduit, so that as the nerves regenerate through the conduit, electrodes become embedded to the regenerated tissue (Fig. 11.3). Previous experiments have demonstrated that the interface was able to record action potentials as early as 8 days after implantation, but since nerve regeneration was highly dynamic in the first 2 weeks, initial neural recordings of the interface were highly variable and unstable; however,

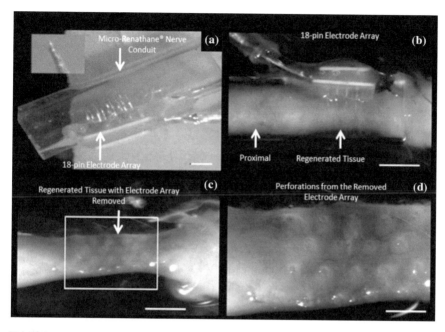

Fig. 11.3 (**a**) Picture of the interface, 18-pin electrode mounted to nerve conduit. (**b**) Regenerated nerve through interface (15 days). (**c**, **d**) Nerve after interface removal. Perforations are left by pin electrode. Scale bar = 2 mm (**a**), 1 mm (**b**, **c**), and 500 μm (**d**). (Adapted with permission from Seifert et al. (2012))

robust regeneration was observed through the interface (Lotfi et al. 2011; Seifert et al. 2012).

Another interface, called the biodegradable nerve regeneration guide (BNRG), used agarose as a substrate (Cho et al. 2008). The BNRG structure comprised of a tube containing several SU-8 microprobes; the microprobe structure consisted of longitudinal perforations embedded with gold electrodes acting as microchannels which guided along nerve regeneration. They were designed to limit the number of axons that neural signals could be recorded from and improve selectivity, while also containing the bipolar grooved electrodes needed for intraneural reading and recording.

An alternative RPNI composed of a free unit of muscle that was neurotized by a transected nerve (Kung et al. 2014). The muscle unit served as the site for electrode implantation and amplified bioelectric signals from the transected nerve, while providing a durable implantation site for the RPNI. Histologic analysis showed that the nerve formed new neuromuscular junctions, and in vivo testing showed that the RPNI remained biologically viable for 7 months and was able to reliably transduce bioelectric signals in this time (Kung et al. 2014). The results of this RPNI point to potential clinical application with prosthetics, but further studies are required for their limited number of electrodes.

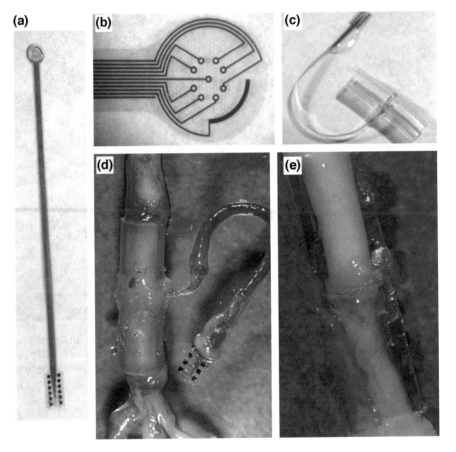

Fig. 11.4 (**a**) Full view of polyimide sieve electrode (**b**) Close up of sieve part (**c**) Sieve electrode inserted into PDMS guide (**d**) regenerated nerve through sieve electrode, 6 months after implantation (**e**) regenerated nerve with PDMS guide open. (Adapted with permission from Lago et al. (2005))

11.3.1 Sieve Electrode Interface

Sieve electrodes are circular substrate structures with perforations known as via holes. Two major substrate materials are silicon and polyimide. Oxidized silicon wafers and a reactive-ion etching process have been used for the fabrication of silicon sieve electrodes. After manufacturing, the sieve electrode was attached on both sides to a PDMS tube serving as the regenerative conduit (Akin et al. 1994). Spinning and curing polyimide resin and photolithograph patterning have been used for polyimide sieve electrodes (Stieglitz et al. 2002). Like conduits, the nerve is sutured so that the axons may regenerate from the nerve's proximal stump through the via holes and connect to the distal stump (Fig. 11.4d). A select number of via holes are embedded with electrodes for stimulation and signal recording (Fig. 11.4b).

The electrodes can stimulate or record action potentials from individual axons or small fascicles which can then be used to control a prosthesis (Edell 1986; Riso 1999; Lago et al. 2007). The extent of the device's functionality depends on the extent of axonal regeneration (Rosen et al. 1990; Navarro et al. 1996, 1998; Wallman et al. 2001; Ceballos et al. 2002). Sieve interfaces have been successfully used in the regeneration of various nerves in rats, with the implant recording and stimulating individual axons (Akin et al. 1994).

Unfortunately, the geometry of sieve interfaces limits their functionality, which has caused them to fall out of popularity. Higher neural action potential amplitudes can be recorded when electrodes are near a node of Ranvier, but one cannot predict where nodes of Ranvier will form after the axons regenerate through the via holes. Sieve electrodes will often record action potentials that travel through extracellular fluid which has low impedance, reducing the amplitude of the recorded action potential which in turn reduces signal to noise ratio (Loeb and Peck 1996). As regeneration occurs and the node of Ranvier shifts, signals recorded by the electrode become inconsistent over time (Shimatani et al. 2003; Lago et al. 2007). Another limitation of sieve electrodes is that the optimal via hole size needs to be large enough to allow easy regeneration and have enough space between holes so that electrodes can be integrated (Zhao et al. 1997). However, this reduces selectivity of the recorded signal by allowing bundles of axons to regenerate through and does not allow single axons to be recorded. One study calculated that an electrode diameter greater than 30 μm was required for ideal nerve recordings (Ceballos et al. 2002; Stieglitz et al. 2002).

11.3.2 Microchannel Interface

The microchannel interface addresses some of the shortcomings of the sieve interface. Similar to conduit and penetrating electrode interfaces, the microchannel interface is composed of an elongated substrate structure (usually silicon, PDMS, or polyimide) with microchannels. The proximal and distal nerve stumps are sutured to cuffs which connect to the interface, and axons regenerate through the microchannels, some of which are embedded with electrodes for recording and stimulating the neurons (Fig. 11.5) (Lacour et al. 2009).

A benefit of using microchannel interfaces has been that the recorded action potential amplitude is no longer dependent on the electrode's proximity to a node of Ranvier. This is because regenerated axons are confined to microchannels made of insulating material. The high impedance of the surrounding microchannel forces the action potential to flow longitudinally along the microchannel, making its way to the recording electrode (Loeb and Peck 1996). The recorded amplitudes are increased as the channel length is increased and microchannel diameter decreased. However, while increasing microchannel length positively affects signal recording, it may negatively affect regeneration. Lacour et al. showed that axons regenerated best up to a length of 1 mm, but reported a sharp decline in regeneration at lengths

Fig. 11.5 Diagram of microchannel interface. Nerve stumps are inserted into PDMS fittings and then sutured to the interface. Implant length may vary from 0.5 to 5 mm. (Adapted with permission from Lacour et al. (2009))

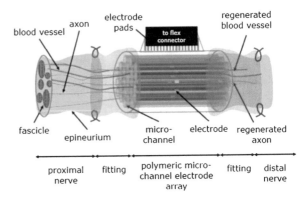

of 2 mm or more (Lacour et al. 2009). Though regeneration was seen at lengths of up to 5 mm, the quality was low with decreased number of axons and myelination. Microchannel diameter reduction could improve recorded amplitude, but axon myelination would decrease. For example, Lacour et al. demonstrated that 100-μm diameter microchannels showed about 4 times as many myelinated axons than the 55-μm diameter microchannels (Lacour et al. 2009).

11.3.3 Rolled Polyimide Interface

The rolled polyimide interface used a combination of PDMS tubes and a rolled polyimide structure (Sun et al. 2008). The polyimide structure was first planarly fabricated with embedded electrodes in the form of strips and was then rolled to form a cylinder with the electrode strips longitudinally oriented (Fig. 11.6a, b). The rolling formed the microchannels as the 2D channels are closed off by the bottom part of the roll's next layer, effectively making it a 3D microchannel array (Fig. 11.6c). Once the polyimide cylinder was formed, it was secured with PDMS tubes to prevent unravelling (Lacour et al. 2008).

This interface model was tested by Fitzgerald et al., who demonstrated that 70% of the polyimide microchannels were well innervated (FitzGerald et al. 2012). Mean axon count per device was 6244, which was similar, and in some cases lower, than what was seen in sieve electrodes (Lago et al. 2005; Ramachandran et al. 2006; Lago et al. 2007). However, a higher percent of the rolled interface's channels contained electrodes (11% compared to 5% in sieve). Stimulation of neural tissue proximal to the implant elicited a muscular response, indicating that axons had successfully regenerated through the axons and reinnervated distal musculature. Responses caused by digital stimulation also pointed to sensory neurons successfully regenerating. These results showed that the rolled polyimide interface has potential but needs further work. For example, due to the rolled nature of the implant, the upper enclosure of a microchannel may be imperfectly sealed which

Fig. 11.6 Major fabrication steps of the rolled polyimide interface. After rolling, structure is secured with PDMS tubes. (**a**) 2D channel array structure with embedded electrodes. (**b**) 2D structure is rolled to form 3D microchannel array structure. (**c**) Micrograph of the interface's cross section. Notice how a layer provides a roof for the prior layer. (Adapted with permission from Lacour et al. (2008))

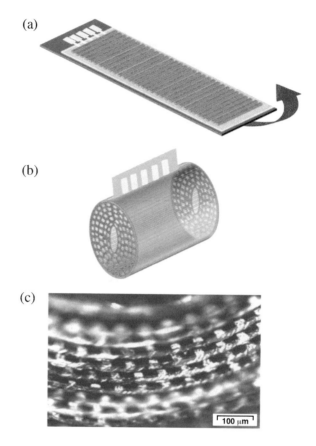

could lead to cross reading and in some rare cases allow neurites to cross into an adjacent channel (Williams et al. 2007; FitzGerald et al. 2012).

11.3.4   Rolled PDMS Interface

An alternative interface addressed the previous sealing problem by introducing a cover layer, which served to better seal each microchannel (Fig. 11.7b). Both the top and bottom layers were made of PDMS, while the walls were made of SU-8 (Fig. 11.7a). The structure formed the microchannels when rolled and the bottom PDMS layer was embedded with electrodes which were used for interfacing with axons that regenerated through the microchannels. In the initial experiment, the embedded electrodes did not perform as expected which prompted a switch to inserting microwires for interfacing. A limitation of the rolling process was that it left an open core at the center of the interface. This limitation was also in the design

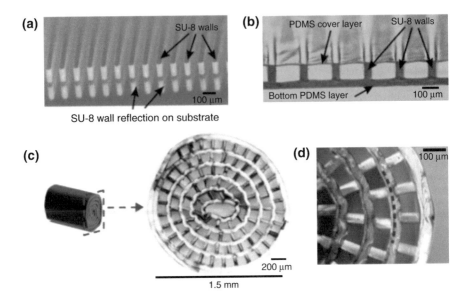

Fig. 11.7 Rolled PDMS with SU-8 wall interface. (**a, b**) Planar form, prior to roll. (**c**) Cross-sectional view of implant (**d**) Magnified view of (**c**); red line shows division between layers. (Adapted with permission from Srinivasan et al. (2015))

developed by Fitzgerald et al., and they implemented procedures to occlude the core; however, occlusion reduced overall regeneration (FitzGerald et al. 2012).

An advantage of this technique was that it was easily scalable to nerves of greater diameter by simply increasing the initial length of the planar structure, which increased the total number of microchannels as well (Faweett and Keynes 1990; Schmidt and Leach 2003). Results verified that the regenerated nerves through this interface were chronically stable and do not cause the formation of neuromas (Carlton et al. 1991; Lindenlaub and Sommer 2000), demonstrating the potential of this technique for application in prosthetics. Since implants for amputees must be functional for extended periods of time and formation of neuromas may not be an issue with this technique, the implants would not be painful to amputees and could allow usage of the prosthetic over extended periods of time (Srinivasan et al. 2015).

11.3.5   Texas Peripheral Nerve Interface

The Texas peripheral nerve interface (TxNI) also made use of microchannels using the technique developed by Lacour et al. (2009), but differed in that it packed wires into a commercially available PDMS tube filled with solidified PDMS to achieve the circular structure and improved dimensional matching. The PDMS structure was soaked in chloroform to expand it, allowing the wires to be removed, leaving behind an array of microchannel structures (Fig. 11.8b). Commercially available

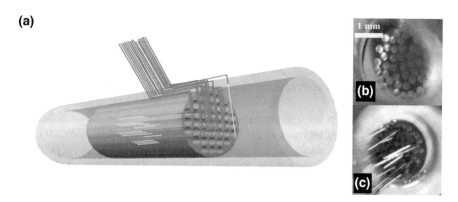

Fig. 11.8 (**a**) Schematic of TxNI; blue wires are used for recording (**b**) 50-microchannel scaffold, no microwires. (**c**) Scaffold from (**b**) with 16 microwires inserted. (From Ajam et al. (2016), open-source)

microwires were fed into the microchannels to record neural signals (Fig. 11.8c) (Ajam et al. 2016). The device could be customized by using different-diameter PDMS tubes and wires in order to match the interfacing nerve. No specialized equipment or cleanroom facility was required for fabrication, and since the materials are commercially available, fabrication could be possible in most labs (Lacour et al. 2010; Minev et al. 2012). During implantation, PDMS tubes were used as suture guides.

Due to the fabrication method, each microchannel was isolated and sealed, preventing cross reading and allowing for higher selectivity (Gore et al. 2014; Kim et al. 2014). The microchannel diameter was 75 μm or greater, since this was observed to improve regeneration in the sieve model (Zhao et al. 1997; FitzGerald et al. 2009). Histological analysis revealed that axons and Schwann cells regenerated successfully in both the 75-μm and 200-μm channels. The microchannels were surrounded by a thin band of tissue, mimicking the compartmentalization of axons seen in healthy cells (Clements et al. 2009; Garde et al. 2009). This tissue could serve to better isolate action potentials from separate microchannels, or it might inhibit action potential amplitudes; further testing would be required. Overall, the device showed potential for application in prosthetics, since it was demonstrated to record action potentials in motion and has a highly customizable nature.

11.4 Conclusion

RPNIs hold exciting potential for applications in prosthetics. As new techniques are developed and refined, signal to noise ratio and selectivity have increased. Should an adequate level of selectivity be reached, it may be possible to process these signals to translate them into movements in prosthetics. RPNIs produce a higher signal to noise ratio than conventional PNIs due to the high axon-electrode proximities.

Additionally, they are well-suited for amputees where the remaining nerves are available for interfacing. Research shows that in long-term amputees, neural pathways for missing limb control remain intact (Sanes et al. 1992; Sanes and Douglas 2000). By transecting residual nerves and allowing them to grow through RPNIs, it may be possible to establish intuitive control of a prosthetic limb using these neural pathways.

PNIs have already been shown capable of interfacing with external devices (Dhillon et al. 2004; Lawrence et al. 2004). RPNIs feature higher selectivity and quantity of individual axons, and this points to that RPNIs have even more potential for interfacing with external devices. Further improvements of RPNIs are needed to push prosthetics further and finally allow successful integration of nerve interfaces with prosthetic technology, introducing a new age of man and machine synergy.

References

Ajam, A., Hossain, R., Tasnim, N., Castanuela, L., Ramos, R., Kim, D., & Choi, Y. (2016). Handcrafted microwire regenerative peripheral nerve interfaces with wireless neural recording and stimulation capabilities. *International Journal of Sensor Networks and Data Communications, 5*(1), 1–5.

Akin, T., Najafi, K., Smoke, R. H., & Bradley, R. M. (1994). A micromachined silicon sieve electrode for nerve regeneration applications. *IEEE Transactions on Biomedical Engineering, 41*(4), 305–313.

Biddiss, E. A., & Chau, T. T. (2007). Upper limb prosthesis use and abandonment: A survey of the last 25 years. *Prosthetics and Orthotics International, 31*, 236–257.

Birbaumer, N., Strehul, U., & Hinterberger, T. (2004). Brain-computer interfaces for verbal communication. In K. W. Horch & G. S. Dhillon (Eds.), *Neuroprosthetics: theory and practice* (pp. 1146–1157). Hackensack, NJ: World Scientific.

Boretius, T., Badia, J., Pascual-Font, A., Schuettler, M., Navarro, X., Yoshida, K., & Stieglitz, T. (2010). A transverse intrafascicular multichannel electrode (TIME) to interface with the peripheral nerve. *Biosensors & Bioelectronics, 26*(1), 7.

Branner, A., & Normann, R. A. (2000). A multielectrode array for intrafascicular recording and stimulation in sciatic nerves of cats. *Brain Research Bulletin, 51*, 293–306.

Branner, A., Stein, R. B., & Normann, R. A. (2001). Selective stimulation of cat sciatic nerve using an array of varying-lenght microelectrodes. *Neurophysiology, 85*, 1585–1594.

Brock, J. H., Rosenzweig, E. S., Blesch, A., Moseanko, R., Havton, L. A., Edgerton, V. R., & Tuszynski, M. H. (2010). Local and remote growth factor effects after primate spinal cord injury. *Journal of Neuroscience, 30*, 9728–9737.

Brown, P. B., Koerber, H. R., & Millecchia, R. (2004). From innervation density to tactile acuity: 1. Spatial representation. *Brain Research, 1011*(1), 14–32.

Carlton, S. M., Dougherty, P. M., Pover, C. M., & Coggeshall, R. E. (1991). Neuroma formation and numbers of axons in a rat model of experimental peripheral neuropathy. *Neuroscience Letters, 131*, 88–92.

Ceballos, D., Valero-Cabre, A., Valderrama, E., Schuttler, M., Stieglitz, T., & Navarro, X. (2002). Morphologic and functional evaluation of peripheral nerve fibers regenerated through polyimide sieve electrodes over long-term implantation. *Journal of Biomedical Materials Research, 60*(4), 517–528.

Cho, S.-H., Lu, H. M., Cauller, L., Romero-Ortega, M. I., Lee, J.-B., & Hughes, G. A. (2008). Biocompatible SU-8-based microprobes for recording neural spike signals from regenerated peripheral nerve fibers. *IEEE Sensors Journal, 8*(11), 1830–1836.

Clements, I. P., Kim, Y.-t., English, A. W., Lu, X., Chung, A., & Bellamkonda, R. V. (2009). Thin-film enhanced nerve guidance channels for peripheral nerve repair. *Biomaterials, 30*, 3834–3846.

Dhillon, G. S., Lawrence, S. M., Hutchinson, D. T., & Horch, K. W. (2004). Residual function in peripheral nerve stumps of amputees: Implications for neural control of artificial limbs. *The Journal of Hand Surgery, 29*(4), 605–615.

Durand, D., Park, H. J., & Wodlinger, B. (2009). Models of the peripheral nerves for detection and control of neural activity. *Conference Proceedings IEEE Engineering in Medicine and Biology Society, 2009*, 3326–3329.

Edell, D. J. (1986). A peripheral nerve information transducer for amputees: Long-term multichannel recordings from rabbit peripheral nerves. *IEEE Transactions on Biomedical Engineering, 33*(2), 203–214.

Edell, D. J., Toi, V. V., McNeil, V. M., & Clark, L. D. (1992). Factors influencing the biocompatibility of insertable silicon microshafts in cerebral cortex. *IEEE Transactions on Biomedical Engineering, 39*(6), 635–643.

Evercooren, A. B.-V., Kleinman, H. K., Seppa, H. E., Rentier, B., & Dubois-Dalcq, M. (1982). Fibronecting promotes rat scwann cell growth and motility. *The Journal of Cell Biology, 93*, 211.

Faweett, J. W., & Keynes, R. J. (1990). Peripheral nerve regeneration. *Annual Review of Neuroscience, 13*, 43–60.

Fischer, W., Björklund, A., Chen, K., & Gage, F. H. (1991). NGF improves spatial memory in aged rodents as a function of age. *The Journal of Neuroscience, 11*, 1889–1906.

FitzGerald, J. J., Lacour, S. P., McMahon, S. B., & Fawcett, J. W. (2009). Microchannel electrodes for recording and stimulation: In vitro evaluation. *IEEE Transactions on Biomedical Engineering, 56*(5), 1524–1534.

FitzGerald, J. J., Lago, N., Benmerah, S., Serra, J., Watling, C. P., Cameron, R. E., Tarte, E., Lacour, S. P., McMahon, S. B., & Fawcett, J. W. (2012). A regenerative microchannel neural interface for recording from and stimulating peripheral axons in vivo. *Journal of Neural Engineering, 9*(1), 016010.

Friedman, B., Scherer, S. S., Rudge, J. S., Helgren, M., Morirsey, D., & McClain, J. (1992). Regulation of ciliary neurotrophic factor expression in myelin-related schwann cells in vivo. *Neuron, 9*, 295.

Garde, K., Keefer, E., Botterman, B., Galvan, P., & Romero, M. I. (2009). Early interfaced neural activity from chronic amputated nerves. *Frontiers in Neuroengineering, 2*, 5.

Gore, R. K., Choi, Y., Bellamkonda, R. V., & English, A. W. (2014). Functional recordings from awake, behaving rodents through a microchannel based regenerative neural interface. *Journal of Neural Engineering, 12*(1), 016017.

Hagbarth, K. E. (1993). Microneurography and applications to issues of motor control: Fifth annual Stuart Reiner memorial lecture. *Muscle & Nerve, 16*, 693–705.

Hess, A., & Young, J. Z. (1952). The nodes of Ranvier. *Proceedings of the Royal Society of London. Series B, Biological Sciences, 140*, 301–320.

Hossain, R., Kim, B., Pankratz, R., Ajam, A., Park, S., Biswal, S. L., & Choi, Y. (2015). Handcrafted multilayer PDMS microchannel scaffolds for peripheral nerve regeneration. *Biomedical Microdevices, 17*(6), 109.

Kathryn Ziegler-Graham, P., Ellen, P., Kenzie, J. M., Patti, M., Ephraim, L., Thomas, P., Travison, G., & Brookmeyer, P. R. (2008). Estimating the prevalence of limb loss in the United States: 2005 to 2050. *Archives of Physical Medicine and Rehabilitation,89*(3), 7.

Kim, Y. T., Haftel, V. K., Kumar, S., & Bellamkonda, R. V. (2008). The role of aligned polymer fiber-based constructs in the bridging of long peripheral nerve gaps. *Biomaterials, 29*(21), 3117–3127.

Kim, B., Reyes, A., Garza, B., & Choi, Y. (2014). A PDMS microchannel scaffold with microtube electrodes for peripheral nerve interfacing. In *The 40th Annual Conference of IEEE Industrial Electronics Society*. Dallas, TX.

Kim, B., Reyes, A., Garza, B., & Choi, Y. (2015). A microchannel neural interface with embedded microwires targeting the peripheral nervous system. *Microsystem Technologies, 21*(7), 1551–1557.

Kovacs, G. T. A., Storment, C. W., & Rosen, J. M. (1992). Regeneration microelectrode array for peripheral nerve recording and stimulation. *IEEE Transactions on Biomedical Engineering, 39*(9), 893.

Kuiken, T. A., Li, G., Lock, B. A., Lipschutz, R. D., Miller, L. A., Stubblefield, K. A., & Englehart, K. (2009). Targeted muscle reinnervation for real-time myoelectric control of multifunction artificial arms. *JAMA, 301*(6), 619–628.

Kung, T. A., Langhals, N. B., Martin, D. C., Johnson, P. J., Cederna, P. S., & Urbanchek, M. G. (2014). Regenerative peripheral nerve interface viability and signal transduction with an implanted electrode. *Plastic and Reconstructive Surgery, 133*(6), 1380–1394.

Lacour, S. P., Atta, R., Fitzgerald, J. J., Blamire, M., Tarte, E., & Fawcett, J. (2008). Polyimide micro-channel arrays for peripheral nerve regenerative implants. *Sensors and Actuators A, 147*, 456–463.

Lacour, S. P., Fitzgerald, J. J., Lago, N., Tarte, E., McMahon, S., & Fawcett, J. (2009). Long micro-channel electrode arrays: A novel type of regenerative peripheral nerve interface. *IEEE Transactions on Neural Systems and Rehabilitation Engineering, 17*(5), 454–460.

Lacour, S. P., Benmerah, S., Tarte, E., FitzGerald, J., Serra, J., McMahon, S., Fawcett, J., Graudejus, O., Yu, Z., & Morrison, B., 3rd. (2010). Flexible and stretchable micro-electrodes for in vitro and in vivo neural interfaces. *Medical & Biological Engineering & Computing, 48*(10), 945–954.

Lago, N., Ceballos, D., Rodriguez, F. J., Stieglitz, T., & Navarro, X. (2005). Long term assessment of axonal regeneration through polyimide regenerative electrodes to interface the peripheral nerve. *Biomaterials, 26*(14), 2021–2031.

Lago, N., Udina, E., Ramachandran, A., & Navarro, X. (2007). Neurobiological assessment of regenerative electrodes for bidirectional interfacing injured peripheral nerves. *IEEE Transactions on Biomedical Engineering, 54*, 1129.

Lawrence, S. M., Dhillon, G. S., & Horch, K. W. (2003). Fabrication and characteristics of an implantable, polymer-based, intrafascicular electrode. *Journal of Neuroscience Methods, 131*, 9–26.

Lawrence, S. M., Dhillon, G. S., Jensen, W., Yoshida, K., & Horch, K. W. (2004). Acute peripheral nerve recording characteristics of polymer-based longitudinal intrafascicular electrodes. *IEEE Transactions on Neural Systems and Rehabilitation Engineering, 12*(3), 345–348.

Lee, S., Carvell, G. E., & Simons, D. J. (2008). Motor modulation of afferent somatosensory circuits. *Nature Neuroscience, 12*, 1430.

Lindenlaub, T., & Sommer, C. (2000). Partial sciatic nerve transection as a model of neuropathic pain: A qualitative and quantitative neuropathology study. *Pain, 89*, 97–106.

Llinas, R., Nicholson, C., & Johnson, K. (1973). Chap. 7: Implantable monolithic wafer recording electrodes for neurophysiology. In *Brain Unit Activity Durig Behavior* (pp. 105–111). Springfield, IL: Charles C Thomas.

Loeb, G. E., & Peck, R. A. (1996). Cuff electrodes for chronic stimulation and recording of peripheral nerve activity. *Journal of Neuroscience Methods, 64*, 95–103.

Lotfi, P., Garde, K., Chouhan, A. K., Bengali, E., & Romero-Ortega, M. I. (2011). Modality-specific axonal regeneration: Toward selective regenerative neural interfaces. *Frontiers in Neuroengineering, 4*, 11–11.

Lundborg, G., Dahlin, L. B., Danielsen, N. P., Hansson, H. A., & Larsson, K. (1981). Reorganization and orientation of regenerating nerve fibres, perineurium, and epineurium in preformed mesothelial tubes - an experimental study on the sciatic nerve of rats. *Journal of Neuroscience Research, 6*(3), 265–281.

Malagodi, M. S., Horch, K. W., & Schoenberg, A. A. (1989). An intrafascicular electrode for recording of action potentials in peripheral nerves. *Annals of Biomedical Engineering, 17,* 397–410.

McNaughton, T. G., & Horch, K. W. (1996). Metallized polymer fibers as leadwires and intrafascicular microelectrodes. *Journal of Neuroscience Methods, 70,* 103–110.

Minev, I. R., Chew, D. J., Delivopoulos, E., Fawcett, J. W., & Lacour, S. P. (2012). High sensitivity recording of afferent nerve activity using ultra-compliant microchannel electrodes: An acute in vivo validation. *Journal of Neural Engineering, 9*(2), 026005.

Munger, B. L., Bennett, G. J., & Kajander, K. C. (1992). An experimental painful peripheral neuropathy due to nerve constriction. 1. Axonal pathology in the sciatic nerve. *Experimental Neurology, 118,* 204–214.

Nagahara, A. H., Merrill, D. A., Copola, G., Tsukada, S., Schroeder, B. E., Shaked, G. M., Wang, L., Blesch, A., Kim, A., Conner, J. M., Rochestein, E., Chao, M. V., Koo, E. H., Geschwind, D., Masilah, E., Chiba, A. A., & Tuszynski, M. H. (2009). Neuroprotective effects of brain-derived neurotrophic factor in rodent and primate models of Alzheimer's disease. *Nature Medicine, 15,* 331.

Navarro, X., Calvet, S., Buti, M., Gomez, N., Cabruja, E., Garrido, P., Villa, R., & Valderrama, E. (1996). Peripheral nerve regeneration through microelectrode arrays based on silicon technology. *Restorative Neurology and Neuroscience, 9,* 151–160.

Navarro, X., Calvet, S., Rodriguez, F. J., Stieglitz, T., Blau, C., Buti, M., Valderrama, E., & Meyer, J. U. (1998). Stimulation and recording from regenerated peripheral nerves through polyimide sieve electrodes. *Journal of the Peripheral Nervous System, 3*(2), 91–101.

Navarro, X., Valerrama, E., Stieglitz, T., & Schuttler, M. (2001). Selective stimulation of the rat sciatic nerve with multipolar polyimide cuff electrodes. *Restorative Neurology and Neuroscience, 18,* 9–21.

Navarro, X., Krueger, T. B., Lago, N., Micera, S., Stieglitz, T., & Dario, P. (2005). A critical review of interfaces with the peripheral nervous system for the control of neuroprostheses and hybrid bionic systems. *Journal of the Peripheral Nervous System, 10,* 229–258.

Nichols, C. M., Brenner, M. J., Fox, I. K., Tung, T. H., Hunter, D. A., Rickman, S. R., & Mackinnon, S. E. (2004). Effect of motor versus sensory nerve grafts on peripheral nerve regeneration. *Experimental Neurology, 190*(2), 347–355.

Pettersson, J., Kalbermatten, D., McGrath, A., & Novikova, L. N. (2010). Biodegradable fibrin conduit promotes long-term regeneration after peripheral nerve injury in adult rats. *Journal of Plastic, Reconstructive & Aesthetic Surgery, 63*(11), 1893–1899.

Ramachandran, A., Schuettler, M., Lago, N., Doerge, T., Koch, K. P., Navarro, X., Hoffmann, K. P., & Stieglitz, T. (2006). Design, in vitro and in vivo assessment of a multi-channel sieve electrode with integrated multiplexer. *Journal of Neural Engineering, 3*(2), 114–124.

Rangasamy, S. B., Soderstrom, K., Bakay, R. A., & Kordower, J. H. (2010). Neurotrophic factor therapy for Parkinson's disease. *Progress in Brain Research, 184,* 237–264.

Rao, S. M., Binder, J. R., Hammeke, T. A., Bandettini, P. A., Bobholz, J. A., Frost, J. A., & e. al. (1995). Somatopic mapping of the human primary motor cortex with functional magnetic resonance imaging. *Neurology, 45,* 919–924.

Riso, R. R. (1999). Strategies for providing upper extremety amputees with tactile hand position feedback - moving closer to the bionic arm. *Technology and Health Care, 7,* 401–409.

Rosen, J. M., Grosser, M., & Hentz, V. R. (1990). Preliminary experiments in nerve regeneration through laser-drilled holes in silicon chips. *Restorative Neurology and Neuroscience, 2,* 89–102.

Rossini, P. M., Micera, S., Benvenuto, A., Carpaneto, J., Cavallo, G., Citi, L., Cipriani, C., Denaro, L., Denaro, V., Pino, G. D., Ferreri, F., Guglielmelli, E., Hoffmann, K.-P., Raspopovic, S., Rigosa, J., Rossini, L., Tombini, M., & Dario, P. (2010). Double nerve intraneural interface implant on a human amputee for robotic hand control. *Clinical Neurophysiology, 121,* 777–783.

Rozman, J., Sovinec, B., Trlep, M., & Zorko, B. (1993). Multielectrode spiral cuff for ordered and reversed activation of nerve fibres. *Biomedical Engineering, 15,* 113–120.

Sanes, J. N., & Douglas, J. P. (2000). Plasticity and primary motor cortex. *Annual Review of Neuroscience, 23*, 393–415.

Sanes, J. N., Wang, J., & Donoghue, J. P. (1992). Immediate and delayed changes of rat motor cortical output representation with new forelimb configurations. *Cerebral Cortex, 2*, 141–152.

Sanes, J. N., Donoghue, J. P., Thangaraj, V., Edelman, R. R., & Warach, S. (1995). Shared neural substrates controlling hand movements in human motor cortex. *Science, 268*, 1775–1777.

Schmidt, C. E., & Leach, J. B. (2003). Neural tissue engineering: Strategies for repair and regeneration. *Annual Review of Biomedical Engineering, 5*(1), 293–352.

Seifert, J. L., Desai, V., Watson, R. C., Musa, T., Kim, Y.-t., Keefer, E. W., & Romero, M. I. (2012). Normal molecular repair mechanisms in regenerative peripheral nerve interfaces allow recording of early spike activity despite immature myelination. *IEEE Transactions on Neural Systems and Rehabilitation Engineering, 20*(2), 220–227.

Shimatani, Y., Nikles, S. A., Najafi, K., & Bradley, R. M. (2003). Long-term recordings from afferent taste fibers. *Physiology and Behavior, 80*(2–3), 309–315.

Srinivasan, A., Tahilramani, M., Bentley, J. T., Gore, R. K., Millard, D. C., Mukhatyar, V. J., Joseph, A., Haque, A. S., Stanley, G. B., English, A. W., & Bellamkonda, R. V. (2015). Microchannel-based regenerative scaffold for chronic peripheral nerve interfacing in amputees. *Biomaterials, 41*, 151–165.

Stieglitz, T. (2004). Considerations on surface and structural biocompatibility as prerequisite for long-term stability of neural prostheses. *Nanoscience and Nanotechnology, 4*, 496–503.

Stieglitz, T., Ruf, H. H., Gross, M., Schuettler, M., & Meyer, J. U. (2002). A biohybrid system to interface peripheral nerves after traumatic lesions: Design of a high channel sieve electrode. *Biosensors and Bioelectronics, 17*(8), 685–696.

Stokols, S., Sakamoto, J., Breckon, C., Holt, T., Weiss, J., & Tuszynski, M. H. (2006). Templated agarose scaffolds support linear axonal regeneration. *Tissue Engineering, 12*(10), 2777–2787.

Suematsu, N., Atsuta, Y., & Hariyama, T. (1988). Vein graft for repair of peripheral nerve gap. *Journal of Reconstructive Microsurgery, 4*, 313.

Sun, Y., Lacour, S. P., Brooks, R., Rushton, N., Fawcett, J., & Cameron, R. (2008). Assessment if the biocompatibility of photosensitive polyimide for implantable medical device use. *Journal of Biomedical Materials Research. Part A, 90*, 648.

Thanos, P. K., Okajima, S., & Terzis, J. K. (1998). Ultrastructure and cellular biology of nerve regeneration. *Journal of Reconstructive Microsurgery, 14*(6), 423–436.

Vallbo, A. B., Hagbarth, K., Torebjork, H., & Wallin, G. (1979). Somatosensory, proprioceptive, and sympathetic activity in human peripheral nerves. *Physiological Reviews, 59*, 919–957.

Wallman, L., Zhang, Y., Laurell, T., & Danielsen, N. (2001). The geometric design of micromachined silicon sieve electrodes influences functional nerve regeneration. *Biomaterials, 22*, 1187–1193.

Wang, K. K., Costas, P. D., Bryan, D. J., Jones, D. S., & Seckel, B. R. (1993). Inside out vein graft prmotes improved nerve regeneration in rats. *Microsurgery, 14*, 608.

Williams, L. R., Powell, H. C., Lundborg, G., & Varon, S. (1984). Competence of nerve-tissue as distal insert promoting nerve regeneration in a silicone chamber. *Brain Research, 293*(2), 201–211.

Williams, J. C., Hippensteel, J. A., Dilgen, J., Shain, W., & Kipke, D. R. (2007). Complex impedance spectroscopy for monitoring tissue responses to inserted neural implants. *Journal of Neural Engineering, 4*, 410–423.

Yoshida, K., Jovanovic, K., & Stein, R. B. (2000). Intrafasicular electrodes for stimulations and recording from mudpuppy spinal roots. *Journal of Neuroscience Methods, 96*, 47–55.

Zhao, Q., Drott, J., Laurell, T., Wallman, L., Lindström, K., Bjursten, L. M., Lundborg, G. R., Montelius, L., & Danielsen, N. (1997). Rat sciatic nerve regeneration through a micromachined silicon chip. *Biomaterials, 18*(1), 75–80.

Chapter 12
Passive RF Neural Electrodes

Katrina Guido and Asimina Kiourti

12.1 Introduction

Implanted brain-machine interfaces (BMIs) have the potential to produce advancements in many fields, including more reliable and more functional prosthetics, detection and prevention of seizures, management of symptoms associated with chronic diseases such as Alzheimer's and Parkinson's, and treatment of mental disorders, as well as improving the current scientific understanding of the brain and consciousness (During et al. 1989; Polikov et al. 2005; Hochberg et al. 2006; Blount et al. 2008; Kipke et al. 2008; Wise et al. 2008). Nonetheless, currently available implantable technologies lack the ability to perform the unobtrusive, chronic monitoring to make such advancements.

The technologies covered in the previous chapters discuss methods in which to connect or interface implants/electrodes to the brain. However, many neglect to discuss the challenges associated with connecting the implant to the external environment (i.e., connecting electrodes from the inside to the outside of the body) to facilitate data transfer. Many of the implantable BMIs currently in use rely on wired connections from neural implants to exterior data recording/transmitting units that perforate the skull for the lifetime of the implant (Wise et al. 2008; Waziri et al. 2009). This setup restricts testing to a clinical environment, limits patient movement, and is prone to infections.

To eliminate perforation of the skull, many groups have turned to wireless data transmission, but this method also has its own unique set of challenges. Employing batteries on the implant to amplify neural activity and to facilitate wireless transmission poses the risk of neuron death caused by heat generation from both the battery

K. Guido · A. Kiourti (✉)
ElectroScience Laboratory, Department of Electrical and Computer Engineering,
The Ohio State University, Columbus, OH, USA
e-mail: kiourti.1@osu.edu

© Springer Nature Switzerland AG 2020
L. Guo (ed.), *Neural Interface Engineering*,
https://doi.org/10.1007/978-3-030-41854-0_12

and the densely packaged implanted electronics (Kim et al. 2007; Dethier et al. 2013). Added to the above, implanted batteries are cumbersome as they increase the size of the implant and require replacement and/or recharging. In response to these so-called "active" implants, passive neural implants were created (Rao et al. 2014; Song and Rahmat-Samii 2017). Using the concept employed by radio-frequency identification (RFID) tags, an external interrogator unit transmits a radio-frequency signal that turns on an implant. The implant then collects, amplifies, and backscatters the recorded neural data to the interrogator where the data is now accessible for post-processing. This type of BMI removes the complexity and the bulk of the power consumption from the implanted device, moving these requirements to outside of the body (the interrogator) instead (see Fig. 12.1). Nevertheless, power is still needed for signal amplification within the implant as is shown by the presence of an integrated circuit (IC) in Fig. 12.1. For such passive implementations, batteries are avoided and radio-frequency (RF) power-harvesting circuits are rather used to rectify, regulate, and store energy. Despite the lack of batteries, heating of neural tissue is still a concern and limits the amount of power that can be supplied to the implant.

Further reducing the power consumption of the implanted device, Schwerdt et al. (2011) and Schwerdt et al. (2012) introduced a new class of fully passive brain implants. This implies implanted devices that have no power requirements (i.e., not just the absence of a battery but also no power harvester, no rectifier/regulator, IC, etc.). Though considered safer, it is worth highlighting that fully passive implants cannot amplify neural signals before transmitting through body tissues, which are associated with high loss. As such, one of the main challenges associated with such fully passive wireless transmission is recording the low-voltage neural signals. The types of neural signals are shown in Table 12.1. With local field potentials (LFPs) as low as 20 μV_{pp}, the minimum detectable signal of a fully passive implant/interrogator system becomes a key design specification. Designing wireless and fully passive implants with sensitivity as high as 20 μV_{pp} has been the focus of several recent works (Lee et al. 2015; Kiourti et al. 2016; Lee et al. 2017). This chapter primarily focuses on this latter class of wireless implants, the fully passive devices.

12.2 Passive Devices

12.2.1 System Overview

Passive devices include those that employ energy harvesters in which the implanted device uses (but does not store) power from the ambient environment to amplify neural signals before transmission back to the interrogator unit. Both harvesting glucose from the body and using inductive power transfer have been suggested for powering BMIs; however, the latter has shown more promise (Rao et al. 2014). Similar to an RFID device, to optimize the efficiency of inductive power transfer, the implanted antenna's termination impedance is modulated and the received power is then converted from RF to DC to power the implanted device. This

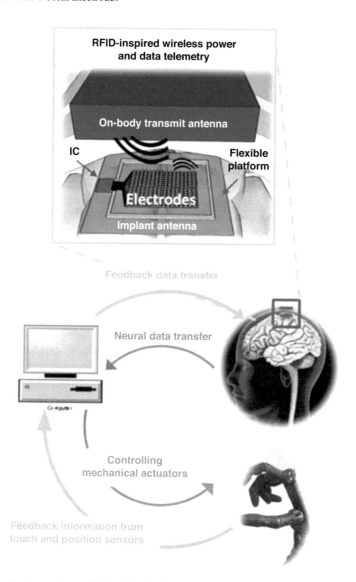

Fig. 12.1 Passive end-to-end BMI with the electrode array and antenna patterned onto a single substrate. Note that the post-processing and a majority of device complexity are located outside of the body. However, wireless power telemetry is still required to operate the implant. (Reproduced with permission from Rao et al. (2014))

modulation is provided by the wireless link. As such, the antennas used to transmit and receive the power as well as the frequency of operation are key components of the system design.

Thus far, implanted loop antennas have been used to achieve such power transfer with the optimal frequency located in the range of a few hundred MHz to a few GHz

Table 12.1 Voltage and frequency range of various neural signals (Muller et al. 2015)

Neural signals	Voltage range	Frequency range
Electroencephalography (EEG)	2–100 μV_{pp}	0.5–50 Hz
Electrocorticography (ECoG)	2–1000 μV_{pp}	1–500 Hz
Local field potential (LFP)	20–2000 μV_{pp}	1–500 Hz
Action potential	20–2000 μV_{pp}	250–10 kHz

(Bjorninen et al. 2012; Rao et al. 2014). Envisioning an end-to-end system to communicate with prosthetic devices, shown in Fig. 12.1, Bjorninen et al. designed an implantable BMI that sits on the surface of the cortex to record ECoG signals. The BMI integrated an 8 × 8 electrode array and the implanted antenna onto the same substrate, patterning both onto the flexible and biocompatible polymer Parylene C using a Pt-Au-Pt layering arrangement.

Several considerations come into play when designing passive brain implants:

- *Implant size.* Placing both the electrodes and the antenna on the same substrate saves space, which is important for clinical considerations when implanting anything into the body.
- *Power efficiency.* Size constraints outlined above, on the other hand, reduce the efficiency of power transfer (Bjorninen et al. 2012).
- *Foreign body response (FBR).* FBR implies the immune system's reaction to a foreign/nonbiological substance in which scar tissue forms around the foreign body. Integration of both the electrodes and the antenna on the same biocompatible substrate helps to avoid such complications.
- *Flexible materials.* Designing the device on a flexible substrate as opposed to a rigid one relaxes the size constraint. A flexible substrate allows the device to conform to the body rather than the body trying to conform to a rigid device (which in part initiates FBR).
- *Frequency selection.* Finite element (FE) simulations using ANSYS high-frequency structure simulator (HFSS) showed that the optimal frequency for link power efficiency for this particular arrangement was 400 MHz. At this frequency, using 6.5 × 6.5 mm² loops with an interrogating loop inner diameter of 15 mm results in a maximum of approximately 0.4 mW reaching the IC for RF-to-DC conversion (Bjorninen et al. 2012).
- *Specific Absorption Rate (SAR).* SAR quantifies the human body's rate of energy absorption when exposed to RF radiation and is given by Eq. 12.1:

$$SAR = \frac{\sigma |E|^2}{\rho} \tag{12.1}$$

Here, σ is tissue conductivity, ρ is tissue mass density, and $|E|$ is the root mean square (rms) electric field magnitude (Cleveland and Ulcek 1999). Expectedly, (inter-) national safety standards limit the maximum allowable SAR levels. In turn, this limits the amount of power one is able to transmit to the implant. As an example, the Federal Communications Commission (FCC) lists the maximum allowable SAR at 1.6 W/kg per 1 g of tissue (for general public expo-sure). SAR levels for passive devices are primarily dependent on the interrogating antenna. To minimize SAR, the interrogating antenna must exhibit a low near electric field as seen in Eq. 12.1 while maintaining a high near magnetic field (high coupling with the implanted antenna) (Song and Rahmat-Samii 2017).

12.2.2 *Interrogator Antennas*

Focusing on optimizing the wireless link, Mark et al. investigated the use of segmented loop antennas as an option for reducing SAR (Mark et al. 2011). To maximize antenna coupling, the interrogator antenna should be larger in size than the implanted one; however, as circumference of the interrogator antenna becomes electrically large (approaching the effective wavelength), current distribution over the antenna becomes nonuniform (Mark et al. 2011). This nonuniform current results in a nonuniform electric field distribution creating "hotspots." To alleviate this issue, Mark et al. divided the loop into segments connected via capacitors. The capacitors adjust for the current phase shift along the antenna, effectively making the current uniform in spite of the large antenna size; however, the capacitors also decrease link power efficiency by increasing power loss over the antenna (Moradi et al. 2013b).

To minimize power loss, the antenna was divided into two segments with four parallel capacitors connecting the segments. Rao et al. also suggested tilting a solid loop antenna (Rao et al. 2014). Tilting reduces SAR by reducing the intensity of the peaks in the electric field incident on the tissue, which are primarily focused at the feed of the antenna. Figure 12.2a shows the solid, 5-segmented loop with each segment joined by a single capacitor and 2-segmented loop joined by four capacitor loops along with the tilted-loop arrangement. Figure 12.2b shows the electric field distributions on the surface of tissue generated by each of the antennas. In Fig. 12.2b, note that though the 5-segmented loop exhibits the best distribution of the electric field in terms of reduction in hotspots, the 2-segmented loop provides the best balance between hotspot reduction and power link efficiency.

Additionally, to maximize power gain of the system, the number of loops of the interrogator antenna can also be increased instead of splitting the ring into segments. By increasing the number of loops of the interrogator antenna to two, Khan et al. (2016) measured an increase of 20 mW in power gain, which equates to an increase of about 2 mW delivered to the implanted device, while staying within safe SAR levels. SAR levels are reduced for this geometry because the antenna feed is moved to the top layer, which is further from the tissue, similar to tilting the antenna.

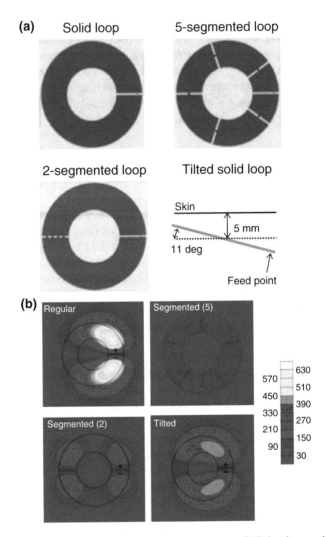

Fig. 12.2 (**a**) Interrogator loop geometries designed to improve SAR levels over the traditional solid loop. (**b**) Electric field distribution in tissue generated by each of the antenna geometries. (Reproduced with permission from Rao et al. (2014))

12.2.3 Implanted Antennas

Given that size of the implanted device is limited, increasing the size of the implanted antenna cannot be used to increase link power efficiency. To increase antenna coupling (and therefore link power efficiency) for a more localized implant (more so than an ECoG recording), Moradi et al. investigated the use of 3-dimensional implanted antennas (Moradi et al. 2013a). Comparing a 1×1 cm^2 planar loop to a $1 \times 1 \times 1$ mm^3 cubic antenna, the cubic antenna performs better, exhibiting an increased maximum power gain. When a magnetodielectric core (SMMDF101) is inserted into the center of the cubic loop, the cubic antenna exhibits an up to 10 dB improvement in maximum power gain over the planar loop. The cubic antenna results in an increased coupling area compared to a planar antenna with the same footprint and results in an increase in size of the current path; the core increases the inductance of the cubic loop, which increases the power gain.

Notably, using a planar structure while increasing the number of loops approaches the performance of a cubic antenna with the same footprint (Song and Rahmat-Samii 2017). Increasing the number of loops to five resulted in a 15-dB improvement in maximum power gain over a single loop but was still 10 dB less than that of the solid cubic antenna.

12.2.4 Disadvantages of Passive Systems

As mentioned earlier in Sect. 12.2, a primary concern with passive devices is SAR. Even though these implants do not have batteries and are therefore passive, they still contain an IC that requires power, raising concerns about SAR levels. Additionally, the IC requires space, adding bulk to the implanted device. Adding an IC to the 1-mm^3 implanted antenna can increase the size of implant by at least 100%. When antennas must be at most on the order of a few millimeters, an additional millimeter or two thanks to an IC poses a challenge in terms of integration into the biological environment.

12.3 Fully Passive Devices

12.3.1 System Overview

Fully passive implants do not rectify or regulate any external power source; thus, all amplification and post-processing occurs at the interrogator. This type of device removes all complex circuitry from the inside of the body, making use of microwave backscattering.

Figure 12.3 shows a block diagram of the basic operation of a fully passive implant employing microwave backscattering. As seen, the system consists of two main components: the brain implant (or neurosensory) and the exterior interrogator. The signal generator in the interrogator unit generates a carrier signal, which the interrogator antenna then transmits to the implanted antenna. The carrier signal (at frequency $f_{carrier}$) acts to turn on the neurosensor by activating the mixer (i.e., diode(s)). Meanwhile, implanted electrodes also connected to the mixer are sensing neuropotentials (at frequency f_{neuro}). The mixer produces third-order products of the two inputs (i.e., $2f_{carrier} \pm f_{neuro}$), which the implant antenna transmits back to the

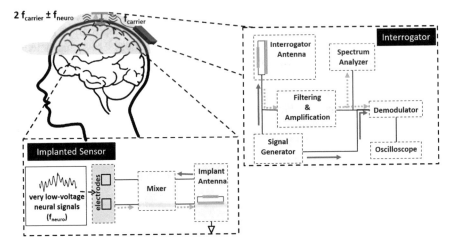

Fig. 12.3 Block diagram of fully passive implant operation. The solid red arrows indicate the path of the carrier frequency, and the dotted green arrows indicate the path of the neural signals

interrogator as amplitude-modulated transmission (hence the term backscattering). Rather than using the first harmonics (i.e., $f_{carrier} \pm f_{neuro}$), the second harmonics (third-order mixing products) are chosen because reflections of the incident carrier signal off tissues and other biomatter creates noise of frequencies primarily centered around the low-voltage, low-frequency neural signals (Schwerdt et al. 2011). These reflections would reduce the sensitivity of the system if first-order products were used. At the interrogator, the third-order products are filtered, amplified, and visualized in the frequency domain via a spectrum analyzer. The neural signal is visualized in the time domain by demodulating the received signal using the initial carrier signal generated by the function generator.

12.3.2 Signal Attenuation and Operating Frequency Selection

Unlike the passive devices aforementioned which use low frequencies and inductive coupling to establish a wireless link, fully passive devices use higher frequencies (on the order of a few GHz) for microwave backscattering. The higher frequencies (smaller wavelengths) allow for smaller antennas; thus, antenna designs beyond inductive coils can be employed (Schwerdt et al. 2011, 2012, 2015; Lee et al. 2015; Kiourti et al. 2016; Lee et al. 2017). However, as frequency of operation ($f_{carrier}$) of the device increases, so does signal attenuation caused by tissue (i.e., the previously mentioned reflections that contribute to noise). In the range of a few GHz, this attenuation is primarily attributed to interactions with the water molecules in the tissues, the complex permittivity of which is modeled via the Cole-Cole equation (Eq. 12.2) (Schwerdt et al. 2011):

$$\hat{\varepsilon} = \varepsilon_0 \left(\varepsilon_r - j\frac{\sigma}{\omega\varepsilon_0} \right) = \varepsilon_\infty + \frac{\varepsilon_s - \varepsilon_\infty}{1 + (j\omega\tau)^{1-\alpha}} + \frac{\sigma}{j\omega\varepsilon_0} \qquad (12.2)$$

In Eq. 12.2, ω is angular frequency; ε_∞ and ε_s are the infinite frequency ($\omega \to \infty$) and static ($\omega \to 0$) permittivity, respectively; τ is the relaxation time; α describes the distribution/spreading and ranges from 0 to 1; and σ is conductivity. The attenuation of the signal is then described by the real part of the complex propagation constant, given by Eq. 12.3, where c_0 is the speed of light in a vacuum, α is the attenuation constant, ε_r is the relative permittivity given by Eq. 12.2, and all other variables are the same as in Eq. 12.2 (Li 2014):

$$\alpha = \frac{\omega}{c_0} \sqrt{\frac{\varepsilon_r}{2} \left[\sqrt{1 + \left(\frac{\sigma}{\omega\varepsilon_0\varepsilon_r} \right)^2} - 1 \right]} \qquad (12.3)$$

The inverse of the attenuation constant (Eq. 12.2) gives the penetration depth of an incident signal. Combining Eqs. 12.2 and 12.3 shows that signal attenuation is increased in tissue as compared to air/free space (in free space $\varepsilon_r \sim 1$) and increases with increasing frequency. Fat, bone, and white matter cause an attenuation of about 1 dB/cm around 1 GHz, and biomatter with an increased water content, such as blood and skin, exhibits a higher attenuation constant (Li 2014).

With this constraint and the losses associated with a given transmit and receive antenna geometry, the optimal transmission frequency can be calculated. For a 5-mm slot receive antenna and a 2.5-mm dipole transmit antenna both with an efficiency of 50% at resonance and assuming a 1-cm^2 beam area, the optimal carrier frequency is ~2–4 GHz with a 4–8-GHz backscatter frequency (Abbaspour-Tamijani et al. 2008).

12.3.3 Proof of Concept Fully Passive Implant with Low Sensitivity

Abbaspour-Tamijani et al. showed the feasibility of such a fully passive neural recorder using: (a) 10 mm × 3.5 mm subresonant slot antenna (milled on 0.50 mm Arlon AR1000 microwave substrate, $\varepsilon_r = 10$) for the implant and (b) dual-frequency (2.4 and 4.8 GHz), dual-port dielectric-loaded waveguide antenna for the interrogator (Abbaspour-Tamijani et al. 2008). The mixer used in the implant was a nonlinear capacitor realized via two diodes connected by a 100-pF chip capacitor. In addition to mixing, the capacitor also serves as a low-pass filter. This filtering prevents RF signals (primarily the carrier frequency) from reaching the input for the electrodes (i.e., the brain). To simulate the electrodes/neural signal, a signal generator was connected directly to the mixer, and a separate signal generator was used to generate the carrier. As this device was a proof of concept, transmission was performed through air with the implant located at the aperture of the interrogator antenna. Using a spectrum analyzer with amplitude modulation (AM) demodulation set to the minimum resolution bandwidth, the minimum detectable signal (minimum signal able to be demodulated) was 10 mV. Abbaspour-Tamijani et al. noted that signals as low as 1 mV still exhibited detectable sidebands, but the spectrum analyzer was unable to perform demodulation; thus, a specially designed low-noise transceiver could improve device sensitivity.

Building on this design, Schwerdt et al. implemented the implanted circuit using two varactors (the mixer) and a bypass capacitor (Schwerdt et al. 2011). The bypass capacitor again acts as a filter, short-circuiting the connection to the electrodes at high frequencies. The varactors act as voltage-variable capacitors, generating the backscattered harmonics.

The implant antenna is a subresonant slot antenna of slot 10×1 mm^2 and, as with the previous example, is designed for the 2.4/4.8-GHz operation. To optimize the antenna impedance (improve conversion gain), two metal-insulator-metal (MIM) capacitors are used, which effectively increase slot length at the carrier frequency and decrease the length at the backscatter frequency. The antenna has near-zero impedance when a DC is applied, preventing any low-frequency signals from radiating outward (Schwerdt et al. 2011). The effectiveness of the implant antenna is in part determined by its ability to backscatter the incident signal, which is described by the effective radar cross section given in Eq. 12.4 (Abbaspour-Tamijani et al. 2008; Schwerdt et al. 2011, 2013):

$$\sigma_{\text{eff}} = 4\pi r^2 \frac{S_{\text{back}}}{S_{\text{inc}}} = G\left(\frac{\lambda^2}{4\pi}\right) D_R D_T e_R e_T \tag{12.4}$$

In Eq. 12.4, r is the distance separating the implant and interrogator; S_{back} and S_{inc} are the magnitude of the Poynting vectors of the backscattered and incident waves, respectively; G is the system conversion gain from incident to backscatter frequency (which is less than 1 since the system is passive); and λ is the carrier signal wavelength in the implant environment (i.e., tissue). D_R and D_T are the directivity of the implanted antenna at the carrier and modulated frequencies, respectively, and e_R and e_T are the efficiencies of the implanted antenna at the carrier and modulated frequencies, respectively.

Maximizing Eq. 12.4 maximizes the power of the backscattered signal. To maximize σ_{eff}, the directivity and efficiency of the implanted antenna at the carrier and modulated frequencies should be maximized by carefully designing the antenna, and the system conversion gain should be maximized by optimizing the implant circuit (i.e., mixer operation and impedance matching). The fabricated implant is shown in Fig. 12.4.

A linearly tapered slot antenna was chosen for the interrogator, providing dual-band (2.4 and 4.8 GHz) operation. Schwerdt et al. chose to fabricate their own demodulator, rather than use one built into a spectrum analyzer. Referring to

Fig. 12.4 Fabricated implant including the subresonant slot antenna, varactors, and bypass capacitor. (Reproduced with permission from Schwerdt et al. (2011))

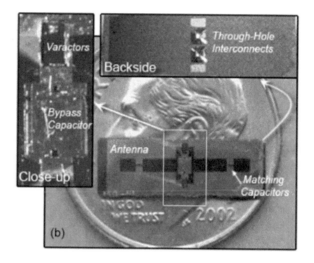

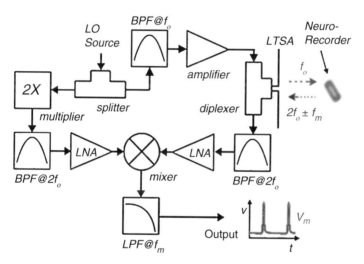

Fig. 12.5 Demodulator design schematic. (Reproduced with permission from Schwerdt et al. (2011))

Fig. 12.5, a signal generator (LO source) feeds the carrier frequency to both the interrogator antenna (LTSA) and a multiplier. The LTSA transmits the amplified signal to the implant, which is then mixed with the neuropotentials and backscattered. The backscattered signal is then filtered around the backscattered frequency, amplified, and mixed with the original signal generator output that has been multiplied, filtered, and amplified. Mixing recovers the neuropotential signals at baseband.

Figure 12.6 shows the results of testing the implant/interrogator system using emulated neuropotentials from a function generator (Fig. 12.6a–c) and using stimulated pulses from the sciatic nerve of a frog (Fig. 12.6d). For both sets of tests, the

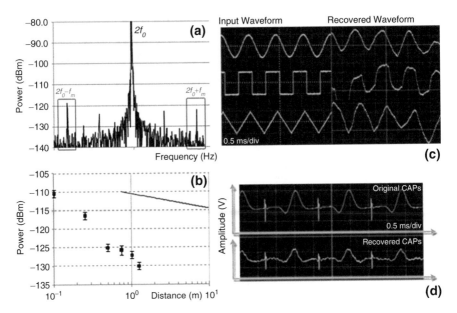

Fig. 12.6 Testing the fully passive device. (**a**) Spectral response with an implant-interrogator separation of 2.5 mm, a sinusoidal emulated neuropotential of 3.4 mV$_{pp}$ at 400 Hz, an input power of 33.1 mW, and a carrier frequency of 2.2 GHz. (**b**) Backscattered power of 50 mV$_{pp}$ at 400 Hz emulated neuropotentials with the implant and interrogator separated by various distances. The black line models the free-space attenuation. (**c**) Comparison of emulated and demodulated sinusoidal, square, and triangular waveforms (from top to bottom) for emulated neuropotentials of 3 mV$_{pp}$ at 800 Hz. (**d**) Stimulated and demodulated neuropotentials of 500 μV$_{pp}$ at 140 Hz from the frog sciatic nerve. (Reproduced with permission from Schwerdt et al. (2011))

antennas were transmitting through air. To simulate detecting a more realistic neuropotential, the implant was connected to electrodes fixed to the sciatic nerve of a dead frog, which was then stimulated by a 0.2-ms biphasic square waveform. The system detected signals as low as 500 μV$_{pp}$ at 140 Hz using signal averaging to improve the signal-to-noise ratio (SNR) (Fig. 12.6d).

To analyze the antennas' operation in the presence of lossy materials, the implant-interrogator system was tested using a multilayer phantom to emulate the human head (the RF properties of skin, bone, dura, gray and white matter can be emulated by mixing water, agar, NaCl, boric acid, TX-151, and polyethylene powder) (Schwerdt et al. 2012). The implant was coated in 4 μm of the polymer Parylene to introduce biocompatibility. With the implant embedded in the dura layer (just below the skull), the interrogator was able to detect neuropotentials simulated by a function generator as low as 6 mV$_{pp}$.

Though less so than with passive implants, SAR is still a concern. Schwerdt et al. used both finite element analysis (FEA) and thermal imaging of the device in the human head phantom to verify that SAR averaged over 1 g was below 1.6 W/kg (Schwerdt et al. 2013). As with passive implants, the interrogator primarily determines SAR, the highest levels of which are concentrated in the outermost layer (skin). The maximum SAR determined via FEA and thermal imagining is 0.216 W/ kg and 0.45 ± 0.11 W/kg, respectively.

12.3.4   Improving the Sensitivity of Fully Passive Implants

With a minimum detectable signal of 500 μV_{pp} at 140 Hz, the device discussed in (Schwerdt et al. 2011) is not sensitive enough to detect the full range of human neural signals (Table 12.1). To begin to move toward more sensitive devices, Lee et al. introduced a wireless, fully passive implant/interrogator system with sensitivity of 200 μV_{pp} at 100 Hz and 50 μV_{pp} at 1 kHz using a similar device design (Lee et al. 2015). To improve sensitivity on the implant side, they used an antiparallel diode pair (APDP) as the mixer, capturing both legs (positive and negative) of the incident carrier signal. Schwerdt et al. did not use both legs and, as a result, experienced increased conversion loss and decreased sensitivity (Schwerdt et al. 2011). When an ideal APDP performs mixing, only odd-order harmonics are produced and no DC terms are produced (i.e., no DC flows toward the brain). On the interrogator side, signals were not demodulated but rather were compared in the frequency domain via a spectrum analyzer.

To maximize the transmission coefficient from interrogator to implant, a low-profile spiral antenna (0.6–6 GHz) was chosen for the interrogator to increase near-field coupling with the implant. An E-shaped patch antenna was chosen for the implant and coated in 0.7 mm of polydimethylsiloxane to maintain biocompatibility. The implanted patch and external spiral are shown in Fig. 12.7a, b, respectively.

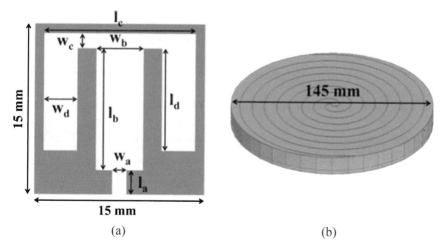

(a)   (b)

Fig. 12.7 Antennas used in the fully passive neurosensing system: (**a**) implanted patch antenna and (**b**) interrogator spiral antenna. (Reproduced with permission from Lee et al. (2015))

The system was validated using a single-layer head phantom made of ground beef with the interrogator separated from the implant by 8 mm. The carrier signal was transmitted with a power of approximately 20 dBm, and neural signals were simulated using a function generator. The system loss and minimum detectable neural signals versus neural signal frequency are shown in Fig. 12.8a, b, respectively. A

(a)

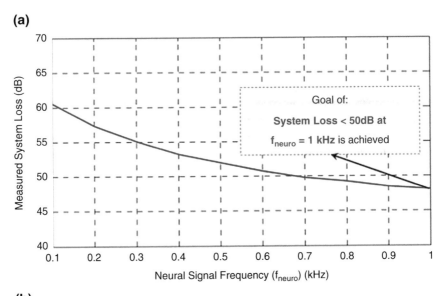

(b)

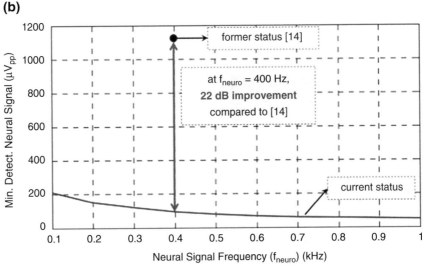

Fig. 12.8 (a) Fully passive neural system loss across the neural signal frequency range. (b) Minimum detectable neural signals across the neural signal frequency range. (Source [14] refers to (Schwerdt et al. 2011), showing the improvement in system sensitivity. Reproduced with permission from Lee et al. (2015))

system loss less than 50 dBm at 1 kHz, as shown in Fig. 12.8a, indicates that neural signals less than 63 μV_{pp} are detectable. Fig. 12.8b also shows the minimum detectable signal of the fully passive device in (Schwerdt et al. 2011) for comparison. Overall, the sensitivity of the system described in Lee et al. (2015) indicates the system is able to record all ECoG signals from 300 to 500 Hz, action potentials from 0.3 to 5 kHz, and LFPs greater than 200 μV_{pp} from 100 to 500 Hz.

To continue improving sensitivity and practicality, the fully passive neurosensing system was further modified to have an up to 20 dB increase in sensitivity and a 59% smaller footprint, as well as incorporate demodulation on the interrogator side (Kiourti et al. 2016). In this case, implant efficiency was improved by using an inductor and capacitor in parallel with the APDP and a single capacitor in place of the two MIM capacitors used previously. The parallel inductor and capacitor provided the ground for the low-frequency neural signals and high-frequency carrier signal, respectively. The single capacitor still served to match the impedance of the antenna to that of the mixer. To reduce the implant size from 39×15 mm^2, which was too large to feasibly be implanted within the brain, the implant was effectively folded in half, separating the antenna and circuit by a ground layer. Folding reduced the implant size to 16×15 mm^2. The folding concept and fabricated implant are shown in Fig. 12.9.

To validate the neurosensing system, a four-layer head phantom (white matter, gray matter, bone, skin) was made using a similar technique as in (Schwerdt et al. 2012) described above. The neural signals were simulated using a signal generator with sinusoidal output. Measured system loss was 40 dB across the 0.1–5-kHz neural signal range, which was 20 dB lower than previously reported (Fig. 12.8a) and indicated that signals as low as 20 μV_{pp} could be detected within this frequency range. In the time domain, neural signals as low as 63 μV_{pp} in the 0.1–5-kHz range were detectable (testing the system reported prior resulted in a time-domain sensitivity of 670 μV_{pp} at 100 Hz). SAR was analyzed using FEA and was shown to be within regulation.

Continuing to improve device feasibility, Lee et al. focused on reducing system sensitivity to 20 μV_{pp} (the lowest possible voltage of LFPs) in the time domain and decreasing the implant and interrogator size (Lee et al. 2017). Efficiency of the implant was improved by replacing the lumped elements with open- and short-circuited stubs, which removed the need for soldering (also improving biocompatibility). To decrease signal path loss, the implant location was moved from under the skull to under the skin (i.e., above the skull). The decrease in implant size was performed by using a higher permittivity substrate material (Rogers TMM13i – $\varepsilon_r = 12.2$, tan $\delta = 0.0019$, versus FR-4 – $\varepsilon_r = 4.6$, tan $\delta = 0.016$) and by using a meandered-arm patch antenna design, increasing the effective length of the antenna without increasing the footprint. The increase in transmission coefficient allowed for a decrease in carrier signal power (6 dBm versus ~20 dBm).

The neurosensing system is intended to be a wearable device, with the external interrogator concealed in a hat or headband, but using a spiral antenna for the external interrogator made the system in (Lee et al. 2015; Kiourti et al. 2016) too large to wear. To begin reducing the size of the external interrogator, the spiral antenna was replaced with an E-shaped patch antenna of footprint 10.1×18.8 mm^2 (98% smaller than the spiral).

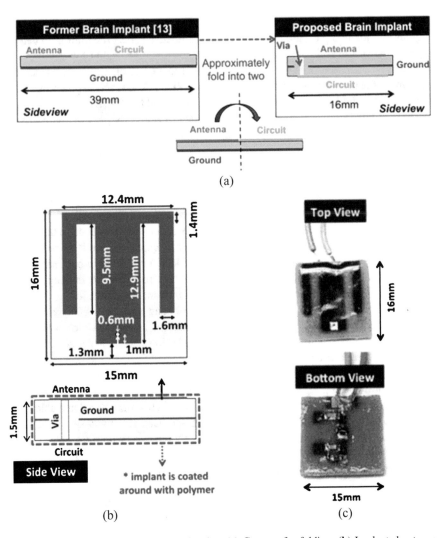

Fig. 12.9 Folding the implant reduces the size. (**a**) Concept for folding. (**b**) Implanted antenna design. (**c**) Fabricated implant. (Reproduced with permission from Kiourti et al. (2016))

These improvements resulted in a system loss of 28 dB across the neural frequency range of 0.01–5 kHz, which is an improvement of about 10 dB on the previous version and means that neuropotentials as low as 20 μV_{pp} are detectable. Figure 12.10 shows various recovered neural waveforms simulated by a function generator outputting a sinusoidal signal of strength 20 μV_{pp}. FE simulations show SAR over 1 g as 0.29 W/kg, which is less than the 1.6 W/kg maximum allowed by the FCC for uncontrolled environment exposure.

Table 12.2 indicates the improvement in sensitivity achieved by the fully passive implants discussed in this section.

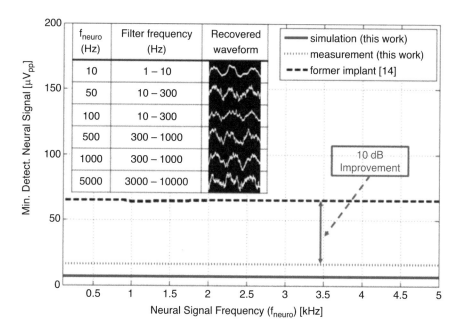

Fig. 12.10 Demodulated time-domain waveforms recovered by the fully passive neurosensing system. The neural waveforms were simulated by a function generator and were 20 μV$_{pp}$, the minimum detectable signal of the neurosensing system. (Reproduced with permission from Lee et al. (2017))

Table 12.2 Comparison of fully passive, wireless neural recording systems

Reference	Number of channels	Footprint	Min. detectable signal (in vitro)	Interrogator-implant separation	Transmission medium	Intended location
Schwerdt et al. (2011)	1	12 × 4 mm^2	3.4 mV$_{pp}$	2.5 mm	Air	–
Lee et al. (2015)	1	39 × 15 mm^2	200 μV$_{pp}$	8 mm	Ground beef	Under skull
Kiourti et al. (2016)	1	16 × 15 mm^2	63 μV$_{pp}$	~15 mm	4-layer phantom	Under skull
Lee et al. (2017)	1	8.7 × 10 mm^2	20 μV$_{pp}$	2 mm	4-layer phantom	Above skull
Schwerdt et al. (2015)	3	~50 × 60 mm^2	700 μV$_{pp}$	3–5 mm	Air	–
Chen et al. (2018)	8	40 × 40 mm^2	20 μV$_{pp}$	2.5 mm	4-layer phantom with pig skin	Above skull

12.3.5 Multichannel Configurations

Thus far, the fully passive systems discussed have had only one recording channel, meaning they can record from a single electrode and a single location in the brain. However, for clinical viability, neurorecorders require the ability to monitor 100 s or 1000s of channels (Schwerdt et al. 2015; Chen et al. 2018). To introduce multiple channels, Schwerdt et al. integrated light-activated switches (photodiodes) into the fully passive device design described in (Schwerdt et al. 2011) to toggle between channels, backscattering a single channel at a time (Schwerdt et al. 2015). For N channels, the device uses N photodiodes and N mixer diodes, one for each channel. Each of the N photodiodes was affixed to an optical filter to allow a certain portion of the visible light spectrum (i.e., a certain color) transmitted by the interrogator (in addition to the carrier frequency) to turn on the photodiode and transmit the modulated neural signal to the interrogator. Simulations predicted the system could detect neuropotentials as low as 10 μV_{pp}, but testing produced a minimum detectable signal of 700 μV_{pp} through air, which they attributed to faults in fabrication. This implementation had a large implant footprint (approximately 50 × 60 mm^2), which would continue to increase as the number of channels increases, and uses visible light which requires cranial windows for in vivo use as visible light is highly attenuated by tissue (Schwerdt et al. 2015; Chen et al. 2018).

To address these issues, Chen et al. modified the design described in (Lee et al. 2017) to introduce multiple channels using a similar concept of light-activated switches (Chen et al. 2018). Rather than facilitate channel selection via visible light, the design utilized near-infrared (IR) light, which is better able to penetrate tissue and therefore does not require cranial windowing for in vivo use (Schwerdt et al. 2015; Chen et al. 2018). Additionally, rather than using N photodiodes and N mixers for N channels, the system uses N photodiodes and one multiplexer for 2^N channels. The transmitted IR signal from N photo-emitters serves two purposes: (1) channel selection as described above for the device in Schwerdt et al. (2015) and (2) providing power for the photovoltaic cell connected to the multiplexer. To facilitate channel selection for eight channels with only three photodiodes, a binary-type selection scheme was used – no light activation represents a zero, whereas light activation represents a one. For example, to record from channel zero, no photodiodes should be activated, whereas to record from channel seven, all three photodiodes should be activated. Channel selection is illustrated in Fig. 12.11.

The implant had a footprint of 40 × 40 mm^2; however, Chen et al. noted that this device was a proof of concept and could be reduced in size by choice of substrate. Testing using a four-layer head phantom with the outer layer being pig skin demonstrated a minimum detectable signal of 20 μV_{pp}, and SAR simulations show the device as having 0.368 W/kg averaged over 1 g.

Table 12.2 contrasts these multichannel devices with the single-channel implementations discussed in Sects. 12.3.3 and 12.3.4.

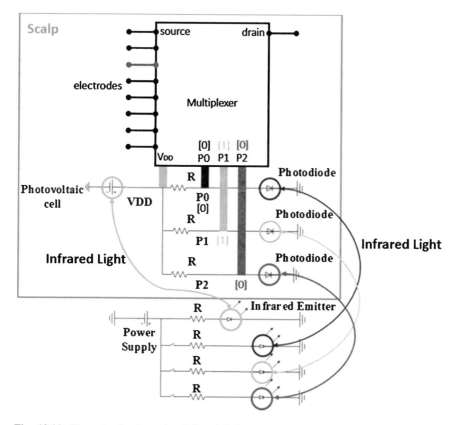

Fig. 12.11 Channel selection using infrared light and three photodiodes for eight channels. (Reproduced with permission from Chen et al. (2018))

12.3.6   Real-World Considerations

Thus far, all of the testing described for fully passive implants uses a function generator to emulate neural signals, but the impedance of a function generator is 50 Ω, whereas the impedance of clinical electrodes falls within the range of tens of kΩs to MΩs (Cogan 2008). This impedance mismatch between the implant optimized for 50 Ω and the high impedance electrodes contributes additional losses, increasing the minimum detectable signal during in vivo recording. Traditionally, battery-enabled and power-hungry operating amplifiers increase the implant's input impedance (Moradi et al. 2013b), but since the devices are fully passive, such methods are not feasible. To begin to move in this direction, Chen et al. introduced a bipolar junction transistor (BJT) into the implant circuit to serve as an impedance buffer and were able to detect neural signals as low as 200 μV_{pp} through a human head phantom using an electrode impedance of 33 kΩ (Chen et al. 2019).

12.4   Conclusion

As the need for more reliable neural interfaces grows, passive RF interfaces have the potential to move such interfaces beyond a clinical or laboratory environment. The devices presented in this chapter do not require wiring through the skin, which increases the risk of infection, nor do they require batteries, the heat generation of which can damage neurons. The passive devices utilize inductive energy transfer to power implanted ICs that can amplify recorded neural signals before transmitting them outside the body. Fully passive devices utilize microwave backscattering and require no implant power. Though these devices eliminate the challenges associated with traditional BMIs, much research is still required to realize these devices in real-world settings. As an example, the interrogator will need to decrease in size to be portable and unobtrusive (e.g., to fit within a hat or headband) and be able to transmit to a separate device with larger memory and post-processing abilities. The implant will also need to be more biocompatible, decreasing or avoiding FBR altogether, as well as integrate electrodes designed for chronic use.

References

Abbaspour-Tamijani, A. et al. (2008). A miniature fully-passive microwave back-scattering device for short-range telemetry of neural potentials. In *2008 30th Annual International Conference of the IEEE Engineering in Medicine and Biology Society* (pp. 129–132). IEEE. https://doi.org/10.1109/IEMBS.2008.4649107.

Bjorninen, T., et al. (2012). Design of wireless links to implanted brain–machine interface microelectronic systems. *IEEE Antennas and Wireless Propagation Letters, 11*, 1663–1666. https://doi.org/10.1109/LAWP.2013.2239252.

Blount, J. P., et al. (2008). Advances in intracranial monitoring. *Neurosurgical Focus, 25*(3), E18. https://doi.org/10.3171/FOC/2008/25/9/E18.

Chen, W.-C., et al. (2018). A multi-channel passive brain implant for wireless neuropotential monitoring. *IEEE Journal of Electromagnetics, RF and Microwaves in Medicine and Biology, 2*(4), 262–269. https://doi.org/10.1109/JERM.2018.2877330.

Chen, W.-C., Guido, K., & Kiourti, A. (2019). Passive impedance matching for implanted brain-electrode interfaces. *IEEE Journal of Electromagnetics, RF and Microwaves in Medicine and Biology, 3*(4), 233–239. https://doi.org/10.1109/JERM.2019.2904024

Cleveland, R., Ulcek, J. (1999). Questions and Answers about Biological Effects and Potential Hazards of Radiofrequency Electromagnetic Fields, *FCC OET Bulletin, 56* (4).

Cogan, S. F. (2008). Neural stimulation and recording electrodes. *Annual Review of Biomedical Engineering*. Annual Reviews, *10*(1), 275–309. https://doi.org/10.1146/annurev.bioeng.10.061807.160518.

Dethier, J., et al. (2013). Design and validation of a real-time spiking-neural-network decoder for brain-machine interfaces. *Journal of Neural Engineering*. NIH Public Access, *10*(3), 036008. https://doi.org/10.1088/1741-2560/10/3/036008.

During, M. J., et al. (1989). Controlled release of dopamine from a polymeric brain implant: In vivo characterization. *Annals of Neurology*. John Wiley & Sons, Ltd, *25*(4), 351–356. https://doi.org/10.1002/ana.410250406.

Hochberg, L. R., et al. (2006). Neuronal ensemble control of prosthetic devices by a human with tetraplegia. *Nature*. Nature Publishing Group, *442*(7099), 164–171. https://doi.org/10.1038/nature04970.

Khan, M. W. A., et al. (2016). Characterization of two-turns external loop antenna with magnetic core for efficient wireless powering of cortical implants. *IEEE Antennas and Wireless Propagation Letters, 15*, 1410–1413. https://doi.org/10.1109/LAWP.2015.2511187.

Kim, S., et al. (2007). Thermal impact of an active 3-D microelectrode array implanted in the brain. *IEEE Transactions on Neural Systems and Rehabilitation Engineering, 15*(4), 493–501. https://doi.org/10.1109/TNSRE.2007.908429.

Kiourti, A., et al. (2016). A wireless fully passive neural recording device for unobtrusive neuro-potential monitoring. *IEEE Transactions on Biomedical Engineering, 63*(1), 131–137. https://doi.org/10.1109/TBME.2015.2458583.

Kipke, D. R., et al. (2008). Symposium Advanced neurotechnologies for chronic neural interfaces: New horizons and clinical opportunities. https://doi.org/10.1523/JNEUROSCI.3879-08.2008.

Lee, C. W. L., et al. (2015). A high-sensitivity fully passive neurosensing system for wireless brain signal monitoring. *IEEE Transactions on Microwave Theory and Techniques, 63*(6), 2060–2068. https://doi.org/10.1109/TMTT.2015.2421491.

Lee, C. W. L., Kiourti, A., & Volakis, J. L. (2017). Miniaturized fully passive brain implant for wireless neuropotential acquisition. *IEEE Antennas and Wireless Propagation Letters, 16*, 645–648. https://doi.org/10.1109/LAWP.2016.2594590.

Li, X. (2014). *Body matched antennas for microwave medical applications.* Karlsruhe: KIT Scientific Publishing.

Mark, M., et al. (2011). SAR reduction and link optimization for mm-size remotely powered wireless implants using segmented loop antennas. In *2011 IEEE Topical Conference on Biomedical Wireless Technologies, Networks, and Sensing Systems* (pp. 7–10). IEEE. https://doi.org/10.1109/BIOWIRELESS.2011.5724339.

Moradi, E., et al. (2013a). Analysis of wireless powering of mm-size neural recording tags in RFID-inspired wireless brain-machine interface systems. In *2013 IEEE International Conference on RFID (RFID)* (pp. 8–15). IEEE. https://doi.org/10.1109/RFID.2013.6548129.

Moradi, E., et al. (2013b). Measurement of wireless link for brain–machine interface systems using human-head equivalent liquid. *IEEE Antennas and Wireless Propagation Letters, 12*, 1307–1310. https://doi.org/10.1109/LAWP.2013.2283737.

Muller, R., et al. (2015). A minimally invasive 64-channel wireless μECoG implant. *IEEE Journal of Solid-State Circuits, 50*(1), 344–359. https://doi.org/10.1109/JSSC.2014.2364824.

Polikov, V. S., Tresco, P. A., & Reichert, W. M. (2005). Invited review response of brain tissue to chronically implanted neural electrodes. *Journal of Neuroscience Methods, 148*, 1–18. https://doi.org/10.1016/j.jneumeth.2005.08.015.

Rao, S., et al. (2014). Miniature implantable and wearable on-body antennas: Towards the new era of wireless body-centric systems [antenna applications corner]. *IEEE Antennas and Propagation Magazine, 56*(1), 271–291. https://doi.org/10.1109/MAP.2014.6821799.

Schwerdt, H. N., et al. (2011). A fully passive wireless microsystem for recording of neuropotentials using RF backscattering methods. *Journal of Microelectromechanical Systems, 20*(5), 1119–1130. https://doi.org/10.1109/JMEMS.2011.2162487.

Schwerdt, H. N., Miranda, F. A., & Chae, J. (2012). A fully passive wireless backscattering neuro-recording microsystem embedded in dispersive human-head phantom medium. *IEEE Electron Device Letters, 33*(6), 908–910. https://doi.org/10.1109/LED.2012.2190967.

Schwerdt, H. N., Miranda, F. A., & Chae, J. (2013). Analysis of electromagnetic fields induced in operation of a wireless fully passive backscattering neurorecording microsystem in emulated human head tissue. *IEEE Transactions on Microwave Theory and Techniques, 61*(5), 2170–2176. https://doi.org/10.1109/TMTT.2013.2252916.

Schwerdt, H. N., Miranda, F. A., & Chae, J. (2015). Wireless fully passive multichannel recording of neuropotentials using photo-activated RF backscattering methods. *IEEE Transactions on Microwave Theory and Techniques, 63*(9), 2965–2970. https://doi.org/10.1109/TMTT.2015.2460746.

Song, L., & Rahmat-Samii, Y. (2017). An end-to-end implanted brain–machine interface antenna system performance characterizations and development. *IEEE Transactions on Antennas and Propagation, 65*(7), 3399–3408. https://doi.org/10.1109/TAP.2017.2700163.

Waziri, A., et al. (2009). Initial surgical experience with a dense cortical microarray in epileptic patients undergoing craniotomy for subdural electrode implantation. *Neurosurgery.* NIH Public Access, *64*(3), 540–545; discussion 545. https://doi.org/10.1227/01.NEU.0000337575.63861.10.

Wise, K. D., et al. (2008). Microelectrodes, microelectronics, and implantable neural microsystems. *Proceedings of the IEEE, 96*(7), 1184–1202. https://doi.org/10.1109/JPROC.2008.922564.

Chapter 13
Wireless Soft Microfluidics for Chronic In Vivo Neuropharmacology

Raza Qazi, Joo Yong Sim, Jordan G. McCall, and Jae-Woong Jeong

13.1 Introduction

Microfluidic neural interfaces hold immense potential for basic neuroscience research and clinical medicine (Chew et al. 2013; Ineichen et al. 2017; Jeong et al. 2015; Minev et al. 2015). In vivo neuropharmacology allows for the delivery of pharmacological agents deep into the brain in order to help dissect complex neural circuits and treat neurodegenerative diseases and brain tumours (Rei et al. 2015). Recent studies show that pharmacological delivery can even be combined with optical intervention for advanced optogenetic and chemogenetic manipulation of the brain (Creed et al. 2015; Jennings et al. 2013; Jeong et al. 2015; McCall et al. 2015, 2017; Stachniak et al. 2014; Stamatakis et al. 2013; Walsh et al. 2014; Williams and

R. Qazi
School of Electrical Engineering, Korea Advanced Institute of Science and Technology (KAIST), Daejeon, Republic of Korea

Department of Electrical, Computer, and Energy Engineering, University of Colorado Boulder, Boulder, CO, USA

J. Y. Sim
Welfare & Medical ICT Department, Electronics and Telecommunications Research Institute, Daejeon, Republic of Korea

J. G. McCall
Department of Anesthesiology, Washington University in St. Louis, St. Louis, MO, USA

Department of Pharmaceutical and Administrative Sciences, St. Louis College of Pharmacy, St. Louis, MO, USA

Center for Clinical Pharmacology, St. Louis College of Pharmacy and Washington University School of Medicine, St. Louis, MO, USA

J.-W. Jeong (✉)
School of Electrical Engineering, Korea Advanced Institute of Science and Technology (KAIST), Daejeon, Republic of Korea
e-mail: jjeong1@kaist.ac.kr

© Springer Nature Switzerland AG 2020
L. Guo (ed.), *Neural Interface Engineering*,
https://doi.org/10.1007/978-3-030-41854-0_13

321

Deisseroth 2013), light-regulated activation of pharmacological agents for high spatiotemporal control of cellular activities (Banghart and Sabatini 2012; Banghart et al. 2013; Kramer et al. 2013), and sophisticated deep brain stimulation for treatment of psychiatric disorders (Creed et al. 2015). Beyond traditional pharmacological agents, there is great potential for the successful delivery of other fluidic agents such as chemogenetic ligands (Burnett and Krashes 2016; Stachniak et al. 2014; Sternson and Roth 2014), gene therapy vectors (O'Connor and Boulis 2015), and antibody treatments for chronic diseases (Sevigny et al. 2016). Therefore, the development of novel, minimally invasive, multifunctional brain-interfacing microfluidic technologies is of the utmost importance for both basic neuroscience and clinical medicine. Despite this need, however, there have only recently been concerted materials and engineering efforts to overcome obstacles inherent in traditional fluid-delivery approaches. The most notable delivery-related challenges derive from the decades-old use of metal cannulas for pharmacological infusions. This conventional method has been effective for basic research to date but is not spatially precise at the scale relevant to brain microcircuits and is not suitable for long-term clinical intervention. The implanted metal tubes extensively damage the targeted brain region and nearby areas, are not well-suited for pairing with concurrent optical or electrical manipulation/observation of neural activity, and severely limit the subject's range of motion due to its tethered operation. In other words, this technology lacks cellular-scale control of drug delivery and the multifunctionality to support wireless, concurrent optical and/or electrical neural stimulation and recording.

To overcome these limitations, innovative efforts combining neuroscience with engineering have been made to develop spatiotemporally-precise tools for in vivo neuropharmacology (i.e. microfluidic neural probe systems) (Canales et al. 2015; Jeong et al. 2015; Minev et al. 2015). Microfluidic neural probe systems greatly minimize neural tissue damage, enable delivery of multiple distinct pharmacological or otherwise fluid agents, and facilitate integration with various other modalities, such as optical, electrical, and/or chemical components within a single implant. Continued advances in materials, microfabrication, and integration technologies have led to unprecedented microfluidic tools with the potential to revolutionize fundamental neuroscience and clinical medicine. In this chapter, we provide an overview of recent advances in microfluidic neural probe technologies for chronic applications.

13.2   Required Conditions for Chronic Microfluidic Interfaces with Neural Tissue

For over a century, metal cannulas have been the standard method in neuroscience for pharmacological infusions into the brain. The basic concept for injecting a fluid into the brain region of interest involves a simple fluidic pump connected to an implanted metal tube (Jeong et al. 2015). This methodology is used in virtually every field of neuroscience. However, these types of regional infusions are limited for the following reasons: (i) the relatively bulky and rigid metal tube (250–500 µm in diameter; E ~ 200 GPa) causes destruction and inflammation of brain tissue, (ii) the sheer size and design of the cannulation system restrict localization of the injection site, and (iii) the tethered tubing that connects to the animal significantly restricts free movement and the ability to receive infusions in more naturalistic environments.

To address these issues for chronic applications, three conditions should be satisfied. First, the probes themselves must integrate well with the targeted neural tissues. This integration relies on the size of the probe, the mechanical properties of the materials used to make the probe, and the biocompatibility of these materials to living tissues. To some degree, these requirements are interdependent (i.e. smaller, flexible probes tend to be more biocompatible (Jeong et al. 2015; Kim et al. 2013; Szarowski et al. 2003)). Secondly, there must be a means to control infusion of the drugs that is both sufficiently small, to enable implantation of the pump hardware, and programmable, to offer on-demand, scheduled, and/or closed-loop fluid delivery. Finally, chronic implantation of a fluid delivery device requires that these first two features operate in a wireless fashion to allow the research subject or patient freedom of movement and mobility to maintain quality of life. This short review attempts to give an overview of recent advances in microfluidic probe designs and considerations that might aid in the development of local fluid delivery devices.

13.3 Compliant Microfluidic Probes for Biomechanical Compatibility

To establish minimally invasive soft, flexible probes for delivery of pharmacological agents, various polymers have been studied to create miniaturized microfluidic implants whose physical properties are well-matched with those of biological tissue in terms of elastic modulus and biocompatibility. Actively investigated materials for this purpose include polydimethylsiloxane (PDMS), Parylene, polyimide (PI), and SU-8. These offer compelling options for numerous biomedical applications due to their biocompatibility, low modulus, and relatively simple fabrication process.

Figure 13.1 shows representative polymer microfluidic probes and their relative moduli compared to those of biological organs and rigid materials such as silicon and stainless steel, which are widely used for conventional cannulas.

PDMS-based microfluidic probes (Fig. 13.1a) are one of the most attractive options because of their low modulus (~1 MPa) and high optical transparency (>95% transmission in visible wavelengths) (Jeong et al. 2015; McCall et al. 2017). Ultrathin, flexible PDMS probes (a total thickness of 50 µm with 10×10 µm^2 fluidic channels) manufactured by soft lithography provide high adaptability to tissue deformation and allow integration of optical elements (e.g. microscale inorganic

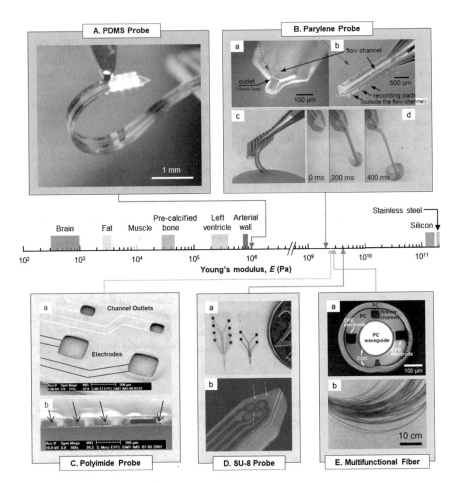

Fig. 13.1 Representative flexible microfluidic probes made of various polymers and their relative moduli compared to those of biological organs and rigid materials (i.e. silicon and stainless steel) used for conventional cannulas. (**a**) An optical image of a PDMS-based optofluidic probe that can deliver multiple fluids and light using an array of microfluidic channels and microscale light-emitting diodes (Jeong et al. 2015). The inset shows mechanical compliance of the probe that can accommodate deformation of tissue. (**b–d**) Polymer microfluidic probes integrated with microelectrodes for simultaneous fluid delivery and electrophysiological signal recording. (**b**) Optical images of a Parylene probe (Takeuchi et al. 2005). (i) A magnified image of the microfluidic channel outlet. (ii) A zoom-out view of the probe showing a fluid channel and recording electrodes. (iii) An image demonstrating flexibility of the probe. (iv) Time-lapse images showing delivery of fluid throughout the microfluidic channel at 0, 200, and 400 ms. (**c**) Scanning electron micrographs of a polyimide probe in (i) top and (ii) side views (Metz et al. 2001). (**d**) SU-8 probe (Altuna et al. 2013). (i) An optical image comparing the size of the entire device to a coin. (ii) A scanning electron micrograph showing the tip of the SU-8 probe. Arrows indicate fluid outlets. (**e**) A multifunctional fibre that integrates microfluidic channels, optical waveguides, and recording electrodes. (i) A scanning electron micrograph capturing the cross-section of the multifunctional fibre, which is made of conductive polyethylene (CPE resistivity, $\rho = 30\ \Omega$ cm), polycarbonate (PC refractive index, $n = 1.58$, Young's modulus, $E = 2.38$ GPa), and cyclic olefin copolymer (COC; $n = 1.52$; $E = 3$ GPa) (Canales et al. 2015). (ii) An optical image showing a bundle of fibre manufactured using thermal drawing (Park et al. 2017)

light emitting diodes [μ-ILEDs]) for versatile operations for a wide variety of applications (e.g. combined optogenetics and in vivo pharmacology, optopharmacology, etc.). Long-term in vivo durability and biocompatibility of PDMS devices verified through integration in different organs demonstrates the potential of PDMS probes for chronic implantation (Abbasi et al. 2012; Chew et al. 2013; Gutbrod et al. 2014; Jeong et al. 2015; Minev et al. 2015).

Figure 13.1b shows another microfluidic probe made of Parylene C (Takeuchi et al. 2005). Parylene possesses many favourable characteristics for use in in vivo pharmacology such as low modulus (2.2 GPa), low water absorption rate, and resistance to chemicals (Noh et al. 2004). Fabrication of fluidic channels using this material relies on thermal bonding of two layers (Ziegler et al. 2006) or a sacrificial structure-assisted process (Takeuchi et al. 2005). The Parylene process is based on vapour deposition, which enables manufacturing of thinner probes (e.g. the thinnest reported is ~20 μm thick (Ziegler et al. 2006)). Polyimide (modulus: 2.5 GPa) and SU-8 (modulus: 4 GPa) are photo-patternable polymers that facilitate creation of fluidic channels in a flexible format as shown in Fig. 13.1c (Metz et al. 2001) and Fig. 13.1d (Altuna et al. 2013), respectively. Although these two materials are not currently part of any devices that have received FDA Premarket Approval, their biocompatibility has been demonstrated by many in vivo studies (Altuna et al. 2013; Lago et al. 2007; Nemani et al. 2013; Rubehn and Stieglitz 2010; Voskerician et al. 2003). Using standard lithography (Altuna et al. 2013; Metz et al. 2001; Minev et al. 2015; Takeuchi et al. 2005; Ziegler et al. 2006) or transfer printing techniques (Jeong et al. 2015; McCall et al. 2017), all these polymer microfluidic probes can integrate with other modalities to enable simultaneous fluid delivery and optical stimulation (Fig. 13.1a) (Jeong et al. 2015; McCall et al. 2017), or fluid delivery and electrophysiological recording (Fig. 13.1b–d) (Altuna et al. 2013; Metz et al. 2001; Takeuchi et al. 2005). This multifunctionality is a fascinating feature that can advance in vivo pharmacology by allowing light-activated drug delivery and/or concurrent analysis of the physiological effects of delivered agents.

Recent developments using thermal drawing have facilitated creation of multifunctional flexible probes (Canales et al. 2015; Park et al. 2017), which resemble conventional optical fibres. These miniature multifunctional fibres (~200 μm in diameter) integrating fluidic channels, recording electrodes, and optical waveguides (Fig. 13.1e) can be manufactured by incorporating multiple materials such as conductive polyethylene, polycarbonate, and cyclic olefin copolymer. Long-term in vivo studies in the brain verify the minimal invasiveness and multifunctionality of these fibres (Canales et al. 2015; Park et al. 2017) demonstrating potential for chronic implantation in other organ systems.

Though soft, flexible microfluidic probes minimize tissue damage when implanted, their initial implantation into intended target regions remains challenging due to high mechanical compliance of the probes. Various approaches to temporarily stiffen probes have been explored, exploiting ultrathin injection-assist needles (Jeong et al. 2015; Kim et al. 2013; Luan et al. 2017; McCall et al. 2013, 2017) and/ or bio-dissolvable polymers such as silk (Lecomte et al. 2015; Metallo and Trimmer 2015), poly-lactic-co-glycolic acid (Park et al. 2016), or maltose (Xiang et al. 2014).

Advanced variations of these approaches and future probe development capable of dynamic stiffness tuning will allow full competence of flexible microfluidic probes for minimally invasive targeted drug delivery.

13.4 Micro-Pumps and Wireless Technologies for Self-Contained Microfluidic Systems

Micro-pumps and wireless modules are the key components for stand-alone implantable fluidic devices to dispense pharmacological agents on demand or in a programmed fashion. The following sections introduce state-of-the art micro-pumps and wireless technologies.

13.4.1 Types of Micro-Pumps

Active fluid pumps are essential to provide temporal manipulation of fluid delivery with a desired flow rate and fluid volume. Micro-pumps integrated with microcontrollers and wireless modules can enable on-demand or programmed delivery of fluid to tissue throughout the microfluidic probe. To do so, these pumps take advantage of various different actuators, the moving components that can drive fluid expulsion. When these components are opened, closed, or otherwise modified (i.e. actuated), the fluid is delivered via the microfluidic probe. Figure 13.2 shows representative mechanical micro-pumps, which are described as follows:

Micro-Pumps with an Activatable Membrane

This type of pump exploits mechanically compliant membrane actuators and check valves to draw fluid from a reservoir and push it out to the outlet. The membrane, integrated with a piezoelectric material (Fig. 13.2a) (Junwu et al. 2005; Liu et al. 2014) or a shape-memory alloy (Fig. 13.2b) (Fong et al. 2015), can be driven to induce bidirectional mechanical pumping motions. This design is suitable when a large volume of fluid needs to be delivered repeatedly from a reservoir.

Thermally Activatable Pumps

Thermally activated pumps (Fig. 13.2c) consist of thermally expandable polymer, micro-heaters, and hemispherical reservoirs (Jeong et al. 2015; Spieth et al. 2012). Joule heating induced by electrical current flow to the heater expands the expandable layer, pumping out fluid from the chamber. This pump can deliver a precise

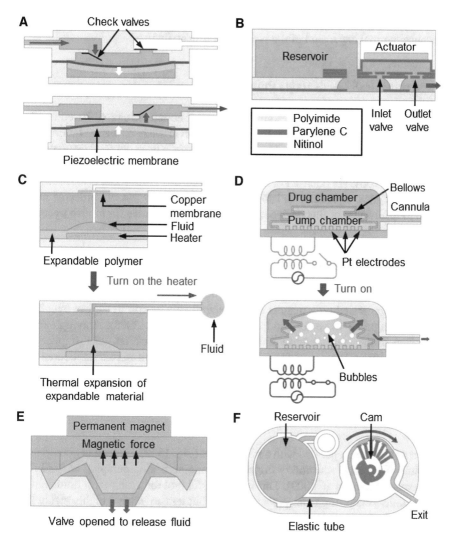

Fig. 13.2 Schematic diagrams of active micro-pumps with different working mechanisms that enable on-demand or programmable fluid delivery when combined with microcontrollers and wireless technology. (**a, b**) Micro-pumps with an actuated membrane. (**a**) A piezoelectric pump consisting of check valves and a flexible piezoelectric membrane, which can drive mechanical fluid pumping via a piezoelectric effect in the membrane induced by an external electric field (Junwu et al. 2005). (**b**) A shape memory alloy pump, which actuates the pump membrane using the shape memory effect of an alloy (e.g. nitinol) in response to temperature changes (Fong et al. 2015). (**c**) A thermally actuated pump based on thermal expansion of an expandable layer by joule heating (Jeong et al. 2015). (**d**) An electrochemically actuated pump, which pushes out fluid by controlling the volume of the pump chamber using an electrolysis effect (Li et al. 2010). (**e**) A magnetically actuated pump controlling fluid pumping via an external magnetic field (Cheng et al. 2008). (**f**) A peristaltic pump using a rotary motor to push out fluid via sequential compression of the fluid tube with mechanical fingers toward the fluid exit (Abe et al. 2009)

fluid volume, predefined by the size of the reservoir, but can only be operated once because thermal expansion of the polymer is irreversible.

Electrochemically Activatable Pumps

Another miniaturized pump design uses electrochemical actuation, based on electrolysis to generate air bubbles from an electrolyte such as water. Figure 13.2d shows an example of an electrochemical pump that can push out fluid by controlling expansion of Parylene bellows using an electrolysis reaction between metal electrodes and electrolyte (Gensler et al. 2012; Li et al. 2010; Sheybani et al. 2015). The non-thermogenic operation and low-power requirements (on the order of microwatts to milliwatts) of this device make it an attractive choice for in vivo operation.

Magnetically Activatable Pumps

Magnetically activated pumps such as that shown in Fig. 13.2e use an external magnetic field generated by a magnet or an electromagnetic coil to control opening of the ferromagnetic valve for fluid delivery (Cheng et al. 2008). Although simple operation of the pumps can facilitate their in vivo application, local implantation may prove difficult because the integrated magnet or coil makes the overall construction relatively bulky.

Motor-Based Peristaltic Pumps

A peristaltic pump operated with a micromotor (Fig. 13.2f) sequentially compresses the fluid-filled tube with rotary fingers, enabling peristaltic pumping of fluid from the reservoir (Abe et al. 2009). This is a powerful tool for a long-term fluid delivery for systems benefitting from peristaltic flow, but less so when a highly stable flow rate is desired.

13.4.2 Wireless Technologies

Integrated wireless modules can minimize tissue damage by eliminating subtle movements of the probe in tissue caused by the tethered tubing. Infrared (IR) (Hashimoto et al. 2014; Iwai et al. 2011; Jeong et al. 2015; McCall et al. 2017) and radio frequency (RF) (Fong et al. 2015; Kim et al. 2013; McCall et al. 2013; Montgomery et al. 2015; Park et al. 2016; Shin et al. 2017) are promising technologies for implementation of wireless implants. The IR approach is simple, offers short-range (1–10 m) control, and can support one-way communication to send a

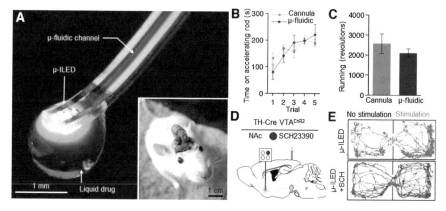

Fig. 13.3 Wireless microfluidic neural probe for in vivo pharmacology and optogenetics. (**a**) An optical image of the flexible optofluidic probe consisting of multiple microfluidic channels and μ-ILEDs, capable of simultaneous pharmacological delivery and optical stimulation. The inset shows a small ($15 \times 15 \times 7$ mm³), lightweight (1.8 g), wireless optofluidic system implanted in the brain and mounted on the head of a freely moving rat. (**b, c**) Activity profiles of mice implanted with wireless- and cannula-based devices. (**b**) Rotarod data showing the equal ability of the animals on the rotating rod as the trial speed increases. (**c**) Spontaneous locomotor activity on a running wheel is not altered by the wireless optofluidic system. (**d**) A schematic diagram illustrating the wireless optofluidic experiment. Infusion of channelrhodopsin-2 fused with enhanced yellow fluorescent protein was made into the VTA, followed by implantation of an optofluidic device filled with SCH23390 in the nucleus accumbens of tyrosine hydroxylase-Cre mice 6 weeks later. (**e**) Traces of mice during real-time place preference control using wireless optofluidics. Optogenetic activation of VTA dopaminergic neurons drives a strong place preference (top), while delivery of SCH23390 (dopamine receptor-1 antagonist) blocks preference behaviour (bottom) (Jeong et al. 2015)

trigger signal to a fluidic device. However, the need for a power supply and "line-of-sight" handicap can limit its applications. On the other hand, the RF technology is more versatile because it can be designed to support bidirectional data transfer and/ or wireless energy harvesting and does not suffer from the line-of-sight problem. However, RF operation can be limited due to susceptibility to RF signal orientation and polarization, therefore requiring careful design. Incorporation of an appropriate wireless technology can significantly empower long-term in vivo implementation of fluidic devices by not only minimizing tissue damage, but also facilitating on-demand or scheduled drug delivery. This section introduces exemplary state-of-the art wireless in vivo microfluidic devices.

Battery-Powered IR Microfluidic Devices

Figure 13.3a shows a wireless microfluidic system that incorporates optoelectronics for in vivo pharmacology and optogenetics (Jeong et al. 2015; McCall et al. 2017). This entirely self-contained, head-mounted system includes an ultrathin flexible microfluidic probe (~50 μm in thickness; stiffness 13–18 N/m; modulus ~1 MPa), micro inorganic LEDs (μ-ILEDs; 100 μm × 100 μm × 6.45 μm in dimension), compact thermally actuated fluid pumps, an IR wireless control module, and small lithium-ion batteries. This system allows untethered, programmable spatiotemporal control of pharmacological delivery and optical stimulation deep in the brain to manipulate neural circuits of interest, therefore holding much potential for many neuroscience and clinical applications involving freely moving subjects. The IR wireless control module interfaces with the optofluidic neural system to enable streamlined operation of the fully self-contained device. The fully assembled device weighs only ~1.8 g; therefore, it does not disturb the movement or behaviour of animals. No gross differences in motor function and physical activity between wireless implanted mice and cannulated mice were found using the rotarod and running wheel tests (Fig. 13.3b, c). One of the most advanced features of the wireless optofluidic system is its ability to simultaneously implement wireless in vivo optical stimulation (e.g. for optogenetics) and pharmacology. In a demonstration experiment (Fig. 13.3d, e), optical stimulation of ventral tegmental area (VTA) dopaminergic neurons was able to drive a real-time place preference (demonstrating the reinforcing nature of this photostimulation) while concurrent wireless delivery of a dopamine receptor-1 antagonist (SCH23390) blocked this behaviour. Such wireless design is desirable to enable unprecedented cell-type and receptor-selective circuit manipulation in freely moving animals.

Fully Implantable RF Optofluidic Devices

Recent advances in wireless energy-harvesting technologies helped engineers to create a miniaturized, fully implantable, and battery-free optofluidic neural system (Fig. 13.4a), which can be operated semi-permanently under tissue (Noh et al. 2018). The device construct includes a stretchable RF energy harvester and an ultrathin, soft, and flexible optofluidic probe (i.e. the same kind of probe as shown in Fig. 13.3a) integrated with a micro-pump for highly localized drug and light delivery. The key component that enables selective control of wireless drug delivery and photostimulation is a stretchable, multichannel RF antenna. The RF antenna is designed to allow selective harvesting of RF energy at a distinct frequency through capacitive coupling between adjacent serpentine metal traces (Fig. 13.4b, c). Due to the non-overlapping nature of the operation channels (1.8 and 2.7 GHz), the harnessed energy at each channel can independently power the micro-pump or μ-ILEDs (Fig. 13.4d, e). This unique antenna design is scalable; thus, more channels can be added for versatile operation. Biocompatibility and biomechanical compatibility of the device are achieved by encapsulating the device body with PDMS and Parylene C, which also help prevent electrical shorts by biofluids. In vivo surgical implantation and proof-of-concept functional testing in a mouse demonstrate its powerful capabilities for seamless implantation as well as wireless manipulation of neural circuits in freely moving animals (Fig. 13.4f).

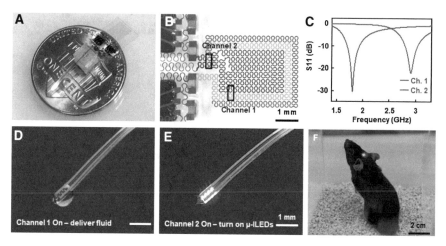

Fig. 13.4 Fully implantable, battery-free optofluidic neural device with a stretchable antenna for RF energy harvesting. (**a**) An optical image signifying the miniaturized nature of the fully implantable optofluidic device (220 mg; 125 mm³) by comparing its size with that of a US penny. (**b**) Top view of the stretchable RF antenna with serpentine-shaped copper traces (false coloured with red, green, and blue) forming two separate channels to enable wireless energy harvesting. (**c**) Scattering parameter S_{11} of the stretchable RF antenna showing distinct resonant frequencies corresponding to channel 1 and 2. (**d, e**) Optical images demonstrating the independent output control capability of the implantable optofluidic device for delivering (**d**) drugs and (**e**) light. (**f**) The fully implantable wireless optofluidic device being operated in a freely moving mouse (Noh et al. 2018)

Smartphone-Controlled Lego Optofluidic Devices

The wireless optofluidic devices introduced in the previous sections (Jeong et al. 2015; McCall et al. 2017; Noh et al. 2018) are limited due to lack of their ability to deliver drugs over long periods of time as well as requirement for special tools for wireless control, which are usually not available in neuroscience labs. Recent development of smartphone-controlled optofluidic devices integrated with replaceable Lego-like drug cartridges (Qazi et al. 2019) resolved these challenges. Figure 13.5a shows a concept of plug-and-play mechanics that helps achieve chronic drug delivery through Lego assembly/disassembly of drug cartridges. After delivering drugs through a cartridge, the clipped harness (that helps achieve hermetic fluidic connection) is removed and then the used cartridge is replaced with a new one before delivering another set of drugs. Integration of μ-ILEDs can allow concurrent delivery of light alongside drugs for advanced chronic optogenetic studies and optopharmacology (Qazi et al. 2018).

The cross-sectional view of the standalone Lego optofluidic system (Fig. 13.5b) highlights its key constituents: (i) a replaceable drug cartridge for chronic drug supply, (ii) a soft optofluidic probe consisting of microfluidic channels and μ-ILEDs for minimally invasive access to neural tissue, (iii) a programmable Bluetooth module to enable wireless smartphone control, and (iv) two rechargeable batteries for continuous power supply. Its compact and lightweight design (~2 g) allows hassle-free

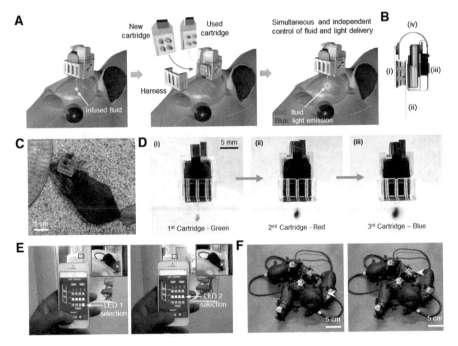

Fig. 13.5 Smartphone-controlled Lego optofluidic device for chronic deliveries of drugs and light with enhanced wireless capabilities. (**a**) Concept illustration for chronic drug delivery and optoge-netics in a freely behaving animal – after delivering the drug (left), the exhausted Lego drug cartridge can be easily swapped with a new one (middle), before delivering the drugs concomitantly with light again (right). (**b**) A sectional view of the Lego optofluidic device highlighting the assembly of its (i) replaceable Lego drug cartridges; (ii) soft, biocompatible optofluidic probe; (iii) wireless Bluetooth circuit; and (iv) rechargeable batteries to form an ultracompact structure. (**c**) An optical image of a freely moving mouse implanted with the ultracompact and lightweight (2 g) Lego optofluidic device showing no effect on its natural behaviour. (**d**) Sequential fluid delivery in a brain phantom (0.6% agarose gel) demonstrating the chronic pharmacological capability of the Lego optofluidic device by delivering (i) green, (ii) red, and (iii) blue aqueous dye solutions using different Lego drug cartridges, respectively. (**e**) Bluetooth-based, smartphone control allowing the user to control the Lego optofluidic device implanted in dummy mouse model (insets) from outside closed rooms without any effect on data reliability as shown through precisely triggered (40-Hz) blue μ-ILED (LED1; left) and orange μ-ILED (LED2; right) from outside a closed door. (**f**) Bluetooth pairing technology further allowing selective wireless triggering and easy target swapping (using the same smartphone) of distinct targets (dummy mice) within a congested group with 100% accuracy (Qazi et al. 2019)

integration in animals as small as mice (Fig. 13.5c) without obstructing their natural behaviour. Figure 13.5d demonstrates their capability to repeatedly deliver multiple distinct fluids to the same target site inside a brain phantom (agarose gel) using different cartridges. Moreover, wireless Bluetooth control through smartphone not only increases wireless operational range, but also enables researchers to control animals from outside behavioural rooms (Fig. 13.5e), thus removing observation effect on animal behaviour studies. Other powerful features include its ability for selective wireless control in a group of multiple animals without any line-of-sight handicap or false triggering (Fig. 13.5f) and scalable closed-loop control. This novel class of optofluidic devices can help uncover basis of various neurodegenerative diseases through in vivo photo-pharmacological manipulations of the same target circuits with high precision over long periods of time.

13.5 Conclusion

Innovation in microfluidic neural probe technologies based on new materials, mechanical concepts, and novel manufacturing/integration approaches is leading to breakthroughs in neuroscience and medicine. Innovative microfluidic probe technologies will establish new horizons for in vivo pharmacology, chemogenetics (Burnett and Krashes 2016; Stachniak et al. 2014; Sternson and Roth 2014), optogenetics, and the relatively new field of optopharmacology (Banghart and Sabatini 2012; Banghart et al. 2004, 2013; Frank et al. 2016; Kramer et al. 2013; Tochitsky et al. 2012), which uses light-regulated control of molecules to enable stimulation or inhibition in specific cells with high temporal and spatial precision. Beyond fundamental research, the microfluidic probe approach also has significant potential in clinical medicine, due to its capability of highly localized pharmacological infusion for targeted therapies without affecting neighbouring neural tissues. Such fascinating characteristics will be invaluable for many clinical applications, including treatment of brain tumours, neural injury, and many neurodegenerative diseases (e.g. Parkinson's, Alzheimer's, etc.). Future multidisciplinary development will lead to compelling advances in neuroscience and clinical applications by enabling advanced, multifaceted functions and capabilities that go far beyond what conventional approaches have been unable to conquer.

Acknowledgments Images in Figs. 13.1, 13.2, 13.3, 13.4, and 13.5 are reproduced with permission from the references cited in the figure captions. Sections 13.1, 13.2 (the first paragraph), 13.4.2.1, and 13.5 are from Sim et al. (2017); Sects. 13.2 (the second paragraph), 13.3, and 13.4 (up to the first paragraph of the Sect. 13.4.2) and Figs. 13.1 and 13.2 and associated captions are from McCall and Jeong (2017), with permissions.

References

Abbasi, F., Mirzadeh, H., & Simjoo, M. (2012). Hydrophilic interpenetrating polymer networks of poly(dimethyl siloxane) (PDMS) as biomaterial for cochlear implants. *Journal of Biomaterials Science. Polymer Edition, 17,* 341–355.

Abe, C., Tashiro, T., Tanaka, K., Ogihara, R., & Morita, H. (2009). A novel type of implantable and programmable infusion pump for small laboratory animals. *Journal of Pharmacological and Toxicological Methods, 59,* 7–12.

Altuna, A., Bellistri, E., Cid, E., Aivar, P., Gal, B., Berganzo, J., Gabriel, G., Guimerà, A., Villa, R., Fernández, L. J., et al. (2013). SU-8 based microprobes for simultaneous neural depth recording and drug delivery in the brain. *Lab on a Chip, 13,* 1422–1430.

Banghart, M. R., & Sabatini, B. L. (2012). Photoactivatable neuropeptides for spatiotemporally precise delivery of opioids in neural tissue. *Neuron, 73*, 249–259.

Banghart, M., Borges, K., Isacoff, E., Trauner, D., & Kramer, R. H. (2004). Light-activated ion channels for remote control of neuronal firing. *Nature Neuroscience, 7*, 1381–1386.

Banghart, M. R., Williams, J. T., Shah, R. C., Lavis, L. D., & Sabatini, B. L. (2013). Caged naloxone reveals opioid signaling deactivation kinetics. *Molecular Pharmacology, 84*, 687–695.

Burnett, C. J., & Krashes, M. J. (2016). Resolving behavioral output via chemogenetic designer receptors exclusively activated by designer drugs. *Journal of Neuroscience: The Official Journal of the Society for Neuroscience, 36*, 9268–9282.

Canales, A., Jia, X., Froriep, U. P., Koppes, R. A., Tringides, C. M., Selvidge, J., Lu, C., Hou, C., Wei, L., Fink, Y., et al. (2015). Multifunctional fibers for simultaneous optical, electrical and chemical interrogation of neural circuits in vivo. *Nature Biotechnology, 33*, 277–284.

Cheng, C.-H., Chao, C., Cheung, Y.-N., Xiao, L., Yang, M., & Leung, W. (2008). A transcutaneous controlled magnetic microvalve based on iron-powder filled PDMS for implantable drug delivery systems. In *2008 3rd IEEE international conference on nano/micro engineered and molecular systems* (pp. 1160–1163).

Chew, D. J., Zhu, L., Delivopoulos, E., Minev, I. R., Musick, K. M., Mosse, C. A., Craggs, M., Donaldson, N., Lacour, S. P., McMahon, S. B., et al. (2013). A microchannel neuroprosthesis for bladder control after spinal cord injury in rat. *Science Translational Medicine, 5*, 210ra155.

Creed, M., Pascoli, V. J., & Lüscher, C. (2015). Addiction therapy. Refining deep brain stimulation to emulate optogenetic treatment of synaptic pathology. *Science, 347*, 659–664.

Fong, J., Xiao, Z., & Takahata, K. (2015). Wireless implantable chip with integrated nitinol-based pump for radio-controlled local drug delivery. *Lab on a Chip, 15*, 1050–1058.

Frank, J. A., Yushchenko, D. A., Hodson, D. J., Lipstein, N., Nagpal, J., Rutter, G. A., Rhee, J.-S., Gottschalk, A., Brose, N., Schultz, C., et al. (2016). Photoswitchable diacylglycerols enable optical control of protein kinase C. *Nature Chemical Biology, 12*, 755–762.

Gensler, H., Sheybani, R., Li, P.-Y., Mann, R. L., & Meng, E. (2012). An implantable MEMS micropump system for drug delivery in small animals. *Biomedical Microdevices, 14*, 483–496.

Gutbrod, S. R., Sulkin, M. S., Rogers, J. A., & Efimov, I. R. (2014). Patient-specific flexible and stretchable devices for cardiac diagnostics and therapy. *Progress in Biophysics and Molecular Biology, 115*, 244–251.

Hashimoto, M., Hata, A., Miyata, T., & Hirase, H. (2014). Programmable wireless light-emitting diode stimulator for chronic stimulation of optogenetic molecules in freely moving mice. *Neurophotonics, 1*, 011002–011002.

Ineichen, B. V., Schnell, L., Gullo, M., Kaiser, J., Schneider, M. P., Mosberger, A. C., Good, N., Linnebank, M., & Schwab, M. E. (2017). Direct, long-term intrathecal application of therapeutics to the rodent CNS. *Nature Protocols, 12*, 104–131.

Iwai, Y., Honda, S., Ozeki, H., Hashimoto, M., & Hirase, H. (2011). A simple head-mountable LED device for chronic stimulation of optogenetic molecules in freely moving mice. *Neuroscience Research, 70*, 124–127.

Jennings, J. H., Sparta, D. R., Stamatakis, A. M., Ung, R. L., Pleil, K. E., Kash, T. L., & Stuber, G. D. (2013). Distinct extended amygdala circuits for divergent motivational states. *Nature, 496*, 224–228.

Jeong, J.-W., McCall, J. G., Shin, G., Zhang, Y., Al-Hasani, R., Kim, M., Li, S., Sim, J. Y., Jang, K.-I., Shi, Y., et al. (2015). Wireless optofluidic systems for programmable in vivo pharmacology and optogenetics. *Cell, 162*, 662–674.

Junwu, K., Zhigang, Y., Taijiang, P., Guangming, C., & Boda, W. (2005). Design and test of a high-performance piezoelectric micropump for drug delivery. *Sensors and Actuators A: Physical, 121*, 156–161.

Kim, T., McCall, J. G., Jung, Y. H., Huang, X., Siuda, E. R., Li, Y., Song, J., Song, Y. M., Pao, H. A., Kim, R.-H., et al. (2013). Injectable, cellular-scale optoelectronics with applications for wireless Optogenetics. *Science, 340*, 211–216.

Kramer, R. H., Mourot, A., & Adesnik, H. (2013). Optogenetic pharmacology for control of native neuronal signaling proteins. *Nature Neuroscience, 16*, 816–823.

Lago, N., Yoshida, K., Koch, K. P., & Navarro, X. (2007). Assessment of biocompatibility of chronically implanted polyimide and platinum intrafascicular electrodes. *IEEE Transactions on Biomedical Engineering, 54*, 281–290.

Lecomte, A., Castagnola, V., Descamps, E., Dahan, L., Blatché, M. C., Dinis, T. M., Leclerc, E., Egles, C., & Bergaud, C. (2015). Silk and PEG as means to stiffen a parylene probe for insertion in the brain: Toward a double time-scale tool for local drug delivery. *Journal of Micromechanics and Microengineering, 25*, 125003.

Li, P., Sheybani, R., Gutierrez, C. A., Kuo, J. T. W., & Meng, E. (2010). A Parylene bellows electrochemical actuator. *Journal of Microelectromechanical Systems, 19*, 215–228.

Liu, G., Yang, Z., Liu, J., Li, X., Wang, H., Zhao, T., & Yang, X. (2014). A low cost, high performance insulin delivery system based on PZT actuation. *Microsystem Technologies, 20*, 2287–2294.

Luan, L., Wei, X., Zhao, Z., Siegel, J. J., Potnis, O., Tuppen, C. A., Lin, S., Kazmi, S., Fowler, R. A., Holloway, S., et al. (2017). Ultraflexible nanoelectronic probes form reliable, glial scar–free neural integration. *Science Advances, 3*, e1601966.

McCall, J. G., Kim, T., Shin, G., Huang, X., Jung, Y. H., Al-Hasani, R., Omenetto, F. G., Bruchas, M. R., & Rogers, J. A. (2013). Fabrication and application of flexible, multimodal light-emitting devices for wireless optogenetics. *Nature Protocols, 8*, 2413–2428.

McCall, J. G., Al-Hasani, R., Siuda, E. R., Hong, D. Y., Norris, A. J., Ford, C. P., & Bruchas, M. R. (2015). CRH engagement of the locus coeruleus noradrenergic system mediates stress-induced anxiety. *Neuron, 87*, 605–620.

McCall, J. G., Qazi, R., Shin, G., Li, S., Ikram, M. H., Jang, K.-I., Liu, Y., Al-Hasani, R., Bruchas, M. R., Jeong, J.-W., et al. (2017). Preparation and implementation of optofluidic neural probes for in vivo wireless pharmacology and optogenetics. *Nature Protocols, 12*, 219–237.

McCall, J. G., & Jeong, J. W. (2017). Minimally invasive probes for programmed microfluidic delivery of molecules in vivo. *Current opinion in pharmacology, 36,*, 78–85.

Metallo, C., & Trimmer, B. A. (2015). Silk coating as a novel delivery system and reversible adhesive for stiffening and shaping flexible probes. *Journal of Biological Methods, 2*, e13.

Metz, S., Holzer, R., & Renaud, P. (2001). Polyimide-based microfluidic devices. *Lab on a Chip, 1*, 29–34.

Minev, I. R., Musienko, P., Hirsch, A., Barraud, Q., Wenger, N., Moraud, E. M., Gandar, J., Capogrosso, M., Milekovic, T., Asboth, L., et al. (2015). Electronic dura mater for long-term multimodal neural interfaces. *Science, 347*, 159–163.

Montgomery, K. L., Yeh, A. J., Ho, J. S., Tsao, V., Mohan Iyer, S., Grosenick, L., Ferenczi, E. A., Tanabe, Y., Deisseroth, K., Delp, S. L., et al. (2015). Wirelessly powered, fully internal optogenetics for brain, spinal and peripheral circuits in mice. *Nature Methods, 12*, 969–974.

Nemani, K. V., Moodie, K. L., Brennick, J. B., Su, A., & Gimi, B. (2013). In vitro and in vivo evaluation of SU-8 biocompatibility. *Materials Science and Engineering: C, 33*, 4453–4459.

Noh, H., Moon, K., Cannon, A., Hesketh, P. J., & Wong, C. P. (2004). Wafer bonding using microwave heating of parylene intermediate layers. *Journal of Micromechanics and Microengineering, 14*, 625–631.

Noh, K. N., Park, S. I., Qazi, R., Zou, Z., Mickle, A. D., Grajales-Reyes, J. G., Jang, K.-I., Gereau, R. W., Xiao, J., Rogers, J. A., et al. (2018). Miniaturized, battery-free Optofluidic systems with potential for wireless pharmacology and Optogenetics. *Small, 14*, 1702479.

O'Connor, D. M., & Boulis, N. M. (2015). Gene therapy for neurodegenerative diseases. *Trends in Molecular Medicine, 21*, 504–512.

Park, S. I., Shin, G., McCall, J. G., Al-Hasani, R., Norris, A., Xia, L., Brenner, D. S., Noh, K. N., Bang, S. Y., Bhatti, D. L., et al. (2016). Stretchable multichannel antennas in soft wireless optoelectronic implants for optogenetics. *Proceedings of the National Academy of Sciences, 113*, E8169.

Park, S., Guo, Y., Jia, X., Choe, H. K., Grena, B., Kang, J., Park, J., Lu, C., Canales, A., Chen, R., et al. (2017). One-step optogenetics with multifunctional flexible polymer fibers. *Nature Neuroscience, 20*, 612–619.

Qazi, R., Kim, C. Y., Byun, S.-H., & Jeong, J.-W. (2018). Microscale inorganic LED based wireless neural systems for chronic in vivo optogenetics. *Frontiers in Neuroscience, 12*, 764.

Qazi, R., Gomez, A. M., Castro, D. C., Zou, Z., Sim, J. Y., Xiong, Y., & Byun, S. H. (2019). Wireless optofluidic brain probes for chronic neuropharmacology and photostimulation. *Nature biomedical engineering, 3*,(8), 655–669.

Rei, D., Mason, X., Seo, J., Gräff, J., Rudenko, A., Wang, J., Rueda, R., Siegert, S., Cho, S., Canter, R. G., et al. (2015). Basolateral amygdala bidirectionally modulates stress-induced hippocampal learning and memory deficits through a p25/Cdk5-dependent pathway. *Proceedings of the National Academy of Sciences of the United States of America, 112*, 7291–7296.

Rubehn, B., & Stieglitz, T. (2010). In vitro evaluation of the long-term stability of polyimide as a material for neural implants. *Biomaterials, 31*, 3449–3458.

Sevigny, J., Chiao, P., Bussière, T., Weinreb, P. H., Williams, L., Maier, M., Dunstan, R., Salloway, S., Chen, T., Ling, Y., et al. (2016). The antibody aducanumab reduces Aβ plaques in Alzheimer's disease. *Nature, 537*, 50–56.

Sheybani, R., Cobo, A., & Meng, E. (2015). Wireless programmable electrochemical drug delivery micropump with fully integrated electrochemical dosing sensors. *Biomedical Microdevices, 17*, 74.

Shin, G., Gomez, A. M., Al-Hasani, R., Jeong, Y. R., Kim, J., Xie, Z., Banks, A., Lee, S. M., Han, S. Y., Yoo, C. J., et al. (2017). Flexible near-field wireless optoelectronics as subdermal implants for broad applications in optogenetics. *Neuron, 93*, 509–521.e3.

Sim, J. Y., Haney, M. P., Park, S. I., McCall, J. G., & Jeong, J.-W. (2017). Microfluidic neural probes: In vivo tools for advancing neuroscience. *Lab on a Chip, 17*, 1406–1435.

Spieth, S., Schumacher, A., Kallenbach, C., Messner, S., & Zengerle, R. (2012). The NeuroMedicator—A micropump integrated with silicon microprobes for drug delivery in neural research. *Journal of Micromechanics and Microengineering, 22*, 065020.

Stachniak, T. J., Ghosh, A., & Sternson, S. M. (2014). Chemogenetic synaptic silencing of neural circuits localizes a hypothalamus→midbrain pathway for feeding behavior. *Neuron, 82*, 797–808.

Stamatakis, A. M., Jennings, J. H., Ung, R. L., Blair, G. A., Weinberg, R. J., Neve, R. L., Boyce, F., Mattis, J., Ramakrishnan, C., Deisseroth, K., et al. (2013). A unique population of ventral tegmental area neurons inhibits the lateral habenula to promote reward. *Neuron, 80*, 1039–1053.

Sternson, S. M., & Roth, B. L. (2014). Chemogenetic tools to interrogate brain functions. *Annual Review of Neuroscience, 37*, 387–407.

Szarowski, D. H., Andersen, M. D., Retterer, S., Spence, A. J., Isaacson, M., Craighead, H. G., Turner, J. N., & Shain, W. (2003). Brain responses to micro-machined silicon devices. *Brain Research, 983*, 23–35.

Takeuchi, S., Ziegler, D., Yoshida, Y., Mabuchi, K., & Suzuki, T. (2005). Parylene flexible neural probes integrated with microfluidic channels. *Lab on a Chip, 5*, 519–523.

Tochitsky, I., Banghart, M. R., Mourot, A., Yao, J. Z., Gaub, B., Kramer, R. H., & Trauner, D. (2012). Optochemical control of genetically engineered neuronal nicotinic acetylcholine receptors. *Nature Chemistry, 4*, 105–111.

Voskerician, G., Shive, M. S., Shawgo, R. S., von Recum, H., Anderson, J. M., Cima, M. J., & Langer, R. (2003). Biocompatibility and biofouling of MEMS drug delivery devices. *Biomaterials, 24*, 1959–1967.

Walsh, J. J., Friedman, A. K., Sun, H., Heller, E. A., Ku, S. M., Juarez, B., Burnham, V. L., Mazei-Robison, M. S., Ferguson, D., Golden, S. A., et al. (2014). Stress and CRF gate neural activation of BDNF in the mesolimbic reward pathway. *Nature Neuroscience, 17*, 27–29.

Williams, S. C. P., & Deisseroth, K. (2013). Optogenetics. *Proceedings of the National Academy of Sciences, 110*, 16287–16287.

Xiang, Z., Yen, S.-C., Xue, N., Sun, T., Tsang, W. M., Zhang, S., Liao, L.-D., Thakor, N. V., & Lee, C. (2014). Ultra-thin flexible polyimide neural probe embedded in a dissolvable maltose-coated microneedle. *Journal of Micromechanics and Microengineering, 24*, 065015.

Ziegler, D., Suzuki, T., & Takeuchi, S. (2006). Fabrication of flexible neural probes with built-in microfluidic channels by thermal bonding of parylene. *Journal of Microelectromechanical Systems, 15*, 1477–1482.

Chapter 14
Gold Nanomaterial-Enabled Optical Neural Stimulation

Yongchen Wang

14.1 Introduction

Conventional electrical stimulation delivers pregenerated electric current through wires to electrodes or microelectrode arrays interfacing the target neurons. Despite its effectiveness and wide use, it is limited by unsatisfactory spatial resolution and invasive surgeries. Noninvasive neural stimulation techniques were, therefore, exploited by directly delivering magnetic field (Hallett 2007), electric current (Paulus 2011), light (Thompson et al. 2014), or ultrasound (Tufail et al. 2010) wirelessly through tissues to the target neurons. These noninvasive platforms avoid surgeries but are still limited by spatial resolution. Optogenetics (Boyden et al. 2005), as an alternative, shows excellent spatiotemporal resolution, but it requires the genetic engineering of neurons which brings technical challenges, safety concerns, and also ethical issues. Thus, it is highly desired to deliver neural stimuli wirelessly, administer neural interfaces without surgeries, and improve the spatial resolution.

An innovative way to realize these goals is to make minimized injectable neural interfaces and target them to neurons as antennae to receive a wireless signal and convert it to a local stimulus capable of modulating neural activity. It is due to their small size, diverse novel physicochemical properties, and specific targeting capabilities that nanomaterials serve as good candidates for nano-antennae (Wang and Guo 2016; Wang et al. 2018). First, nanoparticles can be formulated to be injectable so that the administration of these nanosized interfaces can be significantly less invasive. Second, their unique physicochemical properties enable transductions from wireless signals (light, ultrasound, or magnetic field) to local stimuli (electric signal, heat, mechanical force, etc.) for neural stimulation (Table 14.1). The safety can also be improved by lowering the required power of the wireless signals (Thompson et al. 2014; Eom et al. 2016). Third, the nanoparticles can be distributed

Y. Wang (✉)
Department of Biomedical Engineering, The Ohio State University, Columbus, OH, USA
e-mail: wang.4896@osu.edu

© Springer Nature Switzerland AG 2020
L. Guo (ed.), *Neural Interface Engineering*,
https://doi.org/10.1007/978-3-030-41854-0_14

Table. 14.1 Transduction of nanomaterials from a wireless signal to a local stimulus for neural stimulation (Wang and Guo 2016; Wang et al. 2018)

Nanomaterial	Wireless signal	Local stimulus
Semiconducting nanomaterials	Light	Electric signal
Piezoelectric nanomaterials	Ultrasound	Electric signal
Gold nanomaterials	Light	Heat
Superparamagnetic nanoparticles	Magnetic field	Heat and/or mechanical force
Upconverting luminescent nanomaterials	Near-infrared light	Visible light for optogenetics

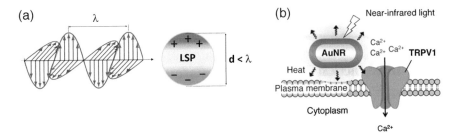

Fig. 14.1 LSPR, photothermal transduction, and their application in neural stimulation. (**a**) LSPR. The electric field of the incident light oscillates conduction electrons over the entire gold nanomaterial having a size (*d*) smaller than the LSPR wavelength (*λ*), generating heat via photothermal transduction. (The figure was adapted with copyright permission from Amendola et al. (2017), © IOP Publishing Ltd. 2017). (**b**) Under the irradiation of NIR light, the heat generated from gold nanorods (AuNRs) can alter plasma membrane's capacitance and/or activate temperature-sensitive ion channels, for example, TRPV1 channels, modulating neural activities (Shapiro et al. 2012; Moran et al. 2004; Nakatsuji et al. 2015). (The figure was adapted with copyright permission from Nakatsuji et al. (2015), © Wiley-VCH Verlag GmbH & Co. KGaA, Weinheim 2015)

in neurons' extracellular space, bound to plasma membrane, surface proteins, and ion channels, and internalized into cytoplasm, enabling cellular- and subcellular-level specificity and spatial resolution (Lavoie-Cardinal et al. 2016).

Among these nanomaterials, gold nanomaterials, which receive light and generate localized heat, are especially intriguing. Such photothermal transduction is due to LSPR (Fig. 14.1a) (Amendola et al. 2017), and the LSPR wavelength can be tailored by controlling the nanomaterials' shape and aspect ratio (Eustis and El-Sayed 2006). For example, cylindrical gold nanorods have near-infrared (NIR) longitudinal LSPR wavelengths within the biological window (650–900 nm) that has the least absorption by blood and water compared to visible and infrared light, so they can be applied to stimulate deeper tissues (Weissleder 2001). The heat generated from these nanosized interfaces activate temperature-sensitive ion channels and/or impact the plasma membrane's capacitance, thereby modulating neural activities (Fig. 14.1b) (Moran et al. 2004; Shapiro et al. 2012; Nakatsuji et al. 2015). Superparamagnetic nanoparticles convert magnetic field to local heat and/or mechanical force modulating neural activities. However, compared to superparamagnetic nanoparticles, gold nanomaterials have higher heat generation efficiency, so they require a significantly lower number of nanosized interfaces and do not require genetic engineering in target cells to express extra temperature-sensitive ion channels (Nakatsuji et al. 2015).

14.2   Critical Factors for Designing Gold Nanomaterial-Enabled Optical Neural Stimulation

There are several critical factors for designing gold nanomaterial-enabled optical stimulation. The general process is that gold nanomaterials targeted to neurons receive light and generate localized heat modulating neural activity. Thus, factors affecting the component(s) of the process should all be considered. First, the properties of gold nanomaterials, including the shape, aspect ratio, surface modification, amount, etc., will affect the LSPR wavelength, biocompatibility, placement, and stimulation outcome. Second, parameters of the laser, including the wavelength, continuous wave or pulsed mode, power, irradiance, duration, etc., will affect the light penetration depth, stimulation outcome, and safety. Moreover, the placement of gold nanomaterials is critical. Gold nanomaterials can be placed in neurons' extracellular space (placement 1 in Fig. 14.2a); immobilized onto a substrate (placement 2); bound to the plasma membrane (placement 3), surface proteins (placement 3'), and ion channels (placement 3''); and internalized into cytoplasm (placement 4) (Wang et al. 2018). Their placement will significantly influence the interface stability, biocompatibility, safety, spatiotemporal resolution, and stimulation outcome. Figure 14.2b gives an example of specific targeting of gold nanomaterials onto surface proteins (Eom et al. 2016), and Fig. 14.2c gives an example of immobilization of gold nanomaterials onto a substrate (Bazard et al. 2017). Furthermore, it is also critical to select the appropriate end points to assess the stimulation outcome, such as indirect end points (calcium transient (Fig. 14.3a) and neurite outgrowth (Fig. 14.3b)), and direct electrophysiological end points (membrane potential and action potential (Fig. 14.3c)). Finally, if the platform is effective for neural stimulation, the potential application on other excitable cells, such as muscle cells, should also be evaluated.

14.3   Development of Gold Nanomaterial-Enabled Neural Stimulation

Initially, dispersed and nonspecifically targeted gold nanomaterials were exploited for development of optical stimulation. Paviolo et al. showed that internalized gold nanorods coupled with continuous-wave laser irradiation of their LSPR wavelength promoted neurite outgrowth (Fig. 14.3b) and triggered intracellular calcium transients in NG108-15 neuronal cells (Paviolo et al. 2013a, b, 2014). Yong et al. found

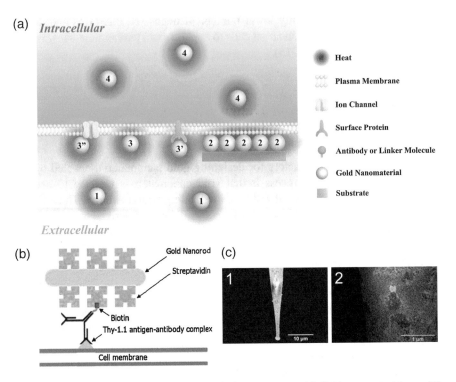

Fig. 14.2 Placement of gold nanomaterials relative to neurons. (**a**) Gold nanomaterials are (1) located in neurons' extracellular space; (2) immobilized onto a substrate; bound to (3) plasma membrane, (3') surface proteins, and (3") ion channels; and (4) located in neurons' cytoplasm (Wang et al. 2018). (The figure was adapted with copyright permission from Wang et al. (2018), © Tsinghua University Press and Springer-Verlag GmbH Germany, part of Springer Nature 2018). (**b**) Specific targeting of bio-conjugated gold nanomaterials to neurons' surface proteins and ion channels. For example, streptavidin-conjugated gold nanorods were specifically bound to Thy-1 antigen in the plasma membrane via its biotinylated antibody (Eom et al. 2016). (The figure was adapted with copyright permission from Eom et al. (2016), © OSA 2016). (**c**) Immobilization of gold nanomaterials onto a substrate. For example, gold nanoparticles were coated onto a glass micropipette via aminosilane linking molecules, as shown in scanning electron microscopy images of a lower magnification (**c-1**) and a higher magnification (**c-2**) (Bazard et al. 2017). (The figure was adapted with copyright permission from Bazard et al. (2017), © The authors 2017)

that gold nanorods coupled with laser pulses of their LSPR wavelength elicited action potentials in rat auditory neurons, while the gold nanorods were distributed extracellularly, on the plasma membrane, and intracellularly (Yong et al. 2014). Eom et al. also provided in vivo evidence that extracellularly distributed gold nanorods coupled with laser pulses of their LSPR wavelength evoked compound action potentials in rat sciatic nerves (Eom et al. 2014). However, the dispersed and nonspecifically targeted gold nanomaterials could be easily washed away by extracellular fluid or internalized into cells, making it difficult to maintain stable interfaces. As a result, the stimulation was inconsistent and variable and even induced cytotoxicity (Paviolo et al. 2013b; Yong et al. 2014).

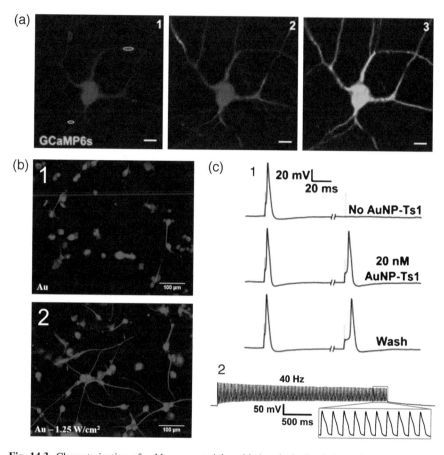

Fig. 14.3 Characterization of gold nanomaterial-enabled optical stimulation using calcium transient, neurite outgrowth, plasma membrane depolarization, and action potential. (**a**) Optical stimulation triggered calcium transients in hippocampal neurons (Lavoie-Cardinal et al. 2016). Gold nanoparticles were attached to neurons' plasma membrane, and an incident NIR laser triggered fluorescence signal increase of a genetically encoded calcium indicator (GCaMP6s) at time points 2 and 3 compared to time point 1 without NIR irradiation. (The figure was adapted with copyright permission from Lavoie-Cardinal et al. (2016), © The authors 2016). (**b**) Optical stimulation induced neurite outgrowth in NG108-15 neuronal cells (Paviolo et al. 2013b). NIR irradiation with an irradiance of 1.25 W/cm² promoted neurite outgrowth in gold nanorod (Au)-treated cells (**b-2**) compared to gold nanorod-treated cells without NIR irradiation (**b-1**). (The figures were adapted with copyright permission from Paviolo et al. (2013b), © Wiley Periodicals, Inc. 2013). (**c**) Optical stimulation evoked plasma membrane depolarization and action potentials in dorsal root ganglion (DRG) neurons (Carvalho-de-Souza et al. 2015). (**c-1**) Ts1 ligand-conjugated gold nanoparticles (AuNP-Ts1) were specifically targeted to the ion channels in DRG neurons. Action potential was evoked electrically (indicated by a blue bar) in both nontargeted and targeted cells, but action potential was evoked optically (indicated by a green bar) only in gold nanoparticle-targeted cells. The nanoparticle binding and optical stimulation were resistant to washing. (**c-2**) In gold nanoparticle-targeted cells, laser pulses of 40 Hz evoked a train of action potentials of the same frequency. (The figures were adapted with copyright permission from Carvalho-de-Souza et al. (2015), © Elsevier Inc. 2015)

Fig. 14.4 Inhibition of neural activity by photothermal effect of gold nanorods under NIR light. Positively charged gold nanorods were electrostatically attached to neurons' plasma membrane, and a train of neural spikes showed inhibition of neural activity during irradiation and restoration upon the removal of irradiation (Yoo et al. 2014). (The figure was adapted with copyright permission from Yoo et al. (2014), © American Chemical Society 2014)

Dispersed gold nanomaterials could be surface modified to be positively charged to bind to neurons' plasma membrane. Interestingly, an excitatory or inhibitory effect was achieved in different works. In the work of Yoo et al., positively charged gold nanorods electrostatically attached to the plasma membrane of rat hippocampal, cortical, and olfactory bulb neurons inhibited spontaneous, electrically stimulated, and epileptic neural activities upon irradiation of laser pulses of the LSPR wavelength, and the neural activities were reinstated upon the removal of irradiation (Fig. 14.4) (Yoo et al. 2014). However, in the work of Nakatsuji et al., the cationic lipoprotein-coated gold nanorods bound to the plasma membrane triggered calcium influxes in HEK293T cells by activating vanilloid receptor 1 under continuous-wave laser (Fig. 14.3a) (Nakatsuji et al. 2015). Meanwhile, they addressed the importance of the surface chemistry of gold nanorods by showing that the cationic polymer-coated gold nanorods triggered calcium influxes through disruption to the plasma membrane.

Bio-conjugated gold nanomaterials could be specifically targeted to the surface proteins and ion channels in the plasma membrane to improve interface stability, stimulation consistency, and spatial resolution. In another work of Eom et al., streptavidin-conjugated gold nanorods were specifically targeted to a plasma membrane protein via its biotinylated antibody (Fig. 14.2b), and laser pulses of the LSPR wavelength triggered action potentials in rat hippocampal neurons (Eom et al. 2016). Compared to NIR stimulation alone, NIR stimulation coupled with these specifically targeted gold nanorods required lower irradiant exposure and showed shorter latency. When these gold nanorods were injected to rat motor cortex, NIR stimulation triggered whisker movement. In the work of Carvalho-de-Souza et al., ligand-conjugated gold nanoparticles were specifically bound to the ion channels in the plasma membrane of rat DRG neurons, and a train of action potentials was evoked upon exposure to the pulsed laser of the LSPR wavelength at the same frequency of the laser pulses (Fig. 14.3c) (Carvalho-de-Souza et al. 2015). This strategy also induced membrane depolarization in acute mouse hippocampal slices ex vivo. In addition, compared to unconjugated counterparts, these ligand-conjugated gold nanoparticles were bound more stably and activated neurons at a lower concentration (Fig. 14.3c).

Other than the LSPR wavelength, an off-resonance wavelength or the double-resonance wavelength can also be applied for optical stimulation with specifically targeted gold nanoparticles. In the work of Lavoie-Cardinal et al., citrate-capped gold nanoparticles were passively attached to the plasma membrane of rat hippocampal neurons, and anti-hemagglutinin antibody-conjugated gold nanoparticles were specifically targeted to the receptors in the plasma membrane (Lavoie-Cardinal et al. 2016). Upon irradiation of laser pulses of an off-resonance NIR wavelength, both placement methods triggered calcium transients and action potentials, but specifically targeted gold nanoparticles reduced nanoparticle aggregation and improved the stimulation reproducibility and specificity. In their work, a high spatial stimulation resolution of several square microns was achieved around an individual nanoparticle or a small group of nanoparticles. In the work of de Boer et al., streptavidin-conjugated gold nanoparticles were targeted to the sugar moieties and proteins in the plasma membrane, and irradiation of laser pulses of the double-resonance wavelength in the NIR range evoked action potentials in acute slices of mouse cortex in vitro and in cortical neurons of mouse visual cortex in vivo and triggered *Hydra* contraction in vivo (de Boer et al. 2018).

Even for the specifically targeted dispersed gold nanomaterials, it is still challenging to maintain stable interfaces regarding their locations relative to the neurons for a sustained period of time (Lavoie-Cardinal et al. 2016). This may even bring further safety concern related to their uncontrolled distribution for in vivo application (Bazard et al. 2017). Besides, gold nanomaterials are not biodegradable, and it is difficult to retrieve them after administration. In order to maintain stable interfaces and increase their retrievability and safety, gold nanomaterials were immobilized to substrates, and the coated substrates could serve as neurodevices for neural modulation (Yoo et al. 2016; Bazard et al. 2017; Lee et al. 2018). Yoo et al. reported that neurons were cultured on the gold nanorod-coated microelectrode arrays and their action potentials could be excited by electric current and inhibited by the photothermal effect of gold nanorods under laser irradiation (Yoo et al. 2016). The gold nanorods and NIR laser combined treatment could inhibit the spontaneous action potentials and the conduction of electrically stimulated action potentials via axons. Lee et al. used gold nanostars as the nano-antennae and demonstrated their use in neural inhibition on rat hippocampal neurons at both the single cell level and network level when coupled with continuous-wave lasers of the LSPR wavelength (Lee et al. 2018). In this work, the gold nanostars were applied to neurons either in their dispersed form or by being immobilized onto the substrate. In both forms, gold nanostars showed improved biocompatibility compared to gold nanorods. In the work of Bazard et al., gold nanoparticles were coated onto a glass micropipette (Fig. 14.2c), and when combined with laser pulses of the LSPR wavelength, it could excite or inhibit the action potentials in SH-SY5Y cells and primary cardiomyocytes depending on the duration of the laser pulse (Bazard et al. 2017). In this work, the excitation and inhibition were studied on single action potentials. Immobilization of gold nanoparticles/nanostars onto a substrate could improve their biocompatibility to neurons (Tran et al. 2017; Lee et al. 2018), enhance the retrievability and safety, and, more importantly, enable control of excitatory or inhibitory effect within the same stimulation platform. Nevertheless, it will also increase the invasiveness associated with the administration of the coated substrate. Another challenge is that only the neurons in the vicinity of the coated substrate can be modulated, so the target neuron population will be significantly limited by the physical geometry of the substrate.

14.4 Challenges and Opportunities

Despite the fact that gold nanomaterial-enabled optical neural stimulation is promising to improve the spatial resolution and lower the invasiveness, it still faces multiple challenges. First, although NIR light penetrates tissues deeper than visible light, the penetration depth of the NIR light within the biological window (650–950 nm) can be only as deep as 1–2 cm (Gao et al. 2004). Such a penetration depth can be very challenging for deep tissue stimulation, for example, transcranial stimulation. In this case, implantation of the light source may still be needed, but this will increase invasiveness. Second, although efforts were made to investigate how the stimulation outcome depended on the laser pulse duration, irradiance, or irradiant exposure in separate works (Yoo et al. 2016; Bazard et al. 2017), no conclusion could universally explain existing findings (Wang et al. 2018). This may be due to the fact that laser is only one of the critical factors, and other factors as discussed in Sect. 14.2 should also be considered. Additionally, the molecular mechanism of how localized heat modulates neural activity remains to be better understood. To realize successful photothermal modulation of neural activity, sufficient heat needs to be generated in the vicinity of the target neurons (Bazard et al. 2017). Thus, the properties, amount, and placement of the nanosized interfaces, the photothermal transduction efficiency, etc. could all affect the modulation outcome. In addition, when assessing the modulation outcome, researchers should be careful, as nanomaterials crossing membrane could cause membrane depolarization that could be falsely regarded as activation (Nakatsuji et al. 2015). Another challenge is that the triggered calcium influxes could induce stress response under an irradiant exposure of 17–51 mJ/cm^2 (Johannsmeier et al. 2018). This could be particularly concerning for long-term repetitive stimulation, as the majority of the studies used a higher irradiant exposure (Wang et al. 2018). For clinical application, when applied on tissues instead of cells, the technique even required notably higher laser powers (Carvalho-de-Souza et al. 2015). Overall, critical factors should be considered to address these challenges in order to develop an effective and safe wireless stimulation platform with good temporal and spatial resolutions. Other than its use in neural stimulation, gold nanomaterial-enabled optical stimulation can also stimulate other cell types, such as astrocytes and muscle cells (Eom et al. 2017; Marino et al. 2017).

References

Amendola, V., Pilot, R., et al. (2017). Surface plasmon resonance in gold nanoparticles: A review. *Journal of Physics: Condensed Matter, 29*(20), 203002.

Bazard, P., Frisina, R. D., et al. (2017). Nanoparticle-based plasmonic transduction for modulation of electrically excitable cells. *Scientific Reports, 7*(1), 7803.

Boyden, E. S., Zhang, F., et al. (2005). Millisecond-timescale, genetically targeted optical control of neural activity. *Nature Neuroscience, 8*(9), 1263–1268.

Carvalho-de-Souza, J. L., Treger, J. S., et al. (2015). Photosensitivity of neurons enabled by cell-targeted gold nanoparticles. *Neuron, 86*(1), 207–217.

de Boer, W. D. A. M., Hirtz, J. J., et al. (2018). Neuronal photoactivation through second-harmonic near-infrared absorption by gold nanoparticles. *Light: Science & Applications, 7*(1), 100.

Eom, K., Kim, J., et al. (2014). Enhanced infrared neural stimulation using localized surface plasmon resonance of gold nanorods. *Small, 10*(19), 3853–3857.

Eom, K., Im, C., et al. (2016). Synergistic combination of near-infrared irradiation and targeted gold nanoheaters for enhanced photothermal neural stimulation. *Biomedical Optics Express, 7*(4), 1614–1625.

Eom, K., Hwang, S., et al. (2017). Photothermal activation of astrocyte cells using localized surface plasmon resonance of gold nanorods. *Journal of Biophotonics, 10*(4), 486–493.

Eustis, S., & El-Sayed, M. A. (2006). Why gold nanoparticles are more precious than pretty gold: Noble metal surface plasmon resonance and its enhancement of the radiative and nonradiative properties of nanocrystals of different shapes. *Chemical Society Reviews, 35*(3), 209–217.

Gao, X., Cui, Y., et al. (2004). In vivo cancer targeting and imaging with semiconductor quantum dots. *Nature Biotechnology, 22*(8), 969–976.

Hallett, M. (2007). Transcranial magnetic stimulation: A primer. *Neuron, 55*(2), 187–199.

Johannsmeier, S., Heeger, P., et al. (2018). Gold nanoparticle-mediated laser stimulation induces a complex stress response in neuronal cells. *Scientific Reports, 8*(1), 6533.

Lavoie-Cardinal, F., Salesse, C., et al. (2016). Gold nanoparticle-assisted all optical localized stimulation and monitoring of Ca2+ signaling in neurons. *Scientific Reports, 6*, 20619.

Lee, J. W., Jung, H., et al. (2018). Gold nanostar-mediated neural activity control using plasmonic photothermal effects. *Biomaterials, 153*, 59–69.

Marino, A., Arai, S., et al. (2017). Gold nanoshell-mediated remote myotube activation. *ACS Nano, 11*(3), 2494–2508.

Moran, M. M., Xu, H., et al. (2004). TRP ion channels in the nervous system. *Current Opinion in Neurobiology, 14*(3), 362–369.

Nakatsuji, H., Numata, T., et al. (2015). Thermosensitive ion channel activation in single neuronal cells by using surface-engineered plasmonic nanoparticles. *Angewandte Chemie International Edition, 54*(40), 11725–11729.

Paulus, W. (2011). Transcranial electrical stimulation (tES-tDCS; tRNS, tACS) methods. *Neuropsychological Rehabilitation, 21*(5), 602–617.

Paviolo, C., Haycock, J. W., et al. (2013a). Plasmonic properties of gold nanoparticles can promote neuronal activity. In *SPIE BiOS*. SPIE.

Paviolo, C., Haycock, J. W., et al. (2013b). Laser exposure of gold nanorods can increase neuronal cell outgrowth. *Biotechnology and Bioengineering, 110*(8), 2277–2291.

Paviolo, C., Haycock, J. W., et al. (2014). Laser exposure of gold nanorods can induce intracellular calcium transients. *Journal of Biophotonics, 7*(10), 761–765.

Shapiro, M. G., Homma, K., et al. (2012). Infrared light excites cells by changing their electrical capacitance. *Nature Communications, 3*, 736.

Thompson, A. C., Stoddart, P. R., et al. (2014). Optical stimulation of neurons. *Current Molecular Imaging, 3*(2), 162–177.

Tran, A. Q., Kaulen, C., et al. (2017). Surface coupling strength of gold nanoparticles affects cytotoxicity towards neurons. *Biomaterials Science, 5*(5), 1051–1060.

Tufail, Y., Matyushov, A., et al. (2010). Transcranial pulsed ultrasound stimulates intact brain circuits. *Neuron, 66*(5), 681–694.

Wang, Y., & Guo, L. (2016). Nanomaterial-enabled neural stimulation. *Frontiers in Neuroscience, 10*, 69.

Wang, Y., Zhu, H., et al. (2018). Nano functional neural interfaces. *Nano Research, 11*(10), 5065–5106.

Weissleder, R. (2001). A clearer vision for in vivo imaging. *Nature Biotechnology, 19*(4), 316–317.

Yong, J., Needham, K., et al. (2014). Gold-nanorod-assisted near-infrared stimulation of primary auditory neurons. *Advanced Healthcare Materials, 3*(11), 1862–1868.

Yoo, S., Hong, S., et al. (2014). Photothermal inhibition of neural activity with near-infrared-sensitive nanotransducers. *ACS Nano, 8*(8), 8040–8049.

Yoo, S., Kim, R., et al. (2016). Electro-optical neural platform integrated with nanoplasmonic inhibition interface. *ACS Nano, 10*(4), 4274–4281.

Chapter 15
Nanomaterial-Assisted Acoustic Neural Stimulation

Attilio Marino, Giada Graziana Genchi, Marietta Pisano, Paolo Massobrio, Mariateresa Tedesco, Sergio Martinoia, Roberto Raiteri, and Gianni Ciofani

15.1   Introduction

The word "piezoelectricity" derives from the ancient Greek, and it literally means "electricity resulting from pressure". Piezoelectric materials, indeed, generate an electric potential when subjected to a mechanical strain (direct piezoelectric effect); conversely, the application of an electric field to a piezo-material induces its own deformation (reverse piezoelectric effect). Since the discovery of piezoelectricity in the late nineteenth century, piezoelectric materials have been exploited in several different devices, for applications ranging from biomedical (e.g., ultrasonic imaging systems) to automotive fields (e.g., air-bag sensors) (Marino et al. 2017).

The rapid development of nanotechnology has reshaped our knowledge in the fields of physics, chemistry, material science, and biology. Nanomedicine refers to the biomedical application of nanomaterials and is one of the branches of nanotechnology, which has been mostly influenced in the course of this scientific/

A. Marino (✉) · G. G. Genchi
Istituto Italiano di Tecnologia, Smart Bio-Interfaces, Pisa, Italy
e-mail: attilio.marino@iit.it

M. Pisano · P. Massobrio · M. Tedesco
Università di Genova, Department of Informatics, Bioengineering, Robotics and System Engineering, Genoa, Italy

S. Martinoia · R. Raiteri
Università di Genova, Department of Informatics, Bioengineering, Robotics and System Engineering, Genoa, Italy

CNR, Institute of Biophysics, Genoa, Italy

G. Ciofani (✉)
Istituto Italiano di Tecnologia, Smart Bio-Interfaces, Pisa, Italy

Politecnico di Torino, Department of Aerospace & Mechanical Engineering, Torino, Italy
e-mail: gianni.ciofani@iit.it

© Springer Nature Switzerland AG 2020
L. Guo (ed.), *Neural Interface Engineering*,
https://doi.org/10.1007/978-3-030-41854-0_15

347

technological revolution (Zhang et al. 2008). Researchers working in nanomedicine recently developed a variety of non-invasive and biocompatible tools capable of remotely delivering specific physical and chemical stimuli in the deep tissue, at single cell level or even with subcellular resolution (Genchi et al. 2017).

In this context, piezoelectric nanomaterials represent a class of nanotransducers, both organic and inorganic, that can be exploited not only for the remote excitation of the neural cells, but, more in general, for the stimulation of the electrically excitable cells, such as cardiomyocytes, osteoblasts, and skeletal myotubes (Marino et al. 2017). Specifically, piezoelectric nanomaterials can be activated with different mechanical energy sources, such as vibrations and acoustic pressure waves in the audible (sound) or non-audible (ultrasound, US) frequencies (Royo-Gascon et al. 2013; Inaoka et al. 2011; Wang et al. 2007).

Concerning US, these pressure waves deeply and safely penetrate soft biological tissues, and are clinically exploited for diagnostic purposes (e.g., sonography). Efficient piezoelectric nanotransducers, such as the ones characterized by barium titanate, are able to generate electric potentials in the order of millivolt and remotely activate neural cells when exposed to US intensities similar to the ones used for sonography (Marino et al. 2015a). Moreover, US waves can be focused into deep tissues through hyperlenses in order to maximize the US intensity in a specific region of the tissue (Zhang et al. 2009). For these reasons, US represents an ideal and safe source of mechanical energy that can be efficiently transduced into biologically relevant electrical cues.

In addition to US, the piezoelectric neural stimulation mediated by acoustic waves in the audible frequencies has been extensively investigated; such studies have been carried out in order to develop a new generation of single-component cochlear implants able to transduce the mechanical waves of sound into electrical signals for the stimulation of the spiral ganglion neurons. These piezoelectric devices have been designed for substituting the functions of the cochlear sensory epithelium, which are compromised in certain types of deafness (Inaoka et al. 2011).

Piezoelectric materials can be exploited not only as actuators for indirect electric stimulation, but, taking advantage of the reverse piezoelectric effect, also as sensors to detect, measure, and accumulate the biomechanical energy developed by single cells and tissues (Nguyen et al. 2012). In this technological framework, piezoelectric nanostructured devices have been incorporated in artificial pacemakers to obtain self-powered battery-free cardiac stimulation systems (Hwang et al. 2015a).

In this chapter, the piezoelectric nanomaterials that have been adopted in the biomedical field will be described and the biological effects of the acute and chronic nanoparticle-assisted piezo-stimulation on neural cells will be reported. Moreover, we will provide a chronological overview of the discovery of the neural activation with this indirect electric stimulation approach, starting from the first experimental evidences to the recent electrophysiological proofs obtained on primary neurons. Finally, in vivo exploitation of the piezo-stimulation approach for activation of the neural system will be presented in the last section.

15.2 Piezoelectric Nanostructured Materials Applied to Nanomedicine

The generation of small electric charges upon the application of mechanical stimuli to piezoelectric nanomaterials is a unique phenomenon in the context of remote stimulation of cells and tissues. Electrical cues are known to foster specific biological responses, and piezoelectric nanomaterials own the ability to act as real "nanotransducers", thus allowing obtaining "wireless" and remote electric stimulation thanks to non-invasive excitation through mechanical sources (usually US or vibrations).

Inorganic piezoelectric nanomaterials can be ceramic or polymeric, with piezoelectric nanoceramics usually showing higher piezoelectric features than polymers. Perovskites (like barium titanate and lead zirconium titanate) and wurzites (like zinc oxide and zinc sulfide) are among the mostly investigated piezoelectric nanoparticles.

Concerning piezoelectric polymers, poly(vinylidene difluoride) (PVDF) and its copolymers show the best piezoelectric features and have been widely investigated to promote cell stimulation, for example on rat spinal cord neurons (Royo-Gascon et al. 2013) and on human adipose tissue-derived stem cells (Ribeiro et al. 2015).

Boron nitride nanotubes (BNNTs), inorganic nanomaterials with structural affinity to carbon nanotubes, have been tested, among others, by our group, and showed beneficial effects as nanotransducers on PC12 neuron-like cells (Ciofani et al. 2013) and on pre-osteoblast human cells (Genchi et al. 2018). In the latter example, they have been used as nanofillers in P(VDF-TrFE)-based scaffolds stimulated with ultrasounds.

Other studies of ours also provided the first direct evidences of piezoelectric stimulation of cell cultures mediated by barium titanate nanoparticles (BTNPs): in particular, experiments have been performed on SH-SY5Y human neuroblastoma cells in the presence of nanoparticles and stimulated with US (Marino et al. 2015a). In this work, BTNPs owing tetragonal crystalline structure, and thus piezoelectric, were tested to demonstrate neuronal stimulation, whereas nanoparticles with cubic crystalline structure (and thus non-piezoelectric) were used as a negative control. A physical model has also been developed to corroborate obtained findings.

Barium titanate nanoparticles, analogously to BNNTs, have been further exploited as fillers to improve piezoelectric properties of scaffolds, giving interesting results on both neuron-like cells (Genchi et al. 2016a) and on human pre-osteoblasts (Marino et al. 2015b).

Finally, it is worth to mention the potentialities of piezoelectric stimulation of cancer cells. It is in fact well-known, as low-intensity electric stimulation represents an alternative treatment able to affect cancer cells without the use of any drugs/chemicals, and to significantly enhance the effects of chemotherapy by reducing multidrug resistance with the impairment of the plasma membrane translocation of P-glycoprotein (P-gp), encoded by the MDR1 gene, the overexpression of which is associated with chemotherapy resistance. Recently, our group provided the first evidences of the efficacy of this antitumor approach mediated by piezoelectric BTNPs

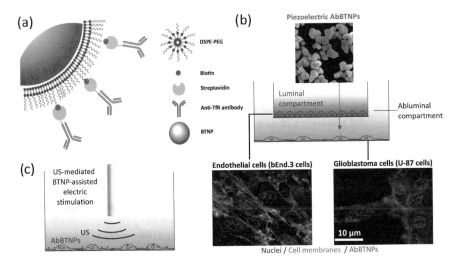

Fig. 15.1 Example of piezoelectric stimulation of glioblastoma cancer cells (U87). BTNPs have been functionalized (**a**) in order to promote the crossing of an in vitro blood-brain barrier model (**b**). Synergic effects of a chemotherapy drug (temozolomide) and indirect electric stimulation promoted apoptosis and reduced proliferation (**c**). (Reproduced with permission from Marino et al. (2019); Copyright Elsevier)

and US, respectively, on breast cancer (Marino et al. 2018) and on glioblastoma multiforme cells (Marino et al. 2019) (Fig. 15.1).

15.3 Nanoparticle-Assisted Piezoelectric Stimulation of Neural Cells

15.3.1 Wireless Nanoparticle-Assisted Modulation of Electrophysiological Activity

Different strategies have been proposed and implemented not only to regulate gene expression and drive cell differentiation, but also to induce short-term effects, in particular to modulate neuronal electrical activity both at the single cell and network level. A neuronal network can be considered as a complex, highly interconnected circuit where signaling is based on the collective effects of electric charges, neurotransmitters, and action potentials.

Stimulation of excitable cells can be provided using electrodes both in vitro and in vivo, and deep electric stimulation plays an important role in many medical treatments for different pathological conditions (e.g., Parkinson's disease, cardiac arrhythmia, chronic pain) (Lonzano et al. 2019; Luan et al. 2014; Keifer et al. 2014). Nanomaterial-based coatings of electrodes and nanostructured electrode surfaces have been shown to improve electrode/cell coupling (lower impedance, higher charge injection capability) and, consequently better stimulation performances in terms of smaller applied voltages, lower power losses and, consequently, less tissue perturbation/damage.

Non-invasive in vivo neural stimulation represents the most effective solution for restoring lost neural functions and correcting neurological disorders in several diseases. Moreover, in vitro neuromodulation allows investigating a wide range of complex phenomena, from neural development to synaptic plasticity. A wireless, spatially resolved and "steerable" stimulation technology would represent a major advancement for in vitro experiments as well (Wang and Guo 2016).

Electromagnetic fields and acoustic waves have both been shown to elicit neuronal responses in vivo and in vitro; however, they both allow poor spatial resolution when targeting regions in the deep brain. In this context, piezoelectric nanoparticles can be used as a localized transducer that is remotely driven by an acoustic control signal and turn it into a suitable neuronal stimulus with subcellular spatial resolution and response time in the millisecond range (Rojas et al. 2018). Cell-type specificity can also be achieved because nanomaterials can be surface-modified and bio-conjugated (Marino et al. 2019).

15.3.2 Ultrasonic Fields for Neuromodulation

US can propagate as compression/rarefaction longitudinal waves in gases, liquids, and biological soft tissues. Soft biological tissues, which contain a large amount of water, behave as liquids, from the point of view of sound propagation; harder tissues, such as bones, allow propagation of shear waves. While propagating, US interacts with matter either by reflection, absorption, or scattering, as electromagnetic waves, even if at a much lower rate. Such interactions attenuate the US intensity and release thermal energy that can locally increase the temperature, generate a stream of fluid, and cause the fast expansion/shrinking of gaseous microbubbles in the fluid (acoustic cavitation). The amount and the type of interaction depend on wave parameters (frequency, wavelength, speed of propagation, and intensity) as well as properties of the liquid or tissue (density and elastic modulus).

Effects of US propagation on biological tissues are at the basis of US medical diagnostic imaging; more in general, effects of US pressure waves on biological tissues, have been studied for safety (US dosimetry (O'Brien Jr. 2007)), therapeutic (tissue ablation (Hesley et al. 2013)), local drug delivery (Carpentier et al. 2016), thrombolysis (Bader et al. 2016), as well as imaging purposes (e.g., US localization microscopy). An extensive description of such reciprocal interactions can be found in Dalecki (2004), while Maresca and colleagues provided a recent review of US biophysics at cellular and molecular levels (Maresca et al. 2018).

For neuromodulation, short pulses at low amplitudes are used in order to minimize thermal effects (i.e., heating) and provide mechanical actuation through radiation force avoiding cavitation. Several in vivo and in vitro studies showed either excitatory or inhibitory neuronal responses to directly applied US. Such results have been recently reviewed (Naor et al. 2016; Blackmore et al. 2019).

Although many models describing the mechanism at the basis of neural response have been proposed, a comprehensive understanding based on experimental evidences requires further investigations on US biophysics at the single cell and single channel level. Such understanding would allow exploiting US-protein interactions in order to increase sensitivity and selectivity to US stimulation by genetically modifying selected neurons to overexpress mechanoreceptors. Such interesting approach, for which the proposing investigators coined the term sonogenetics, has been demonstrated for invertebrates (Kubanek et al. 2018); however, a proof of principle employing mechanoreceptors suitable for mammalians is still to be provided.

Coupling nanomaterials to transduce the US primary stimulus has the potential advantage of a better-defined and, consequently, better-controlled stimulation mechanism, which renders the stimulation more selective.

15.3.3 Indirect Proofs and First Demonstration of Neural Activation upon Piezoelectric Nanoparticle-Assisted Stimulation

Demonstration of the nanoparticle-assisted piezoelectric stimulation was primarily complicated by the vibrations of the electrophysiological electrodes used for monitoring the neural activity during the exposure to acoustic waves. All the traditional approaches of electrophysiological recording, such as intracellular, extracellular, and patch clamp whole cells, suffer from this technical issue.

For this reason, the first proofs of neural stimulation with this approach were fundamentally indirect. Specifically, our group firstly investigated the in vitro development of a neural network during chronic piezoelectric stimulation (Ciofani et al. 2013). In this pioneering work, PC-12 neural-like cells were incubated with BNNTs, and the axonal outgrowth of these cells was monitored in concomitance with the chronic US stimulation ("BNNTs+US"). Results were compared with those obtained from cells incubated with BNNTs but not exposed to US ("BNNTs"), from cells exposed to US without BNNTs ("US"), and, finally, from negative control cultures ("Control"). Interestingly, the combined piezoelectric "BNNTs+US" treatment was able to remarkably promote the development of the neural network with respect to the other experimental classes ("BNNTs", "US", "Control"), both in terms of increased percentage of differentiated cells (+ 15–20%) and of enhanced number and length of β3 tubulin-positive neurites. No significant biological effects were found between the different control conditions ("BNNTs", "US", "Control"). Interestingly, the stimulation tests performed in the presence of non-specific blockers of Ca^{2+} channels (i.e., lanthanum ions ($LaCl_3$)) indicated as the enhanced neural differentiation induced by the chronic "BNNTs+US" stimulation was mediated by the Ca^{2+} influx; this result enforced the hypothesis of an effective indirect electrical stimulation since the intracellular Ca^{2+} elevations are required for the development of PC12 neurites during electric stimulation (Manivannan and Terakawa 1993). An increased axonal outgrowth was also observed in our work (Ciofani et al. 2013), when piezoelectrically stimulating ("BNNTs+US") SH-SY5Y cells, therefore highlighting a good versatility of this nanotechnology-based approach.

Other indirect experimental evidences of the efficacy of the piezoelectric stimulation on neural cells were subsequently collected by other independent groups by using piezoelectric membranes, films, microfibers, and nanofibers (Lee and Arinzeh 2012; Genchi et al. 2016a). As an example, the group of William Craelius developed piezoelectric PVDF films able to transduce mechanical vibration (50 Hz frequency) into oscillating electrical fields; thanks to this indirect electric stimulation, Craelius's group was able to promote the neurite outgrowth in rat spinal cord neurons (Royo-Gascon et al. 2013).

The first direct demonstration of the neural cell activation in response to nanomaterial-assisted piezo-stimulation was subsequently obtained by our group in BTNP-treated SH-SY5Y-derived neurons thanks to Ca^{2+} imaging investigations (Marino et al. 2015a). At 24 h of BTNP incubation, nanoparticles resulted mostly associated to the plasma membranes of these neurons, both at the level of cell bodies and of the neurites. After an acute 5 s exposure to US stimulation, high-amplitude Ca^{2+} waves were evoked only in SH-SY5Y-derived neurons that were previously incubated with piezoelectric BTNPs (tetragonal crystal); no high-amplitude Ca^{2+} waves were observed in US-stimulated cells that were not pre-incubated with BTNPs or that were pre-incubated with non-piezoelectric BTNPs (cubic crystal). This experimental evidence indicated that the cell activation was mediated by the piezoelectricity of the material and not by other non-specific phenomena (e.g., mechanical and thermal). Coherently, the stimulation in the presence of gentamicin, a blocker of mechano-sensitive cation channels, was not able to affect cell activation. The evoked high-amplitude Ca^{2+} waves resulted both tetrodotoxin (TTX) and cadmium (Cd^{2+}) sensitive, therefore indicating as the opening of voltage-gated Na^+ and Ca^{2+} membrane channels was involved, respectively. Finally, experimental evidences were further corroborated by an electroelastic model of the voltage generated by BTNPs when exposed to different US intensities; the generated voltages are in the order of millivolt, values compatible to the ones required for the activation of voltage-sensitive channels.

15.3.4 Electrophysiological Recording of Primary Cultures: Our Results

Despite the difficulties related to electrophysiological recording using tip electrodes, planar multielectrode arrays (MEA) did not result significantly affected during US exposure. In this framework, it was possible to reversibly induce an excitatory

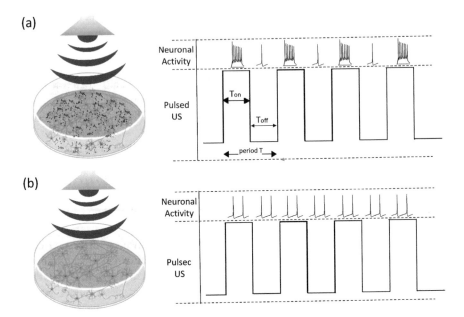

Fig. 15.2 Piezoelectric BTNP-mediated ultrasound stimulation. (**a**) In the presence of piezoelectric BTNPs and low-intensity US, electrical activity of primary neurons is significantly higher with respect to the spontaneous one. (**b**) The US stimulation has no effect on the electrical activity of neurons cultured without BTNPs

response on in vitro networks of hippocampal or cortical neurons with low-intensity ultrasonic pulses in the presence of piezoelectric BTNPs (Rojas et al. 2018). The sketch in Fig. 15.2 describes the working principle of the method: in the presence of piezoelectric nanoparticles (NPs) adsorbed onto the cell membrane, neurons irradiated with an US pulse modify their spontaneous electrical activity. Without NPs or with non-piezoelectric NPs, the same US stimulus does not affect the electrophysiology of the neurons. In the following, we describe the developed procedures for the electrophysiological recording during piezo-stimulation and the main obtained results.

Neurons were extracted and dissociated from the cortex and hippocampus of rat embryos (E18), plated onto a commercial MEA (Multichannel Systems MCS GmbH), and kept in incubator for 20–30 days in order to allow the development and maturation of an interconnected and spontaneously active neuronal network. With the MEA device the correlated electrophysiological activity was easily recorded. Sample preparation, set-up description, and the experimental procedure were previously described (Rojas et al. 2018). Briefly, BTNPs were dispersed in culture medium 12 h before the experiment; BTNPs unspecifically bound the plasma membrane of the cell bodies and of the neurites, as observed by confocal microscopy. In control experiments, primary cultures on MEA were incubated with BTNPs characterized by a cubic crystal structure, thus not showing piezoelectric behavior. Each circular electrode (30 μm in diameter) of the array records the extracellular field potential generated by the cells near it (typically one to three cells). In order to verify the capability of the network to respond to external stimuli and maintain such capability after US exposure, a standard electrical stimulation protocol was performed before and after the US stimulation experiments.

US at 1 MHz frequency and relatively low intensity (~1 W/cm²) was generated using a KTAC-4000 system (Sonopore) and transmitted to the MEA chamber through a thin thermoplastic film coated with a layer of acoustic gel. The pressure field generated by the piezoelectric transducer was experimentally characterized using a chamber with a miniaturize hydrophone (Teledyne RESON, model TC4038). US stimulation consisted of a sequence of US pulses of the same duration as schematically depicted in Fig. 15.2; such periodic stimulation pattern is usually defined by its period and its duty cycle.

Figure 15.3 reports the result of a typical US stimulation experiment in the presence of piezoelectric BTNPs. Raw MEA recordings were processed by using a spike detection algorithm in order to extract the neuronal firing activity. Raster-plot representing with single points each detected spike as a function of the recording electrode (numbered from 1 to 60, in the vertical axis) and of time (horizontal axis) is shown in Fig. 15.3a. Figure 15.3b shows a magnification of the raster plot during piezo-stimulation. The red lines indicate the switching on and off of the US generator. The black trace in Fig. 15.3b represents the trend over time of the instantaneous firing rate, averaged over the 60 electrodes. The raster plot also shows spontaneous

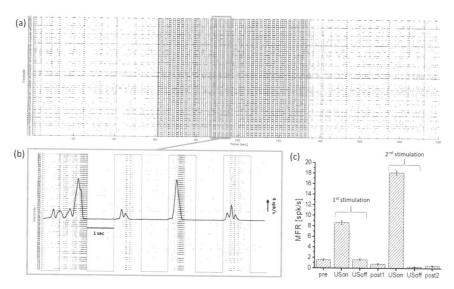

Fig. 15.3 MEA recording of the electrical activity in primary neurons during BTNP-assisted and US-driven piezoelectric stimulation. (**a**) Raster plot recorded from 60 microelectrodes before, during, and after US stimulation. The red line indicates when the US is switched on and off (single pulse duration = 1 s, pulse frequency = 0.5 Hz). (**b**) Raster plot during the train of US stimulations; the superimposed black trace represents the trend over time of the instantaneous firing rate. (**c**) Mean firing rate (MFR) calculated before, during, and after two successive trains of US stimulations

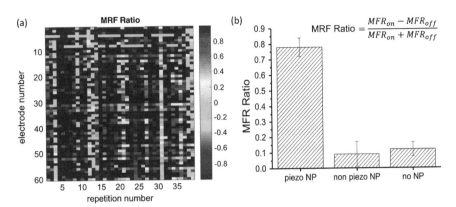

Fig. 15.4 Analysis of the MFR ratio in response to piezoelectric stimulation. (**a**) MFR ratio measured by the recording electrodes of the array in response to different US pulse repetitions in primary neurons pre-incubated with piezoelectric BTNPs. (**b**) Average MFR ratio values obtained by stimulating the primary neurons pre-incubated with piezoelectric BTNPs, pre-incubated with non-piezoelectric BTNPs, or non-incubated with BTNPs

activity before and after the US stimulation. It is possible to appreciate an increase in the firing activity over all the recording area during the stimulation, and, in particular, when the US is switched on (black trace in Fig. 15.3b).

In order to quantify the variation in electrical activity, we calculated the average firing rate (expressed as spikes per second) recorded from the 60 electrodes. Specifically, we calculated both the instantaneous firing rate, defined as the reciprocal of the interval between successive spikes, and the mean firing rate (MFR), defined as the number of spikes in a certain time interval divided by the duration of the interval. In order to calculate the MRF values reported in Fig. 15.3c, we considered 90 s before and after the stimulation ("pre" and "post" intervals) and over 90 consecutive 1-s intervals (total duration = 90 s) during which the US was switched on ("USon" interval) or off ("USoff" interval). Plot in Fig. 15.3c reports the MFR values of two series of piezo-stimulations interspersed with 5 min of recovery interval. The MFR remarkably increases during the US pulses; such increase is both repeatable and reversible.

Moreover, we investigated the MFR ratio, defined as the difference between the MFR during "USon" and "USoff" intervals, divided by the sum of the MFR values during a specific period of time (Fig. 15.4). MFR ratio can vary between −1 (spiking activity only during the "USoff" intervals) and +1 (spiking activity only during the "USon" intervals). Figure 15.4a shows the MFR ratio measured by the recording electrodes of the array in response to different pulse repetitions. The average of the MFR ratios in Fig. 15.4a is 0.78; this remarkably high MFR ratio is a clear indication of the increased activity when the US stimulation is applied in primary cultures pre-incubated with piezoelectric BTNPs. We also observed in a previous work of our group that the induced increase in firing activity during piezo-stimulation depends on the intensity of the generated US (Rojas et al. 2018). The graph in Fig. 15.4b reports the average MFR ratio values obtained by stimulating the primary neurons pre-incubated with piezoelectric BTNPs, pre-incubated with non-piezoelectric BTNPs, or non-incubated with BTNPs. The corresponding values, close to zero, indicate that the US stimulation alone, in our experimental conditions and with the adopted low-intensity levels, does not elicit any relevant response of electrical activity in the neural network. Moreover, since the cubic crystal non-piezoelectric BTNPs are not able to induce any significant increase of the MFR under US exposure, we can affirm that the piezoelectricity of the nanomaterial is required for the transduction of the US pressure stimulus into biologically relevant excitation cues.

In addition to the spiking activity, dissociated cortical or hippocampal cultures display peculiar patterns of electrophysiological activity, named bursts. A network burst is defined as a fast sequence of spikes (at least 5 spikes) and indicates fast re-depolarization at the single cell level and the almost synchronous activation at the network level. Burst activity can be characterized by several parameters, such as the bursting rate (BR), the burst duration (BD), and the number of spikes in each burst (SpkxBurst). All these parameters can be averaged over all the electrodes of the array and over a certain interval of time. We calculated the mean values of these parameters before, during, and after US stimulation in the presence of piezoelectric BTNPs. Figure 15.5 reports the values of the above-mentioned parameters after normalization to their respective maximum values; normalization has been carried out in order to plot them on the same dimensionless scale.

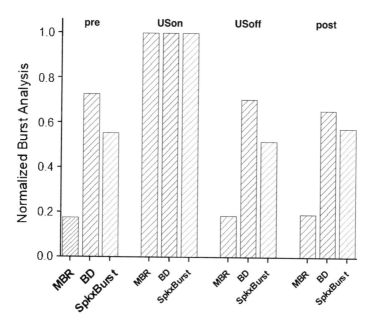

Fig. 15.5 Network burst activity before, during, and after US stimulation for a representative experiment with piezoelectric NPs. Different parameters are plotted for each interval: mean burst rate (MBR), burst duration (BD), inter burst interval (IBI), and number of spikes per burst (SpkxBurst). Mean values are normalized with respect to the maximum during the experiment

The first relevant observation is that US stimulation not only increases spiking activity (MFR) of cultures incubated with piezoelectric BTNPs, but also their bursting activity: during the pulses delivery, the MBR increases about 80% (up to 23.11 ± 0.8 bursts/min). Nonetheless, the MBR increase is limited to the USon intervals. When the US is off, MBR tends to the values measured before (i.e., pre interval) and after (post interval) the stimulation. Consistently with the MBR, also BD and SpkxBurst increase during the "USon" intervals, indicating as the evoked bursts are longer and with a higher number of spikes than the spontaneous ones. To summarize, the analysis of the bursting activity suggests two main observations: (i) the effect of the US stimulation is temporally confined and does not evoke any plastic change at the network level, and (ii) US stimulation models the bursting activity both in terms of number of evoked bursts as well as their shape (i.e., duration). An increased evoked bursting activity (longer bursts with more spikes than the spontaneous ones) can be interpreted as an increased transmission of information throughout the network which is related to the stimulation.

The most intuitive interpretation of the working mechanism of US stimulation mediated by piezoelectric nanoparticles is that the US pressure field deforms the NPs. As a consequence, the NPs, which are interfaced to the cell membrane, generate electric potential differences. A simple analytical model of the mechano-electric transduction mechanism of piezoelectric BTNPs subjected to pressure field has been previously provided (Marino et al. 2015a). The estimated voltage amplitudes generated by the US intensity range used in these cited works and in the experiments presented here ($0.8-1$ W/cm^2) correspond to about 0.2 mV. Since the NPs are interfaced to the cell membrane, such local voltage sources can increase the probability of activation of voltage-gated membrane channels, hence statistically increasing the membrane depolarization and, eventually, the probability that an action potential is generated.

Although preliminary, such results offer an intriguing opportunity to exploit US as an external signal for a highly selective neuromodulation technique based on nanotechnology.

15.4 Piezoelectric Devices for In Vivo Neural Stimulation and Regeneration

In vivo neural stimulation has been shown as fundamental for tissue integrity regeneration and maintenance after trauma, and has promising implications concerning tissue function development and recovery in the case of other pathological conditions (such as genetically derived sensorineural hearing loss, SNHL), including iatrogenic ones (for instance, drug-derived SNHL). Traditionally, it is attained by electrodes which can either externally or internally be applied, but its principal drawbacks consist of electric field attenuation through tissues when external electrodes are used, the high invasiveness of the surgical interventions required for internal electrode positioning/substitution, as well as power management and resupply (Cogan 2011). Wireless stimulation is therefore highly desirable to circumvent these issues, and different approaches have to date been developed, including direct stimulation with ultrasonic waves (Menz et al. 2013; Hertzberg and Volovick 2010) and indirect stimulation mediated by piezoelectric materials (Hwang et al. 2015b). The application of mechanical stimulation to piezoelectric materials intimately interacting with biological environments indeed enables the treatment of deep tissues with high time resolution, though it requires further studies and technological advances in particular in order to improve spatial resolution (Inaoka et al. 2011). In the following, the most important examples of applications of piezoelectric materials to in vivo neural stimulation will be presented, and future directions on the topic will be suggested based on the current technological and nanotechnological opportunities.

Pioneering work on piezoelectric material application to in vivo regeneration of nerves after trauma was conducted by Aebischer and co-workers since late 1980s. In their work, they demonstrated that poled piezoelectric PVDF channels designed for nerve guidance supported transected sciatic nerve regeneration to a higher extent than unpoled (non-piezoelectric) channels in mice. Nerves regenerated in poled channels indeed featured a higher number of myelinated axons than those in unpoled channels, both at early and late stages of regeneration after implantation (Aebischer et al. 1987). In a following study on a rat transected sciatic nerve model, a higher number of myelinated axons were also achieved using positively charged piezoelectric poly(vinylidenefluoride-trifluoroethylene) (PVDF-TrFE) channels compared to unpoled channels. To a lower extent than positively charges one, also negatively charged channels supported better axonal regeneration, thus demonstrating the influence of polarity on neuronal regeneration (Fine et al. 1991).

Over the latest decades, PVDF and its copolymer with trifluoroethylene have also been used for different applications, like sensorineural hearing loss treatment. SNHL is a condition of impaired neural stimulation of the cochlear nerves due to altered or missing cilia on the cochlear epithelium (as an inheritable or acquired disorder). In this concern, Inaoka and co-workers developed and characterized a PVDF membrane based-device equipped with an interdigitated aluminum electrode array in view of its utilization as a self-powered cochlear prosthesis (Fig. 15.6). The multilayered device was designed as a sensor with acoustic/electric signal conversion capability in the absence of battery. Although over-dimensioned compared to cochlear anatomy, the piezoelectric device showed a tonotopic response (in the 6.6–19.8 kHz range in air, and in the 1.4–4.9 kHz in silicone oil). It also generated maximum electrical response from an electrode positioned at the site of maximum vibration amplitude. The device was also scaled down for application in deafened guinea pigs (as rodent models for SNHL), and the PVDF membrane was demonstrated to induce auditory brain-stem responses upon sound stimulation and amplification of the electrical output. In this case, metal electrodes were implanted in the cochlea and the membranes were used externally. When implanted in the *scala tympani* of the basal turn of the cochlea, the device however did not develop an electric output sufficient to activation of auditory primary neurons, which was partially

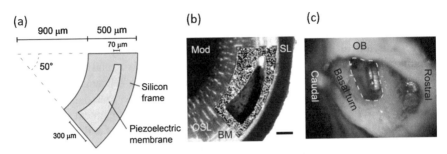

Fig. 15.6 Design of a piezoelectric device based on a PVDF film (**a**) for the treatment of sensori-neural hearing loss in a rodent model (guinea pig). Superposition of the image of an implantable piezoelectric film and the basal turn of the guinea pig cochlea (**b**), where BM stands for basilar membrane, Mod for *modiolus*, OSL for *osseous lamina*, and SL for spiral ligament. A microscopic view of an implanted device in the basal turn of the guinea pig cochlea (**c**). The yellow dotted line highlights an opening in the basal turn of the cochlea (OB stands for *otic bulla*). (Reproduced with permission from Inaoka et al. (2011); Copyright National Academy of Sciences of the United States of America)

ascribed to suboptimal anatomical positioning (Inaoka et al. 2011). Another work from Tona and co-workers aimed at optimizing neural stimulation based on PVDF-TrFE films as electrodes inserted into the *modiolus* of guinea pigs. This enabled a significant decrease in the thresholds of electrically evoked auditory brainstem responses compared with those of electrodes placed in the *scala tympani*. Due to the modest histological alterations detected with long-term analyses, this study represented a further step in the application of piezoelectric films to neural stimulation in cochlear prosthetics (Tona et al. 2015).

Other studies also aimed at applying different piezoelectric materials to cochlear nerve stimulation: for instance, lead-based composites (lead magnesium niobate–lead titanate crystal with saturation polarization, an epoxy composite with saturated polarization, an epoxy composite with unsaturated polarization, and lead zirconate titanate crystal with saturated polarization) were implanted in the *scala tympani* of a feline model for SNHL by Guo and co-workers (Guo et al. 2011). By measuring the maximum decline of hearing thresholds, this investigation demonstrated that the composite supported better hearing ability recovery than the other materials, and supported the evaluation of piezoelectric material performances relying on hydrostatic piezoelectric constants d_h and g_h (Guo et al. 2011).

As another application field of piezoelectric materials, deep brain stimulation (DBS) was demonstrated through a ternary, lead-based composite thin film (lead indium niobate–lead magnesium niobate–lead titanate, also termed PIMT) deposited on a flexible plastic substrate and connected to the primary motor (M1) cortex of a murine model. Upon moderate linear bending, the PIMT film indeed could generate a high current (far above the threshold for real-time DBS of the cortex) and a high voltage that enabled significant forearm movements (1.5–2.3 mm) in anesthetized mice (Hwang et al. 2015a, b).

15.5 Conclusions

This overview of the available literature demonstrates that further investigations are necessary for a realistic application of piezoelectric materials in vivo, as clear indications on neuronal survival and function on mid- and long-terms are still missing. Future studies will have to consider carefully long-term interaction of the materials with the host, and that high piezoelectric performances are ensured along with high safety and proper anatomical site targeting. In particular, lead-based materials of high neurotoxicity (Bressler et al. 1991) should be replaced by more biocompatible compounds (for instance, barium titanate), and targeting should be addressed from different perspectives by implementation of nanotechnology tools, such as nanoparticles, nanotubes, etc., in order to guarantee effective spatial resolution of stimulation from the cellular down to the subcellular level (Genchi et al. 2016b; Salim et al. 2018). Recent literature has clearly shown that piezoelectric nanomaterials and nanocomposites can in particular operate a mechanoelectric signal transduction suitable for (1) opening voltage-gated ion channels on cell membranes (in particular, Ca^{2+} channels), and (2) triggering intracellular signal transduction cascades (Ciofani et al. 2013; Genchi et al. 2016b; Wang and Guo 2016). Since the modulation of these events is involved in neuronal cell communication and survival (Wojda et al. 2008), the role of nanomaterials and nanocomposites (which often show completely different properties compared to their bulk counterparts) in responding to environmental stimulation but also in taking advantage from body motion (heart beating, respiration, etc.)—yet still largely unexplored—will increasingly be a determinant for proper addressing of neural stimulation and regeneration.

References

Aebischer, P., Valentini, R. F., et al. (1987). Piezoelectric guidance channels enhance regeneration in the mouse sciatic nerve after axotomy. *Brain Research, 436*(1), 165–168.

Bader, K. B., Bouchoux, G., et al. (2016). Sonothrombolysis. *Advances in Experimental Medicine and Biology, 880*, 339–362.

Blackmore, J., Shrivastava, S., et al. (2019). Ultrasound neuromodulation: A review of results, mechanisms and safety. *Ultrasound in Medicine and Biology, 45*(7), 1509–1536.

Bressler, J. P., Goldstein, G. W., et al. (1991). Mechanisms of lead neurotoxicity. *Biochemical Pharmacology, 41*(4), 479–484.

Carpentier, A., Canney, M., et al. (2016). Clinical trial of blood-brain barrier disruption by pulsed ultrasound. *Science Translational Medicine,8*(343), 343re2.

Ciofani, G., Danti, S., et al. (2013). Enhancement of neurite outgrowth in neuronal-like cells following boron nitride nanotube-mediated stimulation. *ACS Nano, 4*(10), 6267–6277.

Cogan, S. F. (2011). Neural stimulation and recording electrodes. *Annual Review of Biomedical Engineering, 10*, 275–309.

Dalecki, D. (2004). Mechanical bioeffects of ultrasound. *Annual Review of Biomedical Engineering, 6*, 229–248.

Fine, E. G., Valentini, R. F., et al. (1991). Improved nerve regeneration through piezoelectric vinylidenefluoride-trifluoroethylene copolymer guidance channels. *Biomaterials, 12*(8), 775–780.

Genchi, G. G., Ceseracciu, L., et al. (2016a). P(VDF-TrFE)/BaTiO3 nanoparticle composite films mediate piezoelectric stimulation and promote differentiation of SH-SY5Y neuroblastoma cells. *Advanced Healthcare Materials, 5*(14), 1808–1820.

Genchi, G. G., Marino, A., et al. (2016b). Barium titanate nanoparticles: Promising multitasking vectors in nanomedicine. *Nanotechnology, 27,* 232001.

Genchi, G. G., Marino, A., et al. (2017). Remote control of cellular functions: The role of smart nanomaterials in the medicine of the future. *Advanced Healthcare Materials, 6*(9), 1700002.

Genchi, G. G., Sinibaldi, E., et al. (2018). Ultrasound-activated piezoelectric P(VDF-TrFE) /boron nitride nanotube composite films promote differentiation of human SaOS-2 osteoblast-like cells. *Nanomedicine: Nanotechnology, Biology and Medicine, 14*(7), 2421–2432.

Guo, S., Dong, X., et al. (2011). Properties evaluation of piezoelectric materials in application of cochlear implant. *Ferroelectrics, 413*(1), 272–278.

Hertzberg, Y., & Volovick, A. (2010). Ultrasound focusing using magnetic resonance acoustic radiation force imaging: Application to ultrasound transcranial therapy. *Medical Physics, 37*(6), 2934–2942.

Hesley, G. K., Gorny, K. R., et al. (2013). MR-guided focused ultrasound for the treatment of uterine fibroids. *Cardiovascular and Interventional Radiology, 36*(1), 5–13.

Hwang, G.-T., Byun, M., et al. (2015a). Flexible piezoelectric thin-film energy harvesters and nanosensors for biomedical applications. *Advanced Healthcare Materials, 4*(5), 646–658.

Hwang, G., Kim, Y., et al. (2015b). Self-powered deep brain stimulation via a flexible PIMNT energy harvester. *Energy and Environmental Science, 8,* 2677–2684.

Inaoka, T., Shintaku, H., et al. (2011). Piezoelectric materials mimic the function of the cochlear sensory epithelium. *Proceedings of the National Academy of Sciences of the United States of America, 108*(45), 18390–11839.

Keifer, O. P., Riley, J. P., et al. (2014). Deep brain stimulation for chronic pain. *Neurosurgery Clinics of North America, 25*(4), 671–692.

Kubanek, J., Shukla, P., et al. (2018). Ultrasound elicits behavioral responses through mechanical effects on neurons and ion channels in a simple nervous system. *The Journal of Neuroscience, 38*(12), 3081–3091.

Lee, Y. S., & Arinzeh, T. L. (2012). The influence of piezoelectric scaffolds on neural differentiation of human neural stem/progenitor cells. *Tissue Engineering Part A, 18*(19–20), 2063–2072.

Lonzano, A. M., Lipsman, N., et al. (2019). Deep brain stimulation: Current challenges and future directions. *Nature Reviews Neurology, 15*(3), 148–160.

Luan, S., Williams, I., et al. (2014). Neuromodulation: Present and emerging methods. *Frontiers in Neuroengineering, 7,* 27.

Manivannan, S., & Terakawa, S. (1993). Rapid filopodial sprouting induced by electrical stimulation in nerve terminals. *The Japanese Journal of Physiology, 1,* S217–S220.

Maresca, D., Lakshmanan, A., et al. (2018). Biomolecular ultrasound and Sonogenetics. *Annual Review of Chemical and Biomolecular Engineering, 9,* 229–252.

Marino, A., Arai, S., et al. (2015a). Piezoelectric nanoparticle-assisted wireless neuronal stimulation. *ACS Nano, 9*(7), 7678–7689.

Marino, A., Barsotti, J., et al. (2015b). Two-photon lithography of 3D nanocomposite piezoelectric scaffolds for cell stimulation. *ACS Applied Materials and Interfaces, 7*(46), 25574–25579.

Marino, A., Genchi, G. G., et al. (2017). Piezoelectric effects of materials on bio-interfaces. *ACS Applied Materials and Interfaces, 9*(21), 17663–17680.

Marino, A., Battaglini, M., et al. (2018). Ultrasound-activated piezoelectric nanoparticles inhibit proliferation of breast cancer cells. *Scientific Reports, 8*(1), 6257.

Marino, A., Almici, E., et al. (2019). Piezoelectric barium titanate nanostimulators for the treatment of glioblastoma multiforme. *Journal of Colloid and Interface Science, 538,* 449–461.

Menz, M. D., Oralkan, Ö., et al. (2013). *Journal of Neuroscience, 33*(10), 4550–4560.

Naor, O., Krupa, S., et al. (2016). Ultrasonic neuromodulation. *Journal of Neural Engineering, 13*(3), 031003.

Nguyen, T. D., Deshmukh, N., et al. (2012). Piezoelectric nanoribbons for monitoring cellular deformations. *Nature Nanotechnology, 7*(9), 587.

O'Brien, W. D., Jr. (2007). Ultrasound-biophysics mechanisms. *Molecular Biology, 93*(1–3), 212–255.

Ribeiro, C., Pärssinen, J., et al. (2015). Dynamic piezoelectric stimulation enhances osteogenic differentiation of human adipose stem cells. *Journal of Biomedical Materials Research Part A, 103*(6), 2172–2175.

Rojas, C., Tedesco, M., et al. (2018). Acoustic stimulation can induce a selective neural network response mediated by piezoelectric nanoparticles. *Journal of Neural Engineering, 15*(3), 036016.

Royo-Gascon, N., Wininger, M., et al. (2013). Piezoelectric substrates promote neurite growth in rat spinal cord neurons. *Annals of Biomedical Engineering, 41*(1), 112–122.

Salim, M., Salim, D., et al. (2018). Review of nano piezoelectric devices in biomedicine applications. *Journal of Intelligent Material Systems and Structures, 29*(10), 2105–2121.

Tona, Y., Inaoka, T., et al. (2015). Development of an electrode for the artificial cochlear sensory epithelium. *Hearing Research, 330*(2015), 106e112.

Wang, Y., & Guo, L. (2016). Nanomaterial-enabled neural stimulation. *Frontiers in Neuroscience, 10*, 69.

Wang, X., Liu, J., et al. (2007). Integrated nanogenerators in biofluid. *Nano Letters, 7*, 2475–2479.

Wojda, U., Salinska, E., et al. (2008). Calcium ions in neuronal degeneration. *International Union of Biochemistry and Molecular Biology-Life, 60*(9), 575–590.

Zhang, L., Gu, F. X., et al. (2008). Nanoparticles in medicine: Therapeutic applications and developments. *Clinical Pharmacology and Therapeutics, 83*(5), 761–769.

Zhang, S., Yin, L., et al. (2009). Focusing ultrasound with an acoustic metamaterial network. *Physical Review Letters, 102*(19), 194301.

Chapter 16
Perspectives for Seamless Integration of Bioelectronic Systems in Neuromedicine

Vishnu Nair and Bozhi Tian

16.1 Introduction

The field of medicine has made great strides in improving the quality of life for human beings across the globe. This gradual expansion and maturation of medicine has resulted from contributions of scientists over the centuries who have attempted to understand living systems across different scales. Investigations, from the molecular and cellular levels to organs and whole animal level studies, have revealed how these systems function when in a state of equilibrium (Sung et al. 2013). In the event of deviation from such a state—when these systems are incapable of adapting—the result is either disease, injury, or disorder (Hernandez-Lemus 2012). When such situations arise, the goal of medicine has been to restore humans to the optimal state of equilibrium using short-term or long-term solutions in the case of an acute or chronic condition, respectively (Murrow and Oglesby 1996). However, in order to diagnose a specific medical condition, we need suitable techniques with which to probe a biological system and identify what specific type of non-equilibrium condition exists so as to clinically classify the disorder in an appropriate manner (Wang et al. 2017). Medicine has successfully evolved such techniques to address virtually the entire range of disorders known to mankind. One essential component that has contributed immensely to the field of medical diagnostics are biosensors, tools that form an integral part of medicine as they are a very versatile platform for diagnosing a plethora of human disorders (Bhalla et al. 2016).

V. Nair
Department of Chemistry, The University of Chicago, Chicago, IL, USA

B. Tian (✉)
Department of Chemistry, The University of Chicago, Chicago, IL, USA

The James Frank Institute, The University of Chicago, Chicago, IL, USA

The Institute for Biophysical Dynamics, The University of Chicago, Chicago, IL, USA
e-mail: btian@uchicago.edu

© Springer Nature Switzerland AG 2020
L. Guo (ed.), *Neural Interface Engineering*,
https://doi.org/10.1007/978-3-030-41854-0_16

Beyond diagnosis of a condition or disorder, medicine aims to repair or restore the system to its original state. Medicine is greatly indebted to molecular biology for enabling the use of small molecules—classified as drugs—for treating such disorders (Baig et al. 2016). Though drugs come with their own downsides, such as side effects, low efficacy, and low specificity, they have been the only successful method of treating some of the human disorders (Niu and Zhang 2017). The action of the drug upon the cells is typically mediated through these molecules, which selectively stimulate certain biological pathways, leading to a cascade of chemical reactions that often stabilize a cellular system (which, as a whole, cures the disorder) (Ritter et al. 2008). Cellular stimulation and response to various stimuli have been part of numerous fundamental studies, such as the Hodgkin-Huxley experiment which enabled cell biologists to classify cells as excitable and non-excitable (Hodgkin and Huxley 1952; Cervera et al. 2014). This further enabled electrophysiological research wherein studies utilizing an electrical stimulation generated a deeper understanding of the electrochemical properties of excitable cells (Linaro et al. 2015; O'Shea et al. 2019). These studies supported the development of external and implantable devices for the stimulation of excitable cells. The initial basic devices, such as the automated external defibrillator, vagus nerve stimulator, peripheral nerve stimulator, etc., are some of the most fundamental medical tools for dealing with acute and chronic disorders (Ellison et al. 2017; González et al. 2019; Chakravarthy et al. 2016). Hence, this trajectory of diagnosis and treatment leads to a synergistic relationship between sensing-associated diagnosis and stimulation-assisted treatment (Fu et al. 2016). The demand to further exploit this synergy led to the birth of bioelectronics and biomedical engineering, both of which supply the ever-growing demands of medicine.

The novel aim of bioelectronics is to precisely monitor a clinical condition, supply an appropriate dose of stimulation specific to that disorder, and thus cure the patient. However, depending on the acute or chronic nature of the condition behind the disorder, the time scale for monitoring and the associated feedback for stimulation-assisted treatment would vary. The device would need properties or features that could enable implantation or removal, wireless communication, and engineering design for seamless integration into a biological system (Katz 2014; Zhang et al. 2019). In this chapter, we lay the foundation for the development of nanoscale devices with various modes of stimulation, and the nanomaterial design strategies for the control of stimulation and implantation compatibility, followed by prerequisites and strategies for seamless integration with a focus on neuromedicine.

16.2   Cellular Stimulation

Every cell that forms a part of tissues or organs has a unique role that it plays along with other cells of a similar type, all in an attempt to maintain homeostasis in the human body (Davies 2016). These cells are hence known as the fundamental unit of life and thus are our system of interest (Plopper 2014). Cells are the model system used to investigate different types of stimulation and their after effects, with the goal of advancing our knowledge of bioelectronics.

Cells are a set of complex non-linear systems that cycle in and out of equilibrium and are subject to environmental stresses (Wang et al. 2017); and they are sufficiently robust to handle a certain degree of stress. This cycling in and out of equilibrium is governed by fundamental molecular thermodynamics (Olivier et al. 2016). Thus, a precise understanding of cellular level processes and signaling pathways tells us the nature of the stimulus that would be required to push a cell in and out of equilibrium.

The aim of stimulation is to apply a signal that is recognizable to cells, thus invoking a signaling pathway to push the system out of equilibrium. This would create a non-equilibrium cellular state, the decay of which relies on the stimuli's strength and frequency (Olivier et al. 2016). Depending on these two factors, the cell can either adapt and modify itself or proceed toward apoptosis (Strasser et al. 2000), reflecting a Darwinian-like theory in action. Thus, one initial prerequisite of bioelectronic stimulation is the optimization of stimuli strength and frequency for adaptation—unless the goal is apoptosis.

Post-dosage optimization, the goal is to find the right stimuli transducer. Given that these devices may require implantation and remote control—depending on whether the disorder is acute or chronic—we would need miniaturized devices. Such devices should contain components that can transduce a stimulus to generate a cell-perturbing signal and would be graded with respect to the dosage and frequency of the stimulus. Thus, the transducing material would have a frequency response that would govern its interfacial coupling with cell and output signal magnitude (Fig. 16.1). Nanomaterials are ideal candidates for this role as they have large active surface areas and are electronically active with a frequency response (Lasia 2014; Nair et al. 2018a).

The interfacial coupling of a cell and a nanomaterial results in a junction or cleft (Santoro et al. 2017) through which the transduced signal is scaled, phased, and transferred to a cell (Fig. 16.1). Such cell-material interfaces are chemically well-defined in terms of the surface functional groups that interact with the membrane proteins and the extracellular matrices (ECM) to form active junctions (Cutler and García 2003). Such active junctions are electrically defined using a capacitive element and a resistance in parallel (Fig. 16.1) (Nair et al. 2018a). However, a signal that is transferred via this path has to go through two sets of circuits in series, from a constant phase element (CPE) element in the cleft (Lasia 2014) and the cell membrane (Kandel et al. 2000), such that the effective interfacial impedance here governs the scaling and phasing of the signal that is transferred to the cell (Fig. 16.1) (Sedra and Smith 2000).

16.3 Modes of Stimulation

In order to achieve synergy between sensing and stimulation, we have concluded that miniaturized remote-controlled devices are a necessity. In conjunction with the requirement of signal transduction by active materials in a device for stimulation

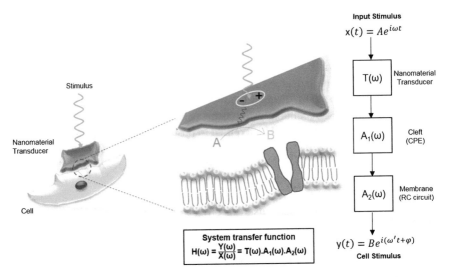

Fig. 16.1 Schematic representation of a cell-nanomaterial interface from a system science perspective. The schematic illustrates how the nanomaterial acts as a transducer with a frequency response $T(\omega)$ to convert the stimulus into a cell-readable signal which is further modulated by the impedances $A_1(\omega)$ and $A_2(\omega)$ of the cleft and membrane, respectively. Thus, from a systems perspective we can define the entire process of signal processing by the nanomaterial transducer and the impedances by a net transfer function $H(\omega)$ given by the product of individual frequency responses in the frequency domain as the net transformation factor. Though the stimulus and cell-readable signal are in the time domain, a frequency domain analysis permits easier analysis as we can consider each component as a set of cascaded systems

and miniaturization, we identify stimuli-responsive nanomaterials as potential candidates (Jiang et al. 2016; Parameswaran et al. 2018; Tortiglione et al. 2017). Nanomaterials or nanoscale interfaces, by virtue of their solid-state structure, have electronic states which evolve in space and time according to the laws of quantum mechanics (Kuno 2011). The development of powerful quantum chemistry computational methods has enabled us to understand how quantum confinement effects impact the electronic structure of nanomaterials and interfaces (Burdett 1996). The electronic structure of a nanomaterial further governs its electronic, optical, magnetic, and surface properties, all of which form an essential set of features that could be controlled by synthetic strategies.

Previous electrophysiology experiments have enabled the classification of cells into excitable versus non-excitable—though the goal of bioelectronic stimulation is to evolve a general strategy for stimulation, as disorders are not specifically confined to excitable cells. Excitable cells have a binary response to an electrical, electrochemical, or thermal stimulus, resulting in a membrane depolarization or hyperpolarization event (Cervera et al. 2014). In stark contrast, non-excitable cells have a continuum response, thus allowing scope for a sub-threshold stimulation which leads to a diversified fate in the responding cells. Thus, the aim of nanomaterial synthesis for bioelectronics is to design materials which can transduce fundamental forms of energy such as electromagnetic waves, heat and sound, to create cell-recognizable stimuli like ionic currents, thermal gradients, and transient molecular species (Carvalho-de-Souza et al. 2018; Rosado et al. 2006). These stimuli, irrespective of their nature, depolarize or hyperpolarize the cell membrane, the consequences of which, though diverse, lead to an unstable non-equilibrium state in the cell (Carvalho-de-Souza et al. 2018; Rosado et al. 2006). We hereby classify the different modes of stimulation as thermal, electrochemical, and mechanical, depending on the form of the final cell receptive signal.

16.3.1   *Thermal Methods*

Heat is a form of energy that can perturb the structure and function of molecules that make up biological components such as proteins, lipids, and nucleic acids (Huang and Lau 2016). Heat that is externally applied to a cell is generally capable of perturbing the lipid bilayer of a membrane and membrane proteins, causing changes in their structure and hence in the membrane capacitance and resistance (Carvalho-de-Souza et al. 2018). The membrane capacitance controls the charge separation via an ionic gradient across the membrane that ultimately provides a measure of the membrane voltage. Thus, we can expect an apparent membrane voltage fluctuation to occur as a thermal stimulus is externally applied to a cell (Carvalho-de-Souza et al. 2018). These membrane voltage fluctuations are read by excitable cells and non-excitable cells in different ways. An excitable cell would depolarize in response to heat only when it causes the membrane voltage to rise above a threshold (Carvalho-de-Souza et al. 2018). On the other hand, a non-excitable cell continuously responds to a membrane potential increase by modulating the ionic concentration inside versus outside the cell to catch up with this perturbation (Cervera et al. 2014). Thus, such a continuous response of non-excitable cells gradually evokes organelle- and genetic-level responses to overcome this change more prominently than excitable cells, leading to sub-threshold stimulation.

In order to introduce such an external thermal stimulus, we should have candidate nanomaterials which can convert an optical stimulus to heat via defect or trap state-mediated carrier recombination or surface plasmon resonance (Streetman and Banerjee 2014; Govorov and Richardson 2007). Alternatively, a magnetothermal effect caused by the release of energy from a magnetization-demagnetization cycle in nanoparticles can also produce such a result (Skumiel et al. 2013). The dosage of thermal stimulus varies according to the fundamental mechanism operating behind the transduction and its efficiency. In case of a semiconductor defect or trap state-mediated carrier recombination, we first need to estimate the defect density using photoluminescence or electrical transport techniques followed by an estimate of generation and recombination rates (Fig. 16.2b). After estimating the rate of recombination, scaling it by the energy gap between the conduction band and trap states would yield a theoretical estimate of the heat generated (though practically there could be multiple dissipation pathways which are underestimated or ignored). In contrast, plasmonic heating deals with the oscillatory dissipation of energy by

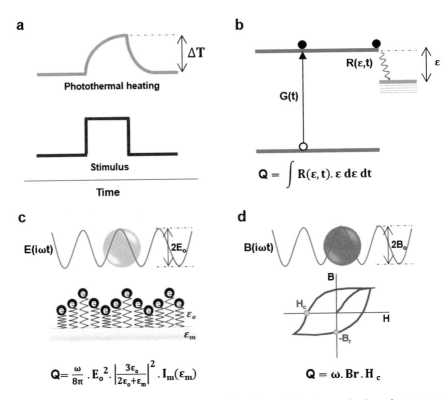

Fig. 16.2 Schematic illustration of different modes of thermal stimulus production using nanomaterials. (**a**) A general schematic showing a standard photothermal response for a constant stimulus pulse. Thermal heating always increases with the application of stimulus and saturates, only to exponentially decay as the stimulus is turned off. The peak of the heat profile measured from the thermal diffusion current can be equated to heat generated Q using the specific heat capacity Cv. (**b**) Illustration of heat generation from semiconductor defect recombination, where G(t) is the carrier generation rate due to photon absorption, $R(\varepsilon,t)$ is the recombination rate from the conduction band into a defect state located energy ε below it and Q is the total estimate of heat generated from this process. (**c**) Illustration of weakly bound surface electrons and dissipating energy from an electrical field according to the Drude-Lorentz model, with Q being an estimate of heat generated. (**d**) Equating the energy stored in a magnetic hysteresis cycle of a nanomaterial being dissipated as heat Q

relatively free electrons at metallic surfaces (Govorov and Richardson 2007). According to the Drude-Lorentz model, surface atoms have an electron cloud which are weakly bound to the nuclei thus making them capable of oscillating in response to electromagnetic radiation (Govorov and Richardson 2007). Depending on the frequency of an electromagnetic wave we could model the electron cloud as an overdamped, underdamped, or critically damped oscillator, ultimately dissipating energy in the form of heat (Fig. 16.2c). On a similar note, a magnetic material placed in an oscillatory magnetic field dissipates energy stored during a magnetization-demagnetization cycle (Fig. 16.2d) (Skumiel et al. 2013). In general, thermal methods focus on the thermodynamic conversion of energy from one form to heat.

16.3.2   Electrochemical and Photoelectrochemical Methods

Free energy (in all systems) is used to take a system toward an equilibrium state. In one such attempt, energy from an electromagnetic field may lead to production of a local charge separation in a material (Streetman and Banerjee 2014). These localized charge separations are non-equilibrium states that have an associated excess of electrostatic potential energy (Streetman and Banerjee 2014). This charge separation, and the consequential electrostatic potential energy, depends on the band structure of the material. Such available free energy in the form of electrostatic potential is precisely what drives electrochemistry at the interfaces (Rajeshwar 2007). An electrochemical reaction thus happens when a local charge separation has significant transient stability in the highest occupied molecular orbital (HOMO) and lowest unoccupied molecular orbital (LUMO) or the valence (VB) or conduction band (CB) of that material (Rajeshwar 2007). Such a stability enables separated charges to be transferred across an interface to a molecular species or material with overlapping density of states (Fig. 16.3a, b) (Rajeshwar 2007). An electrochemical process can polarize the immediate surroundings of an electrode due to charge accumulation on its surface. This polarization can cause membrane depolarization via a capacitive coupling between the membrane and the electrode (Fig. 16.3c) (Rajeshwar 2007). However, if there is an overlap between the density of states of a species in solution and the electrode's HOMO or LUMO as illustrated in Fig. 16.3a or b, then the species could undergo oxidation or reduction (Rajeshwar 2007). The consequences of such a redox reaction or faradaic injection (Fig. 16.3d) are that a species of a specific charge accumulates in the cleft, causing a depolarization or hyperpolarization of cell membranes. Following this, the concentration of ions inside and outside the cell are modified, leading to a new membrane potential governed by the Goldman–Hodgkin–Katz (GHK) equation (Fig. 16.3e) (Kandel et al. 2000). Besides perturbing a concentration gradient, such electrochemical reactions can produce cell-recognizable active molecules through a biochemical process. Cell-recognizable species mainly include reactive oxygen species, reactive nitrogen species, and oxidized or reduced metal ions or clusters (Rosado et al. 2006). Such cell-recognizable molecular species trigger a cascade in the cellular system leading to a membrane potential change, organelle stimulation, or genetic expression (Rosado et al. 2006). A similar effect could also be generated by cavitation effects. Ultrasound-induced cavitation can produce cavitating bubbles which could generate gigapascal and thousands of Kelvin of temperature to generate radical species like peroxides that can stimulate cells in a very similar fashion (Hernández-García et al. 2008). However, a careful control of ultrasound intensity is important to ensure control of the cavitation thus giving electrochemical methods a significant edge over cavitation processes.

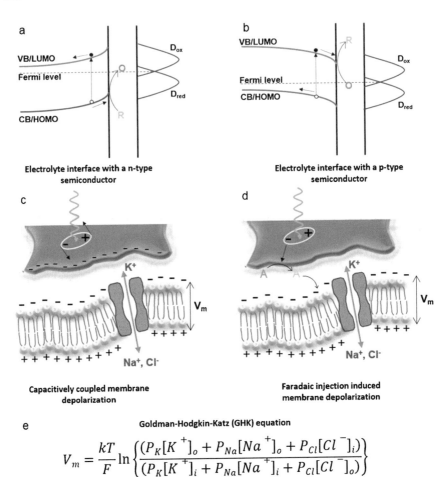

Fig. 16.3 (**a**) Photon absorption and photooxidation process occurring at an n-type semiconductor-electrolyte interface, generating an anodic current. (**b**) Photon absorption and photoreduction process occurring at a p-type semiconductor-electrolyte interface, generating a cathodic current. (**c**) Capacitive depolarization of the cell membrane due to charge accumulation at the nanomaterial-cleft interface, causing charge-induced depolarization of the cell membrane. (**d**) Faradaic reaction-induced redox species which either introduced an accumulation of charged species causing membrane depolarization or a molecular active species which caused receptor binding mediated depolarization. (**e**) The GHK equation describing the variation in membrane voltage with differing concentrations of ions

16.3.3 Mechanical Methods

Compared to either thermal or electrochemical methods, mechanical stimulation operates in a fundamentally different way. This variation arises from the fact that cells have an entirely different machinery with which to sense and respond to mechanical forces. Cells can convert such mechanical cues into intracellular electrochemical signals which lead to a signaling cascade (Epand and Ruysschaert

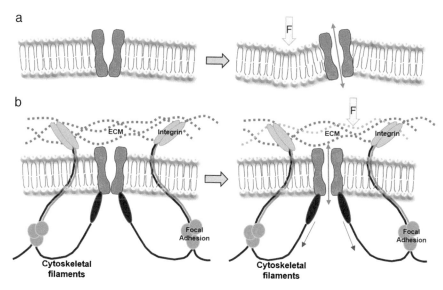

Fig. 16.4 (**a**) Direct force of lipid layer causing mechanosensitive ion channels to activate, open, and produce depolarization. (**b**) Force in ECM transduced via integrin and cytoskeletal filaments, causing the channel to open

2017). This conversion of mechanical forces into electrochemical cues happens at mechanosensors which are classified into three classes. Mechanosensing by the plasma membrane is mediated by caveolin or cilia or microvilli which lead to the inward wrinkling of the plasma membrane (Epand and Ruysschaert 2017). However, mechanosensing can happen through an intracellular or extracellular mechanism as well. In these cases, they are mediated through actin cytoskeleton, glycocalyx, cadherin-rich cell-cell junction, or mechanosensitive ion channels (Epand and Ruysschaert 2017). Among these, the mechanosensitive ion channels are the fastest and thus of extensive interest, though many of these mechanisms could be coupled with respect to each other.

Cells are coupled to the ECM via integrin proteins that extend from the cytosol into the external environment (Kumar et al. 2016). Integrin proteins connect the cytoskeletal filaments to the ECM through the focal adhesions and the phospholipid bilayer (Fig. 16.4b). Thus, any force that the cell receives is via the ECM and goes into the focal adhesions or directly onto the lipid bilayer (Fig. 16.4a, b) (Epand and Ruysschaert 2017). Extracellular forces such as fluid flow shearing, tensile, and traction forces along with intracellular contractile forces are the primary agents which bring about mechanical stimulation. Such forces are transduced by mechanosensitive ion channels present in the cell membrane. These mechanosensitive ion channels physically open or close in response to a force traveling through the lipid bilayer or cytoskeletal filaments. Such channels open or close to produce membrane voltage fluctuations which could lead to depolarization or hyperpolarization of the cell membrane. Thus, magnetic, optical, or piezo-responsive nanomaterials which can exert force on the lipid bilayer or the ECM to transduce a cytoskeletal force are additional modes of cellular stimulation (Epand and Ruysschaert 2017).

16.4   Making Nanomaterial Choices for Seamless Integration

Having explored the different modes of stimulation, we are able to determine a requisite property that each of these nanomaterials should possess in order for a stimulus to be transduced as a cell-recognizable signal. Beyond the functional capability of generating a cell-recognizable signal, we may need to explore other properties that are required for a well-coupled bio-interface. For example, the nanomaterial under consideration should be biocompatible, sufficiently mechanically flexible (like ECM or cytoskeleton), and have the appropriate surface chemistry for cellular recognition (Chen et al. 2017). To further extend the capabilities of this nanomaterial for simultaneous sensing and stimulation with a feedback interlock, it is necessary to use appropriate nanofabrication technologies to design three-dimensional devices capable of interfacing with tissues (Dai et al. 2018). If achieved, this would be the ultimate goal of seamless integration. Hence, our focus is on obtaining a three-dimensional framework of appropriate size—as well as the correct mechanical and electronic properties—capable of carrying out spatiotemporal mapping and stimulation. Existing synthetic nanomaterials have wide-ranging optoelectronic and mechanical properties, thus the task here is to make the appropriate selection from the range of available options. For ideal sensing capabilities, we need to select a material with an electronic structure that has maximum signal-to-noise ratio (SNR) subject to screening limited response at physiological conditions. Similarly, for stimulation capabilities a good understanding of the electronic structure and the subsequent efficiency of signal transduction is important to determine the appropriate dosage. Both stimulation and sensing depend on the electronic structural properties which reside in their dimensionality, chemical composition, and surface chemistry, for dosage and sensitivity, respectively (Jiang and Tian 2018). Thus, the selection process for the choice of nanomaterials could be decentralized on the basis of these fundamental properties. This would ultimately enable us to have electronic components capable of performing simultaneous stimulation and sensing.

Once it is possible to draw up a short list of nanomaterials based on their electronic properties (Alongside information on the appropriate dimension, chemical composition, and surface chemistry). Next, a sub-search within this set has to be performed for mechanical compatibility and biocompatibility. The latter is usually assessed by carrying out a combination of device stability and cytotoxicity testing. This involves checking degradation and its by-products formed from nanomaterials under physiological conditions in human blood plasma or serum, or in vivo testing (Chen et al. 2017; Jiang and Tian 2018). An appropriate quantification and identification of the degradation by-products and immune response like scarring and cell death enable researchers to predict the cytotoxicity. Device degradation by enzymes and pH is an equally important property to be determined through in vivo testing in order to assess device delamination, corrosion, and fractures. On the other hand, mechanical compatibility of a nanomaterial is theoretically predicted by deviation of the persistence length of 1D architectures or nanoscale elasticity modulus in case of 2D or 3D structures, with respect to cellular ECM or cytoskeleton. Nanoscale mechanical property measurement using atomic force microscopy or microrheology allows us to obtain information on mechanical compatibility. Furthermore, the materials for flexible device fabrication and signal readout could be chosen from existing nanofabrication techniques with the goal of selecting materials with properties that facilitate seamless integration. The procedure described here is basically a sequential selection process that allows us to select materials from an existing library, a process that could be efficiently outsourced to machine learning (Fig. 16.5a).

In addition to a smart selection process to find the right set of materials from an existing pool, one could envision design strategies maximized to obtain hybrid materials with the requisite properties. At this juncture, we need to classify materials into hard and soft and attempt to think of hard-soft hybrid materials as a solution to seamless integration. The design of a hard-soft hybrid could be done by embedding a functionally active material into a soft matrix, serving as a mechanically compatible framework for the device. Such a fabrication could be achieved by either using additives in the starting material of the synthesis, by the application of an in situ synthetic process, or by a post-synthetic modification process (Fig. 16.5b). As a result, we have a three-way flexible synthetic strategy by which we can design new hard-soft hybrid materials for seamless integration.

16.5 Toward Seamless Integration in Neuromedicine

The brain may be seen as a gigantic electronic circuit board, composed of a variety of excitable neurons which in turn operate with varying functions. The coherent integration and functioning of the brain is predicated upon the functionality of its neurons (Birmingham et al. 2014). Thus, brain disorders arise from the failure of certain parts of the neural circuitry, resulting in the disruption of coherent functioning. Although acute disorders need prompt medical intervention and are currently out of reach for bioelectronic remediation, chronic disorders are well within the range of what is possible. The prerequisite for addressing chronic disorders, however, is to achieve seamless integration.

With respect to curing a neural disorder that affects organ function, there are two schools of thought regarding stimulation. One is to directly identify the nerve component responsible for the disorder and to stimulate this component in order to address issues regarding a single function of a specific organ (Birmingham et al. 2014). The second is to directly stimulate the organ in a precise way so as to restore the organ function and bypass the brain components (Birmingham et al. 2014). At present, we will focus on the first school of thought in an attempt to remain within the framework of neural interface engineering (Birmingham et al. 2014).

Bioelectronics has evolved significantly over time, allowing for precise targeting and interfacing to modulate neural signaling patterns in the brain. The specificity of modulation for addressing a unique problem associated with a unique part of the body, however, is the first challenge as we move toward seamless integration. The

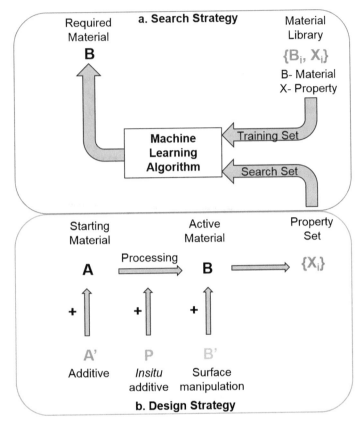

Fig. 16.5 (**a**) Discussion of the selection strategy via machine learning for seamless integration. (**b**) Discussion of a material design strategy via a three-way control to obtain material for biointerface search and optimization

current issue is to decode the entire human neural circuitry with a clear circuit structure-organ function correlation. Deriving such correlation maps of neural circuitry in the brain is a longstanding challenge, though significant progress has been made. We are now able to successfully address diseases such as sleep apnea, hypertension, etc. (Birmingham et al. 2014). However, better imaging and tomography techniques are necessary for fundamental studies involving electrons, X-rays, and optical microscopy, with the ultimate goal of building a neural atlas. The second challenge to seamless integration is to have miniaturized interfaces which are functionally active and well coupled to the cells. To a great extend this depends on the possibility of using the nanomaterial design strategy that was discussed in the previous section. In addition, we would require an appropriate device platform for the device to function wirelessly using electronic components to receive signal, microprocessor-enabled processing of input and output signals—as well as a transmitter for output signals (Dai et al. 2018). The requisite circuitry has been a basic tool of communication electronics and thus is easily accessible among the existing technologies. The third and most significant challenge to seamless integration is to determine the appropriate dosage for treatment (Birmingham et al. 2014). Dosage, in the context of bioelectronic stimulation, means the amplitude and frequency of the stimulus signal to which the cells would be capable of adapting. To date, this has been the most formidable obstacle as there are multiple parameters involved. The primary factors are patient condition and history, material stimulation efficiency, sensor SNR, closed feedback loop coupling, device losses, and failure. The only way to determine the appropriate dosage of therapeutic utility is by performing in vivo experiments in animal models (Birmingham et al. 2014). Once numerous proof-of-concept experiments have been performed and there is a sufficient amount of data, this could be used as a training data set for machine learning. Furthermore, as the training data is used in a machine learning algorithm, it could potentially be used for predicting device design and stimulation dosage for a particular patient. This enables the custom design of wearable and implantable bioelectronics for chronic conditions. However, the long-term effects of these devices should be taken into consideration.

Thus, having broadly discussed the prerequisites for seamless integration and aspects which could help satisfy these demands, we further focus on two promising material strategies.

16.5.1 Mesh Nanoelectronics

Silicon-based nanoscale meshes were designed in order to facilitate three-dimensional devices that have the capability of spatiotemporal sensing and stimulation (Dai et al. 2018). The meshes were designed such that their fundamental architecture permitted an aspect ratio which minimized mechanical mismatch with respect to their biological counterparts. Furthermore, these meshes have arrays of nanowire field effect transistors (FETs) integrated into them in a spatially defined manner so as to enable spatiotemporal sensing or readout. While this design yielded a two-dimensional framework, in order to transform these into a three-dimensional framework, the meshes were mechanically transformed into 3D structures and seeded with cells to yield synthetic tissues. These synthetic tissues could then be loaded into a syringe needle and injected at the requisite location. For the purpose of stimulation, these mesh structures could have individually addressable electrical stimulation devices alongside the FET. This technology holds promise with respect to the goal of seamless integration. Moving forward, we aim to have nanosensors for the detection of biological molecules or analytes. Moreover, the integration of a communication electronic component for wireless input and output would enable us to implement a closed feedback loop for simultaneous sensing and stimulation.

16.5.2 Hydrogel Bioelectronics

Hydrogels are crosslinked hydrophilic polymer networks capable of storing a polar species inside their porous structure (Yuk et al. 2019). Hydrogels have been a material of interest in the field of bioelectronics due to their tissue-like mechanical property and biocompatibility (Yuk et al. 2019). Moreover, hydrogels are not conventional dry electrodes and possess ionic and aqueous components, which may enhance stimulation and sensing performance. Thus, hydrogels represent one class of candidate for bioelectronic devices (Yuk et al. 2019). There are four types of hydrogels used in bioelectronics:

1. Hydrogel coatings and encapsulations: This idea is based on designing devices with conventional metal or semiconductor electrodes, however with a hydrogel coating at the interface of the electrode with the cell. Such a construct reduces the mechanical mismatch.
2. Ionically conductive hydrogels: These are hydrogels that permit fast ion conduction in aqueous condition. Compared to metal electrodes, these hydrogels can have lower junction impedance as they allow ionic conductivity. They can be used independent from conventional electrodes.
3. Conductive nanocomposite hydrogels: Hydrogel frameworks which have a conductive additive that enables electrical conductivity as well as ionic conductivity so as to have even lower junction impedance than ionically conductive hydrogels.
4. Conductive polymer hydrogels: Organic semiconducting polymers possess conductivity due to their conjugated backbone and are also capable of self-assembling into hierarchical structures (Nair et al. 2018b). These properties, as well as the possibility of doping, make them very promising in bioelectronics applications.

Despite possessing far better properties in terms of forming mechanically and electrically compatible junctions, hydrogel bioelectronics still has a long way to go as there are multiple issues when it comes to seamless integration. Hydrogel bioelectronics are soft and have highly hydrated structures which come with long-term stability issues arising from their weak adhesion to devices and dehydration. Moreover, incorporating hydrogels into existing device fabrication techniques with high spatial resolution and complexity is a field of research in itself.

16.6   Conclusions and Outlook

Throughout this chapter, we outlined an overall picture of what is driving all research in the field of bioelectronics. Using fundamental studies involving electrochemical, optoelectronic, and mechanical properties of materials and cells, we are attempting to make sense of what seamless integration should look like. However, due to a lack of elementary comprehension and/or technological innovation, there are still a myriad of unsolved problems in the field. In this chapter, we suggest the technologies and ideas that hold promise to meet the goals of seamless biointegration, despite current limitations.

References

Baig, M. H., Ahmad, K., Roy, S., Ashraf, J. M., Adil, M., Siddiqui, M. H., Khan, S., Kamal, M. A., Provazník, I., & Choi, I. (2016). Computer aided drug design: Success and limitations. *Current Pharmaceutical Design, 22*, 572–581.

Bhalla, N., Jolly, P., Formisano, N., & Estrela, P. (2016). Introduction to biosensors. *Essays in Biochemistry, 60*, 1–8. https://doi.org/10.1042/EBC20150001.

Birmingham, K., Gradinaru, V., Anikeeva, P., Grill, W. M., Pikov, V., McLaughlin, B., Pasricha, P., Weber, D., Ludwig, K., & Famm, K. (2014). Bioelectronic medicines: A research roadmap. *Nature Reviews. Drug Discovery, 13*, 399–400. https://doi.org/10.1038/nrd4351.

Burdett, J. K. (1996). Electronic structure and properties of solids. *Journal of Physical Chemistry, 100*, 13263–13274.

Carvalho-de-Souza, J. L., Pinto, B. I., Pepperberg, D. R., & Bezanilla, F. (2018). Optocapacitive generation of action potentials by microsecond laser pulses of nanojoule energy. *Biophysical Journal, 114*, 283–288. https://doi.org/10.1016/j.bpj.2017.11.018. Epub 2017 Dec 19.

Cervera, J., Alcaraz, A., & Mafe, S. (2014). Membrane potential bistability in nonexcitable cells as described by inward and outward voltage-gated ion channels. *The Journal of Physical Chemistry B, 118*, 12444–12450. https://doi.org/10.1021/jp508304h.

Chakravarthy, K., Nava, A., Christo, P. J., & Williams, K. (2016). Review of recent advances in peripheral nerve stimulation (PNS). *Current Pain and Headache Reports, 20*, 60. https://doi.org/10.1007/s11916-016-0590-8.

Chen, R., Canales, A., & Anikeeva, P. (2017). Neural recording and modulation technologies. *Nature Reviews Materials, 2*. https://doi.org/10.1038/natrevmats.2016.93.

Cutler, S. M., & García, A. J. (2003). Engineering cell adhesive surfaces that direct integrin alpha5beta1 binding using a recombinant fragment of fibronectin. *Biomaterials, 24*, 1759–1770.

Dai, X., Hong, G., Gao, T., & Lieber, C. M. (2018). Mesh nanoelectronics: Seamless integration of electronics with tissues. *Accounts of Chemical Research, 51*, 309–318.

Davies, K. J. (2016). Adaptive homeostasis. *Molecular Aspects of Medicine, 49*, 1–7. https://doi.org/10.1016/j.mam.2016.04.007. Epub 2016 Apr 22.

Ellison, K., Sharma, P. S., & Trohman, R. (2017). Advances in cardiac pacing and defibrillation. *Expert Review of Cardiovascular Therapy, 15*, 429–440. https://doi.org/10.1080/14779072.2017.1329011.

Epand, R. M., & Ruysschaert, J.-M. (Eds.). (2017). *The biophysics of cell membranes* (Springer series in biophysics 19). https://doi.org/10.1007/978-981-10-6244-5_4.

Fu, T. M., Hong, G., Zhou, T., Schuhmann, T. G., Viveros, R. D., & Lieber, C. M. (2016). Stable long-term chronic brain mapping at the single neuron level. *Nature Methods, 13*, 875–882.

González, H. F. J., Yengo-Kahn, A., & Englot, D. J. (2019). Vagus nerve stimulation for the treatment of epilepsy. *Neurosurgery Clinics of North America, 30*, 219–230. https://doi.org/10.1016/j.nec.2018.12.005.

Govorov, A. O., & Richardson, H. H. (2007). Generating heat with metal nanoparticles. *Nano Today, 2*, 30–38.

Hernández-García, D., Castro-Obregón, S., Gómez-López, S., Valencia, C., & Covarrubias, L. (2008). Cell death activation during cavitation of embryoid bodies is mediated by hydrogen peroxide. *Experimental Cell Research, 314*, 2090–2099. https://doi.org/10.1016/j.yexcr.2008.03.005. Epub 2008 Mar 20.

Hernandez-Lemus, E. (2012). Nonequilibrium thermodynamics of cell signaling. *Journal of Thermodynamics, 2012*, 1. https://doi.org/10.1155/2012/432143.

Hodgkin, A. L., & Huxley, A. F. (1952). Currents carried by sodium and potassium ions through the membrane of the giant axon of Loligo. *The Journal of Physiology, 116*(4), 449–472. https://doi.org/10.1113/jphysiol.1952.sp004717.

Huang, R., & Lau, B. L. T. (2016). Biomolecule-nanoparticle interactions: Elucidation of the thermodynamics by isothermal titration calorimetry. *Biochimica et Biophysica Acta, 1860*, 945–956. https://doi.org/10.1016/j.bbagen.2016.01.027. Epub 2016 Feb 3.

Jiang, Y., & Tian, B. (2018). Inorganic semiconductor biointerfaces. *Nature Reviews Materials, 3*, 473–490.

Jiang, Y., Carvalho-de-Souza, J. L., Wong, R. C., Luo, Z., Isheim, D., Zuo, X., Nicholls, A. W., Jung, I. W., Yue, J., Liu, D. J., Wang, Y., De Andrade, V., Xiao, X., Navrazhnykh, L., Weiss, D. E., Wu, X., Seidman, D. N., Bezanilla, F., & Tian, B. (2016). Heterogeneous silicon meso-structures for lipid-supported bioelectric interfaces. *Nature Materials, 15*, 1023–1030. https://doi.org/10.1038/nmat4673. Epub 2016 Jun 27.

Kandel, E. R., Schwartz, J. H., & Jessell, T. M. (2000). *Principles of neural science* (4th ed.). New York: McGraw-Hill, Health Professions Division.

Katz, E. (2014). *Implantable bioelectronics*. Wiley-VCH, Weinheim.

Kumar, A., Ouyang, M., Van den Dries, K., McGhee, E. J., Tanaka, K., Anderson, M. D., Groisman, A., Goult, B. T., Anderson, K. I., & Schwartz, M. A. (2016). Talin tension sensor reveals novel features of focal adhesion force transmission and mechanosensitivity. *The Journal of Cell Biology, 213*, 371–383. https://doi.org/10.1083/jcb.201510012.

Kuno, M. (2011). *Introductory nanoscience: Physical and chemical concepts* (1st ed.). Garland Science, New York.

Lasia, A. (2014). *Electrochemical impedance spectroscopy and its applications*. New York: Springer.

Linaro, D., Couto, J., & Giugliano, M. (2015). Real-time electrophysiology: Using closed-loop protocols to probe neuronal dynamics and beyond. *Journal of Visualized Experiments*. https://doi.org/10.3791/52320.

Murrow, E. J., & Oglesby, F. M. (1996). Acute and chronic illness: Similarities, differences and challenges. *Orthopaedic Nursing, 15*, 47–51.

Nair, V., Ananthoju, B., Mohapatra, J., & Aslam, M. (2018a). Photon induced non-linear quantized double layer charging in quaternary semiconducting quantum dots. *Journal of Colloid and Interface Science, 514*, 452–458. https://doi.org/10.1016/j.jcis.2017.12.034. Epub 2017 Dec 13.

Nair, V., Kumar, A., & Subramaniam, C. (2018b). Exceptional photoconductivity of poly(3-hexylthiophene) fibers through in situ encapsulation of molybdenum disulfide quantum dots. *Nanoscale, 10*, 10395–10402. https://doi.org/10.1039/c8nr01102h.

Niu, Y., & Zhang, W. (2017). Quantitative prediction of drug side effects based on drug-related features. *Interdisciplinary Sciences, 9*, 434–444. https://doi.org/10.1007/s12539-017-0236-5.

O'Shea, C., Holmes, A. P., Winter, J., Correia, J., Ou, X., Dong, R., He, S., Kirchhof, P., Fabritz, L., Rajpoot, K., & Pavlovic, D. (2019). Cardiac optogenetics and optical mapping – overcoming spectral congestion in all-optical cardiac electrophysiology. *Frontiers in Physiology, 10*, 182–196. https://doi.org/10.3389/fphys.2019.00182. eCollection 2019.

Olivier, B. G., Swat, M. J., & Moné, M. J. (2016). Modeling and simulation tools: From systems biology to systems medicine. *Methods in Molecular Biology, 1386*, 441–463. https://doi.org/10.1007/978-1-4939-3283-2_19.

Parameswaran, R., Carvalho-de-Souza, J. L., Jiang, Y., Burke, M. J., Zimmerman, J. F., Koehler, K., Phillips, A. W., Yi, J., Adams, E. J., Bezanilla, F., & Tian, B. (2018). Photoelectrochemical modulation of neuronal activity with free-standing coaxial silicon nanowires. *Nature Nanotechnology, 13*, 260–266. https://doi.org/10.1038/s41565-017-0041-7. Epub 2018 Feb 19.

Plopper, G. (2014). *Principles of cell biology*. Inc: Jones and Bartlett Publishers.

Rajeshwar, K. (2007). *Fundamentals of semiconductor electrochemistry and photoelectrochemistry*. Wiley-VCH, Weinheim.

Ritter J, Lewis L, Mant T, Ferro A (2008) A textbook of clinical pharmacology and therapeutics.

Rosado, J. A., Redondo, P. C., Salido, G. M., & Pariente, J. A. (2006). Calcium signaling and reactive oxygen species in non-excitable cells. *Mini Reviews in Medicinal Chemistry, 6*, 409–415.

Santoro, F., Zhao, W., Joubert, L. M., Duan, L., Schnitker, J., van de Burgt, Y., Lou, H. Y., Liu, B., Salleo, A., Cui, L., Cui, Y., & Cui, B. (2017). Revealing the cell-material interface with nanometer resolution by focused ion beam/scanning electron microscopy. *ACS Nano, 11*, 8320–8328. https://doi.org/10.1021/acsnano.7b03494. Epub 2017 Jul 21.

Sedra, A. S., & Smith, K. C. (2000). *Microelectronic circuits revised edition* (5th ed.). New York: Oxford University Press, Inc..

Skumiel, A., Kaczmarek-Klinowska, M., Timko, M., Molcan, M., & Rajnak, M. (2013). *International Journal of Thermophysics, 34*, 655–666. https://doi.org/10.1007/s10765-012-1380-0.

Strasser, A., O'Connor, L., & Dixit, V. M. (2000). *Annual Review of Biochemistry, 69*, 217–245. https://doi.org/10.1146/annurev.biochem.69.1.217.

Streetman, B. G., & Banerjee, S. K. (2014). *Solid state electronic devices* (7th ed.). Pearson, Prentice Hall India - New Delhi.

Sung, J. H., Esch, M. B., Prot, J.-M., Long, C. J., Smith, A., Hickman, J., & Shuler, M. L. (2013). Microfabricated mammalian organ systems and their integration into models of whole animals and humans. *Lab on a Chip, 13*, 1201–1212. https://doi.org/10.1039/c3lc41017j.

Tortiglione, C., Antognazza, M. R., Tino, A., Bossio, C., Marchesano, V., Bauduin, A., Zangoli, M., Morata, S. V., & Lanzani, G. (2017). Semiconducting polymers are light nanotransducers in eyeless animals. *Science Advances, 3*(1), e1601699. https://doi.org/10.1126/sciadv.1601699. eCollection 2017 Jan.

Wang, L., Haug, P. J., & Del Fiol, G. (2017). Using classification models for the generation of disease-specific medications from biomedical literature and clinical data repository. *Journal of Biomedical Informatics, 69*, 259–266. https://doi.org/10.1016/j.jbi.2017.04.014.

Yuk, H., Lu, B., & Zhao, X. (2019). Hydrogel bioelectronics. *Chemical Society Reviews, 48*, 1642–1667. https://doi.org/10.1039/c8cs00595h.

Zhang, H., Gutruf, P., Meacham, K., Montana, M. C., Zhao, X., Chiarelli, A. M., Vázquez-Guardado, A., Norris, A., Lu, L., Guo, Q., Xu, C., Wu, Y., Zhao, H., Ning, X., Bai, W., Kandela, I., Haney, C. R., Chanda, D., Gereau, R., & Rogers, J. A. (2019). Wireless, battery-free opto-electronic systems as subdermal implants for local tissue oximetry. *Science Advances, 5*(3), eaaw0873. https://doi.org/10.1126/sciadv.aaw0873. eCollection 2019 Mar.

Chapter 17
Voltage-Sensitive Fluorescent Proteins for Optical Electrophysiology

Teresa A. Haider and Thomas Knöpfel

17.1   Introduction

Creating interfaces that smoothly link artificial devices with the living nervous system has been a longstanding major goal of neural engineering. Such interfaces would provide a platform for both monitoring and steering neuronal circuit activity in real time and applications such as neural prostheses and the repair of central nervous system defects.

Optical imaging using genetically encoded voltage indicators (GEVIs) to monitor neuronal activity could be involved in neural interface engineering in two ways: (i) as a tool to better understand the neuronal side of the interface and (ii) as a component of a light-based interfacing mechanism. Here we focus on the former.

To build such a brain-machine interface, it is essential to understand the information processing underlying the function of the neuronal circuits the interface interacts with. One common approach to investigating neuronal circuit function is to record the electrical activity of the neurons in the circuit. Monitoring of the plasma membrane potential, which underlies neuronal electrical activity, is achieved through techniques that range from single-cell electrophysiology to large-scale electroencephalograms. In addition, techniques that indirectly monitor neuronal activity at larger scale, such as functional magnetic resonance imaging, are available. Thus, the established techniques either provide high temporal resolution of single cells (intracellular electrophysiology) or with coverage of large populations of cells at low spatial resolution (EEG) or high spatial coverage (fMRI) with limited

T. A. Haider
Department of Neurophysiology, Center for Brain Research, Medical University of Vienna, Vienna, Austria

T. Knöpfel (✉)
Laboratory of Neuronal Circuit Dynamics, Department of Medicine, Imperial College London, London, UK
e-mail: tknopfel@knopfel-lab.net

© Springer Nature Switzerland AG 2020
L. Guo (ed.), *Neural Interface Engineering*,
https://doi.org/10.1007/978-3-030-41854-0_17

temporal resolution. None of the techniques routinely used today offers the necessary combined spatiotemporal resolution and coverage of large brain areas to investigate how higher brain functions emerge from the activities of neuronal circuits. Understanding of this link between circuit activity and behaviour would be required to optimally steer circuit activity through an interface.

The usage of electrodes for recording activity from neurons across macro−/mesoscopic (cm to mm scale) areas of nervous tissue, comprising hundreds to thousands of cells, is limited by the density at which electrodes or microelectrode arrays can be placed in the tissue and difficulties to separate sources of activity. Overcoming this limitation, optical approaches to neuronal activity monitoring promise to further our understanding of neuronal circuits by enabling the simultaneous recording of up to thousands of neurons. Several decades ago, pioneers of optical imaging started applying voltage-sensitive dyes (VSDs) to nervous tissue. Using fluorescent VSDs, they were able to significantly improve the spatial resolution and coverage of neuronal activity recordings in cell cultures, acute brain slice preparations and living animals. More recently, the optical imaging toolbox has been expanded by genetically encoded activity indicators whose cell subtype-specific expression allows the precise analysis of the contribution of individual neuronal subtypes to neuronal circuit dynamics. Combining the circuit- and network-level monitoring capabilities of genetically encoded activity indicators with the precise timing and targeting of optogenetic actuators is a powerful approach to building a closed-loop, all-optical neuronal interface for neuronal circuits. In this chapter, we summarize the history, mechanisms, main principles and potential future directions of optical activity imaging. Optogenetic actuators are thoroughly discussed in a separate chapter.

We first give an overview of the different approaches to optical activity imaging, including activity-reporting dyes and genetically encoded activity indicators. We then focus on genetically encoded voltage indicators (GEVIs), whose development of readily applicable probes has recently gained momentum. For more detailed reviews on activity-reporting dyes and genetically encoded calcium indicators, we refer to the many extensive reviews of the literature (Baker et al. 2005; Homma et al. 2009; Garaschuk and Griesbeck 2010; Sasaki 2015; Miller 2016). Next, we explain the main components of optical imaging setups. Finally, we discuss current approaches to the analysis of neuronal activity monitored optically. For readers that are eager to get started in GEVI imaging, we provide a short guide on how to choose an indicator.

17.2 Recording of Neuronal Activity

Neuronal activity is based on rapid (ms scale) changes in the plasma membrane potential, caused by the opening and closing of ion channels. Communication between neurons happens mainly through suprathreshold depolarizations of the membrane potential (action potentials), leading to the release of neurotransmitters at the neuron's presynaptic terminals. The released neurotransmitters diffuse across the synaptic cleft and bind to receptors coupled to ion channels at the postsynaptic site of other neurons, where the binding in turn leads to a renewed depolarization of the postsynaptic neuron's membrane potential. If the postsynaptic membrane is depolarized above its threshold, the postsynaptic neuron fires an action potential. This sequence of events can propagate through the neuronal network.

17.2.1   Electrode-Based Approaches

The most direct way of measuring neuronal activity is to measure the fluctuation of the membrane potential of individual nerve cells. The most direct and widely employed method of measuring membrane potential fluctuations or the underlying ionic currents across the membrane is to gain electrical access to a cell's interior by placing a glass-electrode on the plasma membrane and opening a patch of membrane (patch-clamp in whole cell configuration). This approach allows sampling of the membrane potential (current clamp) or of the flow of ions across the membrane (voltage clamp) and yields an excellent temporal resolution of sub-microseconds. However, the upscaling of this delicate procedure to simultaneous recordings from many neurons is practically limited to a dozen cells (Perin and Markram 2013).

17.2.2   Optical Approaches

Optical approaches to measuring membrane voltages are a powerful alternative when the membrane voltages of larger numbers of neurons are of key interest. Traditional voltage-sensitive dyes are (partially) charged, fluorescent molecules with a lipophilic component that locates in the plasma membrane's electric field. These small organic molecules are designed so that their fluorescence emission depends on the electric field to which they are exposed ("molecular Stark effect"). Once the membrane potential changes, the shift in electric field leads to a shift in charge in the voltage-sensing part of the molecule, which leads to a corresponding change in the fluorescence emission spectrum of the reporter part of the molecule. Combined with optical filters, this spectral change can be translated into changes of fluorescent emission intensity at a certain wavelength. These changes in fluorescent emission intensity are tracked with an optical sensor, such as a camera. In comparison to electrode-based approaches, the light-based approach allows for the recording of average membrane potentials across a large number of cells simultaneously at single (point detector) or many sites (camera) within the field of view without the need to gain physical access to the cells' interior.

Given the direct read-out of the plasma membrane potential, VSDs were envisioned and developed early on in optical activity imaging. Pioneering work by Lawrence B. Cohen, Brian M. Salzberg, Amiram Grinvald and Rina Hildesheim led to the synthesis of VSDs that were successfully applied to physiological systems in vitro and in vivo (Grinvald, Salzberg, and Cohen 1977, Cohen, Salzberg, and Grinvald 1978). Notably, VSDs have enabled the study of large-scale cerebral processing in vivo in nonhuman primates (Blasdel and Salama 1986). VSDs respond to changes in the membrane potential within microseconds and can therefore follow even fast electrical events, such as trains of action potentials in brain slices (Vranesic et al. 1994). The two most popular families of VSDs are the fluorescent styryl RH dyes, named after Rina Hildesheim who first synthesized them (Grinvald et al. 1982; Grinvald et al. 1983), and the ANEP (aminonaphthylethenylpyridinium) dyes, developed by Leslie Loew (Fluhler et al. 1985). However, physiologically important but relatively small fluctuations in the subthreshold membrane potential (often in the millivolt range) proved difficult to translate into changes in emission large enough to surpass shot noise. Improvements to the design of VSDs led to improvements in their voltage-sensitivity, linear voltage-emission changes and larger changes in fluorescent intensity. But as advances lagged behind expectations, researchers also measured changes in neuronal physiology associated with neuronal activity that could serve as a proxy for membrane potential changes and provide a better signal-to-noise ratio (SNR). One such indirect measure of activity is the change in cytosolic calcium concentration associated with strong membrane depolarizations. The resting cytosolic calcium concentration is the result of a homeostatic balance between calcium flows across the membrane via calcium channels from the extracellular space and calcium extrusion mechanism. Calcium dynamics are further complicated by storage and release of calcium from internal stores. When a neuron fires an action potential, voltage-gated calcium channels open and the influx of calcium increases the plasma-free calcium concentration. The rise in calcium associated with individual action potentials can be detected with fluorescent calcium indicators that increase their fluorescence brightness upon binding of calcium. Thus, when an increase in fluorescence intensity is detected, the firing of an action potential is deduced. Action potential activity has been monitored with calcium-sensing dyes in vitro (using fura-2 AM, Mao et al. 2001) and in vivo (Stosiek et al. 2003).

For optical measurements, changes in plasma calcium concentrations have several advantages over membrane potential changes. They happen over a larger volume, last longer and can be measured anywhere in the soma. The calcium-indicator dyes now commonly used, such as fura-2 and Oregon Green BAPTA, use a design that combines a calcium-chelating backbone with a fluorescent indicator. Calcium-sensing dyes that monitor calcium with high affinity robustly report even small increases in cytosolic calcium. However, the binding of calcium to calcium-sensing molecules reduces the share of biologically active, free calcium in the cytosol. With high-affinity dyes or high concentrations of the indicator, this buffering can affect cellular functions.

Depending on the application, the calcium- or voltage-sensing dye is either loaded into individual cells via a patch glass-pipette or applied to larger areas of tissue with one of a number of bulk-loading techniques (e.g. incubation in a volume of dye, pressure-injection). Loading of individual cells enables the investigation of calcium or voltage changes in subcellular compartments. Bulk-loading makes it possible to monitor activity in populations of cells with excellent spatiotemporal resolution across large-scale regions of the central nervous system. To this end, the stained tissue is placed under a microscope equipped with a light detector, often a camera, and illuminated with light from an appropriate source. If neurons in the field of view are active, the acquired images show localized changes in the fluorescence emitted by the activity indicator. Depending on the loading technique and resolution of the imaging system, the activity of individual cells or populations of cells can be monitored. This dye-based approach has been successfully applied to many model systems, such as worms, flies, mice and primates. In the last two decades, few additional improvements to voltage- and calcium-sensitive dyes have occurred. While dyes have a place in certain experiments such as in vivo cortical imaging in primates, their broad routine application in other preparations remains limited. To understand neuronal circuit activity, it is necessary to record the full spectrum of neuronal activity: subthreshold and suprathreshold depolarizations and hyperpolarizations. Voltage-sensitive dyes in theory report the whole activity spectrum, but their poor SNR makes high-resolution and high-speed recordings practically difficult to attain. Calcium-sensitive dyes inherently only report on suprathreshold activity. Moreover, behavioural experiments in rodents and primates call for chronic imaging, which is limited by the toxicity of organic dyes and the ability to deliver the dye in a reproducible manner. Dye-based approaches further have the common drawback that they indiscriminately stain all cells and do not allow the experimenter to differentiate between different subgroups of neurons. With the development of readily available transgenic delivery techniques and advances in protein engineering, protein-based neuronal activity indicators have become the main focus of optical activity imaging development. In comparison to dyes, they promise the specific expression of the activity indicator in identified neuronal subtypes, without the interfering background staining seen with dyes. The next section gives a detailed introduction to genetically encoded activity indicators.

17.3 Genetically Encoded Reporters of Neural Activity

Encoded in a neuron's DNA and expressed as proteins, genetically encoded activity indicators are a promising solution to the quest to record neuron subtype-specific activity in circuits. Neuron subtype-specific promoters limit the expression of the indicator to the neurons of that subtype, enabling the experimenter to draw conclusions about the signalling of a particular group of neurons within a circuit or region. Large SNRs have conferred genetically encoded calcium indicators (GECIs) an enduring success in imaging of populations of neurons in living animals. Genetically encoded voltage indicators (GEVIs), which provide a more direct read-out of neuronal activity, have recently experienced considerable improvements. These improvements make them promising candidates to overtake GECIs in experiments in living animals, reporting on the full spectrum of neuronal activity. In this section, we give an overview of common features of GECIs and GEVIs before focusing in on the history, design principles, major applications and current developments of GEVIs.

17.3.1 Genetic Delivery and Cell Type-Specific Expression

Genetically encoded activity indicators depend on the delivery of the DNA coding for the indicator protein to the target neurons. Gene delivery is achieved via viral vectors, most commonly adeno-associated viruses, in utero electroporation or transgenic methods (Knöpfel 2012). Popular viral vectors, such as adeno-associated viruses (AAVs), are injected into the target tissue where they cause only a mild immune response. Cell-type specificity is achieved via cell-type-specific promoters that are only activated if a transcription factor (specific for the chosen cell type) is present in the target cell. Cell class-specific promoters are available for many neuronal groups, such as glutamatergic and GABAergic neurons. After infection, the activity indicator is expressed in target cells and when sufficient protein has accumulated, often after days to weeks, the tissue is ready for functional imaging. Virus production is relatively fast and can easily be adapted for new indicator constructs and model organisms, making viruses a flexible delivery system. The main drawback of viral delivery is the need to inject virus in each experimental animal: under anesthesia, access to the brain or spinal cord is established surgically and then small volumes of virus are injected locally using a stereotactic apparatus, a frame that uses a three-dimensional coordinate system to locate mapped targets within the nervous system. This invasive procedure can damage nervous tissue, especially when deeper brain regions are targeted. Also, even with standardized equipment and maps, injection sites may vary between animals and even the mild local immune reaction can lead to tissue responses (Ortinski et al. 2010). Expression from viral infection can vary with time due to a time lag to the full extent of expression or reduction in expression over time. These factors limit the applicability of viral delivery for experiments that study inflammation or require multiple imaging sessions over longer time-periods. Recently, promising results have been obtained with AAVs optimized for intravenous administration and subsequent spread throughout the whole central nervous system (Chan et al. 2017). For studies of large-scale neuronal activity, this route of administration circumvents the need for stereotactic injections and reduces local disturbances of the tissue. However, the high amount of virus required for systemic delivery currently limits its use. Alternatively, transgenic animals provide stable and reproducible expression of activity indicators. Popular systems are the Cre-lox and the transactivator systems, which limit expression of proteins to identified subpopulations by combining an indicator line with a population-specific promoter driver line. A growing number of selected voltage and calcium indicator lines are available (Madisen et al. 2015). Different cell type-indicator combinations can be achieved by crossing the selected transgenic lines. Currently, transgenic systems are most readily available in fish, flies and mice. Except for rats and non-human primates, where transgenic lines are not as readily available, transgenic animals provide a popular and reproducible means of expressing activity indicators simply by appropriate breeding schemes.

17.3.2 Genetically Encoded Calcium Indicators

Analogous to calcium-sensitive dyes, GECIs report on the changes in calcium concentration associated with action potentials. For details on physiological calcium changes, see Sect. 17.2.2. Given the excellent SNR of GECI variants from the GCaMP family (Nakai et al. 2001), they allow the detection of neuronal firing in many neurons simultaneously. GCaMPs have been successfully adopted for in vivo and in vitro experiments in rodents, flies, worms and other common species studied in neuroscience. Moreover, GECIs expressed in astrocytes have contributed to our understanding of how glial calcium changes relate to neuronal circuit activity (Shigetomi et al. 2010).

However, analogous to calcium-sensitive dyes, GECIs do not report all aspects of neuronal activity. As calcium levels mainly correlate with strong, suprathreshold depolarizations, subthreshold depolarizations and hyperpolarizations in the membrane potential are not readily detected with GECIs. In addition, cytosolic calcium changes are slower than voltage changes. Consequently, the timing of action potentials cannot easily be deduced from calcium signals, and individual spikes in trains of multiple, narrowly spaced action potentials can only be estimated from calcium indicator traces using mathematical deconvolution techniques. GEVIs promise to resolve individual action potentials and record the full range of neuronal electrical signalling. Even though the first GECIs and GEVIs were developed around the same time, GECIs were more rapidly adopted across neuroscientific laboratories, owing to their higher brightness and resulting better SNR. The more slowly progressing development to broad applicability of GEVIs is largely due to inherent differences between measuring calcium versus voltage changes of living cells. Calcium concentration changes occur throughout the plasma, while voltage changes are limited to the plasma membrane, which can accommodate only a fraction of indicator molecules as compared to the cytosolic space. In addition, voltage changes occur at a higher speed than changes in calcium concentrations. The higher camera frame rates that sample these faster changes reduce the number of photons captured per frame, demanding higher fluorescence emittance from GEVIs relative to GECIs.

17.3.3 Genetically Encoded Voltage Indicators

The idea to use a protein that exhibits voltage-dependent fluorescence to study electrical activity in neurons emerged in the 1990s. Scientists hoped for a less invasive and more cell type-specific delivery of the probe in comparison to voltage-sensitive dyes. Multiple GEVI design principles have emerged (Fig. 17.1, Table 17.1). The core principle of voltage-sensing domain-based GEVIs is the modulation of the emission of a fluorescence domain by a voltage-sensitive domain. The discovery of voltage-dependence in the bacterial opsin Archaerhodopsin 3 led the way to the first opsin-based GEVI, Arch (Kralj et al. 2011a). In contrast to voltage-sensing domain-based GEVIs, opsin-based GEVIs sense and report voltage changes through the same domain. Their fast kinetics make opsin-based GEVIs an attractive alternative to voltage-sensing domain-based GEVIs. The broad applicability of opsin-based GEVIs will hinge on reducing the light intensity required to excite them. The demands on a successful GEVI are high: it should report on membrane voltage changes with millisecond speed, show high voltage-dependent fluorescence and be bright to reduce the intensity of light needed and thus reduce phototoxicity. It should further display low bleaching to allow recordings in vivo at timescales relevant to behaviour and not disturb the intrinsic membrane properties of neurons. Moreover, for compatibility with optogenetic actuators, different spectral variants should be available. Currently, no single GEVI exists that fulfils all demands. However, a wide variety of GEVIs optimized for different performance characteristics have been developed. For up-to-date lists of available variants, it is best to consult reviews of the current literature (Lin and Schnitzer 2016, Bando et al. 2019), which are published at regular intervals. This section gives an overview of the different design approaches to GEVIs.

Voltage-Sensing Domain-Based GEVIs

The first GEVIs were generated by inserting a green fluorescent protein (GFP) variant into voltage-gated K^+ and Na^+ channels. For the creation of FlaSh, GFP was attached to the pore domain of a nonconducting mutant of the shaker K^+ channel (Siegel and Isacoff 1997). GFP's position close to the moveable C-terminal end of the pore domain rendered GFP's fluorescence dependent on conformational rearrangements in the K^+ channel (C-type inactivation). SPARC, built according to the same basic principle, used a voltage-sensing domain derived from a Na^+ channel found in rat muscle (Ataka and Pieribone 2002). VSFP1 was the first GEVI to use the conformational changes in the voltage-sensing domain to alter the physical distance of a coupled fluorescence resonance energy transfer (FRET) donor and acceptor pair, translating membrane voltage changes into changes in FRET efficiency. To this end, a voltage-sensing domain from a K^+ channel was coupled to a pair of cyan and yellow fluorescent variants of GFP (Sakai et al. 2001). These first GEVIs were able to optically report membrane-voltage changes; however, GEVIs using voltage-gated ion channels as their voltage-sensing domain share some common limitations, such as a narrow voltage-response range, low fractional fluorescence changes and suboptimal trafficking to the plasma membrane.

In the next generation of voltage-sensing domain-based GEVIs, the voltage-sensing domain of VSFP1 was replaced by voltage-sensitive phosphatases derived from *Ciona intestinalis* (Ci-VSP, Murata et al. 2005) to form the series of VSFP2

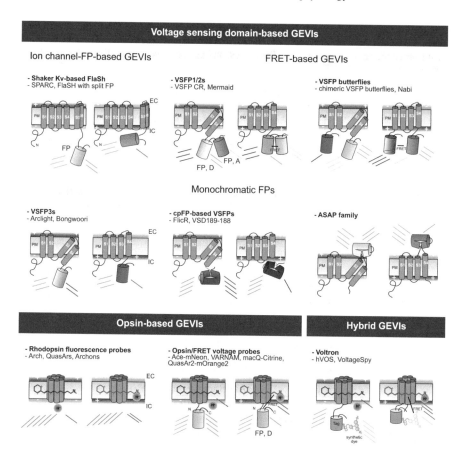

Fig. 17.1 Overview of design principles of genetically encoded voltage indicators (GEVIs). The plasma membrane (PM)-anchoring domains of the GEVI are depicted as numbered barrels (*grey*). Fluorescent protein (FP) domains are attached to the membrane-anchoring domains on the intracellular (IC) or extracellular (EC) side of the membrane. The top two rows show GEVI designs based on voltage-sensing domains (isolated from Kv channels or voltage-dependent phosphatases). GEVIs that work via fluorescence resonance energy transfer (FRET) depend on a pair of FRET donor (FP, D) and FRET acceptor (FP, A) FPs. The voltage domain's conformation directly influences the distance or orientation of the donor and acceptor, modulating the energy transfer efficiency upon PM polarization changes. Voltage-sensitive fluorescent proteins (VSFPs) that use FRET modulation include VSFPs of the 1st and 2nd series, VSFP butterflies and chimeric VSFP butterflies. The second row depicts monochromatic voltage sensing domain-based GEVIs. Monochromatic GEVIs of the VSFP3 type only have one FP, whose fluorescence brightness is modulated by a voltage-sensing domain. Circular permutation (cp) of the FP in cpFP-based VSFP3s and FlicR increases its sensitivity to conformational changes. Similarly, the ASAP GEVI family uses a cpFP, inserted into an extracellular loop of a voltage-sensing domain. Opsin-based GEVIs fall into two groups: rhodopsin fluorescence probes, such as Arch, the QuasArs and the Archons, rely only on the modulation of retinal fluorescence for reporting membrane voltage; opsin/FRET combination probes have an additional attached FP, which is quenched by the retinal in the opsin in a voltage-dependent manner. Ace-mNeon and macQ-Citrine follow this principle. Hybrid GEVIs combine genetically targetable proteins with exogenous small molecules. Hybrid designs comprise opsin-dye FRET (Voltron), FP-dye FRET (hVOS) and photo-induced electron transfer (VoltageSpy). The Voltron design is shown as a representative of this class. Voltron uses a protein tag covalently bound to a linker molecule to bind to an exogenous, synthetic dye that acts as a FRET donor to retinal. (Illustration adapted and updated from Akemann et al. 2015)

Table 17.1 This table lists the most commonly used GEVIs, categorised first by their voltage-sensing principle (voltage domain, opsin, hybrid), second by indicator class. GEVIs are arranged in the order they appear in Fig. 17.1 The corresponding references guide the reader to the first publication of particular GEVIs

Indicator class	Indicator	Reference
Voltage domain-based GEVIs		
Ion channel-FP	Shaker Kv-based FlaSH	Siegel and Isacoff (1997)
	SPARC	Ataka and Pieribone (2002)
	FlaSH with split FP	Jin et al. (2011)
FRET	VSFP1/2s, CR	Sakai et al. (2001), Dimitrov et al. (2007), Lundby et al. (2008), Mutoh et al. (2009), Lam et al. (2012)
	Mermaid	Tsutsui et al. (2008)
	VSFP butterflies	Akemann et al. (2012)
	Chimeric VSFP butterflies	Mishina et al. (2014)
	Nabi	Sung et al. (2015)
Monochromatic FPs	VSFP3s	Lundby et al. (2008), Perron (2009); Perron et al. (2009)
	Arclight	Jin et al. (2012)
	Bongwoori	Piao et al. (2015)
	cpFP-based VSFPs	Gautam et al. (2009)
	FlicR	Abdelfattah et al. (2016)
	VSD189-188	Kost et al. (2017)
	ASAP family	St-Pierre et al. (2014), Platisa et al. (2017), Chamberland et al. (2017)
Opsin-based GEVIs		
Rhodopsin fluorescence probes	Arch	Kralj et al. (2011a, b)
	QuasArs	Hochbaum et al. (2014)
	Archons	Piatkevich et al. (2018)
Opsin/FRET voltage	Ace-mNeon	Yang et al. (2016)
	VARNAM	Kannan et al. (2018)
	macQ-Citrine	Gong et al. (2014)
	QuasAr2-mOrange2	Zou et al. (2014)
Hybrid GEVIs		
FRET-dye	Voltron	Abdelfattah et al. (2018)
Other	hVOS	Chanda et al. (2005)
	VoltageSpy	Grenier et al. (2019)

variants (Dimitrov et al. 2007; Lundby et al. 2008; Mutoh et al. 2009). In addition to its better voltage sensitivity, Ci-VSP's voltage-sensitive domain is structurally simpler, facilitating its transport to the cell membrane. VSFP2s were able to resolve single action potentials in cultured cells (Dimitrov et al. 2007). The Ci-VSP proved a successful scaffold and further GEVIs were developed based on its design:

VSFP3s and ArchLight, whose single fluorescent protein reports voltage changes monochromatically (Lundby et al. 2008; Han et al. 2013); the VSFP butterflies, in which the voltage-sensitive domain is placed between a FRET donor and acceptor, giving them their name (Akemann et al. 2012; Mishina et al. 2014); and Bongwoori, a combination of a pH-sensitive indicator and Ci-VSP, which can resolve action potentials at 65 Hz (Lee et al. 2017). VSFP Butterfly 2.1 reports neuronal activity from acute brain slices with widefield and two-photon illumination (Fig. 17.2). More recently, placing the voltage-sensitive phosphatase from *Gallus gallus* (Gg-VSP) together with a circularly permuted GFP gave rise to the ASAP series of GEVIs (St-Pierre et al. 2014).

Opsin-Based GEVIs

In contrast to voltage-sensing domain-based GEVIs, opsin-based GEVIs sense and report membrane voltage changes via the same protein domain. The molecular backbone is derived from microbial rhodopsins, light-sensitive ion channels that play a role in visual phototransduction. The opsin protein is bound to retinal, its chromophore—a small, light-absorbing molecule. The first opsin-based GEVI, PROPS, reported on electrical spiking in *E. coli*, but could not be optimized to localize to the plasma membrane in eukaryotic cells (Kralj et al. 2011a, b). Following opsin-based designs were based on Archaerhodopsin 3 (Arch; Kralj et al. 2011a, b), whose good localization to mammalian membranes had been established during its use as an optogenetic silencer. Arch's sub-millisecond temporal resolution for neuronal action potentials came with a tradeoff of a laser-induced, hyperpolarizing photocurrent that was strong enough to reduce neuronal firing. Point mutations eliminated the photocurrent but also slowed the probe so that it could no longer resolve action potentials (Kralj et al. 2011a). More recent designs based on Arch have used targeted evolution and rational design to improve its speed, brightness and sensitivity. The QuasArs were combined with a channelrhodopsin actuator for a proof-of-principle of all-optical electrophysiology (QuasArs, Kralj et al. 2011a; Hochbaum et al. 2014). The Arch-derived family of GEVIs is constantly growing. The main challenge to the application of Arch-based probes beyond specialized laboratories is their low quantum yield, requiring high illumination densities that significantly heat the imaged tissue. To improve brightness but retain the opsin's fast kinetics, opsins have been used in FRET pairs with a brighter fluorescent protein (Zou et al. 2014; Gong et al. 2014). In this approach, the opsin quenches the fluorescent protein in a voltage-dependent manner (MacQ-GEVIs, Gong et al. 2014). Limits to opsin-FRET GEVIs are the linker length and challenges in localizing them to the neuronal membrane (Gong et al. 2015; Werley et al. 2017). Employing high illumination and specialized equipment, opsin-based GEVIs have been successfully used to record voltage from single neurons in brain slices and from mammals in vivo (Gong et al. 2015; Werley et al. 2017; Piatkevich et al. 2018).

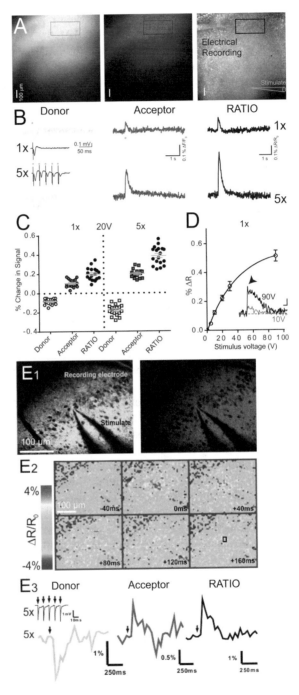

Fig. 17.2 Widefield and two-photon voltage imaging of mouse cortical L2/3 pyramidal neurons in brain slices. The GEVI VSFP Butterfly was expressed in L2/3 pyramidal neurons in mice. Optical and electrical recordings were performed simultaneously. (**a**) widefield images of donor and acceptor fluorescence on the left and in the middle; on the right, a widefield image and illustration

Hybrid GEVIs

Improving GEVI performance, such as speed and brightness, can be tedious because of the complexity of protein expression and the limitations set by amino acid chemistry. Currently available fast GEVIs have time constants around 1 ms. In comparison, commonly used voltage-sensitive dyes have time constants around 1–2 μs (Rohr and Salzberg 1994). In an attempt to circumvent these common challenges, hybrid probes make use of an exogenously applied component that specifically binds to a genetically targeted component. The exogenous component can either fulfil the role of the voltage-sensitive domain or of the modulated fluorophore. For the hybrid voltage sensor (hVOR), researchers express a farnesylated GFP whose fluorescence is absorbed by the synthetic voltage-sensing dipicrylamine (DPA) via FRET (Chanda et al. 2005). DPA moves across the cell membrane in a voltage-dependent manner, rendering GFP's fluorescence voltage-dependent. Even though hVOR's fractional fluorescence changes were larger than other GEVIs available at the time, its applicability is limited by DPA's toxicity and ability to raise the membrane capacitance, influencing population dynamics (Chanda et al. 2005). The hybrid approach has recently seen a revival, with new probes attaching a voltage-sensitive dye-binding domain to an opsin-based domain in a manner that allows for the FRET modulation of the opsin by the exogenously applied VSD. In promising early results, Voltron resolves action potentials in single mouse hippocampal neurons in vivo (Abdelfattah et al. 2018). Drawbacks to hybrid strategies are similar to challenges with voltage-sensitive dyes: the difficulty to reproducibly and uniformly deliver an exogenous substance to living tissue.

Fig. 17.2 (continued) of the placement of the stimulation and recording electrodes during electrical recordings. (**b**) Population activity in response to electrical stimulation to the cortex. Fluorescence traces from the FRET donor (mCitrine) are drawn in yellow, from the FRET acceptor (mKate2) in red. The ratio of the two traces is drawn in black. The top trace shows the population response to a single stimulus (average of 10 sweeps). Below, the local field potential (LFP) recorded with an electrode in response to the same stimulus. The bottom traces show population responses to 5 stimulations (average of 10 sweeps). Below, the LFP in response to the same 5× stimulus. The FRET modulation leads to a decrease in donor intensity and its corresponding increase in acceptor intensity. (**c**) The ratio of FRET donor and acceptor has an improved signal-to-noise ratio in comparison to the single fluorescence signals for both single and five-stimulus recordings (mean and SEM in grey). **d**) The Butterfly VSFP signal increases (calculated as % change in ratio) with increasing stimulus voltage ($n = 5$–8 for each intensity). The inset shows an example recording where the fast (arrowhead) and slow component of the GEVI signal are discernable. (**e**) Two-photon imaging with VSFP Butterfly. (**e1**) mCitrine (left) and mKate2 (right) fluorescence images with the stimulation electrode visible in the lower right and the recording electrode in the upper right corner. (**e2**) A time series of fluorescence ratio images ($\Delta R/R$) recorded at the site of E1. Frames were recorded at 40 ms intervals and colour coded for fluorescence intensity changes (calibration bar shown on the left). At 0 ms, a stimulus (5 pulses, 0.2 ms, 100 Hz) was applied via the stimulation electrode. Note the depolarized areas around the tip of the stimulation electrode at $t = 0$ and the spreading depolarization at $t = 40$ ms. (**e3**) Fluorescence trances (average of 25 trials) from regions of interest (ROIs) in E2. Black arrows indicate the time of stimulation. Inset shows the LFP in response to the same stimulation. (Adapted from Empson et al. 2015)

17.3.4   Choosing a GEVI

Continuous improvements to the original GEVI variants through rational design and targeted evolution have resulted in a broad palette of GEVIs optimized for different performance indicators such as response kinetics, dynamic voltage range, SNR, quantum yield or spectral characteristics. As no single GEVI covers all applications, it is essential to consider individual experimental demands and to choose a GEVI that best fits the particular conditions.

Recording Membrane Voltage In Vivo

If experiments are to be conducted in mammals in vivo, important considerations are biological sources of noise and the required response kinetics and dynamic range for the biological phenomenon to be investigated. Common sources of fluorescence changes that do not directly originate from membrane voltage fluctuations are hemodynamic noise and so-called intrinsic optical signals. Applications based on monochromatic indicators are particularly susceptible to these forms of noise. Ratiometric FRET indicators, which absorb and emit fluorescence at two wavelengths, provide an optical signal that is more robust to biological noise and have successfully been used to record cortical activity dynamics in mice in vivo (Fig. 17.3, Akemann et al. 2012; Song et al. 2018a). For many available GEVIs, there is a tradeoff between fast response kinetics and large SNR. Instead of selecting the fastest available probe, it is advisable to consider the dynamics of the neuronal phenomenon to be studied. If the focus is to be on population responses across wide areas of tissue, it is not necessary to resolve membrane voltage changes with 1–2 ms, as many cortical population phenomena occur at time scales of tens of milliseconds to seconds. In this case, a slower probe with bigger fractional changes in fluorescence can be selected. Similarly, some probes' dynamic ranges are better suited than others for the detection of action potentials or subthreshold events. For example, VSFP Butterfly 1.2 can detect subthreshold events and chimeric VSFP butterfly has been employed for cortical population imaging in awake, behaving mice (Fig. 17.4, Akemann et al. 2012; Song et al. 2018a) .

Combination with Optogenetic Actuators

In all-optical electrophysiology, both neuronal activity monitoring and activity perturbations are achieved using light-based approaches. To combine monitoring membrane voltage with optogenetic control of cellular activity in the same tissue without crosstalk between the systems, it is essential to reduce to a minimum the spectral overlap of the actuator and the reporter. Most optogenetic actuators respond to light in the blue range: wild-type channelrhodopsin has its excitation maximum at 480 nm. So far, most red-shifted GEVIs have been variants of Arch, as Arch itself is

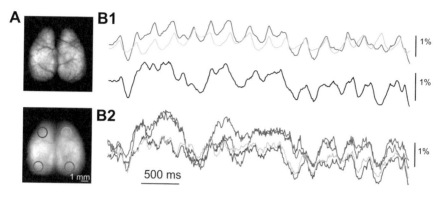

Fig. 17.3 Widefield voltage imaging in living mice using VSFP Butterfly 1.2. The GEVI VSFP Butterfly was expressed transgenetically in cortical LII/III pyramidal neurons in mice. (**a**) Top, dorsal view on the brain of a mouse with a chronically implanted transcranial window over both hemispheres. Lower, outline of four ROIs: (i) top left: left motor cortex (*navy*), (ii) top right: right motor cortex (*cyan*), (iii) lower left: left visual cortex (*red*), (iv) lower right: right visual cortex (*pink*). (**b**) Spontaneous ("ongoing") activity (in absence of sensory stimuli) under light sedation monitored via GEVI fluorescence. (**b1**) Fluorescence traces show the intensity of the FRET donor (*green*) and the FRET acceptor (*orange*) averaged across both hemispheres. The ratio of the two traces is shown in black (lower trace). (**b2**) Ratiometric fluorescence signal of intrinsic activity across the four ROIs outlined in A lower, showing isotopic cortical activity travelling across large distance in both hemispheres. Data collected as described in Akemann et al. 2012 and Carandini et al. 2015. (Adapted from Song et al. 2017)

naturally red-shifted. However, these probes also inherited Arch's low molecular brightness (Kralj et al. 2011a; Hochbaum et al. 2014; Flytzanis et al. 2014). More recently, a directed evolution approach yielded the Archon probes with improved brightness and increased fluorescence changes per action potential (Piatkevich et al. 2018).

17.4 Equipment for Optical Imaging of Neuronal Activity

Optical voltage reporters typically respond with only small changes in fluorescence intensity upon a physiological membrane voltage fluctuation. It requires highly specialized microscopy equipment to record these small optical signals. The excitation light source, shutter, objective, filters and optical sensor are optimized for sensitivity, noise levels and speed. Particular applications, such as mesoscopic or in vivo imaging, can pose additional demands on the system. The ability to detect meaningful changes in signal depends crucially on the degree to which the signal exceeds the noise of the system. The SNR expresses this relationship. Signal detection can be improved by either increasing the amplitude of the signal or decreasing noise. In our application, the signal is the change in fluorescent light intensity in a particular band of the light spectrum, emitted by the activity indicator in response to a membrane potential-altering event of interest. Noise can come from many sources and what is

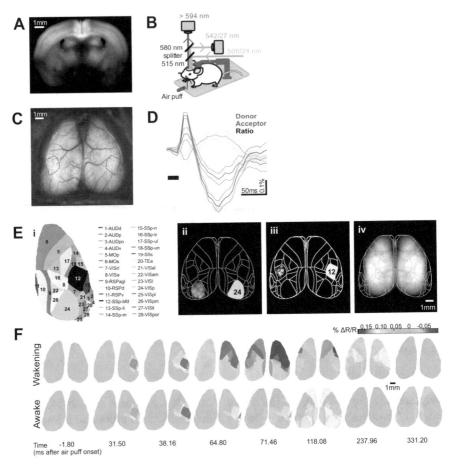

Fig. 17.4 Widefield voltage imaging from defined brain areas in a living mouse. The GEVI chimeric VSFP Butterfly was transgenetically expressed in mice. (**a**) Fluorescence image of a brain slice prepared from a mouse expressing chimeric VSFP butterfly. (**b**) Experimental set-up of a voltage imaging recording in living mice. The mouse with an implanted transcranial window is head-fixated. 500 ± 12 nm light and captured at two wavelength bands (542/27 and > 594 nm) by two synchronized cameras. The whiskers are stimulated using air puffs. (**c**) Dorsal view of the two hemispheres through the chronic, transcranial window. Red outlines mark the primary somatosensory barrel field, which is associated with processing of whisker sensory stimuli. (**d**) Voltage responses in the primary somatosensory barrel field (shown in **c**) in response to an air puff directed at the contralateral whiskers, monitored via dual-emission GEVI fluorescence. Note that FRET donor and acceptor intensity changes are anti-correlated. The ratiometric signal (ratio) is shown in black. Increases in the optical traces indicate depolarization and decreases indicate hyperpolarization. Mean ± SEM shown. (**e**) Brain areas within the cranial window as defined in the Allen Mouse Brain Atlas (abbreviations: 1-AUDd Dorsal auditory area, 2-AUDp Primary auditory area, 3-AUDpo Posterior auditory area, 4-AUDv Ventral auditory area, 5-MOp Primary motor area, 6-MOs Secondary motor area, 7-VISrl Rostrolateral visual area, 8-VISa Anterior area, 9-RSPagl Retrosplenial area-lateral agranular part, 10-RSPd Retrosplenial area- dorsal part, 11-RSPv Retrosplenial area-ventral part, 12-SSp-bfd Primary somatosensory area- barrel field, 13-SSp-ll Primary somatosensory area- lower limb, 14-SSp-m Primary somatosensory area- mouth, 15-SSp-n Primary somatosensory area-nose, 16-SSp-tr Primary somatosensory area- trunk, 17-SSp-ul Primary somatosensory area-upper limb, 18-SSp-un Primary somatosensory

considered noise in one set of experiments could be of interest in another, such as very small fluctuations in the membrane potential. The most relevant noise components in optical imaging of neuronal activity are: biological noise, photon noise and camera noise.

Beyond small fluctuations in the membrane potential that directly contribute to noise, biological noise can stem from physiological movement and fluorescent biomolecules. In vivo, a prominent source of confounding signal components is blood vessels that cross the field of view. They contaminate the signal due to changes in tissue absorbance pulsating with the heartbeat and by the shift of the haemoglobin absorption spectrum associated with changes in its oxygenation level (Ferezou et al. 2016). Strategies to avoid this contamination include fluorescence indicators with far red-shifted absorption and emission spectra and FRET indicators. Photon noise, or shot noise, is a consequence of the discrete, quantized nature of light. Instead of a continuous stream of light, individual photons hit the light detector. Even with continuous illumination, the number of photons (N_{ph}) that hit an individual pixel of the detector varies from frame to frame according to a Poisson distribution. At high light intensities, the variance of the number of incident photons ($\sqrt{N_{ph}}$) from frame to frame is low in comparison to the total number of incident photons. However, at low light intensities, the variance is relatively higher and its contribution to noise as shot noise more disturbing. As a consequence, in a system mostly limited by shot noise, higher illumination increases the SNR $\left(\sim \dfrac{1}{\sqrt{N_{ph}}} \right)$.

Camera noise has temperature-dependent and independent sources. Depending on temperature, electrons in the camera chip are randomly excited and create a dark current. This dark current, also called dark shot noise, can be reduced by cooling the sensor. Temperature-independent components of camera noise originate from electronic signal production (read noise) and sensor design.

The SNR of the optical membrane voltage recording can directly be increased by improving the indicator's brightness, resistance to bleaching and sensitivity to voltage changes. How such improvements can be achieved is discussed in the section on the development of GEVIs. In this section, we discuss the most important

---

Fig. 17.4 (continued) area-unassigned, 19-SSs Supplemental somatosensory area, 20-TEa Temporal association areas, 21-VISal Anterolateral visual area, 22-VISam Anteromedial visual area, 23-VISl Lateral visual area, 24-VISp Primary visual area, 25-VISpl Posterolateral visual area, 26-VISpm posteromedial visual area 27-VISli Laterointermediate area, 28-VISpor Postrhinal area). (ii-iii) Voltage maps evoked by visual (ii) and somatosensory (iii) stimulation (normalized grey scale) registered to the Allen Mouse Brain Atlas (Lein et al. 2007) with outline of cortex areas shown in (i). Evoked responses were thresholded at 10% of peak amplitude. Pixels below the threshold and pixels within areas outside the cortex were set to zero (represented in *black*). (iv) Fluorescence image of registered cranial windows averaged over 5 mice with outline of all cortical areas that project to the dorsal view. Note that lateral cortices are only partially accessed by the cranial window. (**f**) Voltage maps registered to the Allen Mouse Brain Atlas at selected times after stimulation (Wakening condition: $N = 13$, 5 mice; Awake condition: $N = 16$, 5 mice; Grand average shown. (Adapted from Song et al. 2018a)

microscopy components whose optimization yields the biggest improvement to signal quality or whose adaptation extends the range of applications.

Illumination Incandescent lamps, light-emitting diodes (LEDs) and lasers (light amplification by stimulated emission of radiation) are the standard choices for sample illumination. The incandescent light sources commonly used in laboratories, halogen and mercury lamps, emit light at varying intensities across the whole visible light spectrum. To limit the excitation bandwidth, they are paired with excitation filters (band-pass filters). Incandescent light sources are comparatively inexpensive, but the maximum achievable light intensity is lower than with the other options. LEDs emit light within a smaller spectral band than incandescent light sources and a broader band than lasers. For most voltage imaging applications, LEDs offer sufficient illumination intensity and stability. A special application of lasers is two-photon imaging, where a fluorophore is hit with two coincident photons that in combination excite the fluorophore. The advantage of this approach is that the incident photons can have a lower wavelength than photons that individually carry the energy to excite the fluorophore, resulting in less damage to the imaged tissue, which is especially relevant in in vivo preparations. In addition, deeper and more targeted illumination is possible with two-photon imaging. So far, only a small range of voltage indicators is available for two-photon imaging, owing to higher demands on the sensitivity of the probe. These two-photon imaging compatible indicators, VSFP Butterfly 1.2 and ASAP2s, have been successfully applied to record voltage from defined neurons in the brain (Fig. 2E, Akemann et al. 2013; Empson et al. 2015; Chamberland et al. 2017).

Improvements to the SNR with illumination can be achieved by choosing a stable light source and a high illumination intensity, which relatively reduces the prominence of shot noise. However, illumination intensity is limited by bleaching, heating and phototoxicity to the imaged tissue. A shutter blocks the light path to the specimen when the illumination is not needed and therefore extends the maximum possible imaging time. Mechanical shutters can be a source of vibration, i.e. noise, and are therefore often placed off the microscopy table to reduce their impact on the signal.

Microscope and Optical Sensor The basic microscope setup is chosen according to particular experimental needs: subcellular resolution requires high magnification, while mesoscopic imaging across multiple brain regions requires a wide field of view; systems used for in vivo recordings should ideally be portable and ready for chronic recordings; ratiometric indicators require simultaneous dual wavelength detection. Common tradeoffs are imaging speed versus the number of signal photons detected per frame, and field of view size versus magnification. The microscope's objective and filters fundamentally determine how much light emitted by the voltage indicator reaches the optical sensor. Specialized objectives with high numerical aperture (a dimensionless number indicating the range of angles where the objective gathers light) maximize the light collected from the specimen. Excitation filters should be chosen to limit light reaching the specimen to the wavelength used to excite the fluorophore. Superfluous light outside the excitation wavelength can contribute to phototoxicity, sample heating and leaked light to the emission spectral band. Emission filters block excitation light. They should be chosen as to maximally separate the excitation and emission wavelengths and to maximize the fractional change in fluorescence in response to membrane voltage changes. This is commonly achieved by selecting a spectral band where the fractional change is highest and filtering out the emitted light from other wavelengths. High-quality filters that transmit more than 90% of incident light are preferable.

The most common optical sensors in neuronal activity imaging are CCD (charge-coupled device) and sCMOS (scientific [grade] complementary metal–oxide–semiconductor) cameras. Both camera types consist of arrays of light sensors ("pixels") that collect and detect photons with high sensitivity and quantum efficiency. They differ in the way they register and convert the pixel photon count to an electrical signal. Most importantly, CCD cameras typically rely on a single analog-digital (AD) converter and amplifier, whereas sCMOS cameras have one per pixel. As a consequence, each AD converter in a sCMOS camera has more time to process each electron, reducing its read noise and enabling faster frame rates. However, overall CCD cameras have lower noise than sCMOS cameras.

At low light conditions, it is advisable to select the frame rate to fit the time course of the signal of interest and not to image at the fastest frame rate possible with a given sensor. This increases the number of photons collected per frame, thereby improving the SNR. Similarly, binning of pixels, i.e. collecting and reading out photons of adjacent pixels on the sensor together, can improve SNR at the cost of spatial resolution. As discussed above, some camera models can be cooled to reduce dark noise.

Many voltage indicators show a tight dependence of fluorescence emission on ambient temperature. In vitro, the temperature of the recording chamber in the microscope might therefore be regulated with appropriate equipment.

Taken together, careful assembly of optimized microscopy parts is essential for imaging the weak signals of optical activity imaging. As a guiding principle, the signal photon count per frame should be maximized while other sources of light and noise should be minimized, even at the cost of spatial or temporal resolution. As imaging of cellular activity in the central nervous system of freely behaving animals is gaining popularity, so do miniaturized, portable microscopes that can be mounted directly on animals, especially rodents (Piyawattanametha et al. 2009; Ghosh et al. 2011). Adapted lenses and optical fibres that can be inserted deeply into the brain open up the possibility to image activity from previously inaccessible brain areas in vivo, such as the hippocampus or the limbic system. Developments in this regard, increasing spatial coverage while improving signal quality, will vastly expand the applications of functional imaging of neurons.

17.5   Analysis of Optical Neuronal Activity Data

Most often, the output of neuronal activity imaging is an image series, with temporal resolution given by subsequent image frames. How this image series is further processed and analysed is fundamentally determined by the choice of activity indicator. Calcium imaging data are commonly processed with the aim of detecting action potentials on a single-cell level. Voltage imaging data currently often do not have single-cell resolution but offer rich, full voltage-spectrum information across millimetres of nervous tissue. With the goal of closed-loop neural interfaces in mind, fast, accurate and reproducible analysis of signals is essential. This section outlines a typical processing and analysis workflow of neuronal imaging data. As with other aspects of cellular activity imaging, the analysis process is guided by improvements to the modest signal size by reducing noise and variability.

First, image stacks are preprocessed: Raw image stacks are imported to an image processing program, such as ImageJ or Matlab. At this stage, images are often binned by a factor of 2×2 or 4×4 pixels to reduce image size and speed up processing. To improve event detection, stacks can be detrended, either globally or pixelwise, removing bleaching and temperature-related changes in baseline fluorescence. At this stage, movement and drift artefacts can be reduced. Scripts for these applications are widely available and flexible in their input. To improve SNR, multiple trials of stimulus-evoked recordings can be aligned in time and averaged. This process reduces noise and variability, but can also obscure activity components (Alexander et al. 2013).

Next, image analysis depends on the question on hand. Indicators based on the detection of calcium levels result in image series with activity-dependent changes mostly localized to the soma. Analysis focuses on the localization of cell bodies, resolution of overlapping cells and measurement of fluorescence intensity averaged over the cell body per frame. Once this fluorescence intensity trance is extracted, methods well established for electrode-based recordings can be applied. Spike detection, spike sorting and deconvolution to separate narrowly spaced events have been refined and can now be reliable and fast. Using this approach, spiking activity of assemblies of dozens to a hundred neurons has been reported in vivo using calcium indicator dyes (Stosiek et al. 2003) and GECIs (Nguyen et al. 2016). Fluorescence changes from voltage-sensitive indicators are typically targeted to the neuronal membrane. As neuronal dendrites contribute the biggest share of neuronal membrane, the fluorescence changes detected with voltage-sensitive indicators can mostly be attributed to the dendrites, with the notable exception of soma-targeted GEVIs. Consequently, voltage indicators offer a spatiotemporal richness unprecedented by electrophysiological and calcium imaging data. Adding to the complexity, voltage-sensitive indicators report on the full spectrum of membrane potentials, from hyperpolarization to synaptic events and action potential spikes. While spike detection is possible if the resolution is sufficient, reducing voltage-sensitive indicator data to suprathreshold events discards a wealth of information. Existent data offer a glimpse at the intricate and diverse patterns of cellular activity recorded with voltage imaging, including travelling waves, rings, zigzags, stationary bumps and spirals (Urbano, Leznik, and

Llinás 2002; Shimaoka and Knöpfel 2017; Song et al. 2018a). We are at the early stages in the development of the tools to detect and quantitatively describe such patterns (Fagerholm et al. 2018; Muller et al. 2018). Pattern detection could help with reducing the dimensionality of the information contained in the images. Moreover, a uniform classification of dynamic neural activity patterns will be essential for reproducible research and the development of tools, such as neural interfaces, that build upon the detection of complex neuronal activity. While it has been proposed how such classifications could look like (Song et al. 2018b), most likely the majority of patterns are still to be discovered. Methods derived from visual pattern recognition, artificial neural networks and fluid dynamics could inform the development of tools to extract patterns, measure local speeds and track waves and introduce the necessary abstraction in the analysis of activity data (Muller et al. 2018). Throughout the processing and analysis of images, it is of utmost importance to ensure reproducibility of results. To this end, scripts and standardized workflows should be used whenever possible.

Finally, extracting activity profiles of cell assemblies and describing activity patterns of populations are only the basis for putting them into the broader context of their biological function. In this regard, we are only at the beginning of analysing neuronal activity across space and time and making sense of the functional differences between neuronal activity patterns.

17.6   Conclusion and Outlook

In this chapter, we have outlined the history and principles of sensing neuronal activity with fluorescent indicators. We have sketched out how all-optical electrophysiology could be achieved using genetically encoded activity indicators and optogenetic actuators. We have focused on GEVIs, which offer a promising approach to activity imaging by expanding the range of membrane events to subthreshold depolarization and hyperpolarization. We have given an overview of the optical equipment used and how to optimize signal strength. Finally, we have outlined the analysis of data obtained with fluorescent activity probes. While we have stressed promising new developments where appropriate, GEVIs in particular have seen a recent spur in innovative protein engineering approaches and resulting improvements to performance and applicability. Improved GEVIs in the near infrared range promise to soon make all-optical electrophysiology a reality (Kannan et al. 2018; Monakhov et al. 2019). A new hybrid GEVI that combines a genetically targeted voltage-sensitive domain and protein tag with a photostable synthetic dye for high SNR single-trial recordings (Abdelfattah et al. 2018) has redirected interest to the hybrid GEVI approach. Taken together, these advances make a convincing case that the exciting times of neuronal activity imaging are yet to come.

References

Abdelfattah, A. S., Farhi, S. L., Zhao, Y., et al. (2016). A bright and fast red fluorescent protein voltage Indicator that reports neuronal activity in organotypic brain slices. *The Journal of Neuroscience, 36*, 2458–2472.

Abdelfattah, A. S., Kawashima, T., Singh, A., et al. (2018). Bright and photostable chemigenetic indicators for extended in vivo voltage imaging. *bioRxiv.* https://doi.org/10.1101/436840.

Akemann, W., Mutoh, H., Perron, A., et al. (2012). Imaging neural circuit dynamics with a voltage-sensitive fluorescent protein. *Journal of Neurophysiology, 108*, 2323–2337.

Akemann, W., Sasaki, M., Mutoh, H., et al. (2013). Two-photon voltage imaging using a genetically encoded voltage indicator. *Scientific Reports, 3*, 2231.

Akemann, W., Song, C., Mutoh, H., & Knöpfel, T. (2015). Route to genetically targeted optical electrophysiology: development and applications of voltage-sensitive fluorescent proteins. *Neurophotonics, 2*, 2. https://doi.org/10.1117/1.NPh.2.2.021008.

Alexander, D. M., Jurica, P., Trengove, C., et al. (2013). Traveling waves and trial averaging: The nature of single-trial and averaged brain responses in large-scale cortical signals. *NeuroImage, 73*, 95–112.

Ataka, K., & Pieribone, V. A. (2002). A genetically targetable fluorescent probe of channel gating with rapid kinetics. *Biophysical Journal, 82*, 509–516.

Baker, B. J., Kosmidis, E. K., Vucinic, D., et al. (2005). Imaging brain activity with voltage- and calcium-sensitive dyes. *Cellular and Molecular Neurobiology, 25*, 245–282.

Bando, Y., Sakamoto, M., Kim, S., et al. (2019). Comparative evaluation of genetically encoded voltage indicators. *Cell Reports, 26*, 802–813.e4.

Blasdel, G. G., & Salama, G. (1986). Voltage-sensitive dyes reveal a modular organization in monkey striate cortex. *Nature, 321*, 579–585.

Carandini M, Shimaoka D, Rossi LF, et al (2015) Imaging the awake visual cortex with a genetically encoded voltage indicator. Journal of Neuroscience 35:53–63.

Chamberland, S., Yang, H. H., Pan, M. M., et al. (2017). Fast two-photon imaging of subcellular voltage dynamics in neuronal tissue with genetically encoded indicators. *Elife, 6*. https://doi.org/10.7554/eLife.25690.

Chan, K. Y., Jang, M. J., Yoo, B. B., et al. (2017). Engineered AAVs for efficient noninvasive gene delivery to the central and peripheral nervous systems. *Nature Neuroscience, 20*, 1172–1179.

Chanda, B., Blunck, R., Faria, L. C., et al. (2005). A hybrid approach to measuring electrical activity in genetically specified neurons. *Nature Neuroscience, 8*, 1619–1626.

Cohen, L. B., Salzberg, B. M., & Grinvald, A. (1978). Optical methods for monitoring neuron activity. *Annual Review of Neuroscience, 1*, 171–182.

Dimitrov, D., He, Y., Mutoh, H., et al. (2007). Engineering and characterization of an enhanced fluorescent protein voltage sensor. *PLoS One, 2*, e440.

Empson, R. M., Goulton, C., Scholtz, D., et al. (2015). Validation of optical voltage reporting by the genetically encoded voltage indicator VSFP-butterfly from cortical layer 2/3 pyramidal neurons in mouse brain slices. *Physiological Reports, 3*. https://doi.org/10.14814/phy2.12468.

Fagerholm, E. D., Dinov, M., Knöpfel, T., & Leech, R. (2018). The characteristic patterns of neuronal avalanches in mice under anesthesia and at rest: An investigation using constrained artificial neural networks. *PLoS One, 13*, e0197893.

Ferezou, I., Matyas, F., & Petersen, C. C. H. (2016). Imaging the brain in action. In R. D. Frostig (Ed.), *In vivo optical imaging of brain function.* Boca Raton: CRC Press/Taylor & Francis.

Fluhler, E., Burnham, V. G., & Loew, L. M. (1985). Spectra, membrane binding, and potentiometric responses of new charge shift probes. *Biochemistry, 24*, 5749–5755.

Flytzanis, N. C., Bedbrook, C. N., Chiu, H., et al. (2014). Archaerhodopsin variants with enhanced voltage-sensitive fluorescence in mammalian and Caenorhabditis elegans neurons. *Nature Communications, 5*, 4894.

Garaschuk, O., & Griesbeck, O. (2010). Monitoring calcium levels with genetically encoded indicators. *Neuromethods*, 101–117.

Gautam, S. G., Perron, A., Mutoh, H., & Knöpfel, T. (2009). Exploration of fluorescent protein voltage probes based on circularly permuted fluorescent proteins. *Front Neuroeng, 2*, 14.

Ghosh, K. K., Burns, L. D., Cocker, E. D., et al. (2011). Miniaturized integration of a fluorescence microscope. *Nature Methods, 8*, 871–878.

Gong, Y., Wagner, M. J., Zhong Li, J., & Schnitzer, M. J. (2014). Imaging neural spiking in brain tissue using FRET-opsin protein voltage sensors. *Nature Communications, 5*, 3674.

Gong, Y., Huang, C., Li, J. Z., et al. (2015). High-speed recording of neural spikes in awake mice and flies with a fluorescent voltage sensor. *Science, 350*, 1361–1366.

Grenier, V., Daws, B. R., Liu, P., & Miller, E. W. (2019). Spying on neuronal membrane potential with genetically targetable voltage indicators. *Journal of the American Chemical Society, 141*, 1349–1358.

Grinvald, A., Salzberg, B. M., & Cohen, L. B. (1977). Simultaneous recording from several neurones in an invertebrate central nervous system. *Nature, 268*, 140–142.

Grinvald, A., Hildesheim, R., Farber, I. C., & Anglister, L. (1982). Improved fluorescent probes for the measurement of rapid changes in membrane potential. *Biophysical Journal, 39*, 301–308.

Grinvald, A., Fine, A., Farber, I. C., & Hildesheim, R. (1983). Fluorescence monitoring of electrical responses from small neurons and their processes. *Biophysical Journal, 42*, 195–198.

Han, Z., Jin, L., Platisa, J., et al. (2013). Fluorescent protein voltage probes derived from ArcLight that respond to membrane voltage changes with fast kinetics. *PLoS One, 8*, e81295.

Hochbaum, D. R., Zhao, Y., Farhi, S. L., et al. (2014). All-optical electrophysiology in mammalian neurons using engineered microbial rhodopsins. *Nature Methods, 11*, 825–833.

Homma, R., Baker, B. J., Jin, L., et al. (2009). Wide-field and two-photon imaging of brain activity with voltage- and calcium-sensitive dyes. *Philosophical Transactions of the Royal Society of London. Series B, Biological Sciences, 364*, 2453–2467.

Jin, L., Baker, B., Mealer, R., et al. (2011). Random insertion of split-cans of the fluorescent protein venus into shaker channels yields voltage sensitive probes with improved membrane localization in mammalian cells. *Journal of Neuroscience Methods, 199*, 1–9.

Jin, L., Han, Z., Platisa, J., et al. (2012). Single action potentials and subthreshold electrical events imaged in neurons with a fluorescent protein voltage probe. *Neuron, 75*, 779–785.

Kannan, M., Vasan, G., Huang, C., et al. (2018). Fast, in vivo voltage imaging using a red fluorescent indicator. *Nature Methods, 15*, 1108–1116.

Knöpfel, T. (2012). Genetically encoded optical indicators for the analysis of neuronal circuits. *Nature Reviews. Neuroscience, 13*, 687–700.

Kost, L. A., Nikitin, E. S., Ivanova, V. O., et al. (2017). Insertion of the voltage-sensitive domain into circularly permuted red fluorescent protein as a design for genetically encoded voltage sensor. *PLoS One, 12*, e0184225.

Kralj, J. M., Douglass, A. D., Hochbaum, D. R., et al. (2011a). Optical recording of action potentials in mammalian neurons using a microbial rhodopsin. *Nature Methods, 9*, 90–95.

Kralj, J. M., Hochbaum, D. R., Douglass, A. D., & Cohen, A. E. (2011b). Electrical spiking in Escherichia coli probed with a fluorescent voltage-indicating protein. *Science, 333*, 345–348.

Lam, A. J., St-Pierre, F., Gong, Y., et al. (2012). Improving FRET dynamic range with bright green and red fluorescent proteins. *Nature Methods, 9*, 1005–1012.

Lee, S., Geiller, T., Jung, A., et al. (2017). Improving a genetically encoded voltage indicator by modifying the cytoplasmic charge composition. *Scientific Reports, 7*, 8286.

Lein, E. S., Hawrylycz, M. J., Ao, N., et al. (2007). Genome-wide atlas of gene expression in the adult mouse brain. *Nature, 445*, 168–176.

Lin, M. Z., & Schnitzer, M. J. (2016). Genetically encoded indicators of neuronal activity. *Nature Neuroscience, 19*, 1142–1153.

Lundby, A., Mutoh, H., Dimitrov, D., et al. (2008). Engineering of a genetically encodable fluorescent voltage sensor exploiting fast Ci-VSP voltage-sensing movements. *PLoS One, 3*, e2514.

Madisen, L., Garner, A. R., Shimaoka, D., et al. (2015). Transgenic mice for intersectional targeting of neural sensors and effectors with high specificity and performance. *Neuron, 85*, 942–958.

Mao, B. Q., Hamzei-Sichani, F., Aronov, D., et al. (2001). Dynamics of spontaneous activity in neocortical slices. *Neuron, 32*, 883–898.

Miller, E. W. (2016). Small molecule fluorescent voltage indicators for studying membrane potential. *Current Opinion in Chemical Biology, 33*, 74–80.

Mishina, Y., Mutoh, H., Song, C., & Knöpfel, T. (2014). Exploration of genetically encoded voltage indicators based on a chimeric voltage sensing domain. *Frontiers in Molecular Neuroscience, 7*, 78.

Monakhov, M., Matlashov, M., Colavita, M., et al. (2019). Bright near-infrared genetically encoded voltage indicator for all-optical electrophysiology. *bioRxiv*. https://doi.org/10.1101/536359.

Muller, L., Chavane, F., Reynolds, J., & Sejnowski, T. J. (2018). Cortical travelling waves: Mechanisms and computational principles. *Nature Reviews. Neuroscience, 19*, 255–268.

Murata, Y., Iwasaki, H., Sasaki, M., et al. (2005). Phosphoinositide phosphatase activity coupled to an intrinsic voltage sensor. *Nature, 435*, 1239–1243.

Mutoh, H., Perron, A., Dimitrov, D., et al. (2009). Spectrally-resolved response properties of the three most advanced FRET based fluorescent protein voltage probes. *PLoS One, 4*, e4555.

Nakai, J., Ohkura, M., & Imoto, K. (2001). A high signal-to-noise Ca2 probe composed of a single green fluorescent protein. *Nature Biotechnology, 19*, 137–141.

Nguyen, J. P., Shipley, F. B., Linder, A. N., et al. (2016). Whole-brain calcium imaging with cellular resolution in freely behaving Caenorhabditis elegans. *Proceedings of the National Academy of Sciences of the United States of America, 113*, E1074–E1081.

Ortinski, P. I., Dong, J., Mungenast, A., et al. (2010). Selective induction of astrocytic gliosis generates deficits in neuronal inhibition. *Nature Neuroscience, 13*, 584–591.

Perin, R., & Markram, H. (2013). A computer-assisted multi-electrode patch-clamp system. *Journal of Visualized Experiments*, e50630.

Perron, A. (2009). Second and third generation voltage-sensitive fluorescent proteins for monitoring membrane potential. *Frontiers in Molecular Neuroscience, 2*.

Perron, A., Mutoh, H., Launey, T., & Knöpfel, T. (2009). Red-shifted voltage-sensitive fluorescent proteins. *Chemistry & Biology, 16*, 1268–1277.

Piao, H. H., Rajakumar, D., Kang, B. E., et al. (2015). Combinatorial mutagenesis of the voltage-sensing domain enables the optical resolution of action potentials firing at 60 Hz by a genetically encoded fluorescent sensor of membrane potential. *Journal of Neuroscience, 35*, 372–385.

Piatkevich, K. D., Jung, E. E., Straub, C., et al. (2018). A robotic multidimensional directed evolution approach applied to fluorescent voltage reporters. *Nature Chemical Biology, 14*, 352–360.

Piyawattanametha, W., Cocker, E. D., Burns, L. D., et al. (2009). In vivo brain imaging using a portable 29 g two-photon microscope based on a microelectromechanical systems scanning mirror. *Optics Letters, 34*, 2309.

Platisa, J., Vasan, G., Yang, A., & Pieribone, V. A. (2017). Directed evolution of key residues in fluorescent protein inverses the polarity of voltage sensitivity in the genetically encoded Indicator ArcLight. *ACS Chemical Neuroscience, 8*, 513–523.

Rohr, S., & Salzberg, B. M. (1994). Multiple site optical recording of transmembrane voltage (MSORTV) in patterned growth heart cell cultures: Assessing electrical behavior, with microsecond resolution, on a cellular and subcellular scale. *Biophysical Journal, 67*, 1301–1315.

Sakai, R., Repunte-Canonigo, V., Raj, C. D., & Knöpfel, T. (2001). Design and characterization of a DNA-encoded, voltage-sensitive fluorescent protein. *The European Journal of Neuroscience, 13*, 2314–2318.

Sasaki, T. (2015). Probing neuronal activity using genetically encoded red fluorescent calcium indicators. *Optogenetics*, 149–158.

Shigetomi, E., Kracun, S., Sofroniew, M. V., & Khakh, B. S. (2010). A genetically targeted optical sensor to monitor calcium signals in astrocyte processes. *Nature Neuroscience, 13*, 759–766.

Shimaoka, D., Song, C., & Knöpfel, T. (2017). State-dependent modulation of slow wave motifs towards awakening. *Frontiers in Cellular Neuroscience, 11*, 108.

Siegel, M. S., & Isacoff, E. Y. (1997). A genetically encoded optical probe of membrane voltage. *Neuron, 19*, 735–741.

Song, C., Do, Q. B., Antic, S. D., & Knöpfel, T. (2017). Transgenic strategies for sparse but strong expression of genetically encoded voltage and calcium indicators. *International Journal of Molecular Sciences, 18*, 18. https://doi.org/10.3390/ijms18071461.

Song, C., Piscopo, D. M., Niell, C. M., & Knöpfel, T. (2018a). Cortical signatures of wakeful somatosensory processing. *Scientific Reports, 8*, 11977.

Song, M., Kang, M., Lee, H., et al. (2018b). Classification of spatiotemporal neural activity patterns in brain imaging data. *Scientific Reports, 8*, 8231.

Stosiek, C., Garaschuk, O., Holthoff, K., & Konnerth, A. (2003). In vivo two-photon calcium imaging of neuronal networks. *Proceedings of the National Academy of Sciences of the United States of America, 100*, 7319–7324.

St-Pierre, F., Marshall, J. D., Yang, Y., et al. (2014). High-fidelity optical reporting of neuronal electrical activity with an ultrafast fluorescent voltage sensor. *Nature Neuroscience, 17*, 884–889.

Sung, U., Sepehri-Rad, M., Piao, H. H., et al. (2015). Developing fast fluorescent protein voltage sensors by optimizing FRET interactions. *PLoS One, 10*, e0141585.

Tsutsui, H., Karasawa, S., Okamura, Y., & Miyawaki, A. (2008). Improving membrane voltage measurements using FRET with new fluorescent proteins. *Nature Methods, 5*, 683–685.

Urbano, F., Leznik, E., & Llinás, R. (2002). Cortical activation patterns evoked by afferent axons stimuli at different frequencies: An in vitro voltage-sensitive dye imaging study. *Thalamus & Related Systems, 1*, 371.

Vranesic, I., Iijima, T., Ichikawa, M., et al. (1994). Signal transmission in the parallel fiber-Purkinje cell system visualized by high-resolution imaging. *Proceedings of the National Academy of Sciences of the United States of America, 91*, 13014–13017.

Werley, C. A., Brookings, T., Upadhyay, H., et al. (2017). All-optical electrophysiology for disease modeling and pharmacological characterization of neurons. *Curr Protoc Pharmacol, 78*, 11.20.1–11.20.24.

Yang, H. H., St-Pierre, F., Sun, X., et al. (2016). Subcellular imaging of voltage and calcium signals reveals neural processing in vivo. *Cell, 166*, 245–257.

Zou, P., Zhao, Y., Douglass, A. D., et al. (2014). Bright and fast multicoloured voltage reporters via electrochromic FRET. *Nature Communications, 5*. https://doi.org/10.1038/ncomms5625.

Chapter 18
Optogenetics

Aaron Argall and Liang Guo

18.1 History

The first notion that light could influence cellular behavior was shown by Richard Fork in the 1970s, where neurons, when shone with a laser, were stimulated due to their cell membranes being partially disrupted. From this, in 1999 Francis Crick described the needs of neurophysiology to find where a recorded neuron projects, what type of neuron is being recorded, and how to control the firing rate of one or more neurons in a rapid manner, stating "the ideal signal would be light, probably at an infrared wavelength to allow the light to penetrate far enough" (Crick 1999). This line of reasoning concisely frames what optogenetics has become without exactly depicting what optogenetics fully encompasses. A few years later in 2002, Gero Miesenbock showed that three genes from *Drosophila* (NinaE, arrestin-2, and Gα) plus synthetic all-*trans*retinal could allow neurons to be driven by light (Zemelman et al. 2002).

As the first step toward genetically controlling cells with light, Miesenbock's work clearly demonstrated that when expressing the appropriate light-responsive proteins, a neuron could be modulated with light. A slight drawback of this three-gene design is that it requires the administration of an external chemical stimulus, all-*trans*retinal, to kick-start the reaction. Another drawback of this system is spatiotemporal control where it would take many seconds for the system to be turned on and off due to the multiprotein signaling cascade. While ease of use is apparent in in vitro systems, in vivo applications would be complicated by the lack of control

A. Argall
Department of Neurology, The Ohio State University, Columbus, OH, USA

L. Guo (✉)
Department of Electrical and Computer Engineering, The Ohio State University, Columbus, OH, USA

Department of Neuroscience, The Ohio State University, Columbus, OH, USA
e-mail: guo.725@osu.edu

© Springer Nature Switzerland AG 2020
L. Guo (ed.), *Neural Interface Engineering*,
https://doi.org/10.1007/978-3-030-41854-0_18

(a)   **(b)**

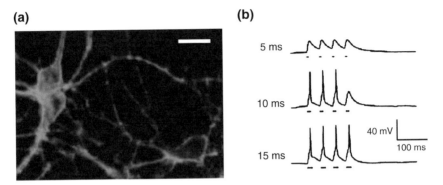

Fig. 18.1 (**a**) First reported expression of ChR2 heterologously expressed in mammalian neurons. (**b**) Electrophysiological traces showing neuronal depolarization upon light stimulation (dashes under each peak). (Adapted with permission from Boyden et al. (2005))

over targeting specific subsets of neurons. A year later in 2003, Nagel et al. demonstrated that a microbial-type rhodopsin named channelrhodopsin could be heterologously expressed in *Xenopus laevis* oocytes and HEK293 cells (Nagel et al. 2003). Cells expressing channelrhodopsin could be depolarized upon light stimulation. This work allowed for a direct translation of a green algae protein to be stably expressed within mammalian cells—what needed to be answered next was whether it could be expressed in neurons and if it could control their activity.

In the beginning of 2005, Karl Diesseroth's team showed for the first time that expression of a light-activated cation channel, channelrhodopsin (ChR2), gene in mammalian neurons could be achieved and the channel could then be stimulated with light to initiate action potentials (Fig. 18.1a, b) (Boyden et al. 2005). Amusingly, the paper was first rejected by *Science* in April 2005, because the journal stated that they had not made any new discoveries other than channelrhodopsins functioned in neurons. A month later, the paper was submitted to *Nature Neuroscience* and was accepted in the end. Shortly after, a slew of papers were quickly published using channelrhodopsins in neurons and nearly a year later the term "optogenetics" was first coined (Boyden 2011). From here, the history catches up with the present and multitudes of papers have been published since, describing specific applications of other type I microbial opsins expressed in various neural systems.

18.2 Biological Mechanism

Optogenetics is a nanobiotechnology that combines genetics and optical methods to allow for targeted, spatiotemporal control of precisely defined events in biological systems. Owing to their unique ability to be stimulated by light, type I and type II opsins are families of light-activated proteins that have been studied since the 1970s but have been adapted to control cellular functions within the early 2000s. Type I opsins, also known as microbial opsins, are seven-transmembrane-domain proteins

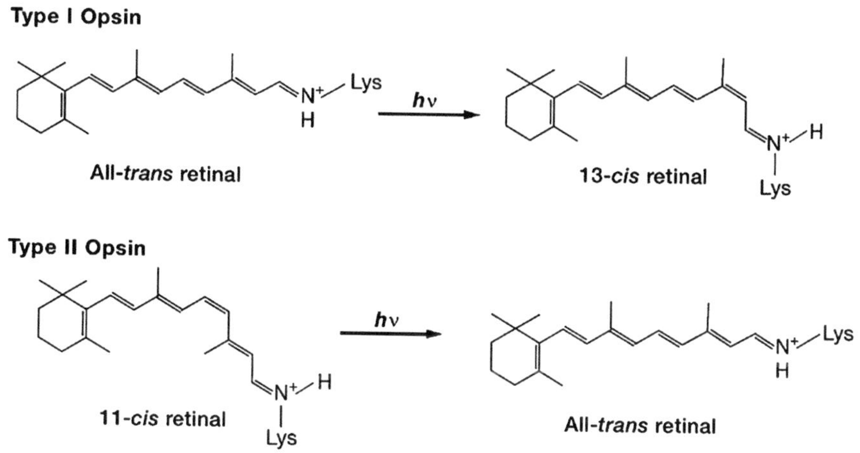

Type I Opsin

All-*trans* retinal   13-*cis* retinal

Type II Opsin

11-*cis* retinal   All-*trans* retinal

Fig. 18.2 Retinal isomerization mechanism in Type I and Type II opsins. Type I: all-*trans* retinal in its dark state, upon photon absorption becomes 13-*cis* retinal (light state). Type II: 11-*cis* retinal (dark state) isomerizes into all-*trans* retinal (light state) upon photon absorption. (Adapted with permission from Zhang et al. (2011))

whose retinal molecule upon light stimulation converts from all-*trans* retinal to 13-*cis* retinal and conducts ions across the cell membrane (Fig. 18.2 top). Type II opsins, also known as vertebrate opsins, are G-protein-coupled receptors whose retinal molecule converts from 11-*cis* retinal to all-*trans* retinal upon light stimulation (Fig. 18.2 bottom). Type II opsins are much slower than type I opsins since their activation results in signaling cascades instead of an immediate flux of ions (Guru et al. 2015).

Several classes of opsins have been implemented in optogenetics: channelrhodopsins, halorhodopsins, bacteriorhodopsins, and optoXRs as shown in Fig. 18.3a–d. These opsins have the capacity to be light activated due to the all-*trans* retinal chromophore embedded within the transmembrane domain, which undergoes photoisomerization to 13-*cis* retinal upon light absorption, causing a conformational change and allowing for the transport of ions or the propagation of intracellular signaling effector proteins. When modulating neurons, a fast response to a stimulus is often desired, as such, type I opsins are typically used. Individually, these opsins can exert an excitatory or inhibitory effect on neural activity either through the movement of ions or through signaling transduction cascades.

18.2.1 Opsin Classes

Channelrhodopsins (ChR), from the green algae *Chlamydomonas reinhardtii*, are a class of blue-light-activated ion channels that pump cations (H^+, Na^+, K^+, Ca^{2+}) across the cell membrane (Fig. 18.3a). Channelrhodopsins are used for neural

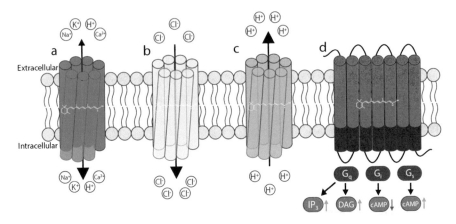

Fig. 18.3 Opsin families in their respective spectral color. (**a**) Channelrhodopsins shuttle cations including calcium, sodium, potassium, and protons leading to neuron depolarization. (**b**) Halorhodopsins primarily transport chloride ions to inhibit the activation of neurons. (**c**) Bacteriorhodopsins shuttle protons out of the intracellular space leading to neural inhibition. (**d**) OptoXRs are chimeric G-protein-coupled receptors/opsins that can elicit diverse intracellular signaling cascades through Gq, Gi, or Gs

activation by increasing the cell's membrane potential and eliciting an action potential (Han 2012). Halorhodopsins (Halo), from the archeon *Natronomonas pharaonis*, are a class of yellow-light-activated ion pumps that drive the transfer of chloride ions into the intracellular space (Fig. 18.3b). Halorhodopsins are typically used for neural silencing by decreasing the membrane potential and preventing an action potential from firing. Bacteriorhodopsins, from the haloarchaea *Halorubrum sodomense*, are a class of green-light-activated ion pumps that facilitate the movement of protons out of the intracellular space and are typically used for neural silencing (Fig. 18.3c). OptoXRs are a unique class of opsins that are an opsin/G-protein-coupled receptor (GPCR) chimera whose intracellular loops of the vertebrate rhodopsin have been replaced with those of the adrenergic receptors (Fig. 18.3d) (Airan et al. 2009). These optoXRs, depending upon the wavelength of light, could elicit signaling responses through Gq, Gi, or Gs (Guru et al. 2015). Gq signals through phospholipase C (PLC) causing the hydrolysis of phosphatidylinositol 4,5-bisphosphate (PIP$_2$) forming diacyl glycerol (DAG) and inositol 1,4,5-trisphosphate (IP$_3$) (Neves et al. 2002). DAG will then signal through protein kinase C (PKC) leading to a range of cellular processes including proliferation, cytoskeletal dynamics, migration, survival, and adhesion. IP$_3$ will bind to a calcium channel on the endoplasmic reticulum (ER) and initiate the release of calcium into the intracellular space causing further signaling cascades to be activated. Gi causes the inhibition of adenylyl cyclase which leads to a decrease in cyclic adenosine monophosphate (cAMP), ultimately leading to repression of cAMP-dependent signaling processes (Rosenbaum et al. 2009). While Gs, alternatively, activates adenylyl cyclase and increases intracellular cAMP leading to the activation of protein kinase A (PKA) and the phosphorylation of downstream targets (Rosenbaum et al. 2009).

18.2.2   Optimizing Opsin Characteristics

In the beginning, these native opsins had baseline kinetics, expression, and wavelength stimulation, progress with mutational screens has developed opsins with faster kinetics, superior expression and cell-type specificity, as well as being spectrally shifted toward infrared. Faster kinetics will allow for increased temporal control over opsin stimulation with a specific wavelength of light, as well as allowing for multiple wavelengths to be used in succession for a quick activation/silencing scheme. Step function opsins (SFOs) are specialized opsins used in paired light-stimulation paradigms where one wavelength of light activates the protein while another inactivates it. The ChR2 mutant C128S exemplifies this unique characteristic by being activated by 470 nm light and inactivated by 560 nm light. Another attribute of ChR2 (C128S) is an unusually long inactivation time, meaning they stay active for longer allowing for only a quick burst of light to activate instead of constant illumination that is necessary for non-step function opsins (Berndt et al. 2009). This is advantageous for in vivo applications where prolonged light stimulation might result in "heating" of surrounding tissue and eliciting unwanted cellular artifacts during recordings. An application of opsins with varying excitation spectra would be to express two different opsins (excitatory/inhibitory) excited with different wavelengths, and expressed in distinct neural circuits to achieve temporally precise two-color stimulation without cross-talk. Two opsins named Chronos and Chrimson are recently discovered channelrhodopsins capable of distinct blue and red-light excitation (Klapoetke et al. 2014). Chronos, a green light (530 nm) channelrhodopsin from *Stigeoclonium helveticum* (ShChR), has the fastest reported kinetics to date with an on/off rate of 2.3 ms and 3.6 ms, respectively. Chrimson, a red-shifted channelrhodopsin from *Chlamydomonas noctigama* (CnChR1), is 45 nm more red shifted than other channelrhodopsins with a peak excitation of 590 nm. When genetically targeted to two distinct neural populations interconnected within a circuit, it is possible to spectrally excite one over the other without excitation crosstalk.

Optimizing the cell-type specificity and expression profile allows for researchers to target only a subset of neurons within an awake animal or in co-culture in vitro models as well as increasing the amount of opsins that are expressed in the membrane or how well they are trafficked in the cell. Accounting for cell-specific expression can be achieved by the promoter by which the opsin gene is controlled. Specific promoters, including neuron-specific enolase (NSE), nestin (NES), and human synapsin (hSyn), can restrict expression within neurons or even within specific subsets of neurons like the promoters: Thy-1 (motor and sensory neurons), tyrosine hydroxylase (dopaminergic neurons) and VGlut1 or CaMKII (excitatory/glutamatergic neurons). Concerning membrane expression, the yellow-light-activated chloride pump halorhodopsin (HR) when tagged with YFP and expressed in neurons under the control of the Thy-1 promoter showed reduced membrane insertion with an increase in intracellular blebs (Zhao et al. 2008). To circumvent this issue, Zhao et al. improved the signal peptide sequence and added an endoplasmic reticulum export signal to HR which eliminated the intracellular blebs and increased membrane expression.

A distinct limitation of optical stimulation is its tissue-penetrating depth. Mechanically inserting an optic fiber/optrode deeper into brain regions is an unfavorable option. As changing the coding sequence of opsins could alter the excitation wavelength, altering an opsin to be stimulated by a different wavelength is of great interest, since the only neural activating opsin is blue-light stimulated which has a low tissue-penetrating depth. Pushing the stimulation spectrum closer to near-infrared will allow for the greatest tissue-penetrating capability and allow for potentially less-invasive optical stimulation. An engineered opsin named ReaChR combines sequences from prior ChR genes that together allow for faster kinetics, red-orange excitation spectra, and increased membrane expression (Lin et al. 2013). Altering the light sensitivity of opsins toward near-infrared is supported by the fact that hemoglobin, in its oxygenated and deoxygenated forms, are the major light absorbers of visible light between 350 nm and 600 nm, making traditional blue-light-activated opsins difficult to implement in in vivo systems. As such, a red-shifted opsin named Jaws, derived from *Halobacterium salinarum* (strain Shark), was engineered to contain point mutations K200R and W214F which increased its photocurrents without impacting the excitation spectra (optimal at 600 nm) (Chuong et al. 2014). Jaws is so far the furthest red-shifted inhibitory opsin allowing for non-invasive transcranial inhibition of neurons in brain regions up to 3 mm deep.

18.3   Optogenetic Expression Systems

In this section, we discuss the various methods to express opsins within in vitro or in vivo conditions, including how opsin genes are introduced either through transgenic or viral delivery.

Transgenic opsin expressing mice models offer a powerful and consistent tool to reproducibly investigate neural circuits, signaling pathways, and behavior (Fig. 18.4a). Utilizing the *Cre-loxP* recombination technology, it is possible to selectively activate or repress specific neuronal subtypes in vivo by crossing one mouse expressing the tissue-specific *Cre* and another mouse expressing the opsin of interest (Fig. 18.4d). *Cre* recombinase is a topoisomerase derived from the bacteriophage P1 that recognizes sites known as *loxP*. *Cre* will bind to two *loxP* sites (typically placed on either side of a stop code or gene), cut the double-stranded DNA, and splice out the sequence in between while DNA ligase will repair the DNA double-strand break (Feil et al. 2009). Depending on the application, *Cre-loxP* can insert the gene of interest, splice out a stop codon upstream of an opsin gene allowing for opsin expression, or can cause a flipping event using other *lox* sites that will rotate the gene of interest from a non-coding orientation to a coding orientation. All three methods can be cell-specific depending on the tissue-specific *Cre*-driver and can be temporally controlled if the *Cre* is inducible upon treatment with a

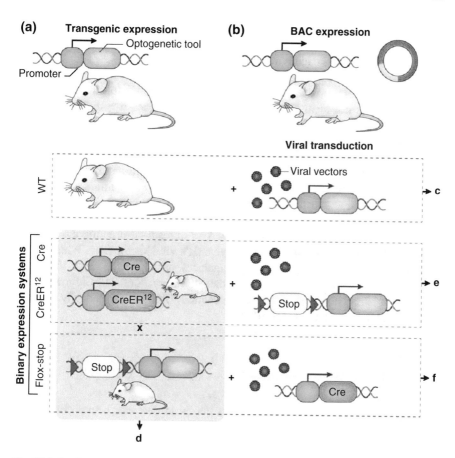

Fig. 18.4 In vivo optogenetic expression paradigms utilizing transgenic mice, bacterial artificial chromosomes (BAC), *Cre* recombinase, and viral vectors. (Adapted with permission from Boesmans et al. (2018))

compound like tamoxifen. Expression and *Cre* efficiency is then measured by the fluorescent reporter gene tagged to the opsin sequence.

Another approach involves the use of bacterial artificial chromosomes (BAC). BACs are large genomic DNA fragments, ranging from 100 to 300 kb in length, that once purified can facilitate the generation of a transgenic mouse via pronuclear injection into a fertilized mouse oocyte (Fig. 18.4b). The oocyte is then transferred into a pseudopregnant dam where the resulting pups are checked for BACs transgene expression. The application of such techniques created the first BAC transgenic mouse line with ChR2, where the ChR2(H134R)-EYFP sequence was inserted in the place of the Vglut2 gene. The construct created BAC transgenic mice expressing Vglu2-ChR2(H134R)-EYFP in glutamatergic neurons within the hindbrain and spinal cord (Hägglund et al. 2010; Zhao et al. 2011). Optogenetic transgenic mice have become increasingly available through The Jackson Laboratory, allowing researchers to create novel transgenic mice models.

Viral delivery can be done either through injection into a specific organ/tissue region (in vivo) or added to culture media (in vitro) (Fig. 18.4c, e, f, yellow shaded area). Use of viruses like lentivirus or adeno-associated virus (AAV) containing an opsin gene allows for sustained, long-term expression. Lentivirus is commonly used for in vitro settings due to its high transfection rate and stable transfection capability. Lentivirus is not ideal for in vivo investigations, since it may elicit stronger immune responses in animals and have off-target integration and expression. In vivo applications are primarily tackled by AAV since its viral genome is more inert, less immunogenic, while also providing stable gene expression. The components for a successful viral delivery strategy include the choice of appropriate virus, plasmid containing a cell-specific promoter, the opsin of interest, and a fluorescent reporter gene. Together these constituents, once transduced into a cell, will be transcribed and translated into the respective gene and protein products. Commercially available optogenetic plasmids for lentivirus and AAV or ready-to-use virus are available through Addgene and University of Pennsylvania's Viral Vector Core. While not the only sources for optogenetic coding sequences or packaged virus, these two resources have a very complete and up-to-date inventory. For further information on optogenetic expression systems, read the section "Targeting genetically encoded tools" found in Boesmans et al. (2018).

18.4   Optical Neural Interfaces

The true power of optogenetics lies in its ability to decipher neural circuits and consequently behavior in awake, free-moving animals. Taking advantage of transgenic mice or stereotactic injection of opsin-containing viral particles, the next step is to have an efficient light-delivery system. One of the most common techniques for optical stimulation is through the use of fiber optic cables inserted into the brain. Aravanis et al. described a setup that used a 200 μm wide fiber optic cable inserted into the brain through a guide, and an LED was placed on one end of the fiber which transmits the light to the desired location (Aravanis et al. 2007). Not only were they able to detect neuronal spiking that matched light pulses, they were able to force an awake mouse to move its whiskers by stimulating the motor cortex neurons. This method resolved two limitations: (1) this allowed the scattering of light in the brain to be minimized, and (2) this allowed the light to be directed onto the photosensitive regions. Doing this ensures a convenient source of power for each LED without the need of a complicated circuit. The only drawback to this method is that the animals are physically tethered.

Recent efforts have attempted to convert the above system into a wireless one. Removing the fiber optic cables would help make the system more feasible. For example, the fibers can be damaged over time or become tangled with other fibers if doing a multi-animal study. In 2011, Wentz et al. described a system that was functionally similar to the above system, but made use of wireless technologies (Wentz et al. 2011). This system used a three-component system comprising a power module, an optics module, and a motherboard. The power module contained a supercapacitor. The optics module was capable of holding up to 16 different LEDs and the motherboard contained the microcontroller to run the design. Additionally, a radio module could be attached to allow for real-time updates to the software. This system intentionally did not need a battery. The alternate use of a supercapacitor complicated the design somewhat, but allowed for increased usage time. The optics module was the only component that was inserted into the brain tissue, while the other modules simply plugged into each other on top of the animal's skull. The motherboard had two important roles: it must (1) accurately control the modulation of the LEDs and (2) ensure that the capacitor did not become insufficiently charged. This wireless system is exciting because it allows for several LEDs, flexibility, reliability, and simplicity in usage while using mostly off-the-shelf material and components.

While optical stimulation allows for high spatial resolution and cell specificity, it can still be improved. Simple fiber optics are easy to construct and implant, but light can only be delivered at the tip (Smedemark-Margulies and Trapani 2013). More complicated methods exist that use 2D and 3D multichannel waveguides, which provide better spatial control. One way of doing this is to use digital micro-mirrors (DMDs) which work by altering the phases of the light that is projected, allowing for constructive and destructive interference. Thus, the desired areas of stimulation can be stimulated without stimulating neighboring cells. These patterns of light can be as complex as needed by using many devices in parallel (Cohen 2016; Mahmoudi et al. 2017).

Optogenetics is not only useful for stimulation of a specific set of neurons, it can also be used as a form of neuronal recording. This is generally done with the use of one- and two-photon microscopy. Neurons are transduced with a fluorescent-based indicator that light up when that neuron is stimulated, which is recorded with use of the microscope (Warden et al. 2014). Genetically encoded neural indicators (GENIs), including calcium, voltage, and even vesicle fusion, can be stably expressed with cell-type specificity (discussed in other chapters). These GENIs, similar to opsins, are stimulated at specific wavelengths of light with their own spectral kinetics that make them appealing options for neural recording. This is an exciting prospect because it will allow researchers to record several magnitudes more neurons than current electrical recordings allowed by multi-electrode arrays. Through the use of simultaneous electrical recording with a transparent micro-ECoG array, an optogenetics-ECoG pair can be used for simultaneous recording and stimulation (Yazdan-Shahmorad et al. 2016). This has thus far been infeasible without the introduction of major electrical artifacts from stimulation or reduced temporal resolution from an alternating stimulation-recording protocol (Mahmoudi et al. 2017; Smedemark-Margulies and Trapani 2013; Yazdan-Shahmorad et al. 2016). For these reasons, optogenetics can allow for unprecedented experimental designs without sacrificing the quality of the recording.

Additionally, optogenetics provides a tool to allow for the mapping of neural activity to anatomical identity. Due to the scattering of light in the brain, one-photon microscopes, which can only provide images up to about 100–150 μm deep, are not as feasible as its two-photon alternative. Because of this, the two-photon microscopes, which can penetrate up to 500 μm, are more common (Deubner et al. 2019). However, one-photon microscopes have several traits that may make it attractive in the long run. For example, they have a naturally higher frame rate than two-photon microscopes and do not suffer from motion-based artifacts.

There are several points of improvement that are currently being researched. For wireless optical stimulation, performance will naturally improve with the development of new technologies for things such as wireless power transfer and supercapacitors (Wentz et al. 2011). Signal processing still has significant areas of improvement such as noise filtering, artifact detection and mitigation, and image processing (Smedemark-Margulies and Trapani 2013). More exotic research is being done on combining optical stimulation with imaging into one package by using the same implanted probe for both tasks, but using different input and output channels (Warden et al. 2014). Finally, while individual neurons can be imaged, there is not a system yet to allow for practical individual neuron stimulation. While there is a system that can stimulate individual neurons using a raster-scanning two-photon microscope, this can only be used on subjects with fixed heads (Fig. 18.5).

18.5 Toward a Brighter Future

Optogenetics has been applied to a range of basic neuroscience questions, such as probing the function of individual neurons within a population, as well as in developing clinical therapies. Optogenetics has specific advantages versus alternative techniques that make it particularly well-suited for the study, control, and modulation of neuronal function. For one, optogenetics can be used to target specific neuronal cell types within a network by expressing light-sensitive proteins selectively in that cell type, allowing the contributions of different cell types to be studied in isolation. Further, these neuronal cells can either be depolarized or hyperpolarized depending on the wavelength of the light being used for stimulation, allowing the activity of the neurons to either be excited or inhibited, and at a millisecond-timescale.

Optogenetics has also been applied to a variety of behavioral and cognitive neuroscience questions within animal models. Several studies have shown that optogenetic stimulation can be used to elicit specific behavioral responses, such as Lima and Miesenböck (2005) who used optogenetics to implicate the role of giant fiber activation in the fruit fly escape reflex, Liu et al. (2012) who used optogenetics to investigate the neuronal network basis of a memory within the dentate gyrus circuitry of the hippocampus in mice, and Jazayeri et al. (2012) who used optogenetics to induce saccadic eye movements in rhesus monkeys. Studies have also shown that optogenetics can be applied for probing neuronal networks, such as Sohal et al. (2009) who used optogenetics to study the role of parvalbumin (PV)-positive cortical interneurons in gamma-range oscillations of brain activity, which had previously been suggested to have a neuromodulatory role in information processing, and

(a)

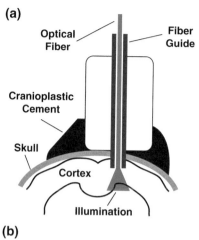
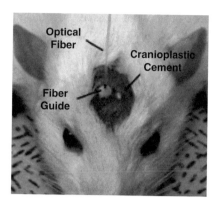

(b)

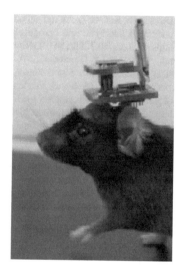

Fig. 18.5 (**a**) Fiber optic cable inserted in the cortex and fixed to skull where the animal is attached to an instrument for light stimulation paradigms. (**b**) Wireless LED module allowing for freely moving animals to be optically stimulated with little detriment to movement. (Adapted with permission from (**a**) Aravanis et al. (2007), (**b**) Wentz et al. (2011))

Piña-Crespo et al. (2012) who transplanted embryonic stem-cell-derived neurons into cultured hippocampal slices and used multi-electrode array recordings in tandem with optogenetic stimulation to show that high-frequency gamma oscillations could be induced in vivo. Similarly, optogenetics has been applied in various animal models relevant to a variety of clinical areas. This includes movement disorders such as Parkinson's, seizures and epilepsy, vision disorders like retinitis pigmentosa, neuropsychiatric disorders like depression and spinal cord injury, to name a few, primarily using rodent models (Erofeev et al. 2016). Though in vivo human experiments have yet to be performed using optogenetics, the results in these animal models have been highly informative for guiding human research for these disorders. In sum, optogenetics has already shown to be quite versatile as a tool for neuroscience, and it is likely that the extent of its applications will only increase in the future.

Acknowledgments This work was supported by the National Science Foundation of USA through Grant # 1749701.

References

Airan, R. D., Thompson, K. R., Fenno, L. E., Bernstein, H., & Deisseroth, K. (2009). Temporally precise in vivo control of intracellular signalling. *Nature, 458*, 1025. https://doi.org/10.1038/nature07926.

Aravanis, A. M., Wang, L. P., Zhang, F., Meltzer, L. A., Mogri, M. Z., Schneider, M. B., & Deisseroth, K. (2007). An optical neural interface: In vivo control of rodent motor cortex with integrated fiberoptic and optogenetic technology. *Journal of Neural Engineering, 4*, S143. https://doi.org/10.1088/1741-2560/4/3/S02.

Berndt, A., Yizhar, O., Gunaydin, L. A., Hegemann, P., & Deisseroth, K. (2009). Bi-stable neural state switches. *Nature Neuroscience, 12*, 229. https://doi.org/10.1038/nn.2247.

Boesmans, W., Hao, M. M., & Vanden Berghe, P. (2018). Optogenetic and chemogenetic techniques for neurogastroenterology. *Nature Reviews Gastroenterology and Hepatology, 15*, 21. https://doi.org/10.1038/nrgastro.2017.151.

Boyden, E. S. (2011). A history of optogenetics: The development of tools for controlling brain circuits with light. *F1000 Biology Reports*. https://doi.org/10.3410/B3-11.

Boyden, E. S., Zhang, F., Bamberg, E., Nagel, G., & Deisseroth, K. (2005). Millisecond-timescale, genetically targeted optical control of neural activity. *Nature Neuroscience, 8*(9), 1263–1268. https://doi.org/10.1038/nn1525.

Chuong, A. S., Miri, M. L., Busskamp, V., Matthews, G. A. C., Acker, L. C., Sørensen, A. T., et al. (2014). Noninvasive optical inhibition with a red-shifted microbial rhodopsin. *Nature Neuroscience, 17*, 1123. https://doi.org/10.1038/nn.3752.

Cohen, A. E. (2016). Optogenetics: Turning the microscope on its head. *Biophysical Journal, 110*, 997. https://doi.org/10.1016/j.bpj.2016.02.011.

Crick, F. (1999). The impact of molecular biology on neuroscience. *Philosophical Transactions of the Royal Society B: Biological Sciences, 354*, 2021. https://doi.org/10.1098/rstb.1999.0541.

Deubner, J., Coulon, P., & Diester, I. (2019). Optogenetic approaches to study the mammalian brain. *Current Opinion in Structural Biology, 57*, 157. https://doi.org/10.1016/j.sbi.2019.04.003.

Erofeev, A., Zakharova, O., Terekhin, S., Plotnikova, P., Bezprozvanny, I., & Vlasova, O. (2016). Future of optogenetics: Potential clinical applications? *Opera Medica et Physiologica*. https://doi.org/10.20388/OMP2016.002.0032.

Feil, S., Valtcheva, N., & Feil, R. (2009). Inducible cre mice. *Methods in Molecular Biology*. https://doi.org/10.1007/978-1-59745-471-1_18.

Guru, A., Post, R. J., Ho, Y. Y., & Warden, M. R. (2015). Making sense of optogenetics. *International Journal of Neuropsychopharmacology, 18*, pyv079. https://doi.org/10.1093/ijnp/pyv079.

Hägglund, M., Borgius, L., Dougherty, K. J., & Kiehn, O. (2010). Activation of groups of excitatory neurons in the mammalian spinal cord or hindbrain evokes locomotion. *Nature Neuroscience, 13*, 246. https://doi.org/10.1038/nn.2482.

Han, X. (2012). In vivo application of optogenetics for neural circuit analysis. *ACS Chemical Neuroscience, 3*, 577. https://doi.org/10.1021/cn300065j.

Jazayeri, M., Lindbloom-Brown, Z., & Horwitz, G. D. (2012). Saccadic eye movements evoked by optogenetic activation of primate V1. *Nature Neuroscience, 15,* 1368. https://doi.org/10.1038/nn.3210.

Klapoetke, N. C., Murata, Y., Kim, S. S., Pulver, S. R., Birdsey-Benson, A., Cho, Y. K., et al. (2014). Independent optical excitation of distinct neural populations. *Nature Methods, 11,* 338. https://doi.org/10.1038/nmeth.2836.

Lima, S. Q., & Miesenböck, G. (2005). Remote control of behavior through genetically targeted photostimulation of neurons. *Cell, 121,* 141. https://doi.org/10.1016/j.cell.2005.02.004.

Lin, J. Y., Knutsen, P. M., Muller, A., Kleinfeld, D., & Tsien, R. Y. (2013). Rea ChR: A red-shifted variant of channelrhodopsin enables deep transcranial optogenetic excitation. *Nature Neuroscience, 16,* 1499. https://doi.org/10.1038/nn.3502.

Liu, X., Ramirez, S., Pang, P. T., Puryear, C. B., Govindarajan, A., Deisseroth, K., & Tonegawa, S. (2012). Optogenetic stimulation of a hippocampal engram activates fear memory recall. *Nature, 484,* 381. https://doi.org/10.1038/nature11028.

Mahmoudi, P., Veladi, H., & Pakdel, F. G. (2017). Optogenetics, tools and applications in neurobiology. *Journal of Medical Signals and Sensors, 7*(2), 71–79.

Nagel, G., Szellas, T., Huhn, W., Kateriya, S., Adeishvili, N., Berthold, P., et al. (2003). Channelrhodopsin-2, a directly light-gated cation-selective membrane channel. *Proceedings of the National Academy of Sciences of the United States of America, 100,* 13940. https://doi.org/10.1073/pnas.1936192100.

Neves, S. R., Ram, P. T., & Iyengar, R. (2002). G protein pathways. *Science, 296,* 1636. https://doi.org/10.1126/science.1071550.

Piña-Crespo, J. C., Talantova, M., Cho, E. G., Soussou, W., Dolatabadi, N., Ryan, S. D., et al. (2012). High-frequency hippocampal oscillations activated by optogenetic stimulation of transplanted human ESC-derived neurons. *Journal of Neuroscience, 32,* 15837. https://doi.org/10.1523/JNEUROSCI.3735-12.2012.

Rosenbaum, D. M., Rasmussen, S. G. F., & Kobilka, B. K. (2009). The structure and function of G-protein-coupled receptors. *Nature, 459,* 356. https://doi.org/10.1038/nature08144.

Smedemark-Margulies, N., & Trapani, J. G. (2013). Tools, methods, and applications for optophysiology in neuroscience. *Frontiers in Molecular Neuroscience, 6.* https://doi.org/10.3389/fnmol.2013.00018.

Sohal, V. S., Zhang, F., Yizhar, O., & Deisseroth, K. (2009). Parvalbumin neurons and gamma rhythms enhance cortical circuit performance. *Nature, 459,* 698. https://doi.org/10.1038/nature07991.

Warden, M. R., Cardin, J. A., & Deisseroth, K. (2014). Optical neural interfaces. *Annual Review of Biomedical Engineering, 16,* 103. https://doi.org/10.1146/annurev-bioeng-071813-104733.

Wentz, C. T., Bernstein, J. G., Monahan, P., Guerra, A., Rodriguez, A., & Boyden, E. S. (2011). A wirelessly powered and controlled device for optical neural control of freely-behaving animals. *Journal of Neural Engineering, 8,* 046021. https://doi.org/10.1088/1741-2560/8/4/046021.

Yazdan-Shahmorad, A., Diaz-Botia, C., Hanson, T. L., Kharazia, V., Ledochowitsch, P., Maharbiz, M. M., & Sabes, P. N. (2016). A large-scale interface for optogenetic stimulation and recording in nonhuman primates. *Neuron, 89,* 927. https://doi.org/10.1016/j.neuron.2016.01.013.

Zemelman, B. V., Lee, G. A., Ng, M., & Miesenböck, G. (2002). Selective photostimulation of genetically chARGed neurons. *Neuron, 33,* 15. https://doi.org/10.1016/S0896-6273(01)00574-8.

Zhang, F., Vierock, J., Yizhar, O., Fenno, L. E., Tsunoda, S., Kianianmomeni, A., et al. (2011). The microbial opsin family of optogenetic tools. *Cell, 147,* 1446. https://doi.org/10.1016/j.cell.2011.12.004.

Zhao, S., Cunha, C., Zhang, F., Liu, Q., Gloss, B., Deisseroth, K., et al. (2008). Improved expression of halorhodopsin for light-induced silencing of neuronal activity. *Brain Cell Biology, 36,* 141. https://doi.org/10.1007/s11068-008-9034-7.

Zhao, S., Ting, J. T., Atallah, H. E., Qiu, L., Tan, J., Gloss, B., et al. (2011). Cell type-specific channelrhodopsin-2 transgenic mice for optogenetic dissection of neural circuitry function. *Nature Methods, 8,* 745. https://doi.org/10.1038/nmeth.1668.

Index

© Springer Nature Switzerland AG 2020
L. Guo (ed.), *Neural Interface Engineering*,
https://doi.org/10.1007/978-3-030-41854-0

Printed in the United States
by Baker & Taylor Publisher Services